The Blockade-Runner *Denbigh* and the Union Navy:

Including Glover's Analysis of the West Gulf Blockade
and Archival Materials and Notes

Dated July 29, 1864, Thomas C. Healy's portrait of the blockade-runner *Denbigh* shows the ship as configured in '64 and '65. Here the *Denbigh* departs Mobile with a full load of cotton bales a week before Admiral Farragut's West Gulf Blockading Squadron ended that port's use by runners. *From a private collection.*

The Blockade-Runner *Denbigh* and the Union Navy:

Including Glover's Analysis of the West Gulf Blockade and Archival Materials and Notes

Robert W. Glover

J. Barto Arnold III
Institute of Nautical Archaeology
Texas A&M University
College Station, Texas

Denbigh Shipwreck Project
Publication 7

INSTITUTE OF
NAUTICAL ARCHAEOLOGY

THE BLOCKADE-RUNNER *DENBIGH* AND THE UNION NAVY: INCLUDING GLOVER'S
ANALYSIS OF THE WEST GULF BLOCKADE AND ARCHIVAL MATERIALS AND NOTES

Library of Congress Control Number: 2015947852
ISBN 978-0-9795874-4-3

Copyright © 2015
Institute of Nautical Archaeology
P.O. Drawer HG, College Station, Texas 77841
http://inadiscover.com/

All rights reserved

No part of this book may be reproduced, stored in a retrieval
system, or transmitted in any form or by any means
without written permission from the Publisher.

Printed in the United States of America

*Cover design by Susanne Van Duyne. Adapted with permission of
Percheron Press / Eliot Werner Publications (http:// www.eliotwerner.com).
Cover illustrations courtesy of Library of Congress, Prints and Photographs Division.*

Contents

Figures 8

Overview of the Blockade-Runner *Denbigh* Shipwreck Project 13

Introduction to the 2015 Edition of Glover 19

Part I. Robert W. Glover's *An Evaluation of the
    West Gulf Blockade, 1861–1865*

Introduction 23
Preface 24
1. Blockaders: Belligerent Power vs. Neutral Rights 25
2. Administering the West Gulf Blockade 33
3. Diplomacy and the Blockade 49
4. Matamoros: Loophole in the Blockade 61
5. The Gulf Trade 71
6. The Galveston Trade 83
7. Blockade-Running: The Economics of the Trade 97
8. Conclusions 105
Bibliography 107

Part II. Archival Material and Notes

Introduction to Part II 117
1. Union Navy Blockade Memoirs 119
2. Union Navy Reports 139
3. Cotton Cargo Jettisoned by the *Denbigh*, 1865 165
4. Galveston Blockading Squadron Deck Logs Regarding the *Denbigh* 195
5. Prize Money Payout for the *Denbigh's* Jettisoned Cotton 227
6. Papers of the Blockade-Runner *Alabama* Prize Case 269
7. Conclusion 433

Figures

Figure 1. The Confederate coast 13
Figure 2. The *Denbigh* 14
Figure 3. Mouth of the Rio Grande 28
Figure 4. Gulf of Mexico 30
Figure 5. Coast of Florida and Alabama 73
Figure 6. Delta of the Mississippi River and Approaches to New Orleans 75
Figure 7. Coast of Louisiana and Texas 80
Figure 8. U.S.S. *Scioto* 124
Figure 9. U.S.S. *Niagara* 134
Figure 10. U.S.S. *Conemaugh* 146
Figure 11. U.S.S. *Sebago* 146
Figure 12. Special Order 111 concerning the *Denbigh's* jettisoned cotton 166
Figure 13. The blockade-runner *Owl* 167
Figure 14. The U.S.S. *Gertrude* prize master's report on the *Denbigh's* cotton 170
Figure 15. The monition in the case of the *Gertrude* vs. Fifty Bales of Cotton 174
Figure 16. Depositions of two crew members of the *Gertrude* 177
Figure 17. Admiralty warrant to "arrest" the cotton picked up by the *Gertrude* 179
Figure 18. The marshal's return, a proof of execution of the arrest and advertisement 180
Figure 19. Advertisement providing public notice of the *Gertrude's* prize case 181
Figure 20. The U.S. marshal reported damaged cotton and recommended quick sale 182
Figure 21. U.S. prize appraisers' report for the *Gertrude's* prize cotton 183
Figure 22. Invoice for services of the appraisers 184
Figure 23. Court order for sale of the *Gertrude's* prize cotton 185
Figure 24. Results of sale of the *Gertrude's* prize cotton 186
Figure 25. Account of the sale of the *Gertrude's* prize cotton 188
Figure 26. Marshal's bill showing the sale expenses for the *Gertrude's* prize cotton 190
Figure 27. *New Orleans Times* bill for advertising sale of the *Gertrude's* prize cotton 191
Figure 28. *True Delta* bill for advertising the sale of the *Gertrude's* prize cotton 191
Figure 29. *Picayune* bill for advertising the sale of the *Gertrude's* prize cotton 192
Figure 30. Expenses of the U.S. prize auctioneer regarding *Gertrude's* prize cotton 192
Figure 31. Docket sheet for the *Gertrude* claim on the *Denbigh's* jettisoned cotton 194
Figure 32. The U.S.S. *Cornubia's* deck log including the recovery of the floating cotton 197
Figure 33. The *Cornubia's* deck log for 20 April 1865 enlarged 199
Figure 34. The *Gertrude's* deck log covering the recovery of floating the cotton 201
Figure 35. The *Cornubia's* deck log covering the destruction of the *Denbigh* 203
Figure 36. Continuation sheet of the *Cornubia's* deck log 205
Figure 37. The U.S.S. *Kennebec's* deck log for the destruction of the *Denbigh* 207
Figure 38. The U.S.S *Seminole's* deck log for the destruction of the *Denbigh* 209
Figure 39. U.S.S. *Princess Royal* 210
Figure 40. The *Princess Royal's* deck log for the destruction of the *Denbigh* 211
Figure 41. The *Princess Royal's* deck log for the destruction of the *Denbigh* enlarged 213
Figure 42. The U.S.S. *Penguin's* deck log for the destruction of the *Denbigh* 215
Figure 43. U.S. Gunboat *Albatross* 216
Figure 44. The *Albatross's* deck log for the destruction of the *Denbigh* 217

Figure 45. The U.S.S. *New London's* deck log for the destruction of the *Denbigh* 219
Figure 46. The U.S.S. *Fort Jackson's* deck log for the destruction of the *Denbigh* 221
Figure 47. The *Fort Jackson*, flagship of the Galveston blockading squadron 222
Figure 48. The *Fort Jackson's* deck log continuation page 223
Figure 49. Entries from the Galveston lookout logbook 226
Figure 50. Prize list or muster roll for the *Gertrude* regarding the prize cotton 240
Figure 51. Treasury ledger for payment of the *Gertrude* crew's prize money 244
Figure 52. Prize list or muster roll of the *Cornubia* regarding the prize cotton 251
Figure 53. Treasury ledger for payment of the *Cornubia* crew's prize money 255
Figure 54. Top part of page of the *Cornubia* prize money document 256
Figure 55. A prize money application form 262
Figure 56. Treasury Department ledger form for prize payment to a *Cornubia* crewman 263
Figure 57. Each page of the prize payment ledger showed the details for two crewmen 264
Figure 58. Treasury Department ledger form for prize payment to a *Gertrude* crewman 265
Figure 59. A page of the prize payment ledger for two *Gertrude* crewmen 266
Figure 60. Bookstore letter to officer of the blockading squadron off Galveston 267
Figure 61. J. H. Nesen, Acting 2nd Assistant Engineer, U.S.S. *Gertrude* 267
Figure 62. Reverse side of the bookstore envelope 268
Figure 63. Crew list for blockade-running merchant Str. *Alabama* 271
Figure 64. Notations on the outer cover of the crew list 272
Figure 65. Reverse side of the *Alabama's* crew list 273
Figure 66. Cargo manifest of the *Alabama* for the return voyage to Mobile 277
Figure 67. The ship's register document of the merchant Str. *Alabama* 279
Figure 68. Four sections of the *Alabama's* register enlarged 280
Figure 69. Attachments to the *Alabama* register for changes of the captain 281
Figure 70. The *Alabama's* articles of agreement, a list of the crew and their pay 283
Figure 71. Reverse side of articles of agreement quoting the terms of employment 284
Figure 72. Crew agreement for additional crewmen recruited in Havana 285
Figure 73. Reverse side of the supplementary crew agreement 286
Figure 74. Customs House clearance to sail from Mobile bound for Havana 287
Figure 75. A customs document issued to the *Alabama* in Havana 288
Figure 76. Informal vouchers for the purchase of small shipments 290
Figure 77. The libel form regarding capture of the *Alabama* 291
Figure 78. Reverse side of the libel form 292
Figure 79. Admiralty warrant for the arrest or seizure of the *Alabama* 292
Figure 80. Reverse side of the admiralty warrant, a form for the marshal's report 293
Figure 81. Order to unload the *Alabama* 294
Figure 82. Appointment of attorney to represent the captors 295
Figure 83. Deposition of Acting Master Wm. Richardson of the *San Jacinto* 299
Figure 84. Deposition of prize master, N. M. Dyer, Actg. Ensign, comdg. the *Eugenie* 303
Figure 85. Deposition and affidavit of N. M. Dyer, commander of the *Eugenie* 305
Figure 86. The prize commissioner reported prize's arrival to the court 309
Figure 87. Interrogatory of Captain Thomas Carrell describing the *Alabama's* capture 315
Figure 88. Chart of Chandeleur Islands and Ship Island 315
Figure 89. The deposition of H. Parker, the *Alabama's* navigator 318
Figure 90. Deposition of Purser Robert Adams of the *Alabama* 319

Figure 91. Bill for the prize ship *Alabama's* arrival fees in New Orleans 320
Figure 92. The inventory of gear and equipment on board the *Alabama* 326
Figure 93. Deposition of Lieutenant Commander Chandler, commanding the *San Jacinto* 327
Figure 94. Deposition of Lieutenant Commander J. N. Quackenbush of the *San Jacinto* 330
Figure 95. Deposition of A. A. Paymaster Asa C. Winter of the *San Jacinto* 331
Figure 96. A group of the *San Jacinto's* officers made a joint deposition 332
Figure 97. Officers of the U.S. Frigate *Colorado* make a claim for a share of the prize 336
Figure 98. Commodore Bell asked Judge Durell to turn the *Alabama* over to the navy 337
Figure 99. Appointment of the appraisers by the marshal 339
Figure 100. Petition and order to transfer the *Alabama* to the navy 341
Figure 101. Letter from prize commissioner to Commodore Bell 342
Figure 102. Testimony of Captains Jenkins and Marchand concerning "signal distance" 346
Figure 103. The report of the appraisers valuing the *Alabama* at $80,000 347
Figure 104. Bill for services of appraisers 347
Figure 105. Attorney for the captors moved for a new appraisal 349
Figure 106. Order for General Stone to appear regarding contempt of court 351
Figure 107. General Stone's statement and admission 353
Figure 108. Captain McClure's testimony regarding goings-on aboard the *Alabama* 355
Figure 109. General Stone's offer to pay the value of the *Alabama* to the court 358
Figure 110. General Stone's letter to Judge Durell pleading the necessity of his actions 362
Figure 111. The U.S. district attorney explained to General Stone the legal impediments 365
Figure 112. Colonel Holabird, Quartermaster, Department of the Gulf, wrote to the judge ... 367
Figure 113. Judge Durell replied to Colonel Holabird with details for the payment 368
Figure 114. Wm. E. Le Roy's testimony on the point of "signal distance" 370
Figure 115. Lieutenant Commander O. F. Staunton's testimony on "signal distance" 370
Figure 116. Judge Durell's order for sale of the *Alabama* and her contents 371
Figure 117. The marshal's return regarding the sale 372
Figure 118. The auctioneer's bill regarding the sale of the *Alabama's* cargo 374
Figure 119. Outside cover of a document reporting advertisement of the *Alabama* auction 375
Figure 120. Funds from the auction of the *Alabama* cargo transferred to the court 376
Figure 121. Appraisal of the *Alabama's* cargo 390
Figure 122. Auction sale of part of the *Alabama's* cargo on 28 Nov. 1863, day one 398
Figure 123. Results of day two of the *Alabama* auction sale 404
Figure 124. *Alabama* auction results of day three 412
Figure 125. The *Eugenie, San Jacinto*, and *Tennessee* were eligible for the prize money 419
Figure 126. Report of the attorney for the captors 421
Figure 127. Invoice for the services of the U.S. district attorney 422
Figure 128. The marshal's invoice for expenses 424
Figure 129. U.S. appraisers' invoice concerning the cargo 425
Figure 130. Invoice for board of two ship keepers 425
Figure 131. Invoice for advertising in *The Era* regarding *Alabama* auction 426
Figure 132. Invoice for advertising in the *Louisiana State Gazette* 426
Figure 133. Invoice for advertising in the *New Orleans Times* 427
Figure 134. Invoice to gauge 10 barrels of alcohol 427
Figure 135. Invoice for producing 500 sale catalogs 428
Figure 136. U.S. marshal's final *Alabama* costs and charges 430

Figure 137. The court's final order in favor of the captors 431
Figure 138. Final list of costs and charges for the *Alabama* prize case 432

Overview of
The Blockade-Runner *Denbigh* Shipwreck Project

by

J. Barto Arnold III
Denbigh Project Director
Institute of Nautical Archaeology

As the Civil War commenced, the newly formed Confederacy found itself short of cash and manufactured goods, while soon its sellable agricultural products were blockaded for the most part. So the South turned largely to Great Britain and a few other European powers as sources of weapons, clothing, tools, and medicines that could be paid for with bales of cotton. Importing goods to and from the Confederate States of America (C.S.A.) was not easy, however, since Southern financial factors had first to establish themselves abroad. Then, too, blockade-running ships and crews had to be bought or otherwise acquired for transporting cotton to nearby neutral ports and merchandise to Dixie (Figure 1).

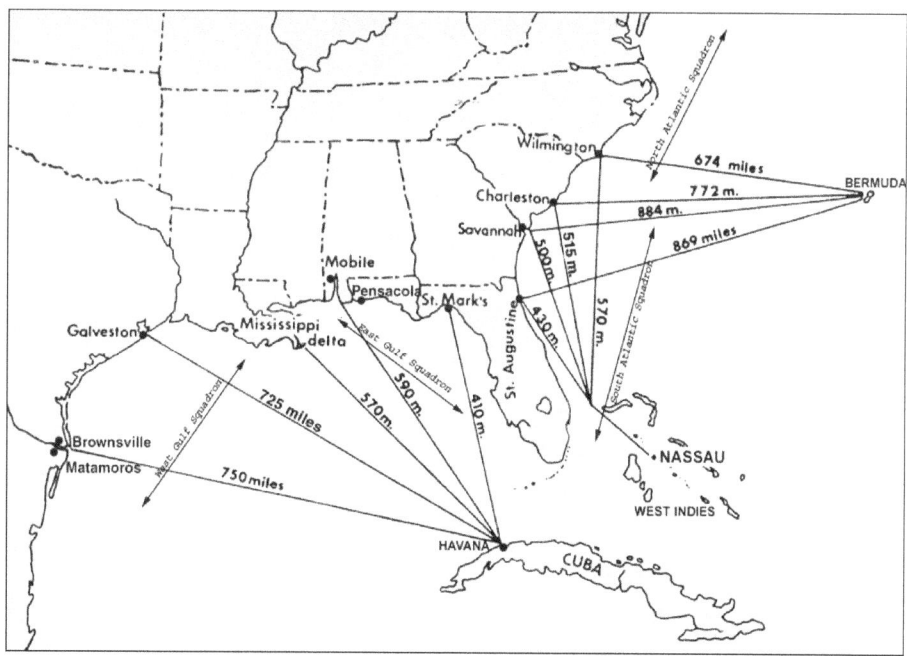

Figure 1. The Confederate coast and blockade-running ports. *Naval Historical Center.*

The present tome considers the activities of the West Gulf Blockading Squadron of the Union navy in taking captive as prizes of war the (mostly) neutral vessels that ran the blockade. It discusses the Union navy's modus operandi and attempts to address the behaviors of the opposing sailors and the how's and why's thereof. Detailed examples are provided for a few particular ships taken off Galveston and Mobile. Archival documents are illustrated.

Mobile and Galveston were the Confederacy's ports of call for the famous and successful blockade-runner *Denbigh* (pronounced "Den-bee"), a shipwreck excavated by the Institute of Nautical Archaeology (INA) located at Texas A&M University. Some of the incidents and documents in the present book reference the *Denbigh* herself and the rest help explain the activities of this ship and her sisters in the runner's trade. Understanding the prize game enhances greatly the understanding of blockade-running.

In the Gulf, the purpose of steam vessels such as the *Denbigh* was to bring in army supplies and manufactured goods, mainly from Havana (Figure 2). The ships then returned through the blockade to Cuba carrying cotton. Smaller and more numerous sailing vessels did much the same but put ashore at secondary and shallower ports, bays, and bayous. The purpose of the Union navy's blockading squadron was to prevent this trade by destroying or (preferably) capturing blockade-runners.

The utility of this book for the *Denbigh* Project is as follows. The most basic context for the *Denbigh* is the 1863–1865 activities of the blockade-runners going to and from Galveston and other western Gulf ports. One important question is how and why the blockaders and the runners did their respective jobs. It was an intricate and complex game of cat and mouse. The operational behavior of both groups was largely influenced by the law of prize. The result is important both to those generally interested in the Civil War and especially to those interested in the history and nautical archaeology of blockade-runners.

To grasp the men of the Union navy's intense interest, one has only to look at the prize money paid to the biggest winners. William Budd commanded the vessel that made the richest capture of the war and received nearly $39,000. As commander of the West Gulf Blockading Squadron, Admiral Farragut's total reached over $56,000. The lowliest landsman on the crew that sank the C.S.S. *Albemarle* received $4,019, while his monthly pay was $12. By 1868, $23,629,627 in prize money had been paid to Union navy officers and men. Prize money was a most important supplement to scanty monthly pay, but those with shore assignments were just out of luck. Congress scrapped the prize money system in 1900 and replaced it with more adequate and appropriate pay rates.[1]

Figure 2. The *Denbigh* in a computer rendering based on a contemporary portrait. Illustration by Andrew W. Hall, *INA*.

The present book is the seventh in a series comprising the *Denbigh* shipwreck excavation results.[2] The first was Captain William Watson's 1892 memoir *The Adventures of a Blockade Runner*. The Watson book was published in the project series because, at the time, we had no firsthand accounts from the *Denbigh*. Watson tells of taking blockade-runners, both sail and steam, in and out of Galveston, and described harrowing experiences and situations much the same as those the *Denbigh* must have encountered. The second book is a new edition of *The Confederate Quartermaster in the Trans-Mississippi* with added material relevant to the *Denbigh*, the army's Cotton Bureau, and blockade-running. The Confederate army was the largest customer of the runners, so a study of the buyers helps one understand

---

[1] Daly, Robert W. "Pay and Prize Money in the Old Navy, 1776–1899," *United States Naval Institute Proceedings* 74(1948):967-971.
[2] See Additional Readings below for listing.

the runners and the cargos they brought in. Third is *Schooner Sail to Starboard* by Block, which chronicles the doings of blockade-runners and their Union navy pursuers in the Western Gulf of Mexico. The INA edition adds memoirs of Union sailors on station with the blockading fleet.

The fourth is *Colin J. McRae: Confederate Financial Agent; Blockade Running in the Trans-Mississippi South as Affected by the Confederate Government's Direct European Procurement of Goods*, which considers mainly what had to be done in Europe to finance and acquire the wanted goods. Part I reprints an excellent study of the business of blockade-running by Charles S. Davis, 1961. Part II discusses the goods involved in the blockade-running trade by publishing fascinating ships invoices as well as customs records. These raw descriptions are supplemented by interesting correspondence between factors, diplomats, and government officials. Book 5, *The* Denbigh's *Civilian Imports: Customs Records of a Civil War Blockade Runner between Mobile and Havana*, gives surprisingly detailed information on the daily life of individually identified families in Mobile. Book 6, *Civil War Blockade-Runners: Prize Claims and the Historical Record*, addresses prize law in effect during the career of the *Denbigh*.

There will be several more books on the *Denbigh* Project: one consisting of the results of archaeological excavation, another presenting the ship's documentary sources, and perhaps others.

The archaeological fieldwork on the *Denbigh* shipwreck (41GV143) began in December 1997. Each year's fieldwork was reported as the project progressed. After 2002, the project proceeded to the phase of artifact conservation, archival research, analysis, and write-up. With conservation completed, the *Denbigh* collection was permanently curated as of 2008 at the Corpus Christi Museum of Science and History.

The *Denbigh* was an iron-hulled paddle steamer. A Liverpool coastal passenger ship built by Laird's shipyard in 1860, she was noted for her speed. As a blockade-runner in the Gulf of Mexico from 1863–1865, she was one of the most successful and famous of the Civil War. The *Denbigh* ran aground entering Galveston in late May 1865 and was destroyed by the Union blockading fleet. To date, several journal articles have been published (see Additional Readings) as well as conference papers on the project.

Table 1. The *Denbigh* at a Glance

Launched:	August 1860
Built:	Laird, Sons & Co., Birkenhead
Length:	182 feet (55.5 meters) between perpendiculars, ~195 feet overall
Beam:	22 feet (6.7 meters), excluding paddle wheels
Depth of hold:	8.7 feet (2.65 meters)
Register tons:	162.69
Sustained speed:	13.7 knots (15.75 mph), third fastest ever at completion
Cost to build:	£10,250 (approximately one million dollars today)
Port of registry:	Liverpool, England
Trips as runner:	13 successful round trips
Cargo capacity:	Approx. 500 bales (225,000 pounds) cotton
Crew:	21 officers and men

Imported munitions and other, often surprising, strategic material meant life or death for the Confederate States of America. Bankers and financiers Erlanger and Schroder and Confederate promoter and merchant Brewer were the owners of the *Denbigh*. Studying the *Denbigh* and her place in the blockade-running supply system was enlightening. In the spring of 1865, the supply system could effectively support the army in the Trans-Mississippi, but the war was by then lost. It was too late. Operations at sea of both the runners and the Union navy were often dictated by the legal strictures of prize law.

It is my recommendation that this and the other books of the *Denbigh* Project Series produced by the Institute of Nautical Archaeology be examined by the reader, for they document the career and demise of a single blockade-runner. Blockade-running was clearly vital to the South's war effort, but for financiers and crew was a dangerous way to earn one's daily bread. For Union navy officers and crews, captured blockade-running ships proved to be sources of wealth, prizes well worth having.

Acknowledgments for *Denbigh* Project Funding

The *Denbigh* Project works under the aegis of the Institute of Nautical Archaeology at Texas A&M University, a nonprofit scientific and educational organization. J. Barto Arnold III is project director and principal investigator; Thomas J. Oertling and Andrew W. Hall were coprincipal investigators for archaeology and history, respectively. The project benefited from the participation of many students and volunteers. The *Denbigh* Project appreciates the kind financial support of the Anchorage Foundation; the Albert & Ethel Herzstein Charitable Foundation of Houston; the Brown Foundation, Houston; the Communities Foundation of Texas; the Clements Foundation of Dallas; the Dana and Myriam McGinnis Charitable Fund of the San Antonio Area Foundation; the Ed Rachal Foundation of Corpus Christi; the Institute of Nautical Archaeology Foundation; the Hillcrest Foundation of Dallas, founded by Mrs. W. W. Caruth, Sr.; the Horlock Foundation, Houston; the Houston Endowment; the Strake Foundation of Houston; the Summerfield G. Roberts Foundation of Dallas; the Summerlee Foundation of Dallas; the Joseph Ballard Archeology Fund of the Texas Historical Foundation of Austin; the Trull Foundation of Palacios; and many individual donors. Finally, the *Denbigh* Project acknowledges its debt to Texas A&M University, Galveston, for generous assistance throughout the course of the project.

Additional Readings

Arnold, J. B. III. "Marine Magnetometer Survey of Archaeological Materials Near Galveston, Texas." *Historical Archaeology* 21.1(1987):18-47.

Arnold, J. B. III. *The* Denbigh's *Civilian Imports: Customs Records of a Civil War Blockade Runner between Mobile and Havana*. College Station, Texas: Institute of Nautical Archaeology, *Denbigh* Shipwreck Project Publication 5, 2011.

Arnold, J. B. III, et al. "The *Denbigh* Project: Initial Observations on a Civil War Blockade Runner and Its Wreck-Site." *International Journal of Nautical Archaeology* 28.2(1999):126-144.

Arnold, J. B. III, et al. "The *Denbigh* Project: Test Excavations at the Wreck of an American Civil War Blockade Runner." *World Archaeology* 32(2001):405-417.

Arnold, J. B. III, et al. "The *Denbigh* Project: Excavation of a Civil War Blockade Runner." *International Journal of Nautical Archaeology* 30.2(2001):231-249.

Arnold, J. B. III, et al. "The *Denbigh* Project: Excavation of a Civil War Blockade Runner." *Bulletin of the Texas Archeological Society* 74(2003):131-140.

Block, W. T. *Schooner Sail to Starboard: The U.S. Navy vs. Blockade Runners in the Western Gulf of Mexico*. College Station, Texas: Institute of Nautical Archaeology, *Denbigh* Shipwreck Project Publication 3, 2007.

Davis, C. S. *Colin J. McRae: Confederate Financial Agent; Blockade Running in the Trans-Mississippi South as Affected by the Confederate Government's Direct European Procurement of Goods*. College Station, Texas: Institute of Nautical Archaeology, *Denbigh* Shipwreck Project Publication 4, 2008.

Nichols, J. L. *The Confederate Quartermaster in the Trans-Mississippi: The Blockade Runner's Texas Connection*. Clinton Corners, N.Y.: Percheron Press, Institute of Nautical Archaeology, *Denbigh* Shipwreck Project Publication 2, 2006.

Powell, G. R., M. C. Cordon, and J. B. Arnold III. *Civil War Blockade-Runners: Prize Claims and the Historical Record, Including the* Denbigh's *Court Documents*. College Station, Texas: Institute of Nautical Archaeology, *Denbigh* Shipwreck Project Publication 6, 2012.

Watson, W. *The Adventures of a Blockade Runner,* reprinted as *The Civil War Adventures of a Blockade Runner*. College Station, Texas: Texas A&M University Press, *Denbigh* Shipwreck Project Publication 1, 2001.

Introduction to the 2015 Edition of Glover

by

J. Barto Arnold III
Denbigh Project Director
Institute of Nautical Archaeology

Over the years of studying and reading about Texas shipwrecks, the present author came across references to Robert Glover's book on the West Gulf Blockading Squadron on numerous occasions. As the work was only available in the form of a 1974 dissertation at North Texas State University (now North Texas University), I passed it by time after time. It was neither easy nor inexpensive to obtain. When I at last had a reason to pursue and study Glover's work, its value became apparent to me as an underwater archaeologist researching in depth a particular Civil War shipwreck like the *Denbigh*.

In planning the series of reports on the *Denbigh* Project, I determined to print Glover's[3] book because it provides vital context addressing the question, "What were the Union blockaders up to on the Texas coast?" Further, publishing Glover makes the study conveniently available to other researchers. One cannot effectively study the blockade-runners without considering the perspective of their opposition, the Union blockading fleet trying diligently to choke off the supplies brought in by the runners and the cotton they took out in order to pay for these supplies.

The present author finds it particularly important and interesting to combine historic overviews like Glover's with illustrating examples of archival documents generated by the activities of both sides. The present book is the seventh in the series of *Denbigh* Project reports, and an example of this sort of combination. I believe this approach can lead to greater understanding for the anthropologist/archaeologist and is also of value, even exciting, for the general reader.

Glover's original 1974 manuscript herein was edited to a very limited extent, and a few notes were added by way of updates.

---

[3] It was interesting for the author to exchange emails and visit with Dr. Glover, who agreed to let us proceed with this plan. I showed him the earlier *Denbigh* publication 2, which included the *Confederate Quartermaster in the Trans-Mississippi* by James Nichols. This was a similar book, including a reprinted study plus *Denbigh*-related archival materials. Interestingly, Nichols, had been Glover's esteemed graduate adviser when he, Glover, was working on his dissertation herein published. (See pp. 16-17 for a listing of *Denbigh* Project books and articles.)

Part I

Robert W. Glover's *An Evaluation of the West Gulf Blockade, 1861–1865*

An Evaluation of the West Gulf Blockade, 1861–1865

by

Robert W. Glover

Introduction

This investigation resulted from a pilot research paper prepared in conjunction with a graduate course on the Civil War. This study (completed in 1974) suggested that the federal blockade of the Confederacy may not have contributed significantly to its defeat. Traditionally, historians had assumed that the Union's Anaconda Plan had effectively strangled the Confederacy. Recent studies which compared the statistics of ships captured to successful infractions of the blockade had somewhat revised these views. While accepting these revisionist findings as broadly valid, this investigation strove to determine specifically the effectiveness of Admiral Farragut's West Gulf Blockading Squadron.

Since the British Foreign Office maintained consulates in three blockaded southern ports and in many Caribbean ports through which blockade-running was conducted, these consular records were vital for this study. Personal research in Great Britain's Public Record Office (now the National Archives) disclosed valuable consular reports pertaining to the effectiveness of the federal blockade. American consular records, found in the National Archives in Washington, D.C. provided excellent comparative reports from those same Gulf ports. Official Confederate reports, contained in the National Archives, various state archives, and in the published *Official Records of the Union and Confederate Armies* revealed valuable statistical data on foreign imports. Limited use was made of Spanish and French consular records written from ports involved in blockade-running. Extensive use was made of Senate and House documents in determining federal blockade policy during the war. The record of the navy's enforcement of the blockade was found in the *Official Records of the Union and Confederate Navies*. The contemporary reports of Union and Confederate governmental officials were found in James D. Richardson's respective works on the Messages and Papers, and in the published diaries of Gideon Welles and Gustavus Fox. Contemporary newspapers and firsthand accounts by participants on both sides provided color and perspective.

In evaluating the performance of the West Gulf Blockading Squadron, a review of the international laws governing blockading was undertaken, emphasizing America's traditional posture regarding the blockades of other nations. Under Gideon Welles, the federal navy became a powerful and efficient force, although the navy's enforcement of the blockade often resulted in serious diplomatic embarrassment, especially from maritime incidents occurring near the mouth of the Rio Grande. Nearby Matamoros, Mexico virtually became an international trade mart for Confederate cotton and imports. However, much contraband trade was conducted through blockaded Gulf ports such as Galveston, Texas.

It is concluded that the West Gulf Blockading Squadron performed only satisfactorily at best. This did not result so much from innate limitations as from outside factors. Among the latter were the open door at Matamoros, the Lincoln administration's diplomatic temerity and national policies that authorized a type of cotton trade with the south. Further, the better vessels were assigned land campaign priorities. The statistics of the cotton trade in this portion of the Confederacy show that cotton exports were significantly high. Most of these exports egressed via Matamoros, but a high percentage exited through blockaded Gulf ports. The fact that 10,000 bales of cotton left the heavily guarded port of Galveston in the last six months of the war indicates the inefficiency of the West Gulf Blockade.

It appears that the West Gulf Blockade was effective enough to create scarcity but never effective enough to seriously interdict the flow of trade. That the Trans-Mississippi Confederacy was largely sustained by imports underscores the blockade's limited effectiveness.

Preface

The genesis of this investigation came in the summer of 1969 with a research paper on the Union blockade of the Texas coast, which disclosed a dearth of published material on the subject. There is no comprehensive narrative of the West Gulf Blockading Squadron, which is due in part to the relatively obscure role of that command in the Civil War.[4] The only published work that came close to the subject is a biography of Admiral David Glasgow Farragut, the most prominent commander of the squadron. Alfred T. Mahan's *The Gulf and Inland Waters* is also disappointingly sketchy.

Once the fragmentary squadron accounts were synthesized into a narrative mosaic, it was also apparent that a precise evaluation of the squadron's effectiveness in checking blockade-running was lacking. The federal accounts indicated that many blockade-runners were captured or destroyed, and this was true. However, the federals had no way of knowing about the runners that got past them. Thus, an accurate measurement of their task necessitated the cross-checking of compiled statistics and accounts, not only from federal and Confederate records, but also from the clearances of vessels through Confederate and neutral Gulf ports. A good index of the effectiveness of the blockade was found in the record of the wartime Texas cotton trade. Although the Matamoros end of this trade had been well publicized, nothing existed concerning the export activities of the Confederate Foreign Supply Office. It was the purpose of this investigation, therefore, to evaluate the effectiveness of the West Gulf Blockading Squadron within a more comprehensive perspective.

---

[4] JBA: This was the case in 1974 when Glover completed his dissertation. The situation was corrected as of spring 2015 when Robert Browning published *Lincoln's Trident: The West Gulf Blockading Squadron during the Civil War*, Tuscaloosa: The University of Alabama Press.

1

Blockaders: Belligerent Power vs. Neutral Rights

The use of naval blockades to obtain strategic goals in warfare has been employed at least since the Peloponnesian War. More recently it has been exercised under the names of quarantine and interdiction. Whatever term is used the purposes remain constant: by the use of military means to confine enemy vessels to their ports and to prevent supplies and provisions from entering the belligerent country by sea.

As the law of nations evolved, the right of a belligerent power to use a blockade became universally recognized. In time, however, the tactical execution of blockades resulted in continuous debate between nations concerning the extent of that right. As the term blockade acquired legal meaning nations attempted to define its more abstract meaning with phrases such as contraband, effectiveness, and neutral rights. Ultimately it was agreed that blockades, to be legally binding and respected, had to be effective, and not paper, or constructed, blockades.[1] What precisely constituted an effective blockade has never been defined for general agreement. In 1818 Prussia and Denmark agreed in treaty that at least two vessels were required at each blockaded port. Earlier a treaty between Holland and the Sicilies had specified that six vessels in a semicircle outside of coastal gun range were effective. Before that France and Denmark had agreed that a blockaded port should be closed by two vessels at least, or by a battery of guns on land. In time admiralty court decisions tended to define effectiveness more flexibly, regarding circumstances rather than numbers as the determining factor. A blockade was effective, they reasoned, if it was dangerous in fact for any vessel to attempt to enter or leave the blockaded port.[2]

Most European powers agreed to the Declaration of Paris of 1856, which established certain rules for the conduct of maritime warfare. An invitation was extended to all nations to accede to the declaration, provided that such nations "shall not in the future enter into any arrangement, concerning the application of the law of neutrals in time of war which does not rest altogether upon the four principals embodied in the said declaration."[3] The four principals of the treaty were:

1. Privateering is and remains abolished.

2. The neutral flag covers enemy's goods with the exception of contraband of war.

3. Neutral goods, with the exception of contraband of war, are not liable to capture under the enemy's flags.

4. Blockades, in order to be binding, must be effective; that is to say, maintained by a force really to prevent access to the coast.[4]

The United States, traditionally sensitive about neutral rights, did not accede to the treaty, feeling that it lacked guarantees for the protection of private property at sea which was not contraband of war. Further, Americans were reluctant to give up privateers, historically their most effective naval weapon. Secretary of State William L. Marcy, arguing the latter point, noted that there was hardly any rule of international law, agreed to in treaty, that nations did not later suspend or modify to suit their wartime needs. Here the issue lay until the Civil War, at which time the United States endeavored to accede to the treaty.[5] In fact the whole role of the United States would change from that of a nation traditionally insisting on maritime rights to one demanding belligerent rights.[6]

Blockades, by their many-sided nature, present hazards for belligerent as well as neutral, because they raise international questions. The Union blockade of the South became one of the four great blockades of modern times–one arising out of the Napoleonic Wars, and the last two arising from the two World Wars. The Civil War blockade was more extensive than the first and based many of its principles on those laid down by Great Britain during earlier struggles, which, incidentally, were furiously rejected by Americans of that time. From 1861 to 1865 it must have been a unique experience for Great Britain to have her own policies applied against her.[7]

During the American Civil War the federal navy undertook this broad role, devoting the majority of its energies to interdicting commercial traffic into and out of Confederate ports. The commercial nature of the blockade understandably involved the rights of neutral countries to trade. Not only did this embroil the United States in a series of diplomatic crises, but it also presented a domestic legal dilemma in interpreting the rules of its own blockade.

From the outset the federal government found itself on questionable legal grounds when President Lincoln declared a blockade of the southern coast. A majority of Lincoln's cabinet preferred an embargo or a suspension of intercourse with the South to a blockade. They argued that the true policy of the government was to close the ports until the rebellion was suppressed. Gideon Welles, Secretary of the Navy, later contended that it was William H. Seward, Secretary of State, who, for good or for evil, had influenced the President in declaring a blockade.[8] Under international law a nation closes its insurrectionary ports and blockades enemy nations. Not wishing to bestow belligerency status upon rebelling Americans, Seward argued that a nation had both belligerent and sovereign rights in suppressing a rebellion and neither need be granted to insurgents.[9] This contradictory position was never recognized by foreign powers.

While the English government recognized the validity of the blockade, many private owners of vessels and cargoes attacked its legality in the prize courts. The administration of the blockade and the conduct of the war rested on the decisions of the Justice Department in these cases. If the courts ruled against the validity of the blockade, southern trade would be open. The Supreme Court would have to determine the question of whether a nation, under international law, could blockade its own ports during an insurrection and whether the President could declare a blockade in the absence of a declaration of war.[10]

It was in the interests of the federal government to postpone final decision in these cases. By this tactic possibly adverse decisions were delayed while the war went on, and English merchants, uncertain of their rights, were more cautious in their trade with the Confederacy. Under diplomatic pressure the Supreme Court finally ruled that the blockade and the capture of prizes were legal. Perhaps more important, the President's use of extensive emergency power was sustained.[11]

The English Rule of War of 1756, later enlarged by the *Essex* case, established the belligerent right to seize cargoes ultimately destined–rather than immediately destined–for enemy shores. Specifically, England ruled that the landing of goods and payment of duties in neutral ports did not interrupt the continuity of the carriage of illegal cargo unless there was an honest intention to dispose of it in the neutral nation. Americans, not recognizing the doctrine, had chosen to interpret this as meaning that if a neutral port was the destination of a neutral ship and cargo that neither could be captured nor condemned. Further, a belligerent could not inquire of the ultimate destination once the cargo was landed at a neutral port. England never accepted America's insistence on free trade and, in fact, had come close to provoking trouble with the Americans during the Crimean War when England announced the doctrine in effect. Treaties followed between the two providing for freedom of neutral trade.[12]

American courts reversed this position during the Civil War. Because of the close proximity of neutral ports to the Confederate shore and the employment of fast blockade-runners, the United States insisted on the right of applying the doctrine of ultimate destination. Two epochal cases establishing this right involved the *Springbok* and the *Peterhoff*. Both cases concerned the doctrine of final destination, but the latter also included the question of blockading neutral ports. The British bark *Springbok* was captured by the U.S.S. *Sonoma* on the high seas while en route from London to Nassau, on 3 February 1863. On evidence that some of the cargo consisted of Confederate military equipage, the New York Prize Court condemned both vessel and cargo.[13] This judgment was foreshadowed in the Supreme Court cases of the *Bermuda* and *Stephen Hart* where both vessels and cargoes had been condemned. Upon appeal the Supreme Court decided in the *Springbok* case that, while the ship was innocent, the cargo, ultimately destined for a Confederate port, was evidence of intent to violate the blockade. Therefore, the cargo was liable to condemnation at any stage of its voyage, as if it had been one continuous voyage.[14]

Throughout the war American courts had to determine the legal intent of both cargoes and ships in prize cases. If a ship's papers proved that it was bound from one neutral port to another the ship might be judged innocent regardless of the nature of its cargo. But if reason existed to believe that the ship ultimately intended to run the blockade, or if efforts had been made to conceal the true destination, it was condemned regardless of how many neutral legs of the voyage the ship might first make. Thus, the question of whether just the cargo was guilty, or both cargo and vessel, depended on what the courts thought was the ultimate destination of the vessel. As for the cargoes, the courts considered that the true test of its intent was whether or not the goods were to be incorporated into the mass of goods for sale in the port of immediate destination. The courts would not accept the premise that a ship would take on contraband cargo for a neutral port. Unfortunately the two maritime offenses of violating the blockade and carrying contraband were not always distinguished by the courts.[15]

The case of the *Peterhoff* involved more complex questions. Union spies in London had determined that contraband was carried on board the vessel, which was bound for the neutral port of Matamoros, Mexico. Alerted federal naval authorities captured her five miles from the Danish West Indies port of St. Thomas. Also seized were British mails, a point that became a source of serious diplomatic problems.[16]

One of the finer points of law to be determined in the *Peterhoff* case was the belligerent right of blockading the mouth of a neutral river, specifically, the Rio Grande. The Treaty of Guadalupe Hidalgo, signed with Mexico in 1848, prohibited the establishment of a blockade at the mouth of the river by either party. The American counsel argued before the Supreme Court that a belligerent had the right to blockade the mouth of the Rio Grande as a natural part of the blockade, since the river was the common boundary with Confederate Texas. Chief Justice Chase ruled that the blockade of Brownsville, Texas–opposite the inland port of Matamoros–did not include the mouth of that river. He contended that neutral commerce, excepting contraband bound for Matamoros, was free.[17]

Justice Chase also advanced the "taint" ruling, as it came to be known. Since the cargo of the *Peterhoff* was of a mixed nature, he ruled that contraband articles contaminated those portions of the cargo not contraband if it belonged to the same owners. He further classified goods as to three types: (1) those that were primarily military, (2) those that could be either military or peaceful according to circumstances, and (3) those that were exclusively peaceful.[18]

During the arguments of this case American counsel also attempted to distinguish between goods unquestionably destined for Matamoros and those suspected of entering Texas. The British countered this, claiming that differentiation of cargo before landing was impossible. Lord Russell, British Foreign Secretary, wrote "that all the goods carried from London might be used in Mexico, and all the goods sent from New York might be transported by land to Texas."[19] The owners of the *Peterhoff* won their reversal in 1866. Even though the war was over, Chase, who was in Lincoln's cabinet before his appointment to the bench, quite likely had lent his legal influence to Lincoln's wartime policy. At least the government had begun to follow the court's opinion three years before it was rendered.[20]

The *Peterhoff* case was one of several of the Matamoros cases that seriously strained diplomatic relations with Great Britain. It was inevitable that neutral Matamoros would become an open door in the blockade, through which contraband and Confederate cotton would flow (Figure 3). High prices for both brought a heavy maritime traffic to the mouth of the Rio Grande. The federal navy in its zeal to curtail this trade–and to share in prize claims–seized vessels passing to and from Matamoros. This resulted in a series of Anglo-American disputes, involving a variety of legal distinctions, arising over the carriage of contraband between neutral ports. When these matters became diplomatically embarrassing to the United States, it often resulted in reprimand or censure of naval officers commanding vessels in the West Gulf Blockading Squadron.

The source of much of the diplomatic sparring during the war was action initiated by the federal government when war became inevitable. One week after the bombardment of Fort Sumter, President Abraham Lincoln "set on foot" a blockade of the ports of the seceded states. The proclamation stated that a competent force would be posted to prevent entrance and exit to the ports. Ships attempting either would be warned by naval officers, and the warnings entered by them on the ships' registers. Thus warned, these ships, if later attempting to violate the blockade, would subject themselves to capture and condemnation in prize courts.[21]

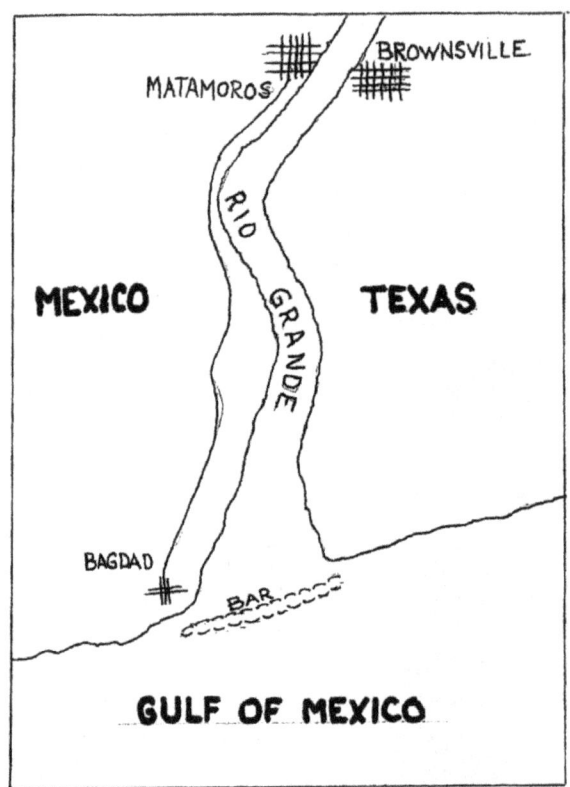

Figure 3. Mouth of the Rio Grande. Copied from 5 Wall (U.S.), 173. *By Glover.*

In proclaiming a blockade the federal government acknowledged by implication two things long recognized by international law: that a war and not an insurrection existed, and that the seceded states were an enemy power.[22] The federal government at no time officially acknowledged either implication, yet it continued to enforce the blockade throughout the war. This contradictory stance resulted in no small amount of diplomatic embarrassment for the Lincoln administration. Privately, cabinet members were often critical of the decision. Secretary of the Navy Gideon Welles confided to his diary that the President should have closed the southern ports, which would have denied the Confederacy belligerency status.[23]

To what extent Seward was the originator of the blockade decision is a matter of conjecture. The British minister to the United States, Lord Lyons, insisted that the administration had come to no decision regarding neutral commerce with the seceded states until the day before the proclamation. Lyons had immediately pressed Seward for definite details on how the blockade was to be carried into effect.[24] On 27 May Seward explained the blockade position to Charles F. Adams, American Minister to Great Britain. The government, he said, had a clear right to suppress insurrection by domestic law, the law of nature, and the laws of nations by the use of blockade. In order to exclude commerce from national ports that had been seized by insurgents, he continued, the use of blockade was a proper means to that end. A few days later he repeated essentially the same argument to Lyons.[25]

Later Secretary Welles attempted to define the American position in a message to congress. Since the commercial laws of the United States could no longer be enforced within the southern ports, he explained, it became necessary for American warships stationed outside the ports to perform this domestic municipal duty. He saw the blockade as a domestic embargo which, however, abridged all foreign rights of trade to neutrals trying to enter those ports.[26]

The official British reaction to the proclamation was guarded. Lyons' response was careful to acknowledge the federal intent to affect a blockade since no de facto blockade existed. Lyons finally got

Seward to comment on the extent of the intended blockade and learned that it included the entire coast from Chesapeake Bay to the mouth of the Rio Grande. When Lyons pointed out that the United States did not have a navy sufficient effectively to blockade 3,000 miles of coast, Seward adamantly asserted that the job would be done.[27]

The blockade posed a serious problem for Great Britain. If she questioned the declaration, a serious rupture between the two could occur. On the other hand, if the blockade was not effective the Americans would be guilty of piracy, which would involve the Royal Navy. The British had anticipated this problem as early as 16 February, when Russell urged Lyons to try to prevent the federals from blockading the Southern coast. To this end Lyons had strongly cautioned Seward on 6 March that any interference with British trade would be ruinous to the North's cause in England. Lyons even hinted at the possibility of British intervention to protect her trade.[28]

Meanwhile, Americans increasingly viewed Britain's non-recognition of the blockade as hostile. To compound the problem, the Confederate government announced its intent to issue letters of marque to privateers. The threat of a maritime war to England's commercial interests prompted that country, as well as most European countries, rapidly to formulate policies. The British policy was stated by Lyons on 4 May, when he notified Seward that America's Civil War was considered a regular war. Thus, Britain recognized the legitimacy of the blockade. Two days later Russell informed Lyons that in maintaining the consistency of this policy, the belligerency rights of the Confederacy would be recognized. One week later Great Britain announced her neutrality, warning her subjects throughout the world not to violate a blockade established by either belligerent.[29]

In view of this perplexing turn of events, the question might be raised as to why the United States did not simply close the southern ports–a recognized measure of suppressing rebellion–rather than blockading the coast. Aside from the reasons attributed to the inexperience of a new administration and the inconsistency of Seward, the answer appears simply to be one of expediency. Faced with a massive rebellion, the federal government felt it was necessary to employ the strongest, most effective means possible in suppressing it. An official blockade would be one of these means. Because of its legal status it promised more benefits, both in domestic enforcement and in foreign compliance. At the same time, by insisting that the war was a domestic rebellion, the United States discouraged foreign diplomatic recognition of the Confederacy.

In spite of the good fortune of the federal government in securing prompt international recognition of their blockade it was apparent to all that the measure was strictly *de jure*. Miserly peacetime appropriations had left the government badly deficient in both number and type of vessels needed to enforce an extensive blockade. Further, three-fourths of these were in foreign ports throughout the world, necessitating weeks of delay in reaching their blockading stations. As of March, 1861, there were forty-two commissioned vessels, plus twenty-seven more available for commissioning. This included all types, logistical and tactical. Of these there were only forty steamers, the best type for blockading duty. Five of these were logistical, nine were not commissioned, and five unserviceable, leaving twenty-one effective steamers to blockade 3,549 miles of coastline. By the end of 1861 there were only sixty-nine vessels of all classes serving in the home squadron.[30]

From the outset federal authorities placed major priority on the southern ports of the Atlantic seaboard. The Gulf coast, having few deep-water ports of any consequence, received a lower naval priority (Figure 4). Pensacola, Florida, had the best harbor with a depth of twenty-two feet. However, a lack of inland communications restricted this port's commercial importance. Mobile, Alabama, though with shallower depth, was a major Gulf port. New Orleans, some one hundred miles up the Mississippi River, was the other major port. Matamoros, Mexico, did, in fact, become a major Confederate port and will be treated separately later.

Secondary ports along the Gulf coast included Galveston, Sabine Pass, Corpus Christi, Velasco, Brownsville, and Indianola, all of which were in Texas. Lake Charles and Brashear City, Louisiana, were inland ports of limited importance, as was Mississippi City, Mississippi. Of all these, Galveston was the third most important port on the Gulf. All suffered from underdeveloped port facilities and lack of inland communications.

The character of the Gulf coast throughout its entirety varies little in depth and appearance. The water is of generally shallow depth for several miles into the Gulf, and the coast is either sandy or marshy. A peculiar aspect of the coast is that in many places it is double-fronted by low, narrow, off-shore islands. Behind these are numerous inland sounds of considerable extent forming internal waterways of bays, channels, rivers, lagoons, and swamps that pierce the land in all directions. Access to these inland waterways was, and is, generally impractical for vessels of any size and draft. Further, storms and currents continuously change the bars, obliterating old channels and creating new ones. The tides, while not mean, were nonetheless adversely affected by the strength and direction of the wind.[31]

By the end of the war the United States had over six hundred vessels in service. The fact that a large percentage of these served in the Gulf attested to the growing importance of that theater of operations. It had early become apparent to the Lincoln administration that a flourishing blockade-running trade all along the Gulf coast would require a major restrictive effort on the part of the Union navy.

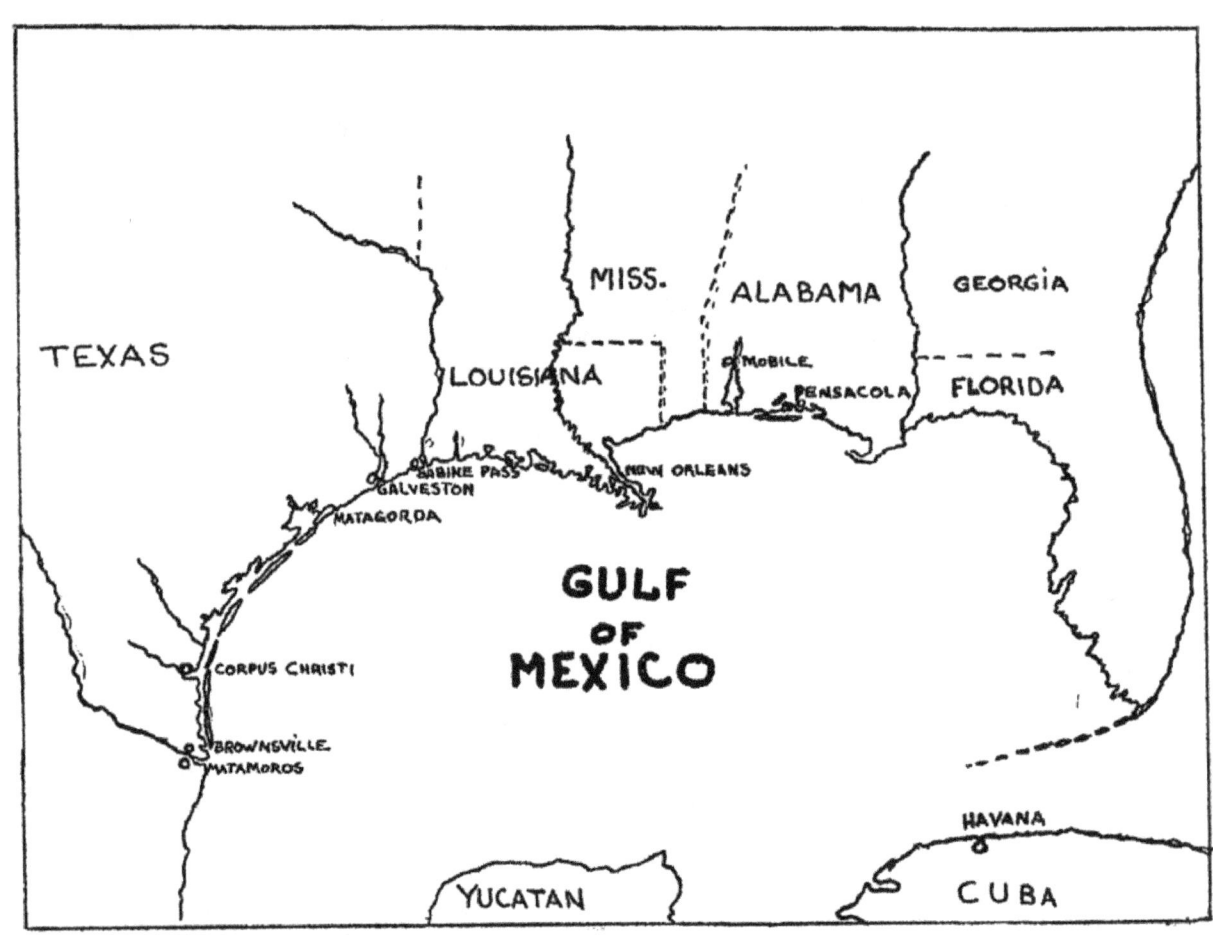

Figure 4. Gulf of Mexico. Copied from Taylor, *Running the Blockade. By Glover.*

Chapter 1 Notes

[1] Editors of the West Publishing Company, *Words and Phrases*, 46 vols. (St. Paul, Minnesota, 1962), 5:790.

[2] Editorial Staff, *American Jurisprudence: War*, 58 vols. (Rochester, New York: The Lawyers' Cooperative Publishing Co., 1948), 56:266-272.

[3] John Bassett Moore, *A Digest of International Law as Embodied in Diplomatic Discussions Treaties and other International Agreements, International Awards the Decisions of Municipal Courts, and the Writings of Jurists, and especially in Documents Published and Unpublished, Issued by Presidents and Secretaries of State, of the United States, the Opinions of the Attorneys General, and the Decisions of Courts, Federal and State*, 8 vols. (Washington: Government Printing Office, 1906), 1:563. Hereinafter cited as Moore, *Digest*.

[4] *Ibid.*; Ephraim Douglas Adams, *Great Britain and the American Civil War*, 2 vols. (New York: Longmans, Green and Co., 1925), 1:37; J. Thomas Scharf, *History of the Confederate States Navy from its Organization to the Surrender of its Last Vessel* (New York: Rogers and Sherwood, 1887), p. 58.

[5] Adams, *Great Britain and the American Civil War*, 1:139-140.

[6] Frank L. Owsley, "America and Freedom of the Seas," in *Essays in Honor of William E. Dodd*, ed. Avery O. Craven (Chicago: University of Chicago Press, 1935), p. 196.

[7] James M. Callahan, *Diplomatic History of the Southern Confederacy* (Baltimore: John Hopkins Press, 1901), p. 167; Katheryn Abbey Hanna, "Incidents of the Confederate Blockade," *Journal of Southern History* 11 (1945):216.

[8] Gideon Welles, *Lincoln and Seward* (New York: D. Van Nostrand and Co., 1874), pp. 122-123, 128.

[9] "Diplomatic Instructions, Great Britain, 1801-1906," Record Group 59, U.S. Department of State, National Archives.

[10] Stuart L. Bernath, *Squall Across the Atlantic American Civil War Prize Cases and Diplomacy* (Berkeley: University of California Press, 1970), pp. 18, 26, 166.

[11] *Ibid.*, pp. 26, 166, Prize Cases, 2 Black (U.S.), 635, (1862); Carolton J. Savage, *Policy of the United States Toward Maritime Commerce in War*, 2 vols. (Washington: Government Printing Office, 1934), 1:90, 446.

[12] Savage, *Policy*, 1:117-118, 132-133; Charles C. Hyde, *International Law Chiefly as Interpreted and Applied by the United States*, 3 vols. (Boston: Little Brown and Co., 1945), 3:2131.

[13] Charles N. Gregory, "The Doctrine of Continuous Voyage," *Harvard Law Review* 24(1911):174.

[14] The *Bermuda*, 3 Wall. (U.S.), 514 (1865); The *Springbok*, 5 Wall. (U.S.), 1 (1866).

[15] Herbert Whittaker Briggs, *The Doctrine of Continuous Voyage*, The Johns Hopkins University Studies in History and Political Science (Baltimore: Johns Hopkins Press, 1926), pp. 44-45, 52-53, 60.

[16] Bernath, *Squall Across the Atlantic*, pp. 63-64, 71-72.

[17] The *Peterhoff*, 5 Wall. (U.S.), 28, 56.

[18] *Ibid.*

[19] Quoted in Hanna, "Incidents of the Confederate Blockade," p. 224.

[20] *Ibid.*

[21] James D. Richardson, ed., *A Compilation of the Messages and Papers of the Presidents 1789–1902*, 10 vols. (Washington: Government Printing Office, 1907), 6:14-15.

[22] *Words and Phrases*, 5:790; James Russell Soley, *The Navy in the Civil War* (New York: Charles Scribner's Sons, 1885), 27.

[23] Howard K. Beale, ed., *Diary of Gideon Welles: Secretary of the Navy Under Lincoln and Johnson*, 3 vols. (New York: W. W. Norton and Co., 1960), 2:79. Hereinafter cited as Beale, *Welles Diary*.

[24] Lyons to Russell, 2 May 1861, as quoted in Savage, *Policy Toward Maritime Commerce*, 1:422-423.

[25] Seward to Adams, 21 May 1861, Seward to Lyons, 27 May 1861, *ibid.*, 1:426-427.

[26] Soley, *The Navy in the Civil War*, 29.

[27] Lyons to Russell, 2 May 1861, Savage, *Policy Toward Maritime Commerce*, 1:422.

[28] Bernath, *Squall Across the Atlantic*, 11; Adams, *Great Britain and the American Civil War*, 1:57-58, 244.

[29] Adams, *Great Britain and the American Civil War*, 1:87, 92; Great Britain, Public Record Office, Foreign Office Record Group 5, Series 2, 755; Bernath, *Squall Across the Atlantic*, pp. 3-4.

[30] Soley, *The Navy in the Civil War*, p. 241; Frank L. Owsley, "America and Freedom of the Seas," p. 197; *War of the Rebellion: The Official Records of the Union and Confederate Navies*, 22 vols. (Washington: Government Printing Office, 1880-1901), Ser. 1, 4:xv-xvi. Hereafter cited as *ORN*.

[31] Alfred Thayer Mahan, *The Gulf and Inland Waters*, vol. 3 of *The Navy in the Civil War*, 3 vols. (New York: Charles Scribner's Sons, 1883), pp. 2-3; W. R. Hooper, "Blockade Running" *Harpers New Monthly Magazine*, 42:105.

2

Administering the West Gulf Blockade

At the beginning of the war very few northerners, including Lincoln and his cabinet, realized the intensity of the task of administering a blockade. When Secretary Welles asked the principal shipping owners of New York for assistance in supplying ships, they advised him that thirty vessels should be sufficient. Welles, although having no professional naval experience, realized more quickly than others what was required. In this respect he had the sound advice of Gustavus V. Fox, the Assistant Secretary, who had eighteen years of naval experience.[1]

In trying to meet the emergency the Navy Department purchased as many merchant vessels as could feasibly be converted to military use. Anything that could carry a gun, including tugs and ferryboats, was purchased and converted as quickly as limited port facilities would allow. Naval yards, inadequate at best, were further overburdened by the loss of the yards at Norfolk, Virginia and Pensacola, Florida. It became necessary for naval officers personally to direct the few hastily assembled unskilled workmen in converting these vessels. As soon as these ships were ready they were assigned to blockading stations.[2]

The lack of naval personnel of all ranks was as serious as the lack of ships. Of the 7,600 seamen on active duty in March 1861, there were but 207 in all the home ports available for immediate duty. Also, 322 naval officers had resigned and tendered their services to the Confederacy. Those remaining who had been born in the South or had married southern girls were suspect, as were those who hesitated in declaring their loyal sympathies. Welles' job of assigning competent, trustworthy officers was further hampered by the lack of a naval retirement policy. Since promotion was based on seniority the highest command ranks were frozen by over age officers.[3] Until Congress rectified this situation, Welles had to assign these officers to flag command, even having to rely heavily on their own recommendation of each other for assignments.

It was largely due to the influence of Commodore Hiram Paulding that Flag Officer William Mervine was assigned to command the Gulf Blockading Squadron on 18 June 1861. Welles, acting against his better judgment, regretted the decision almost from the beginning. Mervin had served fifty-two years on active duty, and while his loyalties were unquestioned, he simply lacked vigorous executive and administrative ability. Welles, more specific in his evaluation, said that Mervine was good for nothing, taking too long in getting to his station, and accomplishing nothing when he reached there.[4]

In selecting Mervine's successor Welles screened assignees more carefully. After several months he knew whose counsel was most trustworthy and was able to narrow his choices to Commanders William W. McKean and C. H. Bell. It was due largely to the recommendations of his confidant, Flag Officer Andrew H. Foote, that he decided on McKean to command the Gulf Blockading Squadron.[5]

As the true proportions of the task of blockading the South became apparent, the decision was made to create a blockade board. This four-man board, headed by the Director of the Bureau of the Coastal Survey, studied the myriad logistical and operational problems of blockading. Quite soundly, they concluded that the most effective means of blockading would be by seizing and occupying key southern ports. These offensive operations would be launched from nearby staging areas. Those harbors too heavily defended could be efficiently sealed off, reasoned the board, by sinking stone-laden hulks in the channel entrances. Ultimately sixteen vessels, dubbed the "rat-hold squadron," were sunk in the ship channel entering Charleston harbor. The experiment failed, however, as new channels washed around the hulks and as British diplomatic pressure discouraged its continuation.[6]

Although Fox's idea of the stone fleet miscarried, his idea for the attack on New Orleans was nothing short of inspirational. The first step in carrying out this phase of the blockade was the seizure of Ship Island, located some one-hundred miles off the mouth of the Mississippi River, to be used as a staging depot. A meeting consisting of Lincoln, General George B. McClellan, Welles, Fox, and Commander David Dixon Porter, was called to present the plans. Lincoln approved the project, but McClellan favored an attack on Mobile first. Finally after deciding on New Orleans, the question of command arose and the name of David Glasgow Farragut was suggested. Porter, then Fox, interviewed Farragut in the following days to ascertain his qualifications for heading the expedition. Satisfied with the answers. Welles ordered Farragut to report to him on 15 December 1861.[7]

At the time he received his order, Farragut was serving on the retiring board at Brooklyn, New York. Welles' decision to use Farragut was not long in being questioned, both by naval officers who noted his advanced years and by congressmen who noted his southern origin. Undaunted by this criticism, Welles notified Farragut on 23 December that the department had decided to divide the Gulf Squadron into two commands shared by himself and McKean. Apparently the decision was made shortly thereafter that rather than divide one squadron two independent squadrons should be organized to blockade the Gulf. Consequently on 9 January 1862 Farragut was notified of his appointment to command the western Gulf Blockading Squadron, whose limits of responsibility began with St. Andrew's Bay, Florida and extended to the Rio Grande. The coasts of Mexico and Yucatan were also included within the cruising limits of the squadron. "As soon as the steam sloop *Hartford* is ready," commanded Welles, "hoist your flag."[8]

Orders to McKean advising him of the reorganization and instructing him to turn over twenty-six vessels to Farragut upon his arrival followed. While Farragut's new command was broad in scope, it was understood from the beginning that his primary objective was the capture of New Orleans. Specific orders to this effect came on 20 January, when Welles instructed him to proceed up the Mississippi River and reduce the defenses guarding the city, seize the town, hoist the flag, and hold until troops arrived. As soon as he finished this seemingly simple assignment he was then to take Mobile Bay. Wishing to impress upon the Flag Officer the importance of his assignment Welles became eloquent: "Destroy the armed barriers which these deluded people have raised up against the power of the United States Government and shoot down those who war against the Union . . . ."[9]

For the next three months Farragut awaited the necessary supplies, vessels, and personnel and devoted the majority of his energies and time to planning the proposed attack on New Orleans. Effectively stopping the illegal traffic along the vast coastline within the squadron's limits would have to await the arrival of more ships. Yet Welles was apparently impatient with his Gulf commanders and could not understand their seeming lack of vigilance in stopping this traffic. Welles was acutely aware of the diplomatic embarrassment caused by repeated blockade-running and had informed Farragut that certain commercial nations wanted to break the blockade. They would excuse their acts, he said, on the lack of effective enforcement of the blockade. He even suggested to Farragut that he should avoid concentrating his vessels around key ports but should spread them evenly along the whole coast.[10] It was apparent that the Secretary lacked a realistic view of the problems existing along the Gulf coast.

Farragut was not long in evaluating the problem and informing the Department. One of the greatest difficulties, he noted, was the shallowness of the Gulf. Light draft blockaders, drawing less than six feet of water were essential to navigate the inlets and rivers and stop the runners at their points of origin. Most of the running, he understood, was conducted by small vessels, some leaving coastal streams incapable of emitting vessels larger than rowboats. It was soon apparent to the new commander that many large steamers from New Orleans were successfully leaving through the guarded passes at the mouth of the river and arriving in Havana. In explaining this dilemma, Farragut told Welles that these runners would lay four or five miles up the river awaiting fog or darkness, at which time they would run out. Having a fast start under a full head of steam, they were soon beyond the reach of the blockaders.[11]

Before this explanation reached the Secretary, he had already heard of the arrival of the Confederate cotton from the American consul in Havana. In a stinging rebuke to Farragut, Welles stated that the Navy Department was surprised and regretful to learn of the arrival of the steamers, having been led to believe that the passes of the Mississippi River were effectively guarded. His note ended with the subtle warning that the Department trusted that it would not hear of this reoccurring. Five days later Welles informed Farragut that information had been received informing him of a constant blockade-running trade being conducted between Sabine Pass, Texas, and Havana.[12] How long it took the Secretary finally to realize that there would never be an airtight blockade of the Gulf coast is a matter of conjecture.

Already international repercussions were being felt over the administering of the blockade at the mouth of the Rio Grande. The blacklisted British steamer *Labuan*, lying at anchor off Boca Chica, one of the mouths of the river, was captured while loading cotton. Captain Samuel Swartwout of the U.S.S. *Portsmouth* found no regular ships papers on board but did find 439 bales of cotton which had been loaded from river lighters during the preceding month. Swartwout determined that she was lying one mile north of the mouth, thus in enemy waters, and was taking on Confederate cotton. After seizing the *Labuan* Swartwout sent her to Ship Island for disposition.[13]

The British Vice-Consul at Matamoros was quick to react to what he labeled an act of piracy, since there was no actual blockade of that part of the Texas coast at the time the *Portsmouth* arrived. He also stated that all the *Labuan's* papers were on deposit in the consulate office and would prove to be in order. As to Swartwout's assertion that the ship was captured in enemy waters, he was probably unaware, suggested the Consul, that the Treaty of Guadalupe Hidalgo declared a neutral zone one nautical league north and south of the mouth of the Rio Grande. Further, he complained of Swartwout's impatient and uncooperative manner when confronted with these facts.[14]

British anger intensified as the incident was reported through diplomatic channels. Lord Russell ordered the admiralty to insure British neutral rights. They, in turn instructed Vice-Admiral Sir Alexander Milne, commanding the North American station, to provide for the security of English vessels and property unquestionably within neutral waters. Milne was to use force only if the foreign office failed to obtain redress from the American government. Lord Lyons presented his country's grievances to Seward on 23 April 1862, impressing upon him the seriousness of the matter and of the potentially dangerous precedent set in the *Labuan* affair. He re-emphasized the fact that the ship had been captured at anchor within the legal limits of the port of Matamoros and that there was no de facto blockade of the Texas coast in the immediate vicinity at the time. His government, therefore, urged the prompt release of the ship and cargo with compensation.[15]

The seriousness of the situation prompted Seward to press the judiciary for a quick settlement. On 21 May 1862, Judge Betts of the New York District Court released the *Labuan* but postponed the matter of compensation. The delay enabled the State and Judiciary Departments to gather substantiating evidence from naval and consular sources near the Rio Grande. Unfortunately, the American Consul in Matamoros was unable to help the government's case. Consular records covering the period had been destroyed in the revolutionary fighting that had broken out in the city. His reports of 1 and 21 March stressed the heavy traffic to and from the Confederacy that came through Matamoros and claimed that the *Labuan* had landed her entire cargo at Brownsville and had been loaded by lighters flying the Confederate flag. Over a year later when the U.S. district attorney pressed the consul for proof of the real nature of the *Labuan's* voyage, the consul had to admit failure. Not only were most of the witnesses to the *Labuan's* cargo scattered or gone, he reported, but those available would not testify because of their interests in the lucrative trade.[16]

Four months to the day after her release, the *Labuan* was again loading cotton from Matamoros, in exchange for New York merchandise. Consul Pierce claimed that several Texas merchants were awaiting her return, and that the ship should be refused clearance in New York. Two days previously he had complained of the absence of U.S. naval vessels to stop this kind of trade. "There are no ships higher on this coast," he testified, "than Vera Cruz."[17]

Both diplomatic necessity and operational limitations accounted for the absence of permanent blockaders near the mouth of the river. Previously, Welles had expressed doubt over the legality of the *Labuan's* capture and had instructed Farragut to make no more captures of vessels off the Rio Grande. He was to do so only if they were on their way to a Texas port and had been duly warned. It appears that with the arrival of British and French warships Commander Swartwout developed a case of jitters. He explained to Farragut the dilemma of having to stop the illegal trade and at the same time avoiding an incident with a friendly nation. Swartwout alleged that the U.S. had no legal right to interfere with the Matamoros commerce anyway.[18]

Farragut was not concerned with diplomatic delicacy. To him Matamoros was not just any neutral port; it was a special port lying immediately adjacent to an enemy whom he was expected to blockade. The letter of the law governing neutral trade was of little use when the spirit of the law was so flagrantly violated. Not only did he defend Swartwout's handling of the *Labuan* incident, but he gave his successor specific instructions to visit and search all vessels going to and from Matamoros. All enemy cargoes leaving Matamoros were to be seized, he emphasized, and those vessels entering or attempting to enter were to be seized if contraband was found on board which was destined for the enemy. "But," he cautioned, "we should [not] interfere with or molest neutrals in their legitimate commerce with Matamoros more than is indispensably necessary to comply with the above instructions."[19]

To Farragut the whole problem was one of practical expedience: let the navy enforce the blockade and let the courts decide the legality of their acts. He expressed his confidence in the court's ability to decide these issues in a letter to the commander of the British naval forces at the Rio Grande, Commander Hugh Dunlop. Farragut reminded his British colleague that the numerous incidents of fraud involving the Matamoros trade necessitated exercising the belligerent right of visitation and search. This was the only way the true nature of cargoes could be determined, he pointed out, especially in the case of the trade on the Rio Grande, which was a common highway for both friends and enemies. Farragut finished his explanation by reminding Dunlop of one of his own country's historic principles: that the flag was no protector of contraband cargo.[20]

Welles, in spite of his earlier diffident views, came to accept Farragut's more obstinate methods of enforcement. The British Foreign Office continued to protest questionable seizures, however, which ultimately resulted in pressure being brought against Welles by Lincoln and Seward. Consequently, on 18 August 1862, the Navy Department issued new orders to its commanders to be vigilant but not to seize vessels in neutral waters. Any seizure had to be proceeded by search only when a vessel was suspected of carrying contraband to the enemy, or by transshipment, or was in some other way violating the blockade.[21]

The whole vexatious problem of Matamoros was one that could be solved by the military occupation of the east bank of the river. Not only would this interdict the cross-river traffic with the Confederacy but it would eliminate the pitfalls of violating international trade. Additionally, and of equal importance to the Union, it would remind the French imperialists, battling for control of Mexico, of the Monroe Doctrine. To Gideon Welles such a move would protect what he believed were large numbers of Union loyalists living in the western part of Texas. For all of these reasons various federal authorities from ships' commanders to cabinet secretaries urged this action throughout 1862. Because of the strategic emphasis placed on the capture of Richmond and the Mississippi River the War Department had to deny the requests of the State and Navy Departments for troops to be used on the Rio Grande.

It was not until 6 November 1863 that several thousand troops occupied Brownsville, Texas and a small portion of the lower Rio Grande. There were too few troops to stop the Matamoros trade which simply crossed higher up the river. Furthermore, the presence of U.S. forces seemed to hinder rather than help American-Mexican relations. Despite American determination to stay out of Mexican internal affairs, there was constant interaction and intrigue between Mexican combatants and American officials. Wisely the federals decided to retire from this turbulent scene. Besides, these groups were needed in upcoming priority campaigns in the East. In midsummer, the federals started demobilizing and by September had abandoned all points except Brazos Santiago.[22]

Welles made it a point to inform Congress of the Brownsville failure in his annual message. He prefaced his report with a statement of the heavy contraband trade at Matamoros, which had been checked, he declared, by the occupation of Brownsville. So effectively had it been stopped, he believed, that the President, by proclamation, conditionally relaxed the blockade of Brownsville. But it was of short duration, he lamented, with the withdrawal of military forces. He reminded them that once again the sole duty of guarding that extensive coast had devolved upon the navy.[23]

Elsewhere along the Gulf coast the federal navy was gradually tightening the blockade. The tactical employment of "flying squadrons," cruising at large in the Caribbean and along the southern coast, was taking effect. This, in conjunction with the courts' application of the doctrine of continuous voyage, was stifling imports into the South. This point was noted by the Confederate Secretary of State, Judah P. Benjamin. Writing to the Confederate emissary in France, he confided that the blockade of southern ports had not interdicted foreign trade; but that the federal cruisers stopping and seizing neutral vessels bound for unguarded ports was damaging. "If Europe, even at this late date, would put an effectual stop to this outrage on its rights of trade with a belligerent we should soon be well supplied with her manufactures . . . ."[24]

Farragut handed the Union a major victory with his capture of New Orleans in April of 1862. By the time this was accomplished the Lincoln administration had realigned its strategic priorities, and Farragut was next ordered to concentrate on reducing the Mississippi River fortifications. Farragut, who felt that the reduction and closing of the major port of Mobile was most important, had to forego these plans and cooperate with the land forces in the occupation of the Mississippi River. Until July of 1863 his control of the vast Gulf blockade was quite remote and its enforcement was left largely in the hands of individual ship's commanders.[25]

For the three months following the occupation of New Orleans Farragut gave his attention to the blockade. Pensacola, which had been evacuated by Confederate forces on 10 May, became his headquarters station. From there he re-outfitted his battered squadron and directed naval operations along the coast until time came for him to move up river against Vicksburg. At this stage his job seemed to be progressing satisfactorily, marred only by the successful daylight entry of the Confederate cruiser *Florida* (*Oreto*) at Mobile.[26]

With the closure of New Orleans and Pensacola, much blockade-running activity shifted to the Texas coast. Galveston had enjoyed a brisk trade in spite of the appearance of the first blockaders in July, 1861. Six months later Commander McKean reported the blockade of the port to be efficient, having recently visited the only vessel on duty there and finding her in fine order. Only three days before this optimistic report was sent, the British Consul in Galveston was forwarding a quite different view. While admitting to the Foreign Office that entrance and exit were not knowingly allowed by the federals, still the blockade could not be considered efficient. This was due, he explained, to the fact that San Luis Pass at the southern end of the island was wide-open, allowing free access to the port via West Bay.[27]

Notwithstanding this easy trade, Confederate authorities knew the poorly defended island could not withstand a determined attack. The attack never came but federal naval forces captured the port quietly in October, 1862, after allowing the Confederate authorities time to evacuate all troops, machinery, government equipment, and ordnance to the mainland. This Union success had been attained by only eight vessels of the West Gulf Squadron.[28]

A few days before the capture of Galveston, Sabine Pass had been captured. Here also, Confederate defenses were almost non-existent, and these were abandoned after an exchange of gunfire. Federal raiding parties cruised at will around Sabine Lake and partially destroyed the railroad bridge between Beaumont and Houston. After taking several prizes in the vicinity, these federal forces cruised unopposed up the nearby Mermentau and Calcasieu Rivers in Louisiana, taking several more prizes near their mouths.[29] Along the south Texas coast Matagorda Bay was entered and patrolled freely by naval forces, which attacked coastal ports from Matagorda to Corpus Christi. Throughout the summer and fall of 1862 federal raiding parties cruised the shallow waters of the elongated bay and captured or destroyed a number of small sailing craft.[30]

On 15 October 1862 Farragut reported–not entirely accurately–to Welles that Galveston, Corpus Christi, Sabine City, and adjacent waters were in federal possession. He was elated that it had been accomplished so easily. Perhaps for the first time Union authorities realized the relative defenselessness of the Gulf coast. It may well have given Farragut a false sense of security as he estimated that only a few soldiers would be required to hold these places. Quite confidently he informed General Butler that one regiment would suffice to hold Galveston. While the Admiral admitted they would be isolated, there was nothing to fear, ". . . with a gunboat inside [the bay] their protection must be perfect."[31]

Fixed with this high-minded optimism the Admiral had little patience with the gloomy news he soon received from Commander W. B. Renshaw, in command at Galveston. In a scathing reply Farragut demanded to know why a third boat crew had been lost to the enemy. But the most distressing thing about the situation on the Texas coast, Farragut stated, was learning that Corpus Christi and Sabine Pass had been abandoned by the blockaders and that Galveston was about to be. "Has it come to this," asked Farragut, "that four well-armed gunboats are to be driven out of a harbor by the report of some reliable person, . . . are you willing that I should make such a statement to the honorable Secretary . . . ?"[32] The Admiral made it clear to Renshaw that he expected the gunboats to hold until the army arrived, and further, he was to reoccupy Sabine City and Matagorda Bay.[33]

Renshaw, infuriated over the Admiral's rebuke, poured out his displeasure to his close friend, Rear Admiral D. D. Porter. Calling Farragut brainless, he related how the Admiral had virtually left him at an unsupported station, facing an imminent Confederate attack, and bereft of supplies. He had related his deplorable situation to "Farraguts," he said, but it was like writing to an ass. Then in a somber vein, he confided that he expected an attack hourly and feared the outcome. "If they come the result will tell how far my opinion was correct."[34]

Farragut himself was not immune from administrative reprimand, which probably resulted from testimony given by Lieutenant George H. Preble. This hapless commander had been relieved of his command for having allowed the C.S.S. *Florida* to get into Mobile. In his own defense, Preble complained of a lack of intelligence disseminated to ship commanders concerning suspected runners. While Preble's complaint did not help his case, it seemed to convince Welles that Farragut had been administratively lax in keeping his subordinates informed. Welles took Farragut to task for poor command communication and closed his letter by reminding him again of the importance of blockading Mobile. These developments had the positive effect of convincing the Department of what should have been evident from the start: that the varied administrative duties of the squadron were too extensive for one man to manage well. Thus, the department ordered specific administrative assignments to be made within the squadron.[35]

As for Mobile, its reduction would have to await the arrival of more ships and large numbers of troops. By now its coastal and harbor defenses were too strong for anything short of a massive combined arms operation to take. Again, federal priorities made heavy demands on both ships and man for other campaigns, notably at Vicksburg. Farragut would have to live with the fact that Mobile was the last port of consequence left open on the Gulf.[36]

All things considered, as 1862 came to a close, the West Gulf Blockading Squadron was accomplishing its mission well. By the first of November 1862, its ships had captured about fifty-four blockade-runners. Two of the most successful ships in the squadron, the *Kanawha* and the *Hatteras,* had taken six prizes each. The majority of prizes had been outward bound with cargoes of cotton, turpentine, resin, rice, cornmeal, flour, sugar, molasses, and lumber. Some cargoes were quite valuable, like that of the *Water Witch*, which was valued at $10,000.[37]

Most of the coast was, to a significant degree, controlled, and although ships continued to run the blockade, federal action had forced a change in their tactics. The U.S. Consul in Havana informed Farragut that the capture of several large steamers had proven so costly that owners were now purchasing schooners and small vessels to lessen the risks. The Consul reported also that there was a drop in the number of arrivals from the ports under Farragut's jurisdiction.[38] The Admiral had every reason to view the coming year with optimism.

However, such optimism proved to be unfounded. Beginning with the earliest hours of the new year, 1863, and continuing periodically throughout the year the squadron suffered a series of disasters and setbacks. January was particularly disastrous, beginning with the loss of Galveston and of four ships of the squadron on the morning of 1 January. This disaster occurred when Confederate forces of Sibley's Brigade stormed across the railroad causeway, supported by an attack by Confederate naval forces. The Confederates had reinforced two riverboats with cotton bales and manned them with several hundred cavalry sharpshooters. Their rifle fire swept the decks of the U.S.S. *Harriett Lane*, killing and driving away the gunners, thereby allowing the Confederates to board and capture her. The gunboat *Westfield* coming up to join battle ran aground and had to be destroyed. The premature explosion of her magazines killed Commander Renshaw and several of the crew. Two support vessels were captured as well as 350 soldiers of the 42nd Massachusetts Infantry–the only reinforcements that arrived to hold the island. The remaining federal vessels steamed away, and the next day Confederate General John B. Magruder officially proclaimed the blockade lifted at Galveston.[39]

Under the law of nations, this fact had to be acknowledged by the chagrined Lincoln administration. Welles was highly critical of Seward's decision in notifying all foreign embassies of the lifting of the blockade, probably because he had not consulted with the Navy Secretary before the decision was made. It was, according to Welles, "an assumption, pregnant with error in which he sometimes indulges."[40] Further, he considered Seward inexcusably ignorant on the subject of blockade. As for the disaster at Galveston, Welles confided that it happened because of a lack of good management and declared that someone was to blame for the failure to garrison the port with an adequate force. Under-Secretary Fox declared the disaster to be the most disgraceful affair in the navy's history. Farragut, who could not tolerate so much as the loss of a skiff, was mortified over the news and moved quickly to convene courts of inquiry. As for Renshaw, all Farragut could say was "bah–he is dead."[41]

The loss of Galveston, along with the casualties sustained, was bad enough, but the shock to federal naval prestige was worse. The affair served thereafter as an object lesson for naval commanders, who cautioned subordinates to remember what happened to the *Harriett Lane*. Farragut was disgusted with the behavior of the commanders at Galveston and openly expressed his feelings. He confided to Welles that the affair had upset his overall plans for naval operations in the Gulf. Particularly was he apprehensive about what the Confederates might do with the *Lane*. "I must try to recover her," he continued, "as she would be a most formidable cruiser if she gets out on account of her speed and battery."[42] To Bell he complained bitterly about how the *Lane's* consort vessels had "basely and most unheroically deserted her in battle." Later he summed up his feelings to Welles: "The capture of the *Harriett Lane* and the abandonment of Galveston was not only the most unfortunate thing that ever happened to the navy but the most shameful."[43]

Farragut was keenly embarrassed over the affair. It is apparent that he was largely responsible for underestimating both Confederate capabilities and the warnings in Renshaw's situation report. Further, the admiral had seriously overestimated the ability of gunboats to hold enemy territory unaided. The Navy Department must bear ultimate responsibility, for they had overburdened Farragut with too broad a command responsibility. It was too much to expect of one man to command military operations against strong objectives like New Orleans and Vicksburg and at the same time maintain an airtight blockade of an extensive enemy shoreline. The Navy Department, or for that matter, Farragut could have assigned a strong deputy commander to be solely responsible for enforcing the blockade. As it was, enforcement was left to the individual ship's officers at their scattered stations. Undoubtedly strict enforcement was impaired by a lack of coordination. Welles, commenting on Farragut's shortcomings, noted that he was good for only one thing at a time and that he was administratively lax.[44] Notwithstanding this noted deficiency, no efforts were made to relieve Farragut of any of his command problems.

More disasters beset the squadron rapidly following the Galveston fiasco. Farragut had first considered returning to Galveston and personally re-establishing the blockade but instead sent his Chief-of-Staff, Commodore Henry H. Bell. The blockade was technically reestablished with Bell's arrival with the gunboat *Brooklyn* on 7 January 1863. No sooner had the new flotilla arrived on station, however, when one of their number, the *Hatteras*, was sunk. She had chased a strange sail out to sea, only to learn too late that it was the Confederate raider *Alabama*. Three days later the squadron lost one of its most promising officers, Lt. Commander H. M. Buckannan, who was killed in an engagement on Bayou Teche. On the sixteenth the C.S.S. *Florida*, whose successful entry into Mobile had cost Preble his job, successfully escaped. Less than two weeks later two Confederate cotton-clad riverboats ran out of Sabine Pass and captured the blockaders *Velocity* and *Morning Light*. In February the squadron lost in action the formidable river gunboats *Queen of the West* and *Indianola*.[45]

As March approached, Farragut was preoccupied with the immediate problems of getting his vessels above the Confederate river defenses at Port Hudson, Louisiana. That he had not attempted this two months earlier, he explained, was due to the disasters at Galveston and Sabine Pass and to the Confederate build-up at Mobile. Thus, when his ships attempted to pass the batteries on the night of 14 March, it was with an inadequate force. Only part of the force passed and one of his biggest ships, the *Mississippi* was destroyed. It was with a sense of failure that he reported these facts to the Navy Department. Less than two weeks later news reached him of the loss of another gunboat, the *Diana* on Atchafalaya Bayou.[46]

The fall of Vicksburg and Port Hudson in July allowed the Admiral to return to New Orleans and to the administrative duties of the squadron. He would have preferred to proceed with the Mobile plans, but orders from the Navy Department brought him to Washington in August. Before leaving, he gave temporary command of the squadron to Bell and briefed him on the situation within the squadron. The Mississippi River, he explained, was under the control of Rear-Admiral Porter. As for the blockade, there were now a sufficient number of vessels on hand. Farragut was optimistic about Confederate weaknesses and predicted that by October or November the federals "will walk into Galveston."[47]

Farragut's dream of entering Galveston in the near future was not realized. The Confederates, having lost nearly their whole coastline once before, would not allow it to happen that easily again. They had endeavored to strengthen their coastal fortifications, especially those at Galveston. Utilizing the splendid ordnance salvaged from federal gunboats along with that purchased from abroad, the defenders forced the blockaders to keep a safe distance, which also had the effect of giving runners more room to get in and out.[48]

Evidence of this strength–or determination to defend–came at Sabine Pass in early September 1863. The administration, for a variety of reasons, mostly political, had determined to develop a military campaign in the southeastern quarter of Texas. Their objective, after securing Beaumont, was Houston. General N. P. Banks with several thousand troops, accompanied by vessels of the West Gulf Blockading Squadron, attempted to force passage of Sabine Pass on 7 and 8 September. The operation was badly managed and ultimately abandoned after the loss of the gunboats *Sachem* and *Clifton*. The whole operation had been stymied by the defense of forty-two cannoneers in a mud fort.[49]

Farragut was in Navy Department headquarters when he learned of the proposed Sabine Pass expedition. He confided his misgivings about the plan to Welles, complaining that the army expected too much of the vessels that would participate. His predictions appeared realized when news of the disaster reached them. The occasion afforded Welles the opportunity to reflect on the good fortune of the Navy Department in the selection of its leaders. Even the President had noted this to Welles, and all agreed that Farragut's appointment was the most fortuitous one made. Welles admitted to his diary that the navy made mistakes but that they were not as conspicuous as those of the army. One of these mistakes, as it developed, had been made by Commander Bell in the Sabine Pass affair. Welles reprimanded Bell for his decisions on the tactical employment of the vessels and for his choice of inexperienced leaders.[50]

Command problems, however, were not the greatest administrative headaches of the squadron. The second year of the war operational requirements had taken a large toll of vessels, not so much in loss to enemy action as to the strenuous wear and tear of blockade duty. Fox noted the generally run-down condition of the West Gulf Squadron and urged DuPont to dispatch some of his ironclads to the Gulf stations as soon as possible. Gulf commanders complained frequently of the poor condition of their vessels, and Farragut claimed that the replacement vessels being sent out were in as bad or worse condition as those sent in for repairs.[51]

The Admiral had often urged the need for expanded naval yard facilities in his sector. In November of 1863 Farragut again lamented the fact that nearly all his steamers were in New Orleans being repaired. "It would be worth a mint," he stated, "to have appliances for repairs at Pensacola." A year later he was so disgusted with the poor maintenance logistics that he told Welles that it was almost impossible to maintain a blockade with vessels that could catch only sailing vessels and schooners. He cited the example of the *Owasco*, which had almost rolled itself to pieces in a gale. Farragut wanted the Department to know of the poor condition of his gunboats which resulted in some of the blockade-runners getting past. "It is a most mortifying fact that we have no vessel with sufficient speed to catch any of the fast steamers that run the blockade," he told Welles.[52] Farragut never got his wish for an adequate Pensacola yard, and Gulf commanders continued to send vessels to New Orleans for repairs. Vessels repeatedly left their duty stations for repairs, and the situation was never alleviated, as evidenced by there being only eighteen vessels on duty along the Texas coast at the war's end.[53]

The Department of the Navy was not altogether at fault in not giving its commanders the best maintenance support. Secretary Welles had urged Congress to establish two modern yards to augment the need for steam vessels and to offset the loss of those yards in southern hands. Self-seeking congressmen tied up Welles' recommendations in deadlocked legislation for the remainder of the war.[54]

By far the most acute logistical shortage facing the squadron was that of personnel. Like the army, naval ranks had been filled with volunteers whose enlistments often expired on or about the same time. When this occurred in the West Gulf Squadron, it often left ships with only skeletal crews aboard. In the summer of 1863 Commander Bell complained to the Department that the squadron was 500 men short, and enlistments were expiring constantly. The gunboat *Pensacola*, explained Bell, was short 103 men out of a complement of 269. He added that it was for lack of men that such a small prize crew had been put aboard the captured *Atlantic*, which was in turn re-commandeered by the captives and sailed into a Texas port.[55]

Subsequently, Bell notified higher headquarters that the ship and crew of the *W. S. Anderson* had to be replaced because of expired enlistments. Even then, Bell complained, those crewmen had been detained beyond their expiration date. Less than two months later Bell reported a near-crisis manpower shortage of 111 officers and 1,286 men. One of the causes of the shortages, Bell noted, was that army enlistment bounties were luring away navy men whose enlistments were up. By August of 1863, Bell had stopped discharging volunteers until their replacements arrived. Sickness and disease compounded the problem, as evidenced by the crew of the *Santee*, which had to leave her Galveston station because so many of her crew were on the sick list due to scurvy.[56]

Gideon Welles was sensitive to these problems and had appealed to Congress for funds for increased enlistments. He cited the case of the *Water Witch* whose understrength crew had been overpowered and captured by a boarding party. Meanwhile, Welles suggested to Bell the expediency of promoting worthy master's mates to ensigns and of utilizing freed slaves to meet his manpower shortages. Further, Bell was encouraged to cannibalize all ships being sent north of any extra supplies and crewmen.[57] Personnel shortages continued to plague the squadron throughout the war.

The squadron also suffered, although not so severely, from periodic shortages of boiler plate, rivets, bar-iron, and steam-maintenance items, as well as food. It was inevitable that serious logistical shortages would occur in supplying ships that were scattered along an extensive coastline. It also seems apparent that the Gulf Squadron was a low-priority command. Notwithstanding this, the navy attempted to keep the Gulf sailors supplied with fresh food. Two ice-packed "refrigerated" vessels were employed in supplying the Gulf squadrons with fresh meat and vegetables. In spite of these efforts, Union crews and soldiers desiring fresh beef staged a large-scale cattle raid below Velasco, Texas. For a while, Confederate authorities in the region suspected an invasion until the true intent of the operation was known.[58]

By the summer of 1863 two unsuspected situations had developed which compounded the problems of administering the blockade. Both of these were related to profiteering, which had grown out of the increase in prices within the Confederacy and in the price of cotton abroad. The federal government invited these problems by enacting laws and making proclamations authorizing trade with certain loyal subjects living within the seceded states.[59] These provisions unwittingly allowed a tremendous amount of illicit trade through the blockade. Especially was this the case where the military departmental commander was one of questionable scruples.

Within a few months after assuming command of the Department of the Gulf, General Benjamin F. Butler was permitting a nefarious trade through the blockade. Farragut first learned of it when he saw an order from Butler to the commanders of blockading vessels to permit the safe passage of the *West Florida*. This vessel, bound from New Orleans, was to exchange her cargo for a load of cotton at Sabine Pass, Texas, and then proceed to Matamoros. Since the owner was loyal citizen with property in New Orleans to ensure her fidelity, explained Butler, the navy was to allow the transaction of contraband cotton. In spite of Butler's permit, the vessel was captured at Sabine Pass by one of Farragut's blockaders and sent north for adjudication. Eventually, the story reached the War Department which strongly rebuked Butler and warned him not to issue any similar permits without specific instructions from the Department.[60]

Individuals within the Confederacy were more explicit about Butler's involvement in this leak in the blockade. The British consul in Mobile explained how Liverpool salt arrived at that and other Confederate points. The salt was purchased in New Orleans by speculators at four dollars a sack, plus one dollar a sack paid to General Butler, who then permitted the salt to be transported to Mobile. Here it brought ninety to one-hundred dollars a sack in Confederate script. On the return trip through the lines, Butler received a rake-off of scarce southern produce, like resin, turpentine, and cotton.[61]

The commercial possibilities were so promising at New Orleans during this period that a French firm offered to supply the Confederate War Department with a million pounds of pork, 200,000 blankets, 100,000 pairs of shoes, and 100,000 yards of flannel in exchange for cotton. Although not mentioning how this was to be allowed, by asserting that the supplies would be delivered to Vicksburg, and the return cotton would proceed unmolested to a French port, it was understood that federal permission was no problem. The Confederate government must have understood it also, and hard-pressed as they were for these goods they declined the offer.[62]

Farragut was more inclined to blame New Orleans customs officials for fraudulent clearances than Butler for permitting this traffic. He complained to the port collector that while he was trying to enforce a blockade, customhouses were clearing vessels daily for Matamoros. Thirty-one such vessels had recently been cleared whose cargoes, the Admiral charged, were for the enemy. Further, there was a considerable amount of smuggling conducted across the coastal lakes of south Louisiana.[63]

With Butler's reassignment, the incidence of fraudulent shipments abated in the Gulf zone, only to recur again in 1864. This time, it seems to have been largely initiated by the Lincoln administration. The scarcity of cotton to feed the mills of New England, coupled with European diplomatic pressure for the American staple, resulted in congressional and executive action aimed at procuring this southern commodity. The Treasury Department was authorized to buy cotton from persons living within the Confederacy, who would have unimpeded passage through the Union lines and blockade in delivering it to treasury agents. The administration also sanctioned the scheme on the grounds that it would lure influential southern planters back into the Union.[64]

The supervising treasury agent at New Orleans, while probably honest, provoked the ire of certain federal officers in his zeal to get the cotton out. In one instance, a ship loaded with cotton consigned to a southerner at Havana was captured off Mobile. Those on board claimed they were headed for New Orleans and presented a permit issued by Agent A. R. Dennison authorizing them to bring out cotton from behind Rebel lines to be delivered to treasury agents. When Dennison was questioned, he declared that he had General Banks' authority for his action and asserted that it mattered not whether the cotton came through a blockaded port as long as it got to New Orleans. Rear-Admiral Bailey of the blockading squadron noted in his report that this was not only trading with the enemy but also was a virtual abandonment of the blockade. When confronted, Banks disavowed all connection with the affair and denounced it roundly.[65]

Admiral Porter was more emphatic in his denunciation of the government's trade regulations, stating that they would enable the South to continue the war indefinitely. In Congress, Senator Ten Eyck charged that the trade had prolonged the war and that some military and naval movements had been made with more intent to seize cotton than to defeat the enemy.[66]

In September Farragut received a copy of an order to General E. R. S. Canby, signed by President Lincoln, directing that Andrew J. Hamilton of Texas be permitted to export cotton from various Texas ports. Hamilton, a Unionist, had fled the state and resided in New Orleans. When General Banks occupied the Brownsville region in force, Lincoln had appointed Hamilton as Military Governor of Texas. Lincoln's order gave Hamilton wide latitude in designating other southern exporters who would, by Hamilton's authorization, be cleared through the blockade.[67]

Upon receipt of this information, Farragut immediately suspected fraudulent intent and protested to Welles, while at the same time he grudgingly notified the blockade commander at Galveston to pass any of Hamilton's cotton. Welles immediately sought out the President and informed him that the executive order conflicted with the blockade. Lincoln, visibly embarrassed by this charge, responded that he saw no harm in it, that Seward had arranged the matter, and that it would probably turn out all right. The President agreed to allow Welles to take the matter up with Seward, which he did several days later. At that meeting, Seward expressed his doubts about the success of the idea saying that Lincoln had endorsed it. Otherwise, Seward saw no reason why this privileged trade violated the blockade. Frustrated in getting the order reversed, Welles confided his caustic opinion of the affair to his diary. Charging unmistakable "rascality" in the cotton order, he first expressed his disbelief that the President could be so easily duped. He suspected that Thurlow Weed, Lincoln's unofficial emissary, had a special interest in the scheme and charged Seward as knowing the fraudulent motive of the order. This was the reason, Welles recorded, why Seward had not discussed the matter with him before acting.[68]

If Welles was unsuccessful in reversing the order, Farragut had his own means of obstruction. Having allowed Hamilton's designees to take a load of cotton out through Galveston, the Admiral refused them re-admittance on the return trip. Explaining his actions to the President, Farragut quoted the letter of the law which made no reference to allowing articles back into the Confederacy.[69] Here the matter laid. While the orders and law were never rescinded, the fact that no further incidents of this kind were reported in the Gulf zone indicates that the scheme was of limited value to all concerned.

Vexed as he was by the schemes of his countrymen to thwart the blockade, Farragut was particularly concerned about the increased efforts of foreign ships to run the blockade. Commensurate with the skyrocketing prices of consumer goods within the Confederacy was an equal price climb for southern cotton in foreign ports. Paradoxically, the blockade which caused this two-way scarcity also increased the profit motive and provided added challenge for running the blockade. While cotton was worth eight and ten cents a pound in the Confederacy, it was bringing eighty-four cents in Liverpool by the end of 1863. One year later the Liverpool quotation reached one dollar and twenty-five cents, and in Boston the staple brought one dollar and ninety cents in greenbacks.[70]

With the development of these economic factors the efforts of foreign and domestic speculators to violate the blockade intensified. Stock companies were created, and the swiftest steamers in Europe were purchased as runners. The risks of being captured were great, but the opportunities for making fantastic fortunes resulted in a flurry of blockade-running activity. For the Confederacy, cotton became a valuable international medium of exchange. Multi-million dollar contracts were made with foreign stock companies to supply the needs of the Confederate government, payment to be made in cotton.[71] All of these factors led to a tripled increase in blockade-running by 1863, but the number of vessels in the West Gulf Blockading Squadron did not increase correspondingly.[72] Furthermore, many blockaders were tied up in special military operations and were of limited use in stopping traffic.

While the increase in blockade-running was occurring, Farragut was again preoccupied with a major military operation. After reporting back to the squadron on 23 January 1864, he immediately began preparation for the attack on Mobile. A quick reconnaissance of the approaches to the bay, however, confirmed his worst fears: the Confederate fortifications had been strengthened and were now quite formidable. Additionally, the Confederates had a small flotilla operating on the bay which included the dreaded ironclad ram *Tennessee*. Confronted with these prospects, Farragut, who had been hoping for a quick campaign, delayed the attack, awaiting the arrival of monitors needed to contend with the ram. He was convinced also that it would have to be a joint operation, necessitating the use of several thousand troops to invest the forts. Unfortunately for Farragut, General Banks' Red River campaign pre-empted all available troops in the area, keeping then occupied for months.[73]

But part of the delay stems from Farragut's own phobia about Confederate ironclads. While he had derided those afflicted with "ram-fever," it is apparent that the *Tennessee* had also assumed super-dreadnaught proportions in his own mind, and he entertained the wildest rumors concerning its intentions. Even his admiring biographer admitted that the Admiral's delay awaiting the arrival of turreted monitors to handle the ironclads was unwarranted.[74]

It was not until early August that preparations were completed for the attack. By then Farragut had sufficient weight of ordnance needed to match that of the forts, as well as having four monitors for the rams. He would learn, as others had seen, that Mobile's greatest defense was in the shallowness of the bay. The bay, which extended for thirty miles between Mobile and its entrances, refused ships of deeper than twelve feet draft for half its length and no more than eight feet of draft for a distance of nine miles from the city. The only approach was a narrow channel running under the guns of Fort Morgan. The other openings to the bay were covered by fortifications, obstructions, or torpedoes.[75]

The details of that famous engagement need not be repeated. Farragut, by his personal bravery and success, won his historical immortality on 5 August 1864. He lost few ships but suffered considerable damage to his fleet in running past the forts, which, for all their impregnability, surrendered, or were abandoned soon afterward. The *Tennessee*, because of a faulty engine, inexperienced crew, and unserviceable ammunition, missed its best opportunities severely to damage the federal ships. Becoming unmanageable, it received the combined battering of Farragut's heaviest ships until compelled to surrender. The victory was widely acclaimed throughout the north though having to share headlines with Sherman's Atlanta campaign. The government was deeply grateful to Farragut, giving him a one-hundred-gun salute at Washington by Executive order. The victory, while complete, proved to be an isolated affair, because once past the fortifications Farragut learned that Mobile was still unapproachable by water. This was due not only to the shallow water but to the fact that the Confederates scuttled an unfinished ironclad in the main channel, completely blocking it. It would be the closing days of the war before Mobile was captured by land forces. For all that, Farragut's victory succeeded in closing the port for Confederate blockade-running.[76]

Command of the West Gulf Blockading Squadron was turned over to James S. Palmer upon Farragut's departure on 30 November. The Admiral had been in poor health for some time; in fact, this had accounted for his six-months leave following the Vicksburg campaign. After Mobile he was apparently worn out and asked to be reassigned, which would, he added, allow others within the squadron to advance. Palmer's assignment was temporary, and the Navy Department ordered Commander H. D. Thatcher to command the squadron in January, 1865.[77]

With the capture of Mobile Bay, most blockade-running activity shifted to the Texas coast. Within a month of Farragut's victory word reached the Navy Department that there were forty blockade-runners at Galveston ready to run out. The intensification of activity along the Texas coast also resulted from the loss of all but one of the Atlantic ports of the Confederacy. Havana replaced Nassau as the busiest transshipment port for the beleaguered Confederacy. The continuing increase in cotton prices infused a certain professionalism into the trade of blockade-running. Clyde-built vessels designed and constructed for the sole purpose of evading the blockade appeared more frequently along the Texas coast.[78]

The blockade commander at Galveston noted the increase in runner traffic near his station and lamented the fact that he had only three ships fast, or sound, enough to catch them. In fact, the activity of the West Gulf Blockading Squadron for the remainder of the war consisted mostly of attempting to stop blockade-running. Yet the bulk of the squadron could not be used to concentrate on this Texas traffic because of the necessity of keeping ships on station along the pacified Gulf coast. Mopping-up operations at Mobile required a large number of vessels there for the remainder of the war. At no time were there more than twenty vessels off the Texas coast; an inadequate number by all accounts. Even after President Johnson lifted trade restrictions along the entire Gulf coast, except Louisiana and Texas, Secretary Welles upbraided Commodore Thatcher for not putting more ships at Galveston to stop the running. He noted, with poorly feigned disgust, that seven steamers had arrived in nine days at Havana from Galveston.[79]

This traffic continued unabated right up to the time Galveston was surrendered by Confederate authorities. Blockade-running ceased only with the general collapse of the Confederacy, and then from causes other than the blockade. On 8 June 1865, Commodore Thatcher notified Welles that the United States flag flew from the Galveston Customs House and that blockade-running had ceased along the Texas coast.[80]

With the close of hostilities, the Navy Department stripped the Gulf squadrons of surplus vessels and merged the remnants into the Gulf Squadron. The squadron consisted of sixteen vessels with headquarters at Pensacola, Florida. It was maintained on Gulf station until May 1867, because of the continuing French designs on Mexico.[81]

Chapter 2 Notes

[1] Alfred Thayer Mahan, *Admiral Farragut*, in *The Makers of American History* (New York: The University Society, Inc., 1904), p. 117; Horatio L. Wait, "The Blockade of the Confederacy," *The Century Magazine* (July, 1899):915.

[2] Wait, p. 915.

[3] *Ibid.*; Soley, *The Navy in the Civil War*, p. 8; *Welles Diary*, 1:19-20, 87-88; U.S., Congress, House, Executive Document No. 1, *Report of the Secretary of the Navy*, 37th Cong., 3d sess., vol. 3, 1862, pp. 24-25.

[4] *Welles Diary*, 1:75-76; 2:116; Soley, *Navy in the Civil War*, p. 123.

[5] *Welles Diary*, 2:116.

[6] James M. Merrill, *The Rebel Shore: The Story of Union Sea Power in the Civil War* (Toronto: Little Brown and Co., 1957), pp. 18-98; Adams, *Great Britain and the American Civil War*, 1:256-257.

[7] Mahan, *Admiral Farragut*, pp. 117-124; *ORN*, ser. 1, 18:4, 5; J. R. Soley, "Early Operations in the Gulf," in *Battles and Leaders of the Civil War*, Robert Underwood Johnson and Clarence Clough Buell, eds., 4 vols. (New York: The Century Co., 1884), 2:13; David Dixon Porter, "The Opening of the Lower Mississippi," *ibid.*, 2:22. Hereinafter cited as *B and L*.

[8] *Welles Diary*, 2:116; *ORN*, ser. 1, 18:4, 5.

[9] *ORN*, ser. 1, 17:56; 18:7-8.

[10] *Ibid.*, ser. 1, 17:38; 18:9.

[11] *Ibid.*, ser. 1, 18:30, 49; Robert M. Thompson and Richard Wainwright, eds., *Confidential Correspondence of Gustavus Vasa Fox Assistant Secretary of the Navy 1861–1865*, 2 vols. (New York: DeVinne Press, 1918),1-301. Hereinafter cited as *Fox Correspondence*.

[12] *ORN*, ser. 1, 18:59, 70; ser. 1, 17:189.

[13] *ORN*, ser. 1, 17:101.

[14] *Ibid.*, 109-111.

[15] Bernath, *Squall Across the Atlantic*, pp. 39-40.

[16] *Ibid.*; Leonard Pierce Jr. to William H. Seward, 1, 21 March 1862, Delafield Smith to Pierce, 30 Sept. 1863, Pierce to Seward, 5 Aug. 1864, U.S. Consular Despatches: Matamoros, Record Group 56, National Archives; Pierce to Seward, 1 March 1862, *The War of the Rebellion: The Official Records of the Union and Confederate Armies*, 130 vols. (Washington: Government Printing Office, 1880–1901), Ser. 1, 9:674. Hereinafter cited as *OR*.

[17] Pierce to Seward, 19, 21 Sept. 1862, Consular Despatches: Matamoros, Record Group 56, National Archives.

[18] Welles to Farragut, 14 March 1862, *ORN*, ser. 1, 18:66; Swartwout to Farragut, 15 March 1862, *ibid.*, 78.

[19] Farragut to Welles, 27 March 1862; Farragut to Hunter, 16 Apr. 1862, *ibid.*, 77-78, 130.

[20] Farragut to Cmdr. Hugh Dunlop, 16 Apr. 1862, *ibid.*, ser. 1, 17:114-115.

[21] Farragut to George H. French, 25 Aug. 1862, *ibid.*, ser. 1, 19:168, 417-418; *Welles Diary*, 1:79-80.

[22] For the interagency correspondence regarding this occupation, see U.S. Consular Despatches: Matamoros, 21 April 1862, 7 Nov. 1863, Record Group 56, National Archives; *OR*, ser. 1, 9:629, 641, 458, 650, 667; *ORN*, ser. 1, 17:446; *ibid.*, 18:38; Hanna, "Incidents of the Blockade," pp. 226-228.

[23] U.S., Congress, House, Exec. Doc. No. 1, *Report of the Secretary of the Navy*, 38th Cong., 2d sess., vol. 6, p. vii.

[24] James D. Richardson, *Compilation of Messages and Papers of the Confederacy*, 2 vols. (Nashville: United States Publishing Co., 1906) 2:376-377.

[25] Fox to Farragut, 24 Apr. 1862, 16 May 1862, *Fox Correspondence*, 1:313-314.

[26] General L. G. Arnold to Maj. C. G. Halpine, 10 Apr. 1862, *ORN*, ser. 1, 18:479-480; Mahan, *Admiral Farragut*, pp. 196-197.

[27] McKean to Wells, 17 Feb. 1862, *ORN*, ser. 1, 17:131; Arthur T. Lynn to Lord Russell, 14 Feb. 1862, Great Britain, Foreign Office Record Group 5, vol. 848.

[28] Special Orders No. 471, 14 May 1862, *OR*, ser. 1, 9:709; *ibid.*, ser. 1, 15:151-152.

[29] *ORN*, ser. 1, 18:217-218, 227.

[30] Lester N. Fitzhugh, "Saluria, Fort Esperanza and Military Operations on the Texas Coast, 1861–1864," *Southwestern Historical Quarterly* 61(1957):9-15.

[31] *ORN*, ser. 1, 19:253-254, 317.

[32] Farragut to Cmdr. William B. Renshaw, *ibid.*, 404.

[33] *Ibid.*, 410.

[34] Quoted in Charles Lee Lewis, *David Glasgow Farragut*, (Annapolis: U.S. Naval Institute, 1941–1943), pp. 149-150.

[35] *ORN,* ser. 1, 19:403.

[36] Mahan, *Admiral Farragut*, p. 196; B. F. Butler to Fox, 2 Nov. 1862; Fox to Butler, 8 Jan. 1863, *Fox Correspondence*, 1:422-423, 446.

[37] Lewis, *David Glasgow Farragut*, p. 137.

[38] R. W. Shufeldt to Farragut, 1 Dec. 1862, *ORN*, ser. 1, 19:386-387.

[39] For the reports concerning the recapture of Galveston, see *ORN,* ser. 1, 19:437-477, 444, 464-477, 836; Bradley S. Osbon, *Handbook of the United States Navy* (New York: D. Van Nostrand Co., 1864), pp. 62-66. J. B. A. 2015 update note: See archaeological report on the *Westfield*. A. Borgens, R. Gearhart, S. Laurence, and D. Jones, *Investigation and Recovery of USS* Westfield *(Site 41GV151) Galveston Bay, Texas*. (Austin, Texas: PBS&J for U.S. Army Corps of Engineers, Galveston, Texas, 2010).

[40] *Welles Diary*, 1:233.

[41] *Ibid.*, 1:220; Fox to Du Pont, 12 Feb. 1863, *Fox Correspondence*, 1:115; Quote in Lewis, *David Glasgow Farragut*, p. 157.

[42] *ORN*, ser. 1, 19:481.

[43] *Ibid.*, 20:157; *ibid.*, 19:493, 440; H. A. Trexler, "The *Harriett Lane* and the Blockade of Galveston," *Southwestern Historical Quarterly* 35(1931):117.

[44] *Welles Diary*.

[45] Bell to Farragut, 12, 24, 26 Jan. 1863, Farragut to Welles, 15, 19 Jan., 2 Mar. 1863, *ORN*, ser. 1, 19:506, 554, 515, 528, 644. J. B. A. 2015 update note: The U.S.S. *Hatteras* wreck site has been located. See J. B. Arnold III and R. J. Anuskiewicz, "USS *Hatteras* Site Monitoring and Mapping," in *Underwater Archaeology Proceedings from the Society for Historical Archaeology Conference*, ed. Paul F. Johnston (The Society for Historical Archaeology, 1995), pp. 82-87.

[46] *Ibid.*, 665-669; J. M. Foltz to Farragut, 22 May 1863, *ibid.*, ser. 1, 20:110.

[47] Farragut to Porter, 15 July 1863, *ibid.*, ser. 1, 20:393; Farragut to Bell, 28 July 1863, *ibid.*, 423-424.

[48] Alwyn Barr, "Texas Coastal Defense, 1861–1865," *Southwestern Historical Quarterly* 45(July 1961):21, 23, 30.

[49] For the Sabine Pass Operations see *ORN*, ser. 1, 20:514-563, 555-563; *OR*, ser. 1, 26 pts. 1 & 2; Osbon, *Handbook of United States Navy*, p. 236.

[50] *Wells Diary*, 1:440-441; *ORN*, ser. 1, 20:538.

[51] *Fox Correspondence*, 1:184, 188; *ORN*, ser. 1, 20:428.

[52] *Ibid.*, 170, 319, 428; Farragut to Fox, 30 Nov. 1863, *Fox Correspondence*, 1:339; *ORN,* ser. 1, 21:726; Lewis, *David Glasgow Farragut*, p. 144.

[53] Cmdr. Thatcher to Welles, 1 Apr. 1865, *ORN*, ser. 1, 22:121.

[54] Richardson, *Messages and Papers*, 6:184-185; Paullin, "Half Century of Naval Administration," *U.S. Naval Institute Proceedings* 39(March, 1913):184-185.

[55] *ORN,* ser. 1, 20:447, 465.

[56] *ORN*, ser. 1, 20:470, 612, 699, 717; Osbon, *Handbook of the Navy*, p. 239.

[57] Paullin, "A Half Century of Naval Administration," p. 192; Welles to Bell, 16 Oct. 1863, 18 Nov. 1863, *ORN*, ser. 1, 20:633, 688.

[58] *ORN*, ser. 1, 20:718, 742; 21:868; Paullin, p. 179.

[59] Richardson, *Messages and Papers*, 6:89-90, 109.

[60] Farragut to Welles, 28 Oct. 1862; Stanton to Butler, 11 Nov. 1862, *ORN,* ser. 1, 19:227, 230-231.

[61] McGee to Russell, 16 Oct. 1862, Foreign Office, Record Group 5, 848:29.

[62] C. A. Barrier to George W. Randolph, 7 Nov. 1862, *OR*, ser. 4, 2:174.

[63] Farragut to G. S. Dennison, 10 Dec. 1862, *ORN*, ser. 1, 19:399-400.

[64] Richardson, *Messages and Papers*, 6:238, 240-241; Ludwell H. Johnson, *The Red River Campaign: Cotton and Politics in the Civil War* (Baltimore: Johns Hopkins Press, 1958), p. 50; A. Sellew Roberts, "The Federal Government and Confederate Cotton," *American Historical Review* 34(October 1927):262-275.

[65] *OR*, ser. 1, 26, pt. 1:670, 702; Roberts, "Federal Government and Confederate Cotton," p. 267.

[66] *OR*, ser. 1, 26, pt. 3:539; *ibid.*, 39, pt. 2:61; Roberts, "The Federal Government." pp. 273-274.

[67] *ORN*, ser. 1, 21:643-644; C. W. Raines, ed., *Six Decades in Texas or Memoirs of Francis Richard Lubbock* (Austin: Ben C. Jones and Co., Printers, 1900), p. 529.

[68] *ORN*, ser. 1, 21:643-644; *Welles Diary*, 2:159-160, 167.

[69] *ORN*, ser. 1, 21:708, 728.

[70] Johnson, *The Red River Campaign*, p. 50; A. Sellew Roberts, "High Prices and the Blockade in the Confederacy," *South Atlantic Quarterly* 24(April, 1925):156.

[71] W. R. Hooper, "Blockade Running." *Harpers New Monthly* 42(Dec. 1870):105; Wait, "The Blockade of the Confederacy," pp. 917-918; Roberts, "High Prices," pp. 159-161.

[72] *ORN*, ser. 21:193.

[73] *Welles Diary*, 2:100; Mahan, *The Gulf and Inland Waters*, p. 218.

[74] On one reconnaissance Farragut saw a ram that he mistook for the *Tennessee*. Admitting that he may have seen one of the smaller rams, he insisted to Welles that the one he saw was at least 300 feet long. A more sympathetic biographer reminds the reader that the Admiral's eyes were weak. Lewis, *D. G. Farragut*, pp. 231-232; *Fox Correspondence*, 1:340-341; Mahan, *Admiral Farragut*, pp. 243-245.

[75] Cridland to Russell, 28 March 1864, Foreign Office, Record Group 5, 970:14; *ORN*, ser. 1, 21:600.

[76] For the naval details of the Battle of Mobile Bay, see *ORN*, ser. 1, 21:397-601; *B and L*, 3:598, 679; 4:99, 106, 345, 385; Mahan, *Gulf and Inland Waters*, 219-245; Mahan, *Admiral Farragut*, pp. 242-291; Richardson, *Messages and Papers*, 6:239-239.

[77] *ORN*, ser. 1, 21:724, 746; *ibid.*, ser. 1, 22:20.

[78] *ORN*, ser. 1, 21:644; Thomas E. Taylor, *Running the Blockade* (London: John Murry Co., 1896). pp. 145-148; Frank Owsley, *King Cotton Diplomacy* (Chicago: Univ. of Chicago Press, 1959), pp. 252-255.

[79] *ORN*, ser. 1, 21:717; William E. LeRoy to Farragut, 2 Nov. 1864, *ibid.*, ser. 1, 22:171, 190, 193.

[80] *Ibid.*, 216-217.

[81] *Ibid.*, ser. 1, 22:217; James Russell Soley, "Closing Operations in the Gulf and Western Rivers," *B and L*, 4:412; Mahan, *Gulf and Inland Waters*, p. 249.

3

Diplomacy and the Blockade

Any discussion of the relationship between the West Gulf Blockading Squadron and diplomacy understandably must be taken within the context of American foreign affairs in general during the Civil War. It is true, however, that many of the most serious diplomatic crises arising during the war grow out of incidents occurring within the zone of operations of that squadron.

The volume of England's maritime trade would, of necessity, account for much of the diplomatic problems of the American State Department. It was the very real prospects of a maritime war that had prompted Britain's decision to declare her neutrality. As previously mentioned the wartime roles of the two powers were exactly reversed from their respective historic stances. Yet in their new positions neither government forgot its historic principles, and both proceeded with a keen eye to their future interests.

Alarmed by the prospects of her commerce being threatened by both Confederate privateers and northern blockaders, the British government ordered Sir Alexander Milne's squadron on the American station reinforced on 1 May 1961. This capable officer-diplomat instructed his subordinates that they were to maintain strictest neutrality between the American belligerents. Most important, they were to refrain from establishing precedents that could weaken British interests in a future conflict. This caution, he stressed, was to be exercised even at the expense of protecting British commerce. These instructions remained the guidelines for the conduct of British naval officers throughout the war.[1]

Anglo-American relations had been seriously tried from the outset of the war by England's neutrality declaration. The declared intentions of the belligerents to wage maritime war prompted England to act quickly. This was the type of war that was most dangerous to British interests, and the type most likely to involve crown subjects the world over. England felt that she had to warn her citizens and clarify her neutrality. It seemed to her a natural course to take in the presence of a de facto war.[2]

The American reaction to this declaration, both public and official, was an unreasonable resentment toward England. The public outcry was evident in American newspapers throughout the land, and the resentment never died, even after the war. Officially, the Lincoln Administration protested the belligerency status extended to the Confederacy by the proclamation. As previously mentioned, Seward and others tried desperately to convince England of the insurrectionary nature of the war. If England had ever believed that the war was only a domestic affair, her proclamation would have been postponed, but the very magnitude of the secession movement convinced her that there was no chance that the United States could subdue the south. Although the sympathy of Lord Lyons, British Minister to America, was clearly with the north, he had to admit to the Foreign Office that all chances of reconstructing the Union were lost forever.[3]

Within a few weeks of the outbreak of war, the United States Justice Department was attempting to define the conflict. This branch was sorely in need of legal definitions to argue the early prize cases coming before them. British ship owners and merchants argued these cases on the grounds that the prizes were taken before Congress met, and since only Congress can declare war, the prizes had been illegally taken. Technically the courts accepted the two dates of President Lincoln's blockade declarations as the official beginnings of the war. The judges argued the justification for this under the President's war-making powers. Subsequently, a variety of Supreme Court cases considered the double-status principle of the Civil War; that it was both a private and a public war. The United States was able to sustain the position of being both belligerent and sovereign, exercising the rights of both.[4]

England never accepted this double-status view, perhaps because they never clearly understood Seward. As Welles contended, this was because Seward relied less on fixed principle than on expedients; trusting to dexterity and skill rather than legitimacy to see him through an emergency. Never was

England more perplexed by the Secretary's behavior than the time Seward entertained the theory that a foreign war, presumably with England, would restore the Union. In Washington, there was no secret of this view, which brought a sharp warning from the British Foreign Office that such blustering demonstration would not be long tolerated. Lyons characterized Seward as arrogant and reckless toward foreign powers, and even Adams, the American Minister to England, found his chief's attitude bewildering. It was not long, however, before it was apparent that Lincoln and not Seward would determine foreign policy. Lincoln's least intention was to hurry into complications over Seward's foreign war scheme.[5] Within weeks Seward's attitude almost completely reversed itself to one of near timidity regarding foreign relations.

As if the blockade was not enough of a threat to England's commerce, the federal government in June took steps to close the southern ports. Lord Russell, England's Foreign Secretary, became more concerned over the prospects of the enactment of the Ports Bill than he was over a blockade that he felt could not be enforced anyway. Following the passage of the bill, a stern official protest was lodged by Russell to Seward stating that England regarded the Ports Bill as merely a paper blockade. So strongly did Russell feel that the bill was detrimental to England's interest that he ordered Milne's fleet reinforced. Even at this it was not England's intent to precipitate war, and Milne was cautioned not to take any steps which might involve hostile action without specific instructions. Russell went on to state his hopes that Lincoln would not act on Congress' authorization to proclaim the closure of Southern ports. In any case, Russell instructed Lyon that friendly relations between the two countries were too important to be endangered by a chance collision or premature action.[6] The Foreign Secretary need not have worried. British diplomacy prevailed, and Lincoln withheld proclaiming the ports closed until 11 April 1865, three days after Lee's surrender.[7]

No sooner did the ports crisis subside because of Lincoln's procrastination than another arose over the federal navy's attempt to block Charleston harbor by sinking a stone fleet in the entrance. England and France lodged formal protests. The sinking of the ships was a manifestation, declared the British, of the despair over restoring the Union since no nation would deliberately destroy facilities through which it derived its prosperity. It was an act, they insisted, that would be perpetrated only against an enemy in revenge. Further, it was a tacit admission of the failure of the blockade by resorting to measures not recognized by the Treaty of Paris.[8]

With the abandonment of this project and a subsiding of tensions, the *Trent* affair exploded, bringing the two countries closer to hostilities than at any other time during the Civil War. Briefly, the incident stemmed from the overzealous actions of Lieutenant Charles Wilkes, an American naval officer, who removed two Confederate diplomats from on board a British mail packet on the high seas. British anger focused on the removal action rather than the time-honored practice of stop-and-search. Throughout the month of December 1861, the British government prepared for war. Commodore Hugh Dunlop, Farragut's English counterpart in the West Gulf, was instructed to abandon the Mexican problem and prepare to attack the West Gulf Blockading Squadron in the event of hostilities. In this event, Dunlop was to attack piecemeal the federal blockaders scattered from Pensacola to the Rio Grande, or, if these vessels had concentrated, they should not be allowed to form a junction with the Atlantic squadrons. Later Milne confided that if war had followed the *Trent* affair he had planned to raise the blockade of the Gulf ports. He would then, in turn, blockade the Northern ports in cooperation with the southern forces, which he admitted would have been allies.[9]

The incident did not completely end here. One year later Admiral Wilkes, now an American national hero, announced his intention of repeating the *Trent* affair. Milne, for once losing his usual diplomatic patience, instructed his officers not to tolerate a repetition of the incident. They were instructed to demand the instant release of any British mail packet so detained. This failing, they were to resort to force to recapture her. The Admiralty revoked Milne's instructions, but Lyons called the attention of Seward to Milne's dangerous threat. Seward and Welles lost no time in directing Wilkes not to carry out his intentions. But as long as Wilkes was assigned to the West Indies Squadron tenseness existed between British and American naval forces.[10]

Evidence of this tenseness occurred when Wilkes guarded too closely the British Bahamas. By now the blockade, and blockade-running, were both gaining system and efficiency. American naval authorities, no longer content to sit outside southern ports, began to cruise the Caribbean and Atlantic trade routes, seeking contraband on the high seas. In fact, Wilkes so closely watched Bermuda that he virtually blockaded that island with his squadron. The British protested this practice and the Lincoln administration acquiesced. Thereafter United States cruisers kept a more respectful distance.[11]

While Anglo-American feeling was still running high over the *Trent* affair, the *Labuan* incident broke. This was the first of the more controversial prize cases, and it is clear that British officials considered these cases most dangerous to neutral rights. In fact, Lyons considered the prize incidents more threatening of hostilities than the controversies over the English-built commerce raiders being used by the Confederates.[12]

One of the finer points of British contention in the *Labuan* case concerned Commander Swartwout's oath requirement of neutral captains. The commander of the *Portsmouth* refused to allow neutral ships to be unloaded off the bar until their captains had guaranteed that the cargoes were destined for Matamoros and not Brownsville. In this, like Wilkes, he had clearly overstepped his authority. However, unlike Wilkes, Swartwout was not lionized by the American public and Congress for his actions. The timely reassignment of Swartwout and his ship to the Mississippi flotilla seems to have served as much to smooth ruffled British diplomacy as it did to add firepower against Ft. St. Phillip.[13]

By now pro-Southern sympathizers in Parliament were chiding the Palmerston Administration for its acquiescence in the prize case decisions. But the greatest outcry from both the British public and press occurred during the *Peterhoff* crisis, in which the most serious point of contention arose over the removal of official British mail from the ship. Lord Lyons informed Russell that the affair had developed into a serious threat of war between the two countries. Later Lyons stated that the situation was more desperate than at any time since the *Trent* affair.[14]

According to Welles, the *Peterhoff* was an advance steamer of a proposed line of packets which were to convey mails and supplies ultimately to Confederate Texas. The American Navy Secretary charged that the whole scheme was deliberately planned to evade the blockade and open communications with the Confederacy via the Rio Grande.[15]

By the time the *Peterhoff* affair occurred Seward had become quite diplomatically evasive with the British. So fearful was he and Lincoln that a war could be precipitated that they usually gave in to strong British demands on hazy points of international law or on questions of unprecedented circumstances. In the midst of the diplomatic controversy over the *Peterhoff* mails a storm of controversy raged over this subject within Lincoln's cabinet. Gideon Welles presented his hawkish, Anglophobe views, arguing that the navy had every right to go through foreign mail captured on blockade-runners in search of pertinent intelligence information. As usual, he blasted the Secretary of State for making concessions to the British and caustically noted that the English complimented Seward on his liberal views.[16]

Having written Seward his views and receiving a totally unsatisfactory reply. Welles branded Seward as supercilious, unstatesmanlike, and prone to illegal acts. He took strong issue with Seward's decision that naval officers must forward all mails captured to their destination speedily without opening or searching them. As for the British complaint that the navy's views were not consistent with the State Department's, Welles contended that the Secretary of the Navy did not take orders from the Secretary of State.[17]

Welles fired another broadside of correspondence at Seward reminding him that Lincoln's instructions were to forward the mail to the nearest British foreign officer for opening. It was understood that if anything was found that could be used as evidence, it was to be turned over to the prize court. Seward's reply to this was equally unsatisfactory to Welles, who noted that the President had been beguiled by the misrepresentations of Seward. Again Welles confided that Seward had managed this by privately bending the President's ear.[18]

Within the next few days Welles did some bending of his own and noted that Lincoln had seemed surprised by and interested in the navy's side of the argument. The Chief Executive admitted that he was confused by the various legal points. At this, Welles proceeded to enlighten the President on the American Law of 1789, which stated that papers found on board prize ships should be sent to the maritime courts. Lincoln answered that the uppermost consideration was to avoid war with England or France, which was threatened by the mail issue. At the next cabinet meeting, Lincoln asked Seward and Welles to remain and present their views again. Seward's thrust was the importance of keeping good relations with England, and that whatever precedent the United States established in the matter would be reciprocated by England in some future war. Welles' argument was that the whole question rightfully belonged to the courts to decide. As for English reciprocity, he flatly did not trust any assumed generosity on her part in the future.[19]

The President procrastinated for the next several months while the two secretaries continued battling over the issue. Welles tenaciously held to the point of view that the prize court was the lawful possessor of the *Peterhoff* mailbag. Ultimately, Seward's argument that any belligerent precedent set by America in this war could be used against her in the future convinced Lincoln. Under mounting British pressure, Lincoln instructed the district attorney of the prize court to forward the *Peterhoff* mail to the British. Welles was convinced that Lincoln's decision in the matter had been made in haste and done without full knowledge of the facts. Throughout the controversy–and the war, for that matter–Welles never once criticized the motives underlying Lincoln's acceptance of Seward's arguments. Nor did Welles give up the battle to reverse the President's policy on captured mail. Ten years later Welles was still arguing that the decision was unfortunately based on the fallacy of Seward's thinking. This he contended was an unauthorized abandonment of a national right which injured the navy and the country. "Law and usage and the practice of nations could not be overturned by a flippant note from the head of a department," he wrote.[20] Indeed, the worst accusation he ever made against Lincoln was his being horrified by the prospects of war with England.[21]

Welles never ceased contending that the *Peterhoff* case set a precedent which defied the right of ultimate destination, thereby virtually leaving Brownsville, Texas, wide open to contraband traffic. He also continued to accuse the British of having resorted to the vilest schemes to evade the blockade and maintain communication with the South, which, he declared, was encouraged by releasing the *Peterhoff* mails. "This fraudulent and evasive traffic was stimulated and encouraged by the assurance which the Secretary of State had given to England, of immunity of the mails." He added, "a great diplomatic error had been committed."[22]

At this stage of the war, the United States appears to have been winning about half of its maritime disputes with Great Britain. Gideon Welles' instructions to naval commanders on enforcing the blockade failed to distinguish between the offenses of carrying contraband to a belligerent and of violating the blockade. It was not surprising that prize courts often restored vessel, cargo, or both to their owners. On the other hand, Lord Russell had publicly acknowledged that there was a notorious illicit trade conducted between England and America, leaving little ground for complaint against the judgments of American prize courts.[23]

At the same time, the *Peterhoff* incident reached its peak the Union had suffered severe military reverses and was in no position seriously to antagonize a foreign power. The President's second annual message, delivered at this point, summarized accurately the diplomatic effect of the blockade. The message regretted that foreign powers had not yet renounced the belligerency status of the Confederacy, which he noted was caused by military reverses and disloyal citizens abroad. He hastened to add that this should not be taken as a complaint toward those governments. The message explained almost apologetically that a blockade of such an extent could not be established without committing some mistakes and injuries upon foreign nations and their subjects. Admitting that the federal blockade had given rise to many complaints of violations, he reminded the world that American rights had also been violated. He had proposed, he noted, that conventions should be established to resolve these maritime issues.[24]

Also, by this stage of the war the British were undecided as to the effectiveness of the blockade. As early as 21 May 1861, Russell had conceded that the blockade might be made effective in view of the small number of Southern ports to be covered. By 15 February 1862, he recognized that the blockade of these ports was such that it was an effective one as recognized by international law. In this he no doubt had an eye toward England's future interests as possible belligerent. Conceding its effectiveness, he nonetheless predicted that blockade-running would be brisk. Three months later he admitted that the blockade had injured the trade and manufacturing of the United Kingdom. By September of 1862, the English acknowledged that every port on the Confederate coast was effectively blockaded.[25]

On the other hand, the British protested that the American doctrine of continuous voyage was a cheap and easy substitute for an effective blockade. Further, they contended that its discriminatory enforcement constituted quasi-hostility toward England. Especially were they disturbed over the American requirement of bonding British shippers passing through American ports bound for Nassau, Havana, and Matamoros. The English maintained pressure on the State Department to admit the fallacy of American regulatory policy against neutral trade. As late as August 1864, Lyons informed Seward that if the blockade could not be enforced without resorting to such irregular methods foreign powers might justifiably consider this as an admission of failure.[26]

Working equally contrary to good Anglo-American relations was a considerable English pro-Southern sentiment. Throughout the war, both public and political pressure was brought to bear on the Palmerston government to actively aid the South. Periodically Parliament would debate whether the Union blockade was effective or not. On one occasion in the House of Commons, a motion was made to declare the blockade ineffective. In the debate that followed Seward was charged with reckless inconsistency, who under the exigencies of the moment totally disregarded international law. The motion failed to carry but debate continued sporadically throughout the war as new prize incidents occurred.[27]

Much of the Parliamentary debate resulted from a very real cotton famine that occurred in England. The *London Weekly* pointed out the need to raise the federal blockade in order to obtain the stockpiles of Confederate cotton for the lack of which English looms were standing still. It noted that nineteen of twenty mills were geared for the longer Southern-grown fiber, which was of the best quality in the world. It closed by calling upon the working classes to get the British government to interpose its power for the preservation of commerce.[28]

Six months later the *Times* ran an article declaring that a terrible cotton dearth was bringing thousands of Britons to destitution. It was as serious as a grain famine, the article noted, as mills were constantly shut down. The article pointed out that in thirteen mill unions unemployment ranged from five to twenty-two thousand people, up to 100 percent more than one year previously. Four months later the *London Herald* noted that the blockade was not hurting the South as much as supposed because they were agriculturally self-sustaining. However, a blockade of the New England states would be ruinous, it continued, if such an intervention move by England and France was necessary to bring about a peace.[29] What effect English public opinion had on Anglo-American diplomacy is academic. But there can be little doubt that such articles as these published in Southern newspapers buoyed the morale of the people within the blockaded Confederacy.

British opinion of these continuing maritime incidents is best exemplified by the attitude of George Coppell, the acting British consul at New Orleans regarding the capture of the *Sir William Peel*. This vessel, captured off Matamoros, had been released by the U.S. District Court, but the case was appealed to the Supreme Court by U.S. authorities. The consul charged that this act was "vindictive" and that the whole case from beginning to end could only be described as a "piratical outrage."[30] In fact, piracy was the common term applied to ships of the West Gulf Squadron by the merchants of Havana, noted the British consul there. He hastened to add that U.S. ships did not interfere with the neutral trade from Havana to Matamoros whenever a British man-o'-war was in the vicinity. He then requested more warships to protect British vessels from what he termed the "illegal manner of [federal] enforcement."[31]

Periodically events occurred at Galveston, which evoked British diplomatic protest. Federal naval commanders soon learned that due to the location of that city in relation to surrounding Confederate fortifications, any shots fired at the forts that went high hit the city beyond. The first bombardment in August 1861, resulted in considerable damage being done to buildings, some of which were uncomfortably close to the various consulates. A storm of protest was lodged by the French, Prussian, and English consulates in Galveston, which caused the shelling to cease altogether. However, with the re-establishment of the blockade in January 1863, Commodore Bell shelled the city for two hours. Again the consulates protested, the English consul complaining to his superiors that his first protest had apparently been ignored by the Foreign Office. The fact that these were the only two bombardments of Galveston throughout the war indicates that the protests through diplomatic channels were not entirely ignored. Just to be on the safe side. However, the British Consul in Galveston systematically raised the Union Jack daily to mark his location for federal gunners.[32]

The raising of the blockade at Galveston with its recapture by Confederate forces created considerable diplomatic concern on the part of the Navy Department. Fox privately confided his fears to Farragut, contending that since the blockade had been physically lifted by enemy action, foreign nations could, if they chose, cause serious diplomatic trouble for the United States. The federal government would be further embarrassed, he added, if the captured *Harriett Lane* left port. What weakened the Union's position, Fox felt, was that the other blockading vessels voluntarily abandoned their stations following the recapture; an act for which those commanders should be court-martialed, Fox added. This abandonment was enough, he declared, to give the Navy and State Departments sufficient diplomatic embarrassment for the rest of the war. Fox's fears were never realized. For unexplained reasons no foreign powers made a diplomatic issue of the incident.[33]

The establishment of the blockade at Sabine Pass had caught four British schooners inside the port. Ultimately, Lord Lyons approached Seward on the subject, stating that he thought the vessels should be allowed to leave. At issue was the point of international law concerning whether official notice had been given to the schooner captains at the time of the appearance of the blockaders. It was months later before Farragut could gather the details and forward them through channels to the State Department. Because the details were hazy on the notification point, Seward decided in favor of Lyons's request and the vessels were allowed to leave.[34]

At the beginning of the war, the British government seriously doubted the federals' ability to establish an effective, and extensive, blockade. So insistent was Great Britain that it did not become a paper blockade, the Foreign Office required all its consuls located in Southern states to submit periodic lists of violations. For the first year and a half of the war, consulates located in Mobile, New Orleans, and Galveston submitted monthly or quarterly reports.[35]

Arthur T. Lynn, consul in Galveston, submitted such lists of ships entering and leaving that port until Confederate authorities stopped supplying names of vessels for security reasons. Thereafter Lynn submitted his estimates of the effectiveness of the blockade. It is apparent from his reports that the blockade of that port was loose at best and non-existent at worst, during that time. Typical of the situation was the report of August 1861. The only interruption of normal trade attributed to the blockade, he reported, was that with the northern states of the Union.

He noted that the foreign trade was slack only because it was seasonal, but estimated the export potential in cotton at 260,000 bales a year.[36] The acting consul at New Orleans sent a similar report in January 1862. Quoting the legal definition of blockades, the Consul let the Foreign Office judge for itself whether the blockade of that port was an effective one under the circumstances:

> But from the fact that there is a perfect rage here at the present time for running the blockade (among) merchants and others emboldened by the success of those vessels, who are quietly chartering, buying, loading, and getting ready for sea almost every available vessel.[37]

While meticulous records were not kept by American, Confederate, or British officials after the fall of New Orleans, the consular reports from the Gulf region all testified to the inefficiency of the blockade. In November of 1861, Lyons had reported to Russell that the blockade, while not a paper one, failed to intercept the majority of commerce entering or leaving. He had to add, however, that it was seriously interrupting trade. All these reports were carefully scrutinized by both British and French officials with the view of re-evaluating their strict neutrality positions. Apparently to some extent their degree of neutrality depended on the effectiveness of the blockade.[38]

Another diplomatic goal that Seward attempted to gain was accession to the Treaty of Paris. Five years previously, the U.S. had rejected accession, but with Confederate privateers taking Union prizes, accession would benefit the Union. Understandably Great Britain was the principal foreign power with which Seward negotiated. The negotiations failed for several reasons, but primarily failure was due to Lord Russell's fear of American bad faith arising out of interpretations over the treatment of privateers.[39]

Seward, for all his timidity, was not above exerting diplomatic pressure against the English on occasion. Such was probably the case in his advocacy of the act authorizing the President to issue letters of marque and reprisal. From the beginning of the war private ship owners, for one reason or another had deluged the government with offers of volunteer ships to help the navy. Not until the English-built Laird rams became a menace did the administration seriously consider the use of privateers. If the regular navy could not capture the Laird-built *Alabama* private enterprise might take a hand. Despite this logic, the whole plan may have been merely a gesture by Seward to induce the British to prevent the sailing of Confederate vessels under construction at Birkenhead.[40] If so, it worked.

England was not the only foreign power the American State Department had to convince of the legitimacy of its blockade policies. Generally, French opinion and, therefore, resistance coincided with British. Napoleon III candidly expressed his policy to one of his generals as the Mexican situation developed in 1862:

> It is to our interest that the United States be powerful and prosperous, but it is not at all to our interest that it should control the entire Gulf of Mexico, should dominate from there the Antilles and South America and should be the sole distributor of the products of the New World.[41]

The American prize court decision in the *Springbok* case was criticized in the Memorial *Diplomatique*, an organ of the French Foreign Office. The paper strongly contended that there could be no capture of neutral property bound from one neutral port to another on its assumed ultimate destination. Legally, it argued, neutral property bound for neutral ports could not even be classified as contraband, since that term had meaning only as it applied to belligerent ports. Further, a few articles found on board that might be considered contraband could not subject an entire cargo and vessel to condemnation.[42] For all this, France still accepted the doctrine of ultimate destination.

The cotton famine in France, though not as pronounced as in England, nonetheless resulted in strained diplomacy with the United States. In 1862 Thurlow Weed, one of Seward's unofficial representatives felt that France was considering turning against the United States because of the need for cotton. The French Minister for Foreign Affairs on two occasions told Weed emphatically that France must have cotton and that France looked to the North to get it for her one way or another. This situation probably underlay the North's reasons for purchasing southern cotton, a policy to which Gideon Welles and military commanders objected.[43]

It was an incident involving the capture of French cotton that led to a serious rift in Lincoln's cabinet. Seward, responding to French pressure over the seizure, approached Welles for details of the incident. Welles, apparently not having the details, replied that if it was a naval capture the courts would decide the issue. In this, he was joined by Attorney-General Edward Bates, who impatiently demanded to know what authority either the State or Executive Departments had to interfere with a legal matter. Seward, infuriated, stammered that the Attorney-General should have to be responsible for diplomacy also, informing him of the difficulty of trying to prevent war with foreign powers. Welles, for the first time, felt that maybe they had been too rough on Seward, and the next day wrote him a soothing note complimenting his zeal. He ended by gently reminding him of the proper procedure in handling the French claim.[44]

The French, however, were in no position to press America too strongly over diplomatic issues. The ambitions of Louis Napoleon in attempting to establish a puppet regime under Maximillian in Mexico brought France into direct conflict with the American Monroe Doctrine. Seward did not press the French closely on the matter for fear of driving them into an alliance with the Confederacy. Interestingly enough, the French blockaded the Mexican coast at the same time the federals were blockading the south Texas coast. In the interest of diplomacy, no French ships blockaded the mouth of the Rio Grande, nor any area within ten miles of it. In this respect, Matamoros enjoyed an exceptionally heavy traffic, both for Mexican revolutionaries as well as for Confederates.[45] It further complicated the job of the federal navy in deciding for whom contraband munitions bound for Matamoros were actually intended.

It was with the Spanish, however, that the U.S. came closest to an actual shooting incident. The steamer *General Rusk*, re-christened the *Blanche*, left Matagorda Bay with 800 bales of cotton, bound for Havana. Several days out she was sighted by the West Gulf blockader *Montgomery*, which gave chase. Unable to make Havana, the *Rusk* turned toward shore hoping to reach port inside Cuban territorial waters. Undaunted, Commander Hunter of the blockader entered the waters, firing at the *Rusk*. In desperation, the runner captain ran his ship ashore within a dozen miles of Havana. Spanish authorities, who had been watching on shore, boarded the *Rusk*, accepted the jurisdiction of ship and cargo, and hoisted a Spanish flag over the vessel. Despite the Spanish claim Hunter sent a boarding party to the grounded runner, disregarded the Spanish authorities' protests, and claimed the *Rusk* as a prize. In the confusion that followed, a fire broke out which ultimately destroyed vessel and cargo.[46]

Spanish officials in Havana were outraged and immediately dispatched warships to catch or sink the *Montgomery*. Fortunately for U.S. diplomacy, Hunter had already put out to sea, barely missing, an almost certain shooting incident. The seriousness of the affair was evident to the Americans, and Seward instructed the new American Minister to Spain to assure that government that justice would be done. Spanish diplomatic protests were strong, demanding satisfaction, the removal of Hunter, and reparations of over $300,000. Seward's reply which was based largely on naval reports denied responsibility for any personal loss because the ship was, in fact, a blockade-runner. Hunter's indiscretion was another matter. Charged with violating Spanish territorial waters and with scandalous conduct, he was found guilty of the first and innocent of the second count. Lincoln dismissed Hunter from the service amid pleas for reconsideration by fellow naval officers. Many of them felt that Hunter's only crime was overzealousness and that he should not have been made a diplomatic scapegoat.[47]

The Spanish were not entirely satisfied with the outcome of the affair and shortly thereafter announced that their territorial waters were extended to a six-mile marine limit surrounding Cuba. Although Spain patiently explained that modern ordnance improvements necessitated increasing the traditional cannon-shot limit, the action was obviously a diplomatic reprisal over the *Rusk* incident. The fact that the limit specifically applied to Cuban and no other Spanish territorial waters thinly disguised Spain's desire to encourage the Havana blockade-running trade.[48]

Seward was seriously considering recognizing Spain's claim to the six-mile limit in the hopes of detaching her from other European powers. Seward's proposal brought the usual condemnation from Welles, who considered such recognition as a surrender of American rights. Nor was Lincoln anxious to cede this point to the Spanish. Eventually, the matter was submitted for arbitration to the King of Belgium, provoking Welles to inquire acidly of Seward if the King was an authority on international law.[49]

It is apparent that the American blockade declaration was accepted by foreign powers from the beginning, in spite of the fact that for the first two years of the war it was more de jure than de facto. While occasionally questioned seriously by the maritime powers, the blockade was recognized throughout the war. As the war continued the blockade became de facto, a point also recognized by foreign powers. It also seems apparent that England, the largest and most involved maritime power, submitted to America's liberal interpretations of the rules of blockading.[50]

Welles always felt that America had lost the diplomatic war, at least with England. He quoted an English authority writing on the subject during the war who was amazed at how submissive American wartime diplomacy was. The Englishman warned his countrymen that the real fear was not that England would be offered too little but that she would accept too much. "A humiliating commentary," wrote Welles, "on our diplomacy, by an English writer of no mean ability."[51]

Many of the British felt that it was just the other way around, especially in the case of continuous voyage. The London *Times* presented the case for British public consumption: "blockade is by far the most formidable weapon we possess. Surely we ought not to be over ready to blunt its edge or injure its temper."[52] The official view was presented by Sir Randall Palmer, Solicitor General:

England has as strong an interest as any power in the world in understanding what she is about, when she is invited to take a step that hereafter may be quoted against herself and may make it impossible for her, with honor or consistency, to avail herself of her superiority at sea.[53]

From the events of World War I, it appears that never was England's foresight better rewarded.

The United States was painfully aware of the future as well as the immediate consequences of its blockade. Seward had convinced Lincoln that "it is also obvious that any belligerent claim which we make during the existing war will be urged against us as an unanswerable precedent when [we] may be ourselves at peace."[54] In the ticklish diplomacy of the Civil War, America more than once sacrificed an advantage in the interest of legal precedent, treaties, or future interests. The question of trade with the neutral port of Matamoros was a case in point.[55]

American diplomacy with Mexico was exceptionally frustrating since Mexican political affairs were in a constant state of flux. Although the Juaristas, who were in control of Matamoros in 1861, were supposedly pro-federal, they not only encouraged and abetted trade with the Confederacy but also protested certain acts of the United States consul over American interests. When U.S. customs agents in New York began restricting suspected clearances for Matamoros, Matias Romero, the Mexican minister, protested. He charged that it was a treaty violation and a virtual extension of the blockade to Mexico. He argued that the booming Matamoros trade could not be attributed solely to Texas markets. He explained that overland, as well as oceanic trade routes to Mexico from the North, had been closed by the war, which forced all exports to the Atlantic shipping centers. Nonetheless, the United States refused to relent, apologizing for the inconvenience imposed but explaining that unregulated commerce would be too risky for the Union.[56]

In November 1862, President Lincoln proclaimed an embargo on all arms exports, a move which should have silenced some of the Matamoros controversy. However, with the French invasion of Mexico, the Juaristas were deprived of all their available ports except Matamoros. The American dilemma is appreciated when it is seen that had she continued her policies she would be trading in non-contraband goods with Mexican ports controlled by the French, whom she opposed, and denying aid to the Juaristas, whom she favored. Realizing this, the United States relaxed its export restrictions to Matamoros.[57]

Perhaps the most careful summation of the state of American diplomacy during the Civil War is found in Lincoln's annual messages. While admitting that serious maritime problems had arisen from the blockade, he noted that these differences had been accommodated in a spirit of frankness. He emphasized the impartiality of the judicial branch in handling prize cases, a point which foreign powers conceded to a degree. As several Southern ports were captured and re-opened for non-contraband commerce, Lincoln urged foreign powers to trade legally through these ports. He reminded them that this was preferable to pursuing a contraband trade through ports that were closed by a lawful and effective blockade.[58] Here too, the logic would be conceded but for the higher profits of the contraband trade.

Chapter 3 Notes

[1] James P. Baxter, III, "The British Government and Neutral Rights," *American Historical Review* 34(October 1928): 10-11.
[2] Adams, *Great Britain*, 1:111.
[3] *Ibid.*, 1:113; James G. Randall, *Constitutional Problems Under Lincoln* (Urbana: University of Illinois Press, 1964), p. 62.

4 Randall, *Constitutional Problems*, pp. 49-52, 71, 295, 374; Bernard, *The Neutrality of Great Britain During the American Civil War* (London: Taylor Printers, 1870), p. 161.
5 Adams, *Great Britain*, 1:124-127; Welles, *Lincoln and Seward*, p. 47.
6 Adams, *Great Britain*, pp. 246-252; Baxter "British Government," pp. 14-15.
7 Welles, *Lincoln and Seward*, p. 129.
8 Scharf, *History of the Confederate Navy*, p. 663.
9 Baxter, "British Government," pp. 16-17.
10 *Ibid.*, p. 19-20.
11 Francis B. C. Bradley, *Blockade Running During the Civil War* (Salem, Mass.: The Essex Institute, 1925), p. 33.
12 Bernath, *Squall*, p. 2.
13 *Austin State Gazette*, 22 February 1862, 14 August 1862. For full details of the *Labuan* incident see citations in Chapter 2. For a memoir including service on the Rio Grande by an officer on the *Portsmouth*, see herein Part II, Chapter 1.
14 Bernath, *Squall*, pp. 156-159.
15 Welles, *Lincoln and Seward*, p. 121.
16 Gideon Welles, *Lincoln and Seward*, p. 90; *Welles Diary*, 1:269.
17 *Welles Diary*, 1:270.
18 *Ibid.*, 1:271-275.
19 *Ibid.*, 1:277-279; Welles, *Lincoln and Seward*, pp. 89-89.
20 Welles, *Lincoln and Seward*, p. 87.
21 *Welles Diary*, 1:281, 282, 284-287, 300, 302; Baxter, "British Government," p. 9.
22 Welles, *Lincoln and Seward*, pp. 114-118.
23 Briggs, *Doctrine of Continuous Voyage*, pp. 48-49.
24 Richardson, *Messages and Papers*, 6:126-127.
25 Adams, *Great Britain*, 1:263-264; *OR*, ser. 4, 2:1028; Baxter, "British Government and Neutral Rights," p. 12.
26 Bernath, *Squall*, pp. 15-17; Baxter, p. 18.
27 Adams, *Great Britain*, 1:95, 267-269.
28 As quoted in *The Galveston Weekly News*, 14 January 1862.
29 *The Times*, 19 April 1862, as quoted in Galveston *Weekly News*, 11 June 1862; *London Herald*, 16 September 1862, as quoted in *The Galveston Weekly News*, 22 October 1862.
30 Coppell to Lyons, 6 June 1864, Coppell to Russell, 16 June and 27 August 1864, Foreign Office, America, Record Group 5, 1183.
31 James T. Crawford to Russell, 16 March 1862, Foreign Office, Spain, Record Group 72, 1041, 1141.
32 Arthur T. Lynn to Russell, 12 January 1863, Foreign Office, America, Record Group 5, 909:5; Walter Lord, ed., *The Fremantle Diary Being the Journal of Lieutenant Colonel James Lyon Fremantle, Coldstream Guards, on his Three Months in the Confederate States* (New York: Capricorn Books, 1960), p. 56.
33 Fox Correspondence, 1:324-326.
34 Welles to Seward, 11 December 1862, Welles to Farragut, 30 December 1862, *ORN*, ser. 1, 19:402-403, 435.
35 Owsley, *King Cotton Diplomacy*, pp. 252-253.
36 Lynn to Russell, 8 August 1861, Foreign Office, America, Record Group 5, 788:25.
37 George Coppell to Russell, 8 January 1862, Foreign Office, America, Record Group 5, 848:6.
38 Owsley, *King Cotton Diplomacy*, p. 251; Adams, *Great Britain* 1:254-259.
39 Adams, *Great Britain*, 1:140-141, 169.
40 Welles, *Lincoln and Seward*, pp. 145-164; Baxter, "The British Government," pp. 26-28.
41 Quoted in Hanna, "Incidents of the Confederate Blockade," p. 215.
42 Bernath, *Squall*, p. 87.
43 Roberts, "The Federal Government," pp. 262-280.
44 *Welles Diary*, 2:106-107.
45 Corpus Christi *Ranchero*, 24 September 1863.
46 Galveston *Weekly News*, 29 October 1862; *ORN*, ser. 1, 19:279-281.
47 Bernath, *Squall*, pp. 101-104.
48 Welles, *Lincoln and Seward*, pp. 167-171.
49 *Welles Diary*, 1:170, 460; Richardson, *Messages and Papers*, 6:179.
50 Baxter, "The British Government," pp. 29-29.
51 Welles, *Lincoln and Seward*, p. 129.
52 London *Times*, 10 February 1862.

[53] Quoted in Samuel Flagg Bemis, *A Diplomatic History of the United States* (New York: Dodd, Mead and Co., 1936), pp. 377, 596.
[54] "Papers Relating to Belligerent and Neutral Rights," *American Historical Review* 34(1928–1929):87.
[55] Hanna, "Incidents of the Blockade," p. 217.
[56] U.S., Cong., *House Executive Documents*, 39th Cong., 1st sess., No. 1, part 3, p. 535; Hanna, "Incidents of the Blockade," pp. 222, 224-225.
[57] Hanna, "Incidents of the Blockade," p. 226.
[58] Richardson, *Messages and Papers*, 6:179, 243-245.

4

Matamoros: Loophole in the Blockade

Before the war Matamoros, Mexico, was a small border town of a few thousand inhabitants. Its importance in the commercial affairs of the region was limited. As an inland port, it received the cargoes of fewer than a dozen ships a year. Within a few months after the establishment of the blockade, the city experienced a boom as traders from all over realized the commercial opportunities afforded by the nearby Confederacy. Within four years, the city had a cosmopolitan flavor and had a population of 40,000 to 50,000, most of whom were newcomers. There were anywhere from fifty to two hundred vessels anchored off the bar on any given day throughout the war.[1] At the peak of the war the city became, as one observer said, "a great center of commercial activity, rivaling the trade of New York or Liverpool."[2]

Added to this polyglot atmosphere were the civil strife of Mexican politics and the closeness of the American belligerents. The state of affairs is best indicated by an eyewitness report made shortly after the arrival of federal forces in 1863:

> Things are most certainly working on the Rio Grande. What with Banks and his Yankees, Videl and his traitors, and Cortinas with his *pronunciamentos*, the phase of events have shifted there in one short week with all the quickness and all the fancifulness of the kaleidoscope. On the one hand the Yankees hold the Rio Grande on this side. On the other the "Rojas" who are a political party confined to the state of Tamaulipas, indeed mainly to the city of Matamoros with Cortinas, a daring skillful and unscrupulous bandit, at their head, are setting their wheel within the wheel most beautifully in motion. Ruiz is in prison. Lopez is ditto. Vidaurri is at a distance and the Prince of Mischief has nothing to check him.[3]

The lower Rio Grande must have presented a strange sight, when, on another occasion, U.S. troops were at Brazos Santiago, Confederates at Brownsville, the French at Matamoros, and the Juaristas encamped nearby. At one time, the campfires of all four could be seen along the banks of the river.[4]

Because of the bar off the mouth of the river and the twisting shallow water within the river, traffic between Matamoros and the ships was confined to small river lighters. There was an overland road between the city and the point of land nearest the bar which carried a lesser amount of wagon freight. On the point of land was the fishing village of Bagdad, which had originated as an American supply point during the Mexican War. It too was caught up in the boom and became a busy, squalid trading center.

A reporter for the Ft. Brown *Flag* described conditions at Bagdad in January 1863, listing forty-three vessels at anchor, plus sixteen schooners in the river. He noted that two cotton presses, one of northern manufacture, were in constant operation. The city of Bagdad was described as consisting of one house and twenty jacals, built of straw, sticks, and rushes. He claimed that along the single street there were twenty restaurants, a similar number of barrooms, and in every room a mountebank. As for commercial activity, he noted that there was a large amount of goods on board ship as well as cotton stockpiled on shore, all awaiting transportation by lighters. These lighters, he added, were restricted by weather conditions from regular activity.[5]

Three months later a British observer described the town as a miserable village consisting of a few wooden shanties. Stockpiled nearby for an immense distance were endless bales of cotton waiting to be loaded on some seventy vessels outside the bar. Other observers noted that upon approaching Bagdad there were to be seen a great number of things that looked like black hillocks which proved, on closer inspection, to be piles of goods covered with tarpaulins.[6]

As was to be expected the British were quick to exploit the Matamoros market. Trade flourished to such an extent that by 1863 a steamship service was begun with London. Commercial houses in Liverpool and Manchester shipped "super cargoes" to Matamoros to be exchanged for cotton. To hasten the transaction, British agents went to San Antonio to buy cotton and hire transportation. They could purchase cotton for five and six cents a pound, but costs to Bagdad would increase it to perhaps forty cents. Even at its lowest price, cotton sold in Liverpool yielded twenty per cent profit and usually much more. One report claims that in one year's time eighty to eighty-five thousand bales of cotton were exported at Matamoros.[7]

During the early stages of the Rio Grande traffic, all lighters were Confederate-owned, and they plied between the ships and Brownsville. With the arrival of the blockaders, these lighters took Mexican registries. After Matamoros was declared a free port its commercial traffic increased at the expense of Brownsville. The U.S. consul there reported in March 1862, that Matamoros was then the great thoroughfare to the Southern states, with goods coming in and cotton stockpiled on the east bank ready to go out.[8]

The consul, Leonard Pierce, continued to inform the State Department of the increasing traffic, reporting the arrival of ships by name and noting their contraband cargoes which were transshipped to Brownsville. In June of 1862, Pierce reported that Confederate authorities were seizing all cotton coming from the interior, paying par value for it in script, and then selling it across the river for specie. Three months later he reported that large contracts had been entered into between Confederates and agents in Mexico for supplying the Southern armies for the coming winter. The supplies were to be paid for in seized cotton, he added. Repeatedly Pierce complained to the State Department about the absence of a federal ship off the mouth which would, he believed, stop much of the contraband traffic. In December 1862, Pierce declared that not only supplies arrived through Matamoros for the Confederacy, but also official dispatches to and from Richmond came through there. He further informed Seward that "Large quantities of cotton are shipped to Europe, some of which belongs to the 'Ladies Gunboat Fund' of Texas for equipping privateers."[9]

Three months later Consul Pierce was again lamenting the absence of a blockader, noting that there were seventy-four vessels awaiting discharge, among which, were twelve American registries. All of these ships, he claimed, contained supplies for Texas. Pierce did not appreciate the delicate diplomatic reasons underlying the absence of federal vessels at the mouth of the Rio Grande. He would learn soon that before a warship could be sent there the order first had to receive State Department approval.[10]

In reality, the various secretaries of Lincoln's cabinet had devoted considerable time to the perplexing problem of Matamoros. Quite possibly it was the urgency of the consular reports that prompted the decision to occupy the Brownsville side of the river with military forces. Throughout the summer of 1863 Welles attempted to impress upon Seward this necessity. Seward appears to have been unimpressed until he learned that there were possible French designs upon Texas. Still Seward's desires to occupy some part of Texas were thwarted by General Halleck's complaisance and War Department troop allocations. Welles continued to remind Seward of the heavy flow of Matamoros trade to the Confederacy. He declared that the blockade would continue to be ineffective until federal forces occupied western Texas.[11] Even when the federal occupation was accomplished in December of 1863, the contraband trade continued.

By the fall of 1862 the Matamoros trade had already reached boom conditions. The price of cotton in specie reached thirty-eight cents a pound and imported articles were eagerly purchased at escalating prices. These factors brought into the city more speculators, merchants, agents, as well as more cotton and goods. By the Spring of 1863 the Matamoros consumer goods market was glutted due to over speculation and the rising costs of transportation to and from interior points in Texas. The wagon teamsters demanded freight payment in specie only. Further, they could only haul when weather and water conditions allowed them to cross 300 or more miles of sandy, semi-arid land. Since road conditions required ten oxen or six mules to pull one wagon, there was a constant scarcity of both teams and wagons to meet the burgeoning Rio Grande trade. Because of these transportation limitations Matamoros merchants had to depend on local consumption which diminished in 1863.[12]

The generally depressed civilian market conditions in and around Matamoros had the effect of backing up maritime traffic in the roadstead. One reporter claimed that there were some seventy vessels with cargoes estimated at $3,000,000 awaiting unloading. He noted that in Brownsville there was no scarcity of any article of luxury and that this city was probably the only place in the Confederacy where gold and silver circulated as the only currency.[13] As specie became scarce transactions tended toward the barter system, payment generally being made in three-quarters goods and one-quarter cash. Cotton prices dropped to a low of twenty cents a pound, and the exchange ratio of Confederate currency to specie was six-to-one. By September 1863, the cotton quotations were back up to twenty-five cents specie, but the trade was uncertain. Also, bad roads and lack of grass and water restricted freight traffic. By October Confederate script was up and transportation costs down, but even then costs were high enough to keep local merchants from selling into the interior of the state.[14]

These conditions were further worsened by the changes in Mexican political affairs. Matamoros was periodically caught up in revolutionary fighting, and each new commandant had his own trade regulations to enact. Something of a cotton panic ensued in that city when rumors circulated that General Cortinas was going to prohibit all foreigners from trading in the city. Under Cortinas' regime, a double import-export duty was levied on goods coming through Mexico. Upon landing, goods were taxed twelve-and-one-half percent and taxed the same rate upon leaving for Texas. His successor, Cobos, levied a tax of twenty dollars per bale of cotton to support his army.[15] Thus, costs of goods were exorbitant even if they never reached the interior.

Despite these conditions a reporter noted that Brownsville was crowded with merchants and traders from all parts of the world, and the sidewalks were blocked with goods. He noted the American involvement by saying that Boston and New York were familiar trade names therein. In exchange, he said, for Texas hides, wool, and cotton, the Yankees sent back goods, clothing, powder, shot, nitro, Phosphorus, Sulphur, percussion caps, and other explosives. They would send rifled cannon, he continued, if it was as easy to do as shipping rifle powder. In fact, Brownsville was generally understood to be neutral since "Yankees" came and went frequently.[16]

The fact that New York, New Orleans, and other American cities contributed to the Confederate economy through Matamoros was of continuing concern to federal authorities. Yet it was federal authorities who were encouraging much of this trade. As early as July 1861, the Secretary of the Treasury was granting clearance papers from Now York for Matamoros, lending official sanction to a contraband trade conducted by England and France.[17]

The figures of the New York customhouses revealed that in 1861 there was only one arrival at New York from Matamoros, but there were twenty arrivals in 1862 and seventy-two arrivals in 1863. In the quarter ending 18 March 1864, there had been thirty-two arrivals. Clearances from New York to Matamoros had been even heavier, showing an aggregate tonnage of 35,000 tons cleared. The exports of specie from New York to Matamoros showed a similar trend: for the six months ending 1 January 1862 there had been exported $1,408,953, and for the same period ending 1 January 1863 there had been exported $31,785,708.[18] Pierce complained in May 1863, that Texas agents in Matamoros were converting state indemnity bonds into money and merchandise which were used to purchase supplies in New York and Europe. "Every dollar paid for them by New York merchants will aid the Rebel cause," he warned.[19]

Included in these shipments from New York were large cargoes of military equipment. Though ostensibly for the Mexican combatants, undoubtedly nine-tenths of these shipments entered Texas and Confederate service. Texans were appreciative of these northern imports and noted the irony that some of the gunpowder used to capture the 42nd Massachusetts Regiment at Galveston had been manufactured in Boston the previous year.[20] General Hamilton P. Bee, Confederate commander at Brownsville, claimed that he could procure as many carbines and pistols, shipped from New York, as were needed. A New Orleans newspaper reporter claimed that one merchant house in Matamoros had received a consignment of percussion caps in New York and had sold them to a man in Brownsville, and that other contraband was regularly supplied in like manner from northern ports. Early in 1865 exports from New

York to Mexico were reported at over one million dollars a week and exports for the week ending 18 February had amounted to $1,700,000.[21]

A congressional investigation of the New York Custom House in 1864 by the Committee on Public Expenditures revealed some interesting facts concerning the New York trade. Particularly the cases of the ships *Banshee* and *Jasper*. Both cases concerned the ownership of cotton shipped from Matamoros to the mercantile house of Smith and Dunning in Burling Slip, New York. The purpose of the investigation was to determine the true ownership of the *Banshee's* cotton. The evidence left little doubt that the cotton belonged to a Confederate agent who had shipped through a Matamoros citizen. One of the members of the house of Smith and Dunning swore that he knew nothing of certain mysterious entries of large sums in the company's books, and throughout he displayed an amazing ignorance of the operation of this company. Nothing definite was proved in the investigation because Smith and Dunning's New York contact man fled to Canada just before the case was brought before the committee. It was brought out in the testimony that a member of the company had once declared that within the year Smith and Dunning had made a million dollars for the Confederate agent as well as commissions for themselves.[22]

The *Jasper* was chartered in 1863 by a Columbus, Ohio, firm which shipped cotton to the North and goods to Matamoros. The investigation showed that a New York law firm had used $3,500 as "grease" money to get the vessel cleared from New York. The two Matamoros agents for the firm were said to have done about $200,000 worth of business as middlemen since the beginning of the war. Cotton, wool, and skins were sent to New York in exchange for coffee, flour, starch, soap, and other goods. The testimony further revealed that papers of another suspicious transaction by the Burling Slip firm had been seized and placed in the hands of the United States District Attorney, E. Delafield Smith. However, just before the case was called Smith mysteriously disappeared, and the case had to be dropped.[23]

Trade between New Orleans and Matamoros was not as free as the New York trade because of the whim, inexperience, or degree of integrity of military and port officials in the Crescent City. Under mounting complaints from Farragut, Banks, and E. R. S. Canby New Orleans customs collectors began discontinuing clearances for steamers from that city to Matamoros. Apparently the only effect this had was to shift trade from New Orleans to northern ports. In spite of the restrictions imposed by federal officialdom in New Orleans, large shipments of Confederate uniforms and Enfield rifles entered Matamoros from that port before the discontinuances.[24] As the war continued, the State Department required all American ship captains arriving in Matamoros to sign an affidavit. By this, they certified that their goods landed in Mexico were sold in good faith to Mexican citizens for local consumption and were in no way intended for Confederate use.[25] Apparently this did little to stop the contraband trade.

An American admiralty court judge of the Civil War suggested that more goods were imported into the Confederacy by Americans than by foreigners. He contended that more secrecy was observed by merchants living in New York engaged in this business than by those actually involved in running contraband to Matamoros. Horace Greely may have approached the truth, the judge felt, when he said "that the ideals and vital aims of the South were more generally cherished in New York than in South Carolina and Louisiana."[26] The business activities at Matamoros would not refute such an assertion.

The Confederate government was quick to realize the purchasing power of its chief staple and the importance of utilizing Matamoros as a supply route. The Confederacy dispatched to Mexico Juan A. Quinterro, one of its most capable agents. Being a Cuban by birth, a citizen of Texas, a loyal Southerner, and sometime resident of Mexico, he was admirably suited for the job. It became Quinterro's responsibility to direct the trade between Mexico and the Confederacy and to maintain friendly relations with the various political factions of Mexico. He soon won the cooperation of Santiago Vidaurri, the leading politician of the north Mexican state of Tamaulipas. When Vidaurri obtained control of the state of Tamaulipas, Matamoros was opened to the Confederacy.[27]

In the fall of 1862 the Confederate Secretary of War, Judah P. Benjamin, ordered Major Simon Hart to Texas to purchase army supplies from Mexico. Within two weeks of his arrival, he had reportedly bought over a million dollars' worth of supplies. So available were Mexican sources of supply that

Hart assured General H. H. Sibley not to concern himself about supplies, because Sonora would supply all his immediate needs. By January 1864, the head of General Kirby-Smith's Cotton Bureau informed the Confederate Secretary of the Treasury that Mexico remained the chief source of supplies for the Trans-Mississippi Department.[28]

As early as the summer of 1862 Confederate authorities began controlling the export of cotton across the Rio Grande. General Theophalous Holmes ordered that no cotton would be exported except by permits issued by General Bee, the Confederate commander, in Brownsville. Bee issued his own explanatory orders, emphasizing that only those engaged in the traffic that were providing for the necessities of the people and the army would be issued permits. Planters were required to sign affidavits attesting that goods received for their cotton would not be used for speculation or re-sale. Further, he ordered the cotton held on the Texas side until the equivalent value of goods had been imported.[29]

From the beginning of the war Confederate authorities began buying cotton at fixed prices, paying either in script or cotton bonds. Both instruments were, incidentally, promissory notes and cost the government nothing until maturity date which was two years after the war. Since most planters were landlocked, their best market became the Confederate government, which obtained vast quantities of cheap cotton. This cotton was shipped abroad, via the blockaded southern ports and Matamoros, and was consigned to Fraser, Trenholm and Company of Liverpool (and New York and Charleston). In Europe, Confederate agents purchased large stores of supplies for their hard-pressed government, paying by checks drawn on Fraser, Trenholm and Company.[30]

Captain Caleb Huse, one of the most successful Confederate agents, was commissioned by his government two days after the bombardment of Ft. Sumter to proceed to Europe. Once there his insistence on procuring quality firearms, and having to compete with federal agents on the same mission, delayed the arrival of his first shipment of arms until February 1862. By then only about 15,000 small arms had arrived in the Confederacy. However, during the first twenty months of his stay he managed to buy and ship a total of 131,129 stands of arms. By the spring of 1864 the arms shortage within the Confederacy had eased considerably. By then domestic production was supplementing that which came through the blockade.[31]

After federal forces captured the Mississippi River, the Trans-Mississippi Confederacy was virtually isolated from the central Confederate government at Richmond. General Edmund Kirby-Smith, with headquarters at Shreveport, was given broad, autonomous military and civilian powers. One of the appointments he made was that of Major J. P. Minter, selected to be the Confederate purchasing agent in Europe for the Trans-Mississippi Department. General Smith asked for and received approval for Minter being placed on an equal footing with Huse and other agents abroad. In the months that followed, Minter contracted for large, shipments of military supplies that arrived at the mouth of the Rio Grande.[32]

One of the problems in the Trans-Mississippi Confederacy was the multiplicity of agents buying cotton on behalf of the Confederate civil government as well as for branches within the army. This was further compounded by purchases made by various generals and by the state of Texas Military Board. All of this ultimately required a myriad of regulations that brought increasing official control over the sale and export of cotton through Matamoros. Because of the confusion and the competition between the various purchasing agents, General Smith created a centralized Cotton Bureau in August of 1863. All cotton procurement, save that made by the state, was coordinated under this department. A survey made by this new bureau shortly after its creation showed an inventory of 111,673 bales on hand in the Trans-Mississippi Department.[33]

Headquarters of the Bureau were in Shreveport, but most all transactions were conducted through a branch office in Houston, known as the Texas Cotton Office. Lt. Colonel W. J. Hutchins was in charge of this office with a competent force of Texas businessmen assisting him. Hutchins supervised the activities of all Confederate cotton agents in Texas. Specifically, the office was to purchase half of the owners' cotton, giving exemptions from impressment for the remaining half. The cotton was to be purchased in certificates redeemable in cotton bonds or their equivalent as determined by the Confederate Congress.[34] Ideally, the Bureau was to make uniform and just rules, satisfactory to planters and all concerned. Planters were allowed to export as much for themselves as they transported to the border for the government.[35]

The expected civilian complaints accompanied the increased Confederate involvement in the cotton business. An article appearing in the *Ranchero* typified civilian sentiment. Commenting on the surprise of the New York Chamber of Commerce upon learning of the heavy Matamoros cotton trade, which rendered the blockade ineffective, the editor noted that the Confederate cotton orders were making the blockade effective. He noted the irony of both governments working toward the same end stopping the trade across the Rio Grande. He singled out cotton order no. 28, which would, he said, destroy confidence and trade as well as raise prices and bring misery upon the poor. Besides being monopolistic, he continued, it was a violation of rights, a defying of the constitution, and it "stunk" of despotism. "Let trade seek its own channel and good will result."[36] Free enterprise would further suffer when shortly thereafter Confederate Generals Magruder and Bee were authorized to impress cotton to meet contractual obligations.[37]

Despite civilian criticism, limited military impressment continued. In July 1863, three British ships, the *Sir William Peal, Sea Queen*, and *Gladiator* were anchored off the mouth of the Rio Grande waiting to off-load valuable military, stores. The cargo of the *Gladiator* consisted of virtually every type of army equipment except munitions. Confederate cotton was to be exchanged for the cargoes at the rate of thirty cents a pound. Even under the impressment plan it was another month before Quartermaster officers in the area had accumulated enough cotton to pay off the *Gladiator* and the *Sea Queen*.[38]

These same transactions were reported by Consul Pierce, who substantiated that the *Peel* and the *Sea Queen* were under consignment to Milmo and Company of Monterrey, Mexico, to supply the state of Texas. As for cargoes, he knew personally of large quantities of vegetables and army blankets that had been unloaded for the Confederates. He, too, was aware of General Bee's order of 13 July which prohibited the crossing of all cotton, while the Confederates were awaiting the arrival of a large quantity of the staple within the next sixty days. He also had seen the Confederate order for impressing one-fifth of all cotton within the vicinity for immediate shipment abroad.[39]

In fact, the Cotton Bureau continued to be hard-pressed to accumulate enough cotton to meet their contractual commitments abroad. One reason was that the Texas Military Board offered better terms to suppliers on both ends than the Confederate government could. Such competition continued until an agreement was reached between General Smith and Governor Pendleton Murrah.[40]

Meantime, the Matamoros trade flourished, despite the regulations imposed by the respective authorities. The Confederate authorities limited the impressment of cotton to that owned by speculators and contractors. It was apparent that it was this class who enriched themselves from the Matamoros trade. In November and December of 1862, it was commonly discussed in London commercial circles that contraband articles delivered at Matamoros would yield a one-hundred percent profit, shared between shipper and agent.[41]

By 1863, the Confederate government had made many contracts for supplies with brokers abroad, who drained off tremendous amounts of cotton while making fantastic profits for themselves. The competent Confederate purchasing agent in England, Colin J. McRae, urged his government to adopt stringent cotton controls to stop this drain on southern resources. He cited a classic example of profiteering in which the Confederate government agreed to pay $1,836,000 for goods delivered at Matamoros, the cost of which was only $650,000. Payment was to be made with 9,180,000 pounds of "middling-fair" cotton, which when sold in Liverpool at the current quotations of sixty cents a pound, would yield $4,590,000. Thus, the contractor would gross over four-and-one-half million dollars on an investment of about $651,000 on non-contraband goods conveyed between two neutral ports.[42]

McRae went on to point out that one of the big evils of the system was the cost-plus arrangement which allowed invoice prices to be manipulated, producing even greater profits. He predicted that purchases made only by official Confederate agents would strengthen the government's credit abroad as well as conserve cotton resources. To emphasize his recommendations, he cited a published market statement which showed that the port of Liverpool alone had received 100,000 bales in the previous ten months, worth $20,000,000. This figure, he reminded, was double the amount of the celebrated Erlanger loan which was secured with 260,000 bales of cotton.[43]

Due to the urgings of McRae and others, the Confederate government eventually enacted stringent controls on all imports and exports. Because of too few necessities and too many luxuries being imported, the government ordered that one-half the tonnage on all inward and outward-bound vessels be reserved for government use. The owners of such vessels were bonded, and the Confederate portions of the freight charges would be paid in cotton or coin at a fixed exchange rate. In this manner one-half of every outward bound cargo of cotton (and other items) would apply to Confederate accounts overseas while imports thereafter would consist of more military supplies and fewer trinkets.[44]

Confederate purchasing agents along the Rio Grande continued to experience difficulties in stockpiling cotton to meet commitments. A lack of adequate funds was a major cause. Moreover, as the war progressed, Confederate currency depreciated so rapidly that people became increasingly reluctant to accept it in payment for goods. Transportation shortages and costs contributed to this problem, also. Even before the war freight rates were so prohibitive that it was unprofitable to transport heavy articles any great distance. In 1858 rates were quoted at twenty cents per ton per mile.[45] During the war, civilian teams and teamsters were hired, impressed, or conscripted, depending on their circumstances. A driver was exempted from conscription if he supplied his own wagon and team. However, he would still haul at Confederate prices. These measures still fell short of the need, and in July 1864, General Smith added the Transportation Corps to his many bureaus. As if teamsters did not have enough troubles, they were occasionally highjacked by bandits, some of whom were women who demanded a share of the cotton load.[46]

Despite the obstacles, high prices, and logistical shortages, the Matamoros trade flourished. Cotton brought unheard of prices; the specie price alone rose from sixteen cents a pound in August 1862, to ninety cents by November 1864. By 1865, it had reached one dollar and twenty-five cents per pound. According to one source, the lucrative war-trade produced at least ten millionaires in Matamoros and Brownsville.[47] Observers claimed that in early 1865 there were from one to two-hundred vessels off Bagdad. During the winter of 1864 another observer was amazed at the change that had come over the city during the war. He noted that many fine brick houses had been or were being built and that every available nook in the city was filled with goods. Small rooms in the business district rented for up to $150 a month, he noted.[48]

So vital was this trade to Confederate existence that the Brownsville area commander refused to reduce his forces, as ordered, to meet an expected federal invasion at Galveston. He explained that he needed all available forces to protect the Matamoros trade.[49] Union authorities verified this trade, one claiming that in a short period, every description of property was imported into Mexico from all parts of the world. Of these imports, he added, fully nine-tenths entered the Confederacy, and he knew personally of the importation of revolvers and percussion caps.[50] In February 1865, a federal officer stationed at Brazos Santiago, requested additional troops with which to recapture Brownsville. This would, he contended, prevent the "immense contraband trade continually carried on between Matamoros and Brownsville."[51]

Supplies brought in from Mexico were not equally distributed throughout the Trans-Mississippi Department. Most stayed in Texas. The other districts, more distant from the source and physically suffering from enemy actions, were practically destitute by war's end. A Union spy touring the department in the spring of 1865 claimed Arkansas to be literally "starved out." Louisiana, he claimed, was a little better off but could not support an army. On the other hand, Texas was "full to repletion."[52] One evidence of this repletion was the heavy influx of emigrants from Arkansas and Louisiana into Texas during the last year of the war.

Because of a paucity of sources it is virtually impossible accurately to gauge the magnitude of the Matamoros trade. The chief ordnance officer in the Trans-Mississippi Department was answerable only to General Smith. He was given complete control of ordnance purchasing and made no report whatsoever of his expenditures. However, a sample indication of the flow of goods purchased through Matamoros by the Cotton Bureau as of December 1864, is as follows:

5,000 reams of paper and stationery	230,000 yds. mixed cotton goods
$100,000 in medicines	77,000 yds. satinet
$15,000 in commissary stores	67,000 yds. army cloth
23,000 Rifles	25,000 yds. flannel
10,000,000 Percussion caps	22,000 pairs socks
200,000 lbs. gunpowder	18,600 blankets
200,000 lbs. nitro	18,000 yds. mosquito net
2 five-pounder guns and carriages	64,000 pairs shoes and boots
30,000 saddlers and shoemakers stores	200,000 yds. bagging
2,400 pairs pants	245,000 lbs. bale rope
32,650 shorts	2,400 pieces Mexican bagging[53]

That many of the goods purchased by the Cotton Bureau during its year of operation were actually entering the Confederacy is indicated by the report of Captain F. W. Lynch of the Quartermaster Department. Lynch, located at Brownsville, reported goods arriving in such quantity that he could predict that a six months' supply for the Trans-Mississippi would be available by March 1865. For the quarter ending December 1864, Lynch had forwarded to San Antonio and Alleyton, in south Texas, the following shipments:

600,000 lbs. army stores	72,000 lbs. bar and rod iron
19,700 pairs shoes	100,000 lbs. chains, axes, hdwe., etc.
14,000 pairs blankets	4 cases lint
14,000 yds. gray army cloth	8 cases thread
8,000 yds. gray satinet	2 cases needles, thimbles
80,000 yds. brown drilling	200 ozs. quinine
50,000 yds. sheeting	200 lbs. blue mass
12,000 yds. tent ducking	100 ozs. morphine and other medicines
20,000 woolen shirts	50 cases stationery
20,000 yds. flannel	12,000 reams paper
50,000 lbs. leather	2,000,000 percussion caps[54]
10,543 lbs. of sheet iron for mess kettles	

The flow of all classes of war material from and through Mexico continued throughout the war. Despite the lack of complete statistics on these importations, there can be little doubt that they virtually comprised the entire source of supply for the Trans-Mississippi Confederacy. It is equally impossible to measure the importance of these imports, although the statement made by Colonel W. J. Hutchins of the Cotton Office, is probably accurate: "many of the guns and most of the powder which gained the [Confederate] victories at Mansfield and Pleasant Hill and won the campaigns in Louisiana and Arkansas" were imported.[55]

In the face of many obstacles Confederate agents in Mexico and Europe succeeded in procuring for the Trans-Mississippi Confederacy the materials which made possible the continuance of its struggle. A student of the Confederacy has suggested that the first great campaign of the war was not fought by armies. It was, he suggests, a commercial campaign fought by agents of the federal and Confederate governments and having for its aim the control of European sources of war supplies.[56] It cannot be argued that the South won the campaign nor can it be refuted that the South was continually on the receiving end of a valuable flow of supplies. The trade across the Rio Grande undoubtedly had a significant influence on the life and success of the Trans-Mississippi Confederacy.[57]

There is evidence that in March and April 1865, Confederate authorities planned to discontinue the Matamoros trade. The prohibitive expense of transporting cotton across Texas, especially from the eastern regions of the state where it was grown, had created a noticeable slack in the Matamoros trade. The Foreign Supply Officer in Shreveport ordered his purchasing agent in Houston not to send any more cotton out by Matamoros because of excessive expenses. From now on they would trade through Galveston and other Texas ports where blockade-running continued to be successful.[58]

Chapter 4 Notes

[1] *OR*, ser. 1, 17:403; Robert W. Delaney, "Matamoros, Port for Texas during the Civil War," *Southwestern Historical Quarterly* 63(1955):473-487; Sherrill L. Dickeson, "The Texas Cotton Trade During the Civil War," (Master's Thesis, NTSU, 1967), pp. 12-13.
[2] Moore, *Digest*, 7:716.
[3] *Houston Tri-Weekly Telegraph*, 18 November 1863.
[4] Hanna, "Incidents of the Blockade," p. 228.
[5] *Corpus Christi Ranchero*, 15 January 1863.
[6] Lord, *Fremantle Diary*, p. 16; Hamilton Cochran, *Blockade Runners of the Confederacy* (New York: Bobbs-Merrill Co., 1958), pp. 17-20.
[7] *New York Herald*, 18 February 1864; Hanna, "Incidents of the Blockade," p. 222.
[8] Leonard Pierce to Seward, 1 March 1862, U.S. Consular Despatches; Matamoros, Record Group 56; *OR*, ser. 1, 9:674.
[9] Pierce to Seward, 8 June, 27 Aug., 19 Sept., 23 Dec. 1862, U.S. Consular Despatches: Matamoros, Record Group 56.
[10] Pierce to Seward, 23 Dec. 1862, and enclosures of 25 April, 12 May 1863, *ibid*.
[11] *Welles Diary*, 1:334-335, 387-391; *OR*, ser. 1, 26, pt. 2: p. 405.
[12] *Corpus Christi Ranchero*, 19 Mar. 1863, 23 July 1863; *Texas State Gazette* (Austin) 10 June 1863; Lord, *Fremantle Diary*, p. 23.
[13] *Fort Brown Flag*, 11 Mar. 1863, as quoted in *Ranchero*, 19 Mar. 1863.
[14] *Texas State Gazette*, 10 June, 17 1 June, 24 June, 2 Sept., 28 Oct. 1863; *Corpus Christi Ranchero*, 28 May 1863.
[15] General Dana to Gen. Banks, 21 Dec. 1863, *OR*, ser. 1, 26, pt. 2: 414-415; *Texas State Gazette*, 23 Dec. 1863; Cochran, *Blockade Runners*, pp. 20-21; Annie Cowling, "The Civil War Trade of the Lower Rio Grande Valley," (M.A. Thesis, University of Texas, 1926). p. 87.
[16] *Texas State Gazette*, 1 July 1863.
[17] *Welles Diary*, 1:590.
[18] *New York Journal of Commerce*, as quoted in *Texas State Gazette*, 4 March 1863; Delaney, "Matamoros, Port for Texas," p. 479; U.S., Congress, *House Committee Reports*, 38th Cong., 1st sess.. Report no. 111.
[19] 31 May 1863, U.S. Consular Despatches: Matamoros, Record Group 56.
[20] Houston *Tri-Weekly Telegraph*, 12 February 1863.
[21] *ORN*, ser. 1, 20:741-742; *ibid*., 15:1014; Delaney, "Matamoros, Port for Texas," pp. 479-480; E. Merton Coulter, "Commercial Intercourse with the Confederacy in the Mississippi Valley, 1861–1865," *Mississippi Valley Historical Review* 5;377-395.
[22] U.S., Congress, House, Report of the Committee on Public Expenditures, *House Executive Documents*, 38th Cong., 2nd sess., no. 25, pp. 2-36.
[23] *Ibid*.
[24] Delaney, "Matamoros, Port for Texas," pp. 481-482; Pierce to Seward, 30 Sept. 1863, U.S. Consular Despatches: Matamoros, Record Group 56.
[25] Pierce to Seward, 26 January 1865, *ibid*.
[26] Quoted in Charles Cowley, *Leaves from a Lawyer's Life Afloat and Ashore* (Lowell, Mass.: Penhallow Printing Co., 1879), pp. 112-113.
[27] Samuel Bernard Thompson, *Confederate Purchasing Operations Abroad* (Chapel Hill: University of North Carolina Press, 1935), pp. 107-108.
[28] Randolph to Hart, 14 November 1862, *OR*, ser. 1, 15:866; William Diamond, "Imports of the Confederate Government from Europe and Mexico," *Journal of Southern History*, 6(1940): 499-501; W. A. Broadwell to Christopher Memminger, *OR*, ser. 1, 53:955.
[29] *Houston Tri-Weekly Telegraph*, 1 Dec. 1862.

[30] Caleb Huse, *The Supplies for the Confederate Army, How They Were Obtained in Europe, and How Paid For*, reprint (Houston: Deep River Armory, Inc., 1970), pp. 22-23.

[31] Diamond, "Imports of the Confederate Government," pp. 474-475, 486.

[32] General Smith to Jefferson Davis, 21 November 1863; *OR*, ser. 1, 27, pt. 2:1064, 1074.

[33] General Orders, 3 August 1863, *OR*, ser. 1, 26, p. 5. 2:535-537; Broadwell to Memminger 26 Dec. 1863, *ibid.*, 536; Broadwell to Smith 15 Mar. 1864, *ibid.*, 53:971-974.

[34] U.S., Congress, Senate, "Cotton Sold to the Confederate States," *Senate Documents*, 62nd Cong., 3rd sess., 8:308-309; *OR*, ser. 1, 26, pt. 2:480-482; William T. Windham, "The Problem of Supply in the Trans-Mississippi Confederacy," *Journal of Southern History* 27(1961):159.

[35] Florence Elizabeth Holladay, "The Extraordinary Powers and Functions of the General Commanding the Trans-Mississippi Department," (M.A. Thesis, University of Texas, 1914), pp. 122-128.

[36] *Corpus Christi Ranchero*, 2 April, 11 June, 30 July 1863.

[37] Yancy to Bee, 2 July 1863; *OR*, ser. 1, 53:100-101.

[38] Hart to Seddon, 13 July 1863; Bee to Magruder, 17 August 1863; *OR*, ser. 1, 53:877-878, 892; James L. Nichols, *The Confederate Quartermaster in the Trans-Mississippi* (Austin: University of Texas Press, 1964), pp. 56-57.

[39] Pierce to Seward, 6, 21 July 1863, U.S. Consular Despatches: Matamoros, Record Group 56.

[40] Charles W. Ramsdell, "The Texas State Military Board, 1862–1865," *Southwestern Historical Quarterly*, 27 (1924):271.

[41] Tom Lea, *The King Ranch*, 2 vols. (Boston: Little Brown and Co., 1957). 1:189-190.

[42] Charles S. Davis, *Colin J. McRae; Confederate Financial Agent*, Confederate Centennial Studies No. 17 (Tuscaloosa: Confederate Publishing Co., Inc., 1961). Also, see McRae to Memminger, 7 Oct. 1863, *OR*, ser. 4, 2:982.

[43] Richardson, *Messages and Papers of the Confederacy*, 2:586-587; *OR*, ser. 4, 2:982-983.

[44] *OR*, ser. 4, 3:80-82; Amended Regulations, 6 February 1864, Henry Sampson Papers, Rosenburg Public Library, Galveston, Texas. Hereinafter cited as Sampson MSS.

[45] Cowling, "Civil War Trade," p. 70.

[46] Nichols, *Confederate Quartermaster*, pp. 68-69; Windham, "The Problem of Supply," p. 153.

[47] Lea, *The King Ranch*, 1:192, 445a; Dickeson, "The Texas Cotton Trade," p. 132.

[48] *Houston Tri-Weekly Telegraph*, 30 Dec. 1864; Cowling "The Civil War Trade," p. 114.

[49] Slaughter to Magruder, 30 January 1865, *OR*, ser. 1, 48, pt. 1:1353-1355.

[50] E. D. Etchison to Gen. Canby, 27 Feb. 1865, *ibid.*, pp. 1045-1049.

[51] Gen. R. B. Jones to Gen. Canby, 28 Feb. 1865, *ibid.*, p. 1005.

[52] *Ibid.*, pt. 2:398-403; Windham, "Problem of Supply," pp. 167-168.

[53] Houston *Daily Telegraph*, 9 January 1865; *Texas Republican* (Marshall), 20 January 1865.

[54] Cited in Agnes Louise Lambie, "Confederate Control of Cotton in the Trans-Mississippi," (M.A. Thesis, University of Texas, 1915), pp. 88-89.

[55] Cited in Thompson, *Confederate Purchasing Operations*, p. 117.

[56] Nathaniel W. Stephenson, *The Day of the Confederacy* (New Haven: 1919), p. 49; Diamond, "Imports of the Confederate Government." p. 502.

[57] Thompson, *Confederate Purchasing Operations*, p. 126,

[58] Maj. A. M. Jackson to Lt. Col. C. T. Christensen, 25 Apr. 1865, *OR*, ser. 1, 48, pt. 2:230; Captain W. C. Black to Henry Sampson, 23 March 1865, Sampson MSS.

5

The Gulf Trade

Marcus W. Price, former director of photographic research in the National Archives at Washington, has compiled voluminous statistics covering the blockade-running activity along the southern coasts during the Civil War. His findings indicate that the efficiency of the federal blockade of the Gulf ports was more thoroughly tested than that of the Atlantic Squadrons guarding the blockade-running ports of the Carolinas. During the period from 20 April 1861, the day following Lincoln's blockade proclamation, and 4 June 1865, the day before Galveston was occupied by federals, there were 2,960 violations or attempted violations of the Gulf blockade. This was a daily average of two attempted violations as compared to 1.5 for the Carolina ports.[1] A violation is considered to be any one-way attempt to run the blockade.

Price's statistical tabulations appear to be exhaustive and definitive, certainly from the standpoint of recorded sources. At that, his findings must represent the known or minimum number of violations. As Price notes, the extant records of Gulf port customs houses do not show all business conducted during the war. Upon arrival, the masters of vessels were required to dispatch messengers to the nearest collection district so that revenue officers might be sent to perform their duties. In many instances, captains entering and clearing such ports dispensed with the required amenities, leaving no record of their violating the blockade. Confederate customs authorities noted that there was a class of small light draft vessels, of from six to twenty tons burden, engaged in trade between Gulf coast and Caribbean ports. A. B. Noyes, collector of customs for the District of St. Marks, Florida, reported that these vessels ran into the numerous small streams on the coast and freighted their merchandise cross-country, neither having entered nor cleared vessels or merchandise nor having paid any duties. Included in this class of offenders were those numerous and nameless individuals who engaged in the coasting trade, flitting about in skiffs, schooners, and fishing smacks, hauling a bale or two or perhaps forty, between bayous and fishing villages. As Noyes commented, "there seems to be a general opinion among this class that they are expected to pay no duties during the continuance of the blockade."[2] Their part in violating the blockade will perhaps never be known, but judging from contemporary reports, one suspects their numbers were legion.

For the first several months of the war, the Confederacy deliberately embargoed their cotton exports. It was commonly known that eighty-four percent of the world's supply of cotton was produced by the seceded states. Confederate strategists reasoned that the sudden disappearance of cotton from the world markets would paralyze the economic and social welfare of those countries dependent upon the staple. In both New England and old England, great fortunes were invested in huge mills for the spinning and weaving of cotton, and in both places large numbers of employees were dependent on the cotton crop for their livelihood. Southern diplomacy was organized around the assumption that the withholding of cotton from the mills of Europe, even for a short duration, would so disrupt their domestic affairs that both England and France would intervene militarily to ensure the continued supply of this commodity. Thus, Southern independence would be obtained. As a result, when the blockade was first proclaimed Southerners hailed it as an act playing directly into their hands.[3]

With the dual imposition of the blockade and the embargo, only a limited number of attempts were made to run the blockade in the first months of its existence, when it could have been done with relative ease. England, the South's largest customer, noted the sharp decrease in cotton imports. In the six months ending with May 1862, only 11,500 bales had been received in their ports. This was only one percent of the previous year's figure for the same period. It was hardly the blockade that accounted for this, as in many places the only evidence of a blockading fleet was a decrepit ex-ferryboat, mounting a few jury-rigged cannon.[4]

Despite the embargo, there appears to have been a relatively heavy Gulf trade the first year of the war. The success of this trade is explained largely by the fact that the United States Navy was inadequately prepared both in ships and experienced personnel effectively to blockade the long Gulf coastline. Often, for considerable periods of time, many of these ports remained open while others were so poorly covered that runners showed little concern. The summary of the Gulf trade for 1861 shows that there were thirty-four steamers and 397 other types of craft involved. The steamers attempted 375 violations while the other vessels attempted 1,348. Of these, the steamers succeeded in 371 attempts and the others succeeded 1,293 times. The percentage of successful runs for each class of vessel were ninety-nine and ninety-six per cent respectively.[5]

After the fall of New Orleans, only Mobile and Galveston were left open as the best and largest ports for steamer traffic. This type traffic gradually decreased until the latter part of 1864. R. W. Shufeldt, the American Consul in Cuba, reported in December 1862, that there was a marked reduction in steamer arrivals. It was his opinion that attempts to run the blockade in large steamers had proven so costly that the traffic was trending toward small schooners in order to lessen the risks.[6]

While Shufeldt's observations were correct, his reasoning was in error. The slackening of steamer traffic and the utilization of schooners and sloops in the Gulf trade resulted from necessity rather than desire. When New Orleans fell, there were so few Gulf channels left that were deep enough to admit steamers that the bulk of steam traffic shifted to the Carolina coasts. Yet it was soon apparent, as the blockade tightened, that the employment of sailing vessels in the trade was highly risky. Nevertheless, some veteran captains preferred the centerboard schooners over steam ships. This yacht-like sailing vessel had an adjustable centerboard, or keel, that could be raised or lowered as circumstances required. With center board down the craft was maneuverable and clean sailing, yet in shallow water, so characteristic of the Gulf coast, the board could be retracted allowing a draft of only five feet when the ship was fully loaded. Further, these craft were capable of uncannily disappearing before one's eyes in broad daylight by stripping sails and turning the vessel at right angles to a blockader.[7]

If some preferred sailing vessels, the truth was that their lack of speed made them easy prey for the federal steamers that were constantly being added to the blockading squadrons. They were further limited by their cargo capacity which minimized the importance of their attempts to run. Despite the grisly statistics, captains persisted in their attempts to run sailing vessels through Farragut's blockade. At least 216 such vessels were engaged in traffic through Gulf ports in 1863, and not less than 119 engaged the next year. This, in spite of the fact that in 1863, 136 of 329 attempts to run had resulted in the loss of the vessel.[8]

Following Lincoln's proclamation, federal naval officers displayed considerable uncertainty and confusion as to exactly what constituted a violation of the blockade. Some of the early instructions to them declared that key Gulf ports were under the general provisions of the proclamation but not specifically closed. Ships must be allowed to enter unless they were carrying contraband of war.[9]

An article appearing in the Montgomery, Alabama, *Weekly Mail* reported that the schooner *Sara Cole* and an English vessel had arrived at Pensacola, Florida, on 2 May 1861. The ships had been boarded by the federals, inspected, and then permitted to enter port. The English master had been told that he was lucky to have arrived when he did since the blockade was to commence on 11 May. The article added that ". . . There are nothing but Englishmen now in port. The last Yankee left yesterday morning, with every inch of canvas spread. He has been terribly scared for two or three weeks. . . ."[10]

Following instructions from Gideon Welles, formal notice of a blockade at Pensacola was communicated to Confederate authorities at that port on 13 May 1861. The captain of the United States frigate *Sabine* dispatched a note under a flag of truce stating that "Foreign vessels will be allowed two weeks from this date to settle their accounts and sail. After the expiration of that time, they will be detained if they attempt to leave the harbor and dealt with according to the laws against violating blockades."[11] On that same day, the blockaders off Pensacola prohibited the British ship *Perthshire* from entering port but volunteered the information to her master that Mobile was still open.[12]

Pensacola was never a very active blockade-running port. This was due to the fact that the federals retained control of Fort Pickens on the tip of Santa Rosa Island, which commanded the entrance to the harbor (Figure 5). In spite of the frowning guns of the nearby federal fort, limited running was conducted at night and, for some vessels, via the inlet between island and coast. The British Consul at Havana noted that from 1 January to 14 March 1862, only two vessels, the *Al Jones* and the *Essayons*, had arrived from Pensacola. Both were schooners of ninety-four and twenty-two tons burden respectively and carried cargoes of turpentine. Shortly after this the federals occupied the port, entirely stopping the running activity. On 19 November 1864, President Lincoln proclaimed the port open to non-contraband traffic, "[Pensacola] having sometime been in the military possession of the United States."[13]

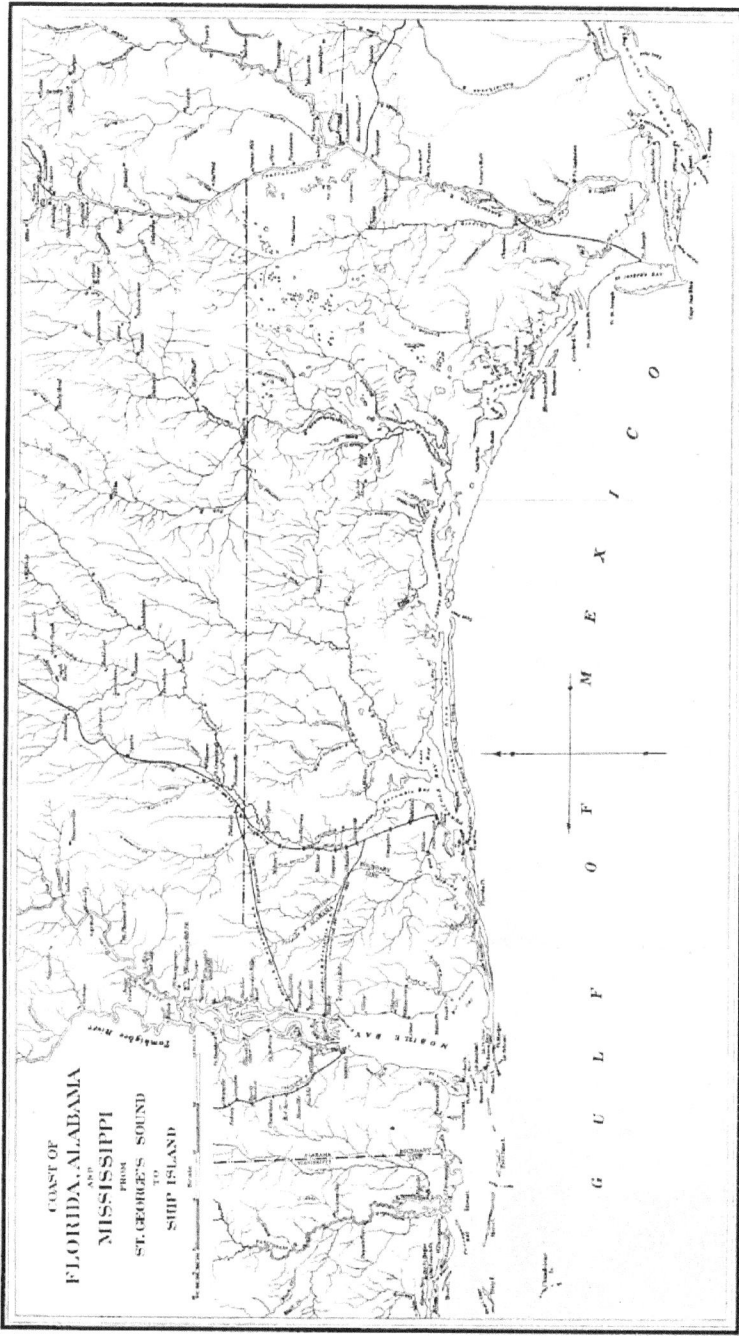

Figure 5. Coast of Florida and Alabama. *ORN*, ser. 1, 19:3.

At the beginning of the war authorities in New Orleans had also been in the dark regarding the exact date of the official blockade of that port. The British Consul there informed the *Daily Picayune* on 17 May 1861, nearly a month after the proclamation, that he had not received any official notice respecting the date it was to be established off the mouth of the Mississippi. He had heard through official channels, however, that neutral vessels would be allowed fifteen days to leave port after its actual imposition, whether with or without cargoes and whether the cargoes were shipped before or after the actual blockade commenced. He added that it was rumored that the blockade of the Port of New Orleans would be effectually established on or about 25 May 1861. The rumor proved to be true.[14]

Foreign consuls in blockaded Gulf ports had been apprehensive over the possibility that ships belonging to subjects of their nations, or containing cargoes consigned to foreign houses or individuals, would not be able to enter or leave port before the blockade became effective. Their concern was well founded. On 17 May 1861, there were in the Port of New Orleans seventeen American, ten British, five Bremen, three Spanish, two Oldenburg, and one Italian vessel. By June 6, the number had increased to nearly thirty, all stranded because of the low stage of water on the bar. The British Consul reported that the majority were bound for Liverpool with an aggregate cargo valued at not less than one million pounds sterling. To complicate matters, the New Orleans Towboat Association had given public notice that, inasmuch as vessels being towed over the bar had been fired upon by the blockader *Brooklyn*, the Association's tugs would suspend services until assured of safe passage while towing vessels that had a perfect right to leave.[15]

Wishing to resolve the dilemma, the worried British, French, Spanish, and Bremen consuls, accompanied by a manager of the Association, called on the captains of the *Brooklyn* and *Powhatan*, then lying off the mouth of the Mississippi River. The delegation was received courteously and learned to their surprise that formal notice of the blockade had been given by Captain Poore of the *Brooklyn* to a Confederate major at the mouth of the river. By this, they learned that the expiration date of formal notification was to be 10 June. The British Consul asked for and was granted additional time to allow vessels to be towed down river and over the bar. The captains agreed and set 14 June as the deadline within which vessels must put to sea. The federals also acknowledged that towboats were to be regarded in the light of the sails of the vessels, or as pilots, to be treated as neutrals. It was understood that certain towboats that carried guns would be considered as privateers, and not be exempted.[16]

As for the vessels that had cleared but had been unable to cross the bar, the captains ruled that while their instructions did not allow them to extend the time of departure, they would assume the responsibility and allow some latitude. This provision, of course, was contingent upon those vessels making every effort to get over the bar with no partiality being shown by the towboat company in taking over vessels owned by the South. Under this agreement tugboat operations began and twelve vessels put to sea. At least one ship, the *Senator Jken*, was still hard aground on the bar on 18 June. Apparently she was later floated, and there is no record that her departure was interfered with.[17]

While the consuls were busy arranging official clearances, the citizens of New Orleans conducted their own shipping affairs with a business-as-usual regard for the blockade. Vessels arrived and departed with regular frequency. If they were warned off Pass a l'Outre where the *Brooklyn* was stationed or off South West Pass by the *Powhatan* they simply ran for the bayou-bays to the south of the city or for the lakes to the north. New Orleans newspapers published their arrivals, clearances, and schedules of departure as they had in peacetime. A typical issue reported that the steamer *California* and the schooners *Fanny, Sea Drift, General Surprise, Olive Branch, Balcares, Mohongo, Southern Independence*, and *Virginia John* had arrived on the previous day and the steamer *Oregon* on 2 June. It also announced the clearance on 3 June of the barks *Teresita, Valentina, Antonio*, and *Julia* for European ports, and of the schooners *Sara Burr* and *D. F. Keeling* for Mobile and Vera Cruz respectively.[18]

Eventually, the reality of the blockade became apparent to the citizens of New Orleans. On June 25, the *Picayune* reported that a United States warship had anchored off Atchafalaya Bay the previous day and had interrupted communications with Texas via Berwick's Bay (Figure 6). "While we are remaining in fancied security and boasting of the impossibility of the invasion of our shores by the enemy," it declared, "a fleet takes possession of Ship Island and the Sound and cuts off our communications with Mobile, while another of the enemy's ships blockades the mouth of the Atchafalaya." It demanded, "Where is our fleet of gunboats and why are they permitted to lie idle when there is plenty of opportunity for then to exercise their bravery . . . ?"[19] The same issue reported that several shots had been fired at the mail boat *Oregon* and the steamer *J. D. Swain* and that both vessels, bound for Mobile, had returned to New Orleans. "On account of the blockade of the lakes," it stated, "the suspension of the Mobile mail line is announced."

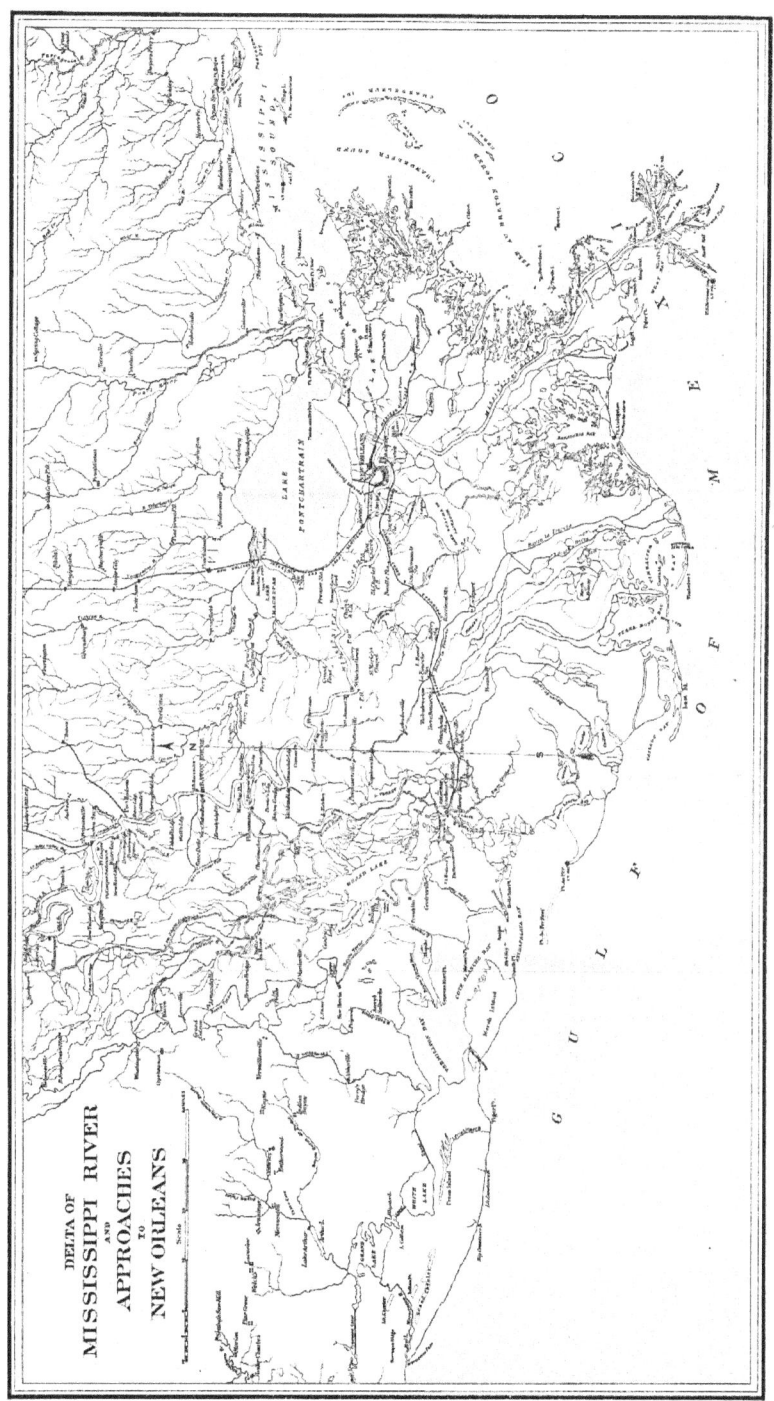

Figure 6. Delta of the Mississippi River and Approaches to New Orleans. *ORN*, ser. 1, 20:4.

In spite of this state of affairs, the record shows that entrances and clearances to and from New Orleans to Mobile continued with almost daily regularity. Arrivals from Confederate ports like Galveston, Velasco, Indianola, Brownsville, Apalachicola, and Havana, as well as departures for these ports, were not infrequent. Indeed, the volume of business carried on at the port of New Orleans was heavy, considering the blockade, until its capture by Farragut in May 1862. During the single month of September 1861, there were twenty-four clearances. Shortly thereafter, the citizens of New Orleans sent an agent to Tampico, Mexico, to establish monthly mail connections at that port with British mail packets.[20]

As previously mentioned, British consulates were directed by the Foreign Office to submit periodic reports of blockade-running activity. On 8 January 1862, acting Consul George Coppell left the following notation regarding the New Orleans trade:[21]

> 30 November 1861–Steamer *Vanderbilt*, flying C.S. flag left N.O. arrived Havana 3rd Dec. without opposition. 387 bales cotton, 100 bbls. turpentine; left Havana arrived Sabine City, Texas, 29 Dec. U.S. Yacht *Dart* did not stop her. Brought in 91,000 lbs. cannon powder and other munitions and articles; preparing another trip. Name as permanent British register is *Black Joker*.
> 2 December 1861–Steamer *Calhoun*, 651 bales, left N.O. arrived Havana 6 Dec.–news brought by British sloop *Margaret* [which] came in through bayou from Havana.
> 14 December 1861–Steamer *Pizarro*, 493 bales, left Berwick Bay for Havana, 160 miles out developed leak and returned. No trouble from blockade.

Coppell's quarterly report for the period ending March 1862, shows an even greater traffic and this on the eve of the capture of New Orleans:[22]

> –Steamer *Victoria* left N.O. for Havana, 15 Jan., 1862, via river, 535 bales–returned from Havana [by] Barataria Bay 20 Feb., [with] 700 kegs, 5 barrels and 23 boxes gunpowder, 616 bags coffee, assorted cargo of drugs, tin, copper, 125 stand of arms, percussion caps, shot, etc.–provisional British register.
> –Steamer *A. I. Whitmore*, left N.O. 1 Feb. for Havana, with 490 bales, arrived 23rd. Return voyage arrived Berwick Bay 11 Mar. entered at custom house 21 Mar. with cargo of 200 boxes gunpowder, 400 sacks coffee, 4 cases of arms and assorted cargo. Provisional British registry, renamed *Fox*.
> –Steamer *Vanderbilt* sailed from Sabine Bay, Texas for Havana, 15 Feb., with 399 bales, arrived Havana 20 Feb.
> –Steamer *Florida* left N.O. 9 Feb., with 839 bales, by river, arrived Havana on 23rd; returned to a Florida port apparently not blockaded with 70,000 lbs. gunpowder, 550 sacks coffee, 500 revolvers, 2,000 rifles, 900 cases merchandise and miscellaneous goods, total cargo estimated at $500,000. Owners live in New Orleans.
> –Steamer *Elizabeth* arrived at Grand Caillon, this state, 16 Feb., from Nassau with ten tons gunpowder, 400 muskets, coffee, leather, and sundries–a permanent British register.
> –Schooner *Lilly* arrived Berwick Bay 19 Mar. from Havana with lead, copper, zinc, etc., etc.,
> –Schooner *Hermosa*, left Nassau Feb. 4, arrived Grand Caillon 12 March with 2,400 bushels salt, coffee, tin, printing paper, etc., etc.,–permanent British register.

James T. Crawford, the British Consul in Havana, noted the arrivals and departures for the Foreign Office during the same quarter. In addition to the *Victoria*, *Lilly*, *Whitmore*, and *Florida*, which had arrived from New Orleans, he also named seven other vessels, omitted by Coppell, that arrived from the same city.[23] Only the *Vanderbilt* was unaccounted for by Crawford, but considering that the ship carried Confederate colors, had a permanent British registry, and sailed under two names, Crawford may have been more confused than negligent in not noting her arrival.

It is also apparent that much of the New Orleans trade was conducted through Lake Salvador, Berwick, and Atchafalaya Bays, and almost exclusively so once the lower Mississippi became clogged with Farragut's flotilla. Railroad and overland routes between the Crescent City and Bayous and lakes that emptied into coastal bays afforded an easy flow of commerce to and from the city. Federal naval officers were not long in discovering this fact, and thus extended the blockade to cover this lacy network of waterways. As was to be expected, the gunboats carried too much water effectively to stop the traffic, and Farragut asked the Navy Department to send the first available vessels of less than eight feet draft for bay service.[24]

Even after the fall of New Orleans there was a continuous trade flourishing on the coastal bays and inland waterways of south Louisiana. While captures were frequent, the trade was never stopped, and this activity constitutes another obscure chapter of blockade-running. Especially was this the case since much of the bay commerce was lateral traffic, more accurately smuggling than running.[25]

The blockade of Mobile Bay went into effect technically on 28 May 1861. The harbor was soon cleared of all foreign vessels that intended to leave safely. The first blockader was the U.S.S. *Powhatan* commanded by Captain David D. Porter. Belatedly, Porter informed James Magee, acting British Consul at Mobile, to get all British registries to sea immediately. Yet Magee reported to the Foreign Office on 29 July that so far he had neither seen nor heard of any official notification that the Port of Mobile was blockaded. United States warships were lying some four to seven miles off Fort Morgan, at the mouth of the bay, he wrote,

> But such is the effect of the blockade that steamboats and small vessels are regularly running between this port and New Orleans. Although it does happen [that] a U.S. war vessel is occasionally by to stop them; but as this vessel does not regularly blockade, and frequently is not to be seen, the intercourse between Mobile and New Orleans is not regularly interrupted.[26]

Mobile was second to New Orleans in importance as a major Gulf port and became pre-eminent in blockade-running activity after the fall of the latter in the spring of 1862. For the federal navy, blockading duty off Mobile Bay was difficult due to the many entrances to the harbor that could be used by light draft runners. Blockading vessels were also obliged to lie at a distance from shoal water because of their draft. The distance was further necessitated by the need for additional depth when violent gales blew up suddenly. Having sturdy ships with heavy ground tackle enabled blockaders to ride out severe winter southeasters without being once driven aground according to those stationed off Mobile.[27]

For blockade-runners, such foul weather, if not too severe, afforded the best opportunities for running. But foul weather or fair, the Port of Mobile experienced a busy contraband trade until the fall of 1864. Consul Magee was fairly diligent in reporting blockade violations conducted by at least twenty-one vessels. Seven of the ships were either captured or grounded and destroyed to prevent capture, representing a loss of one-third of the vessels. All except one, the steamer *General Miramon* were powered by sail, which accounts for the high loss ratio.[28]

After an unexplained lapse of three months in his reporting, Magee resumed submitting details of blockade-running activity in the fall of 1862. Of all the schooners that had been mentioned in his previous reports, the *Wide Awake* was the only familiar name mentioned in December. He did note with some detail the arrival of the British steamer *Alice* (*Matagorda*) with munitions and her departure one month later with 830 bales of cotton. Two months later the *Alice* was safely back in Mobile with a full cargo of powder, blankets, and assorted contraband. He added that the American steamer *Yorktown* had also run out safely with 830 bales of cotton.[29]

The success of these and other steamers in the Mobile trade was noted by Farragut. He confided to Fox that it was impossible to prevent steamers from running the blockade on dark nights. "They are past you at twelve or fourteen knots," he explained, "while our ships are lying still." Frustrated by the success of the runners, he ordered a court of inquiry for Lt. Woodworth for allowing the *Cuba* and the *Clio* to run out safely for Havana. Later he lamented the decision to give priority to the army operations against Vicksburg, "otherwise, I could take Mobile today with wooden ships," he declared.[30]

The success of the federal blockade at Mobile was further retarded by the necessity of dispatching all except a bare minimum of vessels to the Mississippi. Further, the steamers that were left on Mobile station were in sad condition. Commander R. B. Hitchcock reported that the gunboats *Supply* and *Oneida* had collided. The *Oneida* was so badly crippled that she could make only three knots "with the safety valves made fast." Under these circumstances, Hitchcock warned, the Department should not expect too much of the Mobile blockade. "We have to have vessels to cruise off the line at Mobile," he explained, "but we cannot take them off channel guard." He indicated the difficulty of the task under the best of conditions by declaring that if he was a runner he would come out the first dark night "and not mind all the vessels you could put off here." Lie ended his gloomy report by echoing a common complaint of West Gulf Blockade commanders: "We are so far off, I am afraid you forget us. Norfolk and Du Pont seem to stop all [good vessels] coming this way."[31]

In the spring of 1863 an issue of the *Galveston Weekly News* reported that blockade-running at Mobile was quite active because of the heavy import business there at that time of year.[32] If this was true, the increased traffic coincided with the general intensification of blockade-running throughout the South. As noted heretofore, this increased traffic began in early 1863 due to the shortage of cotton in Europe and high prices for consumer goods within the Confederacy.

At the peak of the spring rush to get into the running business, the Mobile Committee of Safety protested to the Alabama governor about the conversion of riverboats into blockade-runners. The chairman of the committee reported that no less than six of the river steamers were fitting out to carry cotton to Cuba. It was their opinion that the boats were more vital to harbor and river commerce and if lost could not be replaced. Facilities did not exist for the building of new boilers, he explained, though admittedly, hulls might in time be constructed. Further, these boats, if retained, could serve emergency uses in transporting troops or provisions on the rivers in the event the railroad line to Montgomery was captured. After citing several other emergency strategic uses for the boats, the chairman closed with the best argument of all: "Being of light draft for these waters, if captured, they would form a large fleet of transports and would greatly aid the enemy in transporting troops and munitions of war along the coast, in the bays, and bayous about the entrance of the bay."[33]

This intensification of trade at Mobile is clearly reflected in Magee's reports for 1863. In one five-week period from 12 March to 20 April, there arrived in Mobile eleven vessels, five of which were steamers. During the same period, there were seventeen departures from that port. On 1 April, Magee reported that, even with the increased cotton shipments, total exports were less than the amount that had been destroyed in New Orleans on the night of 23 April 1862. From 22 April to 25 May there were only seven recorded arrivals and thirteen departures. Traffic either slowed from 21 May to 5 August, or Frederick J. Cridland, Magee's replacement, was not as careful in recording arrivals and departures, as only ten departures and three returns were reported for the period. According to Cridland, several steamers were captured in September, which virtually suspended trade with Mobile for some weeks. If his fourth quarter returns are any indication, this was true, as only three arrivals and two departures were noted.[34]

The capture or destruction of these steamers engaged in the Mobile trade apparently did cause a serious setback in the Mobile maritime traffic. Major General Dabney H. Maury, C.S.A., in reporting to General Joseph E. Johnston the loss of the steamers *Fanny* and *Alabama* on 12 September 1863, stated that a fine cargo of government supplies had been lost on *Fanny* and an even more valuable one on *Alabama*. These two steamers, he related, had carried 450 bales of cotton to Havana where they brought thirty-six cents a pound. "They may be regarded as the last of the blockade runners," he wrote dejectedly, "as they were the best of them."[35] As events proved they were neither the best nor the last of the steamers that entered and cleared the port of Mobile, but their elimination seriously hurt for some time the volume of trade done there.

Federal sailors learned in time that often outward-bound blockade-runners made use of cotton saturated with turpentine to build a full head of steam before dashing through the blockaders. Runners preferred anthracite coal, as it made less smoke to betray their presence, but this fuel was scarce and was used only in emergency circumstances. For the remainder of the trip, the common British coal was used though its tell-tale smoke would draw any federal cruisers in the vicinity. On several occasions blockaders gave chase to Mobile runners by following their smoke trail at night. The smoke would gradually increase in volume, indicating that the victim was firing her boilers to the fullest. After perhaps several hours of unrelenting pursuit the blockader would suddenly come to the end of the line of smoke and find that the runner had vanished. Federals later learned from captured captains that when pursued, they gradually increased the volume of smoke until it was dense. Then with enough distance for the rest of the vessel to be hidden by darkness the runner would close its dampers, shut off smoke entirely and change course at right angles, leaving her bewildered pursuers in the dark.[36]

Schooners engaged in the runner trade, while leaving no smoke to be spotted, were strictly at the mercy of prevailing breezes. Even so, some schooners, like the *Joe Flanner* made many round trips between Mobile and Havana. The *Flanner* had run the blockade sixteen times before her luck ran out. Her captain disclosed to federal officers that his success was due to his coming out on foggy or very dark nights with all sails lowered and simply drifting through the blockaders on the current. He had seen the federal flagship on several such occasions, but the low hull and slender spars of his schooner had been invisible to those on the gunboats.[37]

And so it went at Mobile as long as the port remained open. There were active and slack periods until August 1864, when Farragut carried the outer defenses of Mobile Bay, which virtually eliminated blockade-running at this port, although a few runners did get in occasionally.[38]

With the exception of Galveston, Sabine Pass, Texas, was perhaps the Gulf port next in importance in the Gulf contraband trade. Although the port was of secondary importance, continuous references were made by both federal and Confederate officials throughout the war alluding to the traffic there. Confederate customs records for the port have not survived the years to leave a record of the volume of trade. Mention is made of both steamer as well as schooner traffic entering and clearing the port. Because of the initially weak defenses of the bay, the port was controlled by federal vessels from September 1862, to January 1863. At other times, no trace of blockading vessels were to be seen off the port for days at a time.[39]

As related heretofore, the blockade of Sabine Pass was physically raised by the heroic exploits of the Confederate riverboat "cottonclads" *Uncle Ben* and *Josiah Bell*. The port is most commonly remembered for the lopsided Southern victory over General Banks there in October 1863. A Confederate artilleryman stationed at the port remembered the night that William L. Yancey, Confederate Minister to England, was landed through the surf near the pass. Upon being picked up by a Confederate patrol on the beach, Yancey explained that he was returning from England but that the other Gulf ports were so closely guarded that their landing at Sabine Pass was the only recourse.[40]

The inland port of Lake Charles, Louisiana, located some sixty miles east of Sabine Pass, was one of many small ports of limited commercial activity. Agricultural production for that region was not extensive, and most export commodities were lumber and timber by-products. Seven local lumber mills comprised the chief industrial effort for the parish. Commerce was restricted to all but small schooners by the shallow depths of the lakes and river that connected the port to the sea. One such vessel, captured by the federals, was held to be so insignificant that it was used only for target practice off Galveston. Nonetheless, the federals considered the Calcasieu trade important enough to station two blockaders, when possible, off the mouth of the river.[41]

Velasco, Texas, probably handled as much contraband tonnage as Sabine Pass, though fewer steamers visited the port. The Brazos River, which empties into the Gulf at that point drained a region of considerable cotton production. Since the river was broad and relatively straight, schooners could navigate inland, serving the cotton centers of Brazoria and Columbia. Throughout the war, a constant small craft trade continued through Velasco, and the port was often sought out by blockade-runners because it was not consistently guarded by the federals. Nearby, the mouth of the San Bernard River was unguarded entirely. No doubt the federals felt that the four-foot bar and lack of coastal habitation eliminated any need for guarding the mouth. The truth was that the San Bernard drained the same rich cotton counties, and small trading vessels frequently entered the river.[42]

William Watson, the centerboard schooner captain, found upon approaching Velasco, that the usual bar existed off the mouth and that a pilot was required to guide vessels through the crooked, six-foot deep channel. While ascending the river he met two schooners whose decks were piled so high with cotton that the vessels appeared in danger of capsizing. At Columbia, he found three or four blockade-runners discharging or awaiting cargoes. Apparently a large amount of unbaled cotton was loaded loose into the hold of Watson's schooner and then compressed. In their efforts to stuff as much as possible of the precious staple on board, they pressed the bottom out of Watson's vessel.[43]

From the uppermost tip of Matagorda Bay to the mouth of the Rio Grande, a distance of roughly 250 miles, a string of narrow, offshore islands parallels the Texas coast (Figure 7). The string is broken here and there by passes which enter the inland bays, where numerous coastal villages and towns became active trading centers during the Civil War. The narrow water passage between islands and coast afforded a natural canal for an active shallow-draft coasting trade. The importance of this waterway in laterally funneling contraband goods was noted by Major C. G. Forshey, Confederate Corps of Engineers, in February 1862. "The salt from the lakes below Corpus, and the vast lead importations from Mexico, and coffee and other articles now coming from Matamoros," he informed the authorities, "pass up this route and the cotton and sugar going out in return pass out this way." He then added, "a constant run of small craft is visible from outside."[44]

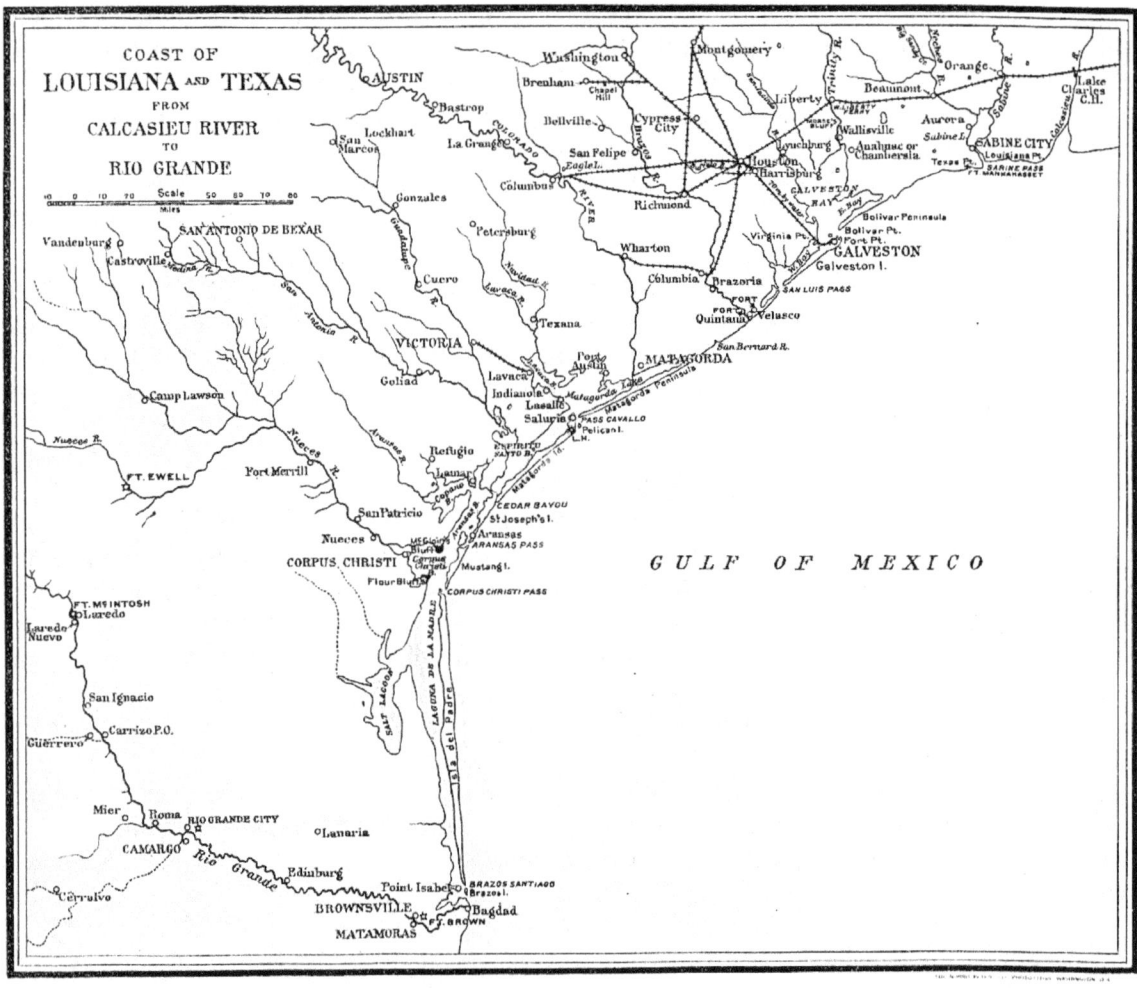

Figure 7. Coast of Louisiana and Texas. *ORN*, ser. 1, 17:3.

Five months later the Confederate Provost Marshall at Matagorda reported that in the previous six months some 3,850 bales of cotton had been shipped, and 16,000 bushels of salt had been received at the port. The federal navy had noted this activity and seriously interdicted it with a series of raids and cutting out expeditions beginning in the summer of 1862. For the next year, the presence of federal naval forces in the bays and off the ports appreciably stifled the inland trade.[45] The federals lessened their harassment when ships were reassigned following the disasters at Galveston and Sabine Pass, and the contraband coasting trade resumed. The extant shipping vouchers and freight receipts indicate that a considerable volume of cotton made its way to Brownsville and Matamoros by the inland passage. The evidence indicates that cotton was shipped on schooners and sloops in small lots from Matagorda, Texas, to Corpus Christi, Penascal, and Flour Bluffs on the south Texas coast. From these points, it was freighted overland to the Rio Grande, thus avoiding the risk of being captured by federal blockaders near the mouth of the river.[46] With the onset of Banks' Rio Grande expedition, the south Texas coasting trade was again interdicted, this time completely. Occupied not only were the lower reaches of the river but also various passes on the offshore islands. This situation existed from November 1863, to the following summer. As expected, as soon as the expedition was recalled the coasting trade resumed, continuing until the close of the war.

That cotton which found its way south along the Texas coast did not necessarily stop at Matamoros. Any Mexican Gulf port would do, especially for smaller craft that preferred hugging the coast to braving the uncertainties of the open Gulf. There were foreign buyers and brokers in Tampico and Vera Cruz, ready to buy, or to exchange, goods and war equipment for the white staple. Belize, British Honduras, was quite active as a foreign port for Confederate commerce, according to the U.S. Consul there.[47] These and other convenient neutral ports were sought by ships running out of the Confederacy. Once inside port, ships were safe, and their cargoes shipped out in neutral bottoms from neutral ports found their way safely into the burgeoning cotton trade.

Chapter 5 Notes

[1] Marcus W. Price, "Ships that Tested the Blockade of the Gulf Ports, 1861–1865," *The American Neptune* 11(1951): 262-290; 12(1952):229-258; 13(1953):154-161; 15(1955):97-131.
[2] *Ibid.*, 11:278.
[3] For a full treatment of the South's dependence on cotton to exert diplomatic pressure, see Owsley, *King Cotton Diplomacy*; also, A. Sellew Roberts, "High Prices and the Blockade in the Confederacy," *South Atlantic Quarterly* 24(1925):154-163.
[4] Roberts, "High Prices," p. 155; Adams, *Great Britain*, 2:9.
[5] Price, "Ships," 11:263-264, 290.
[6] Owsley, *King Cotton Diplomacy*, p. 251; Shufeldt to Farragut, 1 Dec. 1862, ORN, ser. 1, 19:386-387.
[7] William Watson, *The Adventures of a Blockade Runner*, (London: T. Fisher Unwin, Publisher, 1888), pp. 145, 155, 294. JBA: Watson's book reprinted as *Denbigh* Shipwreck Project Publication 1. See herein p. 17.
[8] Price, "Ships," 11:262-263.
[9] ORN, ser. 1, 4:156, 162.
[10] Quoted in Price, "Ships," 11:264.
[11] Welles to Capt. Charles H. Poor, ORN, ser. 1, 4:167-168; *The New Orleans Daily Picayune*, 18 May 1861.
[12] Quoted in Price, "Ships," 11:264.
[13] Crawford to Russell, 14 Mar. 1862, Foreign Office, Record Group 72, Spain, 1041; Richardson, *Messages and Papers*, 6:230.
[14] *The Daily Picayune*, 18 May 1861; Price, "Ships," 11:264; ORN, ser. 1, 4:187.
[15] Bernard, *Neutrality of Great Britain*, p. 231; *The Daily Picayune*, 4 June 1861; Price, "Ships," 11:265.
[16] Bernard, *Neutrality*, p. 281.
[17] *Ibid.*; Price, "Ships," 11:266.

[18] *The Daily Picayune*, 4 June 1861.

[19] *Ibid.*, 25 June 1861.

[20] Price, "Ships," 11:267-268; Franklin Chase to Seward, 24 Jan. 1862, *OR*, ser. 1, 9:641-642.

[21] Coppell to Russell, 8 Jan. 1862, Foreign Office, Record Group 5, America, 848:325.

[22] Coppell to Russell, 24 March 1862, *ibid.*

[23] Crawford to Russell, 14 March 1862, Foreign Office, Record Group 72, Spain, 1041.

[24] *Fox Correspondence*, 1:337.

[25] For some indication of this bay traffic see *OR*, ser. 1, 48, pt. 1:153, 1426.

[26] Scharf, *History of the Confederate Navy*, p. 535; Magee to Russell, Foreign Office, Record Group 5, America, 848; Price, "Ships," 11:269.

[27] Wait, "The Blockade of the Confederacy," p. 920.

[28] Magee to Russell, 2, 6, 29 Jan., 3, 11, 25 Feb., 5, 17, 27 Mar., 12, 23 Apr., 7, 22 May 1862, Foreign Office, Record Group 5, America, 848; Crawford to Russell, 14 Mar. 1862, *ibid.*, Record Group 72, Spain, 1041; Price, "Ships," 12:52-59.

[29] Magee to Russell, 26 Aug., 29 Sept., 16 Oct., 26 Dec. 1862, Foreign Office, Record Group 5, America, 848.

[30] *Fox Correspondence*, 1:318-312, 322.

[31] *Ibid.*, 2:490-491.

[32] *Galveston Weekly News*, 15 April 1863.

[33] Committee of Safety to John Gill Shorter, 25 March 1863, *OR*, ser. 4, 2:462.

[34] Magee to Russell, 1 Apr. 1863, William Porter (acting Consul) to Russell, 20 Apr., 25 May 1863, Cridland to Russell, 5 Aug., 31 Dec. 1863, Foreign Office, Record Group 5, America, 908.

[35] *ORN*, ser. 1, 20:596; Price, "Ships," 11:270.

[36] Wait, "Blockade of the Confederacy," p. 923.

[37] *Ibid.*

[38] Maury to S. Cooper, 14 August 1864, *OR*, ser. 1, 52, pt. 2:723-724; Frank Vandiver, *Confederate Blockade Running through Nassau, 1861–1865; Letters and Cargo Manifests* (Austin: University of Texas Press, 1947), p. xxxvii.

[39] Arthur T. Lynn to Russell, 14 February 1862, Foreign Office, Record Group 5, America, 848; *Galveston Weekly News*, 15 Apr. 1863.

[40] *OR*, ser. 1, 15:237; Reminiscence of E. I. Kellie, undated Port Arthur, Texas, newspaper clipping found in vertical file, Port Arthur Public Library.

[41] Stuart Alfred Ferguson, "The History of Lake Charles Louisiana" (M.A. Thesis, Louisiana State University, 1931), p. 34-35; Grade Ulmer, "Economic and Social Development of Calcasieu Parish, 1840–1912," *Louisiana Historical Quarterly* 33(1955).

[42] Watson, *Adventures of a Blockade Runner*, p. 39; *OR*, ser. 1, 9:616; Captain Forshey to A. T. Lynn, Oct. 1861, Foreign Office, Record Group 5, America, 848.

[43] Watson, pp. 44, 48-52, 64, 79.

[44] *ORN*, ser. 1, 17:165-166.

[45] Its R. H. Chinn to R. L. Upshaw, 12 July 1862, *ORN*, ser. 1, 17:724-725; Chinn to R. M. Franklin, 22 July 1862, *OR*, ser. 1, 17, pt. 2:728-729; *ibid.*, ser. 1, 15:182.

[46] Assorted papers, William Pitt Ballinger collection, University of Texas Archives. Hereinafter cited as Ballinger collection.

[47] A fair indication of this running traffic is reflected in: U.S., Department of State, *Consular Despatches*, Havana, 45, 46; Tampico, 6, 7; Vera Cruz, 8, 9; Belize, 1, 2, 3.

6

The Galveston Trade

The importance of Galveston as a trading port for the Trans-Mississippi Confederacy has been somewhat de-emphasized by Civil War historians. This oversight may in part be accounted for by the fact that blockade-running activity through this port was relatively light, albeit continuous, until the capture of Mobile Bay in August 1864. Until then, it was obviously more efficient for blockade-runners to ply the shorter trade routes between Havana and Confederate ports to the east of Galveston. Ships that felt compelled to trade as far west as Galveston were better off to extend their voyages to Matamoros, where there was little or no threat of federal capture. With the intensification of traffic at Galveston in the winter of 1864–1865, it was soon apparent that the heavy blockade of that port could easily be violated. The effectiveness of the blockade, and therefore, the limited trade through the port prior to the closure of Mobile Bay, appears to have been based more on fear than actuality.

Peacetime commerce through Galveston had been heavy, considering the frontier nature of the state. Of over 400,000 bales of Texas cotton produced in 1860, 300,000 of these had passed through Galveston bound for other ports. This export was estimated at being worth fifteen million dollars in gold. Although cotton was the largest single commodity exported, corn, wheat, wool, lumber, hides, and other agricultural products were shipped in large quantities. In the quarter ending December 1860, over fifty-thousand hides alone had been shipped from Galveston.[1]

Upon the secession of Texas, Arthur Lynn, British Consul in Galveston, forwarded his views on this event to the Foreign Office. It was his conviction, as it was of most British diplomats at the time, that the United States was incapable of restoring the seceded states to the Union short of war. He expressed popular Southern sympathies in stating that the Republican party regarded slavery under a higher law than the Constitution and also in noting that party's belief that an irrepressible conflict faced them and slaveholders. He added that coercion would not work, either. As for the commercial outlook, the prospects were bright, he felt, since the Confederate States would probably adopt a low tariff and free trade. This, he contended, would strengthen peaceful separation and should increase the cotton production. He concluded that there was no fear of a shortage of cotton, "unless war occurs."[2]

War came to Galveston on 2 July 1861, with the appearance of the federal blockader, *South Carolina*. Lookouts atop the Hendley building ran up a red flag to sign all ships as well as to alert Confederate defenders of the city. After dropping anchor outside the bar, the *South Carolina* was approached by Confederate officers under a flag of truce. The Southerners were received courteously and were handed a written notification that Galveston was blockaded.[3]

This was true only in a technical sense, as ships of all classes entered and left the port in defiance of the single blockader. Observers atop Hendley's notified Lynn on 7 August, that the steamer *Texas Ranger* and the schooners *Hope* and *Relief* had entered the main channel, while at San Luis Pass, at the other end of the Island, schooners and sloops had been entering and leaving regularly. Further, there was an unbroken communications system operating between Galveston and Sabine Pass and New Orleans, conducted in small sailing boats of two to eight tons burden. Lynn added that of the twelve vessels that had been captured by the *South Carolina* since the blockade began, eight had been stopped in the first three days and the rest within ten days of the arrival of the federal ship. None of these vessels, Lynn contended, were notified that a blockade existed until they received shots across their bows.[4]

Lynn informed the Foreign Office on 23 September that, while there had been no intermission of the blockade at Galveston, the federal force had never consisted of more than one vessel, assisted by an unarmed schooner of seventy-five tons. This force, he said, had maintained a position commanding the main channel, but San Luis Pass had never been blockaded though it was occasionally visited by the schooner. He stated that the position of the blockader was such that vessels entering by the main channel would have to pass under her guns. "But vessels of not more than six feet draught could have ingress or egress by the shore channels, east and west of the main channel, should they elect to run out or approach these channels during the night." Then he added, "all the channels to the westward of Galveston are yet open."[5] Apparently this situation existed at least until February, 1862.[6]

In spite of the meager Confederate shore defenses, Federal sailors soon learned to keep a respectful distance from the island. On one occasion the schooner *Sam Houston* while returning from a periodic visit to San Luis Pass was fired upon by the island's south battery receiving some damage to her rigging. The *South Carolina* cleared for action and began returning fire. Many of the townspeople flocked to the beach to watch the show despite the sixty-eight-pound solid shot that impacted among them. The gunboat was hit severely enough to draw off after a short engagement.

While most of the blockade-runners engaged in the Galveston trade during the first two years of the war were small schooners and sloops, at least one-steamer was active. This was the *General Rusk*, a side-wheeler that had been one of the Morgan Line before the war. She was seized by state authorities in Galveston harbor upon secession of the state and, for a while, served as a makeshift gunboat, posted off Bolivar Point.

Ultimately her defensive uselessness was recognized, her guns removed, and she was fitted out as a blockade-runner. In early 1862 the *General Rusk* ran the blockade to Cuba where her destruction caused a serious international incident between the United States and Spain.[7]

By May 1862, the situation had changed drastically with the appearance of a sizeable naval force intent on capturing Galveston. Confederate defenses were so weak that General Hebert issued special orders to his battery commanders to be ready to spike their guns if the city was attacked since they would be unable to hold for long. On May 19, Captain Henry Eagle of the blockading flotilla sent word to all foreign consulates in Galveston to remove themselves and their families before the bombardment of the city, which was scheduled to commence four days later.[8]

The bombardment came amid the loud protests of the consuls, but the expected attack to take Galveston did not come for five months. Had the federals known of Hebert's eagerness to spike his guns and leave at the first serious demonstration, the Galveston trade would have stopped that much sooner. The navy's reluctance to move against the island is explained only by Captain Eagle's overestimation of Confederate strength. However, the trade was seriously stifled, due to the formidable array of gunboats, which increased in number each month off the Galveston bar. Small schooners and sloops continued in the trade, but judging from federal captures, traffic was heavier at Sabine Pass, Velasco, and elsewhere on the Texas coast throughout 1862.[9]

Maritime traffic at Galveston showed little increase in 1863 judging from newspaper as well as federal accounts of arrivals, departures, and captures. Clearly, the dynamics of the Matamoros trade overshadowed all other maritime activities on the Texas coast for the first three years of the war. Judging the traffic by the incidence of federal captures of craft in the vicinity of Galveston, indications are that there was a remarkable increase in the Galveston trade in 1864. This increase was tempered by the fact that virtually ninety-five percent of the craft were small sailing vessels. Even as late as October of 1864, Arthur Lynn explained that he had long since discontinued his returns of the Galveston trade since there "was absolutely no trade to report." Admitting that small quantities of cotton had been shipped on small craft, the large bulk of cotton produced in the state had been "exported by the way of Matamoros, crossing the Rio Grande at several points," he said.[10]

Lynn added an interesting note in explaining that the shortage of imported manufactured goods had "obliged" the people of Texas to establish manufactories of their own. Looms for the production of cotton cloth, as well as tanneries were in operation, he reported, and he had heard of two buildings being constructed for the manufacture of cotton and woolen cloth. Further, the Confederates had erected at Marshall, Texas, works for smelting, casting, and rolling iron for military purposes. "Should the state remain free from hostile invasion," he predicted, "the people by their industry will be soon able to supply all their necessities."[11]

During the federal intermission at Galveston, there was concern on the part of the islanders that they would be shut off from all supplies. Since they depended on the mainland for most consumer goods, the federal commander was prevailed upon to keep open the railroad between Galveston and Houston. His decision to do so may have been a costly one, as the Confederate attack force that helped retake Galveston used this route of approach. Otherwise, when Galveston was under Confederate control, consumer goods were about as plentiful as anywhere within the Confederacy. The city was fortunate in having first chance at goods that did run the blockade there. But the supply of coal for manufacturing gas was soon exhausted, and the townspeople had to rely on candles for illumination. To this inconvenience was added the burden of having to pay fifty cents a pound for tallow with which to make candles. Prices of produce rose correspondingly. Lynn's periodic returns show corn prices ranging from seventy-eight cents a bushel in April 1862, to $2.02 by the following October. Flour rose from $7.56 per hundred pounds to $24.06 during the same time. Middling cotton, however, because of the ever-fluctuating Liverpool advices ranged between six and nine dollars per hundred pounds. In spite of wartime shortages Colonel Fremantle noted that the ladies of Galveston were very well dressed, "considering the blockade."[12]

Fremantle described Galveston as being desolate looking during his visit in May 1863, and said that many of the houses were vacant, and others showed the effect of indiscriminate shelling. He described Magruder's feverish activities to strengthen the island's defenses, not only with new earthworks constructed under the supervision of Colonel Sulokowski, formerly of the Austrian army but of the bay defenses as well. The Britisher was informed, or, deliberately misinformed, that the bay was now blocked with pilings, torpedoes, and other obstructions. Four miles offshore he could see the ever-present blockaders, at that time consisting of three gunboats, an "ugly paddle-steamer," and two supply vessels.[13]

The war, for all its impositions upon the lives of Galvestonians, apparently did not seriously impair the merchant class. Fortunately, the records of several prominent cotton merchants who transacted business from Houston and Galveston offices have survived, which attest to a flourishing trade that existed throughout the sixties. Houston was the cotton center of Texas, which received for processing and compressing the bulk of fiber produced in the state. This is where the factors, with their Liverpool connections, bought, sold, and shipped through Galveston preferably, or Brownsville, if necessary, the white staple. At the beginning of the war when the bogus blockade at Galveston caused the trade to languish there, factors, agents, European buyers, and cotton all focused at the Rio Grande.

One of the more prominent merchants engaged in the Galveston trade was Thomas W. House of Houston. The firm name of "T. W. House, Plantation Commissary and Wholesale Grocery," belied the other ventures of the firm, including cotton and wool factoring, and merchandising a complete line of dry goods and hardware. This successful enterprise led to the establishment of a bank and also provided the means for acquiring a sugar plantation and a 70,000-acre cattle ranch. He helped organize the Houston and Galveston Navigation Company, which operated a steamboat service between the two places.[14]

During the war House's cotton wagons plied the roads to the Mexican border. Documents indicate that he either owned or had interests in a string of blockade-runners, including the schooners *Charles Russell*, *Clarinda*, *Chaos*, and *Tiger*. He was also a trustee of the captured federal gunboat *Harriet Lane* and held the first lien on the runner *Pelican*, which had been purchased with the *Lane's* cotton cargo. Often on squally evenings, House could be seen on the cupola atop his Galveston home scanning with a glass the federal blockaders. Next morning he would survey the hostile vessels again and if any were missing he could rightly presume they were chasing his blockade-runners.[15]

Under the half-cargo regulations of the Confederate government, House occasionally provided the government's half of cotton if their own sources were depleted. One such contract exemplifies the working relationship between House and Confederate authorities. On 5 January 1865, Henry Sampson, the Confederate Treasury agent at Houston agreed to purchase half the cotton cargo then on board the British blockade-runner *Will-o' the Wisp*, docked at Galveston. House was designated "agent" for the steamer and was to be reimbursed by the Confederate agent at Havana for the freight charges on the government's 300 bales. The freight rate, set by regulations, stipulated that five pence sterling per pound be paid for middling cotton and a "corresponding rate for other grades." The contract further stipulated that any or all portions of the cargo that were captured, lost, jettisoned, or otherwise destroyed by the public enemy, would be compensated for by the Confederate government. That is, the owner's half would be reimbursed, either in coin or in kind. House's commission on this contract, was apparently forty-seven bales–just under eight per cent.[16]

When all was ready, the *Will-o' the Wisp*, ran out but was soon spotted and chased by blockaders. In order to elude her captors, she not only changed course for Vera Cruz but also jettisoned 131 bales of cotton, which probably comprised her dock load. Ship and remaining cargo reached Vera Cruz safely, and when the facts were made known, House presented his claim to Sampson for sixty-five and one-half bales. Pragmatist that he was, House took his reimbursement "in kind"–33,251 pounds of C.S. cotton. It was probably a wise choice; the date was 19 May 1865.[17]

Yet, judging from House's transactions as reflected in his correspondence with Liverpool buyers, the end of the war brought no noticeable change in his fortunes. There was still a market for the staple after the war as there had always been, and there were ships available for transporting civilian cotton. The only indication that the Confederacy had ceased, at least in House's case, was the falling cotton quotations from Liverpool.[18]

Another prominent Texas merchant engaged in the Galveston-Matamoros cotton trade was John Milton Swisher. Swisher had participated in the War of Texas Independence and later served as a Ranger in the Mexican War. After distinguishing himself as a soldier, he devoted his next years to establishing successful business enterprises in the state. The firm of J. M. Swisher and Company, at one time or other, included, banking, merchandising, and cotton factoring. During the Civil War, he was a member of the Texas State Military Board, which duty ultimately brought him into competition with the Confederate Treasury Department for cotton grown in the state.

The J. M. Swisher and Co. balance sheet for the years 1863–1866 indicated that the firm received large consignments of goods through the international merchandising firm of Droege, Oetling and Co. of Matamoros. There is no indication that the ships *Malabar*, *Raleigh*, *Wilhelm*, *Cornella*, and *Maid of the Mill*, which brought his consignments, encountered any difficulty with the federal blockaders. Although the single-entry figures are incomplete, they indicate that the firm's total income over consignment indebtedness was a comfortable 105 percent for 1863. As in House's case, the end of the war brought no interruption in the operations of the firm.[20]

An interesting contractual arrangement entered into between John S. Sydnor, a Houston merchant and financier, and John A. Sauters, a cotton commission merchant, provides a glimpse into the mechanics of civilian cotton transactions in Civil War Texas. Under the agreement, Sydnor was to advance Sauters $100,000, with which to buy cotton, transport it to Matamoros, or Galveston, ship to European markets, and pay all freight, commissions, and other expenses. It was agreed that once the cotton was sold Sydnor was to receive five thousand dollars dividend on the advancement, plus twelve percent interest from Sauters for the amount expended. All profits remaining after these arrangements had been paid were to be divided, share and share alike, between the two parties.[21]

That commercial cotton transactions were carried on at all is nothing short of miraculous when one considers the complexities of Confederate cotton regulations. It was not that any one plan was so prohibitive of civilian trade as it was the uncertainty caused by conflicting authorities and amended regulations. These were especially complicated during the earlier stages when a rife competition arose between Confederate agents representing different governmental agencies, who, in turn, competed with private traders. The Richmond government had assumed that by sending Major Hart to Texas, vested with broad and exclusive powers of buying supplies, and selling cotton, the problems would be alleviated by being confined to this one officer.[22]

For a while, it appeared that Hart would be able to stabilize the situation and reestablish Confederate credit across the Rio Grande. However, jealousies arose between Hart and his associates, General Bee and his associates, and Captain Bloomfield, acting under Magruder's orders. Civilians took advantage of the confusion and used every trick to evade the conflicting orders. The net result was that the public became generally outraged with the whole procedure and came to regard all government involvement in the cotton trade as dishonest.

It was at this time that the British steamers *Sir William Peel*, *Gladiator*, and *Sea Queen* arrived, laden with badly needed war material which had been contracted for with cotton. There were insufficient amounts of cotton on hand to exchange for the cargoes, which Magruder needed immediately. Bee, therefore, issued orders impressing twenty percent of all cotton arriving at, or en route to, the Rio Grande. Hart felt that these transactions rightly belonged within the scope of his own broad instructions and wrote Secretary Seddon about his problems. To Kirby-Smith, Hart complained that Magruder's Quartermaster officers had impressed cotton that had been procured by his own agents. By August 1863, Confederate government cotton necessary to meet commitments was so scarce that Kirby-Smith, in desperation, issued orders to impress one-half the cotton in the department.[23]

One month later these orders were rescinded following a storm of protest from planters and merchants. For the most part, they expressed their willingness to provide whatever cotton was needed, as long as the matter was handled by consistent, trustworthy government agencies. After much planning and further reports to Richmond the Cotton Bureau was created, which, in its one year of existence, somewhat re-established the government's credit, centralized cotton procurement, and imported large quantities of military supplies.[24] However, public reaction to the Cotton Bureau so intensified during the year of its existence, that Richmond ultimately placed its operations under Treasury control, to coincide with its overall "New Plan."[25]

The Honorable P. W. Gray became, under the new scheme, the virtual Secretary of the Treasury of the Trans-Mississippi Department. On 12 January 1864, Gray issued instructions supplementing Treasury regulations, which governed overland commerce with Mexico. Under these instructions, Captain W. C. Black was assigned to duty in the Treasury Agency in Marshall, Texas, in charge of what came to be called the Foreign Supply Office. In this capacity, he had charge of purchasing and exporting cotton and purchasing foreign supplies for the army and government. Assisting Black as sub-Treasury agents were T. C. Twichell at San Antonio, Perry Nugent at Goliad, and Henry Sampson at Houston. Twichell and Nugent were authorized to purchase and export cotton overland to Mexico while Henry Sampson was authorized to do the same by running the blockade at Galveston.[26]

After the Treasury Department assumed control of Confederate imports and exports, the exact function of the old Cotton Bureau became obscure. Gray's authority replaced that of Broadwell, yet Broadwell remained head of the Bureau, at least in title. W. C. Black assumed the duties that had been performed by Hutchins, and finally, as shipping and receiving agent, Sampson, a civilian, was given the responsibilities normally carried out by Captain C. G. Wells of Magruder's staff. This transition was gradual and seems to have been completed in late 1864. Further, the changes probably served to produce a more efficient bureaucratic operation, as they tended to centralize efforts under the Treasury Department.[27]

The expected arrival of government steamers at Galveston in early 1865 caused the Treasury agency to redouble its efforts to stockpile cotton to be ready to go out as soon as the steamers arrived. Captain Black appealed through channels to General Smith asking for the use of cotton still under control of the Cotton Office. Thereupon, Colonel Broadwell ordered the Cotton Office in Houston to turn over, unconditionally, to the Treasury Department any cotton belonging to the Bureau. Also, Broadwell ordered a large amount of cotton that was on its way to Waco to be redirected to Navasota for the use of the Treasury. Later advices instructed Sampson not to keep separate shipping accounts on Cotton Office cotton, since all government shipments, under the sea regulations, would be made for the account of the Treasury Agency. By January 1865, the Cotton Office had apparently become a bureau without the means to function.[28]

The efforts of Gray, Black, and Sampson to accumulate, process, and ship out cotton through Galveston affords an interesting view into the working of an obscure branch of the Confederate Treasury. This sub-bureau was called the Foreign Supply Office, and the extant papers relating to it concern themselves largely with instructions to Sampson on proper bookkeeping. Indeed, Sampson's paperwork must have been maddening. Besides being the Treasury agent responsible for providing the government's half of all cotton cargoes exported, and being the receiving officer for incoming invoices, he was also the general agent for the Confederate Produce Loan for the state of Texas, as well as the Treasury agent for all tithe cottons. Furthermore, all his accounts payable, besides having to be kept separate for each Treasury function, had also to be maintained in each medium of exchange–Confederate currency, specie, and exchange certificates.[29]

The cotton that Sampson shipped out through the Galveston blockade was consigned either to Major Helm in Havana or directly to Fraser, Trenholm and Company in Liverpool. This English firm credited the account of P. W. Gray and either purchased supplies and shipped then to Texas or just bought the cotton outright, issuing bills of exchange drawn on themselves. These bills of exchange virtually became legal tender in Texas, the Treasury Department purchasing cotton and other supplies with these endorsed bills. At least in the Houston area sterling exchanges circulated freely.[30]

If insufficient amounts of Treasury cotton were on hand to fill out cargoes, Sampson was authorized to purchase cotton from private owners, paying in coin or sterling exchanges. At first, Gray advised Sampson to avoid paying this way for fear that if cargoes of cotton were lost, Fraser-Trenholm might not honor their own exchange certificates. He predicted that if a few shipments got through safely then "the Treasury scheme will be placed on a good footing." For the time being, Sampson was advised to use Treasury and tithe cotton that were most accessible to Houston.[31] Black's preliminary fears proved to be unfounded, and exchange certificates were readily accepted from the Treasury by Texas cotton suppliers.

The Foreign Supply Office continued to buy what cotton it needed from civilian suppliers when their own resources were depleted. Since, under the regulations, no ship could clear port unless half its cargo contained Treasury cotton, private shippers were induced to sell half their cargoes to the Treasury. If shippers complied, they received an exemption on freight charges for their half of the cotton to Havana. On 21 December 1864, Black suggested to Sampson that in order to encourage shippers who were reluctant to accept fixed government prices, he might pay an additional ten cents a pound, "to be paid as charges on the property by consignor." It should be understood, Black continued, that if the cargo was captured the owner could present a claim against the Confederate government for these uncollected charges. To help clear away some of the red tape, private shippers of government cotton were to be cleared by special orders and would not be required to post bonds.[32]

The Foreign Supply Office found it to their best interest to purchase half cargoes of private cotton. Black reasoned, quite logically, that such shippers had a much greater interest in the total cargo getting safely through the blockade. Especially did he recommend this where sailing vessels were involved. Schooner shipments were unsafe at best, he said, but there was greater security "in their having a large interest in the entire cargo."[33] Later, the absence of Treasury cotton was so acute that in order to dispatch vessels as quickly as possible, the Foreign Supply Office made exceptions to the regulations. Apparently, some private shippers had offered to give one-fourth of their cotton cargoes, rather than sell half cargoes, to the Treasury agency, just to facilitate matters. Black advised Sampson to take this type of proposition unless he could do better.[34]

Despite the emphatic regulations, some private shippers were continually trying to circumvent the law and carry out entire cargoes on their own account. The evidence in one case indicates that the mercantile firm of King and Company, of Galveston, contracted to purchase the captured federal gunboats *Wave* and *Granite City* and operate them on joint account with the government. James Sorley, formerly with the Cotton Office, and at that time, customs officer for the port of Galveston, apparently informed Gray that the deal had been made with Colonel Broadwell and had General Kirby Smith's endorsement. Eventually, Gray saw the contract and found that the conditions therein had been misrepresented by Sorley. What was worse, Gray learned that Sorley had cleared the *Granite City* with its entire cargo for private account. That Sorley may have been duped into believing that the cargo was jointly owned, according to King and Co., and had allowed it to clear port is indicated. At any rate, Gray considered it a grave error on Sorley's part and wrote him a scathing letter ordering him never to allow any cargoes to clear port unless one-half their cargoes had been laden by Treasury agents–not by King and Co., nor other officers. What the outcome of the incident was, if any, is unknown, but Gray considered it serious enough to call it to the attention of Generals Kirby Smith and Magruder.[35]

It was such questionable practices as this that led Sampson to balk at authorizing the clearance of the steamer *Pelican*, a vessel jointly owned by the government and civilians represented by T. W. House. Sorley informed Sampson that the steamer was load with 900 bales of cotton, half of which had been laden by Magruder's Chief of Quartermaster, Major S. M. Stone, for government account. House had supposedly offered to the Treasury half of the civilian portion, but Sampson declined, claiming half the entire cargo for the Treasury, or none. Sorley tried to point out to Sampson that the regulations did not apply to vessels owned jointly, and since the government already owned half the cargo, only the civilian half was subject to the regulations. The whole complicated affair is explained simply enough: the Treasury Agency refused to recognize Magruder's authority in shipping cotton.[36]

The difficult task of procuring and transporting cotton was no less hectic for the Treasury Department than it had been for preceding agencies. Much cotton sat idly at depots in the interior, waiting to be removed. The shortage of land transportation caused the Treasury agency to establish river transportation on the Sabine. Pulaski, located in Panola County, was chosen as the depot for processing and shipping of cotton grown in the heartland of the Trans-Mississippi Department. This river port was located near Treasury headquarters at Marshall, yet not so high on the river as to prohibit low water navigation. On 18 January 1865 Black informed Sampson that 300 bales were on their way down to Orange, Texas, at the head of Sabine Lake. Sampson was to have bagging and rope available to put this cotton in shipping condition.[37]

Beginning in September 1864, Black had employed ten to fifteen people at Pulaski in putting the accumulated cotton in order to ship. Later, he complained of their slow pace, claiming that the workers were only repairing and rebaling about 100 bales a week. By March, there was a sizeable stockpile of cotton on hand at Pulaski when the redoubtable riverboat *Uncle Ben* started down river with 590 bales. In the vicinity of Logansport, Louisiana, she was discovered to be on fire. Efforts to save her were futile, and soon vessel and cargo were completely destroyed. Precious as the white staple was to Confederate operations, it was the loss of the vessel, as Black noted, that was most serious.[38]

Sampson was sending as many riverboats up to Pulaski as he could contract for, but even so they were not removing the cotton rapidly enough. The *Roebuck* carried down 375 bales, followed by the *Sunflower* laden with 309 bales. At this rate, Black complained, it would be a considerable time before the depot cotton was removed. One month later Black admitted his failure to remove the Pulaski cotton. Apparently, the riverboat owners were unwilling to take full loads, perhaps remembering the fate of the *Ben*. Further, civilian cotton consignments were competing for space on board. Even Black's offer to pay an additional ten cents a pound "freight charges" to Beaumont failed to produce the desired results. As an alternative, Black suggested to Sampson that he try to utilize the Trinity River to remove cotton, paying for it in cotton.[39]

One of the most acute logistical shortages facing the Treasury was that of bagging and rope needed to bale cotton. The planters of Texas, being further removed from the source of these supplies, were particularly handicapped, and their tithe cotton coming into the depots was in various stages of packing, which resulted in considerable loss. Because of the wastage incurred in transporting, the Treasury Department agreed to supply the bagging and rope to the planters for their tithe cotton. Much of the space on incoming runners at Galveston was devoted to these articles. During the first three months of 1865 ten shipments, totaling 73,350 yards of bagging and 99,117 pounds of rope, entered Galveston.[40]

During the winter of 1864–1865, the tithe cotton was slowly accumulating all over the state. Agents collected the precious staple at county depots and then awaited the means to move it either to Houston, Goliad, or San Antonio. That collected nearest the last two places made its way overland to Matamoros, while that produced in the eastern half of the state was directed to Houston. An inventory of the 1864 tithe cotton, made on 1 May 1865, showed that Sampson's agency had on hand 3,575,856 pounds. A large portion of this was scattered throughout the state awaiting removal to the collection points.[41] Records are incomplete on how much of this actually reached Houston for shipment through the blockade, but Wolston's press in Houston processed 1,116 bales for the Treasury.[42]

In spite of the arrival of more federal warships at Galveston as the war diminished elsewhere, blockade-running not only increased at the port but continued to be highly successful. Two weeks after

Lee's surrender Black commented on how easily the Galveston blockade was run. "The enemy is not in our way," he said, "except to make us pay high profits on our goods."[43] He might have added that the blockade was just effective enough to support the high prices received for Texas cotton. Indeed, a popular toast in Havana among runner captains went:

> The Confederates that produce the cotton;
> The Yankees that maintain the blockade and keep up the price of cotton,
> The Britishers that buy the cotton . . .
> Here is to all three and the long continuance of the war and success to blockade runners.[44]

Among those toasting their own success was the daredevil runner Captain William Watson. Upon his arrival at Galveston in December 1864, he noticed several runners in the harbor and four or five others at the docks awaiting or loading compressed cotton. A short time later he reported that notwithstanding a federal blockading force of twelve to fifteen armed steamers the number of blockade-runners that passed in and out of Galveston was almost incredible. It was his opinion that the traffic here was heavier than had developed at any of the Atlantic ports. A common belief among those engaged in the Galveston trade, he declared, was that owners could afford to give the Yankees a prize every now and then to encourage a tight blockade, which supported the high price of cotton.[45]

One of the federal sailors stationed off Galveston estimated that as of February 1865, there were nineteen steamers regularly engaged in the Galveston trade.[46] Another federal report in April 1865, coming from the Havana end of the trade, stated that as many as twenty-five different steamers were engaged. This observer insisted that nearly all were successful and mentioned among the more important ones, the *Owl*, *Luna*, *Fox*, *Denbigh*, and *Pelican*. While admitting that the traffic was decreasing, it was not, he insisted, due to any danger from the blockading fleet, "but from the scarcity of cotton at Galveston."[47]

Skippers of blockade-runners might have contested the statement that the Galveston blockade posed little danger. Occasionally, runners came in riddled by shellfire, as in the case of the *Susanna*. This large, fast steamer cleared Havana for Galveston in October 1864, with a cargo of arms and ammunition. Just as she spotted the first federal blockaders off the port, her machinery malfunctioned. During the delay for repairs, a gunboat noticed the derelict and started after her. After getting the engines going again, the skipper of the runner realized that their limited coal supply ruled out trying to escape out to sea and decided to run the federal gantlet. She came under fire from three cruisers and sustained damage to her funnels and bow, but kept up full steam and reached the harbor safely. Eighteen days later she succeeded in running out, but some distance from Havana, her machinery again broke down. A federal cruiser took her in tow as a prize, though it is doubtful that the loss was a crushing blow to her owners as this was her fifteenth trip.[48]

The steamer *Zephine* was another privately owned runner engaged in the Galveston trade, reportedly the property of Charles Morgan of New York. She appears to have entered the trade after changing her Philadelphia registry from *Marian*. The vessel had a known capacity of 1,400 bales and was commanded by one of the most experienced runner captains in the trade. She made at least four trips from September 1864, until the following spring and was never captured. The federal consul in Havana reported that the *Zephine* had cleared Havana with a cargo costing $80,000 and had returned with a cargo valued at $300,000. On this one round trip, the combined value of the cargoes exceeded the value of the vessel by $260,000.[49]

The blockade-runner *Owl* had the unique distinction of being commanded by one of the most famous Confederate captains in the war. Captain John Newland Maffitt had formerly commanded the Confederate commerce raider *Florida*, the ship whose in-and-out exploits at Mobile had caused Farragut so much embarrassment. Maffitt and the *Owl* had been engaged in the Charleston trade until that port was closed. On his first run into Galveston he made it safely through the sixteen blockaders only to run aground on Bird Island Shoals, at the entrance to the harbor. A Confederate naval crew, manning the captured *Diana*, steamed out to help the *Owl*, which had come under intense shelling. With this assistance, Maffitt and his crew finally got the *Owl* afloat. At the docks, Maffitt was welcomed by an enthusiastic crowd of townspeople, many of whom had witnessed the drama from their rooftops.[50]

Grounding on sand bars was a common occurrence for runners engaged in the Galveston trade. The waters surrounding the island were shallow, and all vessels were at the mercy of the tricky current, prevailing winds, and "northers." Since most runners were usually moving fast, and often under conditions of poor illumination, they frequently missed the channel entrances. Once aground they were soon spotted and shelled by the gunboats, but if lucky, they were re-floated before sustaining severe damage. One vessel whose luck finally ran out was the *Will-o' the Wisp*.

This vessel was one of those fast but weakly constructed Clyde steamers built especially for the running trade. She had made several trips in and out of Wilmington, North Carolina, one of which had almost ended in her destruction. How many trips she later made in the Galveston trade before her last, in which T. W. House contracted to carry out Treasury cotton, is not known. On her return run, she was chased by a blockader as she approached the island at dusk in a storm. Apparently trying to make the "swash" channel which paralleled the coast, she was blown aground about eight miles from the city. The notice of public auction of the salvageable wreck gives some interesting details about the vessel:

> The fine steamship *Will-o' the Wisp* as she now lies stranded, together with her tackle, apparel, [*sic*] furniture, etc. The said ship is built of iron, 1/16 inch, her length 238 feet, breadth of beam, 23 feet, 8 floats to each wheel, and 6 brass brushes to each float; one donkey engine (with pump), two deck pumps . . . in water tight compartments, one pair osolating [*sic*] engines, 2 cylinders 49 inches, length of stroke 2 feet, 2 inches, normal horsepower 180, 4 tubular boilers.
>
> All of the above machinery would be invaluable to various manufactures in operation, or about to be put into operation in the state . . . . Specie or cotton will be received in payment of purchase.[51]

There were other steamers active at Galveston, like the *Fox* whose narrow escape into port afforded Galveston spectators one of their most exciting chases. For several miles, two federal cruisers maintained a parallel chase with the *Fox*, shelling her all the way. Upon nearing the harbor the entire fleet fired at her, but notwithstanding four hits she made port safely. The *Lizzie*, a sister ship to the *Will-o' the Wisp*, was also active at Galveston. Her most noteworthy asset was her speed, which on her trial run set the unheard of record of twenty-two statute miles per hour.[52] Even though these sophisticated runners were numerous in the Galveston trade, schooners like the *Rob Roy* continued active in the trade. That they did their part to aid the beleaguered Confederacy is evidenced by the following manifest of Watson's schooner: 200 Enfield rifles, with bayonets and accoutrements, 400 Belgium muskets with bayonets, 400 assorted firearms, 400 cavalry sabers, 6 cases saddlery and accoutrements, 25 boxes ammunition, 1 large box cavalry currycombs, as well as a lot of several bales of blankets, assorted lots of clothing, boots, shoes, hardware, tea, coffee, cheese, spices, thread, needles, brandies, and wine.[53]

Often runner captains, planning to run out at night, would climb to the top of some Galveston building at sunset and chart the positions of Union blockaders. Upon coming out, the runners would attempt to put a distance of ten miles behind them without being detected before daybreak, in which case there was little chance they would be caught. If the runners were discovered they would try to run for it, keeping close to the Gulf shore of the island. Often, deckloads of cotton were jettisoned, but if escape became impossible, the vessel would be beached. The next day bales of cotton would be salvaged all along the beach.[54]

On the morning of 20 February 1865, federal sailors on board the sloop-of-war *Lackawanna* were alerted when the lookout cried that a steamer was close aboard. Despite the fact that the small sidewheel runner had sped by close, and in daylight, she was a mile past by the time the blockader could turn around. Throughout the day, the runner gradually pulled away and lost the federal ship at night. Notwithstanding, the blockader set sail for Moro Castle light in Havana, and to their surprise, spotted the steamer the next morning. The runner, upon discovering her relentless pursuer, began jettisoning her deckload of cotton, not only to lighten her but hoping that the federals would stop to fish out the valuable cargo. The blockader pressed on, however, consuming three tons of coal per hour and working the firemen in thirty-minute reliefs to maintain maximum speed. The runner, not having the crew to keep reliefs, gradually lost speed as its firemen fell from heat exhaustion. When overhauled, the runner proved to be the *Isabel* bound for Havana with 600 bales of cotton.[55]

By December 1864, the efforts of Colin J. McRae to involve the Confederate government directly in blockade-running were beginning to bear fruit. At least four government vessels reached Galveston in time to aid materially the efforts of the Foreign Supply Office. By then, Black, the head of the office, was able to inform Henry Sampson in Houston to expect their arrival soon. McRae, Black explained, had completed contracts with Mr. C. K. Prioleau of Fraser, Trenholm, and Company and with J. K. Gilliat and Company, to obtain four vessels to run on government account exclusively. The first two vessels, the *Lark* and the *Wren*, supplied by Prioleau, had departed from Liverpool on 15th and 30th November, respectively. Black described them as light-draft steamers, capable of carrying 350 bales on five feet of water and 700 bales on six feet.[56]

The *Curlew* and *Plover*, the Gilliat and Company vessels, were to depart about the middle of December, Black reported and were admirably suited to the Galveston trade "as they were strong and staunch with great speed." He described them as being larger than the two Trenholm ships, capable of carrying 700 bales on six feet of water and 800 to 1,000 bales on seven feet. Black further informed Sampson that the terms of the contract required that all government cotton be consigned to these two suppliers as top priority shipments as soon as possible. He closed by cautioning Sampson to keep the details confidential concerning the status and names of the vessels.[57]

The details were known and reported soon enough by William T. Minor, the United States Consul-General in Havana, who described the ships as ". . . new steamers of a class very much superior to those heretofore engaged in the contraband trade,"[58] He identified the *Lark* and *Wren* and stated that they were of 800 bale cargo capacity and very fast. *Curlew* and *Plover* never entered the trade by those names, but were probably those named as *Evelyn* and *Flamingo*, by the consul. These vessels were also quite fast, according to the consul's intelligence reports, and had a 1,000-bale capacity–the same as reported confidentially by Black. Whatever their names, the four government steamers made a total of sixteen successful runs in and out of Galveston. Both *Lark* and *Wren* completed six round trips each and none of the four were ever captured.[59]

The government steamers were handling Treasury accounts so well that by the end of April Black was advising Sampson not to buy the famous *Denbigh*, which the latter wished to add to the government's squadron. ". . . It is to be hoped that you will not have so large a fleet of vessels to attend to in the future," Black stated, ". . . our own vessels–four in number–can do the work for us and with more profit to the Government, than so many outsiders."[60]

Interestingly, only two days after Black's confident forecast for the future of Treasury blockade-running, General Magruder was advising General Kirby-Smith that Galveston would have to be abandoned. It was Magruder's conviction that the island would be untenable in the face of a determined enemy attack, no doubt a sound enough tactical conclusion. To emphasize his views he declared that blockade-running at Galveston was almost at an end. This, he insisted, was due to Judge Gray's "injudicious tenacity" in carrying out the Treasury regulations, which, he said, had been sufficient a year earlier, but were unworkable at the present. Gray's inflexibility, Magruder declared, had already driven away most of the steamers.[61] Treasury records, however, show that steamers continued to enter and leave safely until the third week in May 1865. It must be remembered that Magruder's control of cotton transactions had been subordinated to the Treasury for over a year.

The total number of vessels engaged in the Galveston trade throughout the entire war will probably never be known. There is, however, a fairly accurate record of the number engaged during the peak six-months period from November 1864 to 19 May 1865. Fortunately, the Treasury ledger kept in Sampson's office survived the years and discloses the names of ships and the amount of export duty paid by their agents on outgoing Treasury cotton. Even so, these figures are minimal, as the ledger is incomplete. Working solely with these sketchy figures it can be concluded that there were at least thirty-seven vessels engaged in the Galveston trade during this period. These sixteen steamers, twenty schooners, and one sloop violated the blockade a total of fifty-two times. Perhaps most significantly, the total government cotton exported, upon which duties were paid and listed, amounts to 10,115 bales, and this represented only the Confederate government's half of all exported cotton.[62]

Aside from the four government steamers purchased and the bagging listed heretofore, there is little record of the material and supplies that reached the Trans-Mississippi through Galveston. An inventory of Treasury property still on hand that had not been looted or destroyed by mobs was submitted to federal authorities on 22 June 1865, as follows:

One roan horse	–	For use of Produce Loan
One letter press	–	Do Do
Office blanks, envelopes, etc.	–	Do Do
Lot gunny sacks for corn	–	Do Do
Lot coal at Galveston	–	In charge of A. P. Lufkin
Four (18) pounder steel guns	–	Left in warehouse by Lt.
Twenty-four Artillery Wagon wheels	–	Von Phul, Ord. Officer.
Twelve Ammunition chests	–	Do Do
Eight Limbers	–	Do Do
Six axles	–	Do Do
seventy-five boxes, shell	–	Do Do
One-hundred loose shell	–	Do Do
Four artillery saddles	–	Do Do
	–	
Tithe and Treasury cotton on hand 26th day of May, 1865, in the hands of Sub-Agents throughout the state. A report will be furnished to the U.S. Treasury Agent	–	All of which so far as heard from has been plundered and carried away by mobs of armed men.
(certified correct)		Henry Sampson Gen'l. Agent Produce Loan[63]

The Foreign Supply Office, as perhaps the last agent of the Confederate government, officially ended on 22 June. The Galveston trade, upon which this agency had depended, continued with but the briefest interruption for federal occupancy. Then, it was business as usual, and a thriving maritime commerce resumed immediately upon the surrender.[64]

Chapter 6 Notes

[1] Thomas North, *Five Years in Texas* (Cincinnati: Worley Brothers, 1870), p. 157; U.S., Department of Commerce, Bureau of the Census, *Eighth Census of the United States, 1860: Agriculture*, p. 149; Lynn to Russell, 1 March 1861, Foreign Office, Record Group 5, America, 848.
[2] Lynn to Russell, 14 February 1861, Foreign Office, Record Group 5, America, 848.
[3] *Corpus Christi Ranchero*, quoting the *Civilian Extra*, 2 July 1861.
[4] Lynn to Russell, 7 August 1861, Foreign Office, Record Group 5, America, 848.

[5] Lynn to Russell, 19 October 1861, *ibid.*

[6] Lynn to Russell, 2 February 1862, *ibid.*

[7] Ben C. Stuart, "Scrapbook," clipping from *Galveston Daily News*. For complete details of the *Rusk* incident, see Chapter 4.

[8] Special Orders No. 471, *OR*, ser. 1, 9:709; Eagle to Hebert, 19 May 1862, *ibid.*, p. 711.

[9] Compilation of statistics quoted in Price, "Ships;" also, see *OR*, ser. 1, 15:151; *ORN*, ser. 1, 20:469; *Galveston Weekly News*, 1 May, 29 May, 9 July, and 7 Oct. 1862.

[10] Lynn to Russell, 22 October 1864, #12, Foreign Office, Record Group 5, America, 970.

[11] *Ibid.*

[12] Robert Hayes, "The Island and City of Galveston," 1879, typescript, Rosenburg Library, Galveston, p. 514; Lynn to Russell, 3 April, 7 July, 1 October 1862, Foreign Office, Record Group 5, America, 848; Lord, *Fremantle Diary*, p. 57.

[13] Lord, *Fremantle Diary*, pp. 55-57.

[14] Walter P. Webb, ed., *The Handbook of Texas*, 2 vols. (Austin: University of Texas Press, 1964), 1:843.

[15] Unknown to House, 1 December 1864; Frederick Huth and Company to House, 1 February 1865, T. W. House papers, University of Texas Archives. Hereinafter cited as House MSS; unsigned, undated deposition on ownership of *Lane*, Ballinger MSS; Webb, *Handbook*, 1:843.

[16] Contract of sale, with attached Receipt of Charges Certificate, Sampson MSS.

[17] *Ibid.*

[18] Assorted House Correspondence, House MSS.

[19] Webb, *Handbook of Texas*, 2:699.

[20] Consignments balance sheet of J. M. Swisher and Co., Ballinger MSS.

[21] Contract agreement, 15 February 1864, Ballinger MSS.

[22] The summary of cotton control in Texas up to the final Treasury regulations is taken from the report of W. J. Hutchins to "The Senators and Representatives in Congress from the State of Texas," 20 November 1864; a nineteen page booklet, "News" Print, Houston, catalog item 71-0043, Rosenburg Library, Galveston.

[23] Hutchins MSS, p. 2; Maj. George Williamson to Kirby-Smith, 26 October 1863; Seddon to Kirby-Smith, 29 October 1863; W. A. Broadwell to Memminger, 28 January 1864, official true copies of original orders, Ballinger MSS.

[24] For a more detailed report of the Cotton Bureau see Chapter 4.

[25] The Confederate plan of control of foreign commerce is also found in Chapter 4.

[26] "Amended Regulations" and "Additional Regulations Under the Overland Commerce with Mexico," Treasury Broadsides, Marshall, Texas, 12 January 1865, Sampson MSS; Nichols, *Confederate Quartermaster*, p. 78.

[27] Gray wrote Sampson on 11 November 1864, "it is too complicated for you to prepare cotton for shipment, pay expenses and keep accounts and for another to ship and open new accounts . . . [you] are authorized to attend to the whole business of shipments by sea . . ." Sampson MSS.

[28] Black to Sampson, 15 December 1864, 1, 2 January 1865; Kirby-Smith to Gray, 27 December 1864, Sampson MSS.

[29] Gray to Sampson, 11 November 1864; Black to Sampson, 5 January 1865, Sampson MSS.

[30] Gray to Sampson, 11 November 1864.

[31] *Ibid.*

[32] Black to Sampson, 15, 21 December 1864, Sampson MSS.

[33] Black to Sampson, 13 February 1865, Sampson MSS.

[34] Black to Sampson, 19 April 1865, Sampson MSS.

[35] Grey to Sorley, 14 April 1865; Gray to Sampson, 19 April 1865, Sampson MSS.

[36] Sorley to Sampson, 20 April 1865, Sampson MSS.

[37] Black to Sampson, 18 January 1865, Sampson MSS.

[38] Black to Sampson, 13 February and 28 March 1865, Sampson MSS.

[39] Black to Sampson, 26 April 1865, Sampson MSS.

[40] Gray to Sampson, 31 January 1865; "Abstract of Baling Materials Received by Henry Sampson," Sampson MSS.

[41] "Statement of Tithe Cotton received by Sub-Agents of the Treasury Department, Houston, 1 May 1865," Sampson MSS.

[42] "Weights and Receipts of Cotton Stored at Wolston's Press, Houston by Hy. Sampson, Treasury Agent, Foreign Supply Office from 31 August 1864 to 25 May 1865," Sampson MSS.

[43] Black to Sampson, 26 April 1865, Sampson MSS.

[44] Watson, *Adventures of a Blockade Runner*, pp. 304-305.
[45] *Ibid.*, pp. 258-260, 304-305.
[46] William F. Hutchinson, "Life on the Texan Blockade," *Soldiers and Sailors Historical Society: Personal Narratives*, ser. 3, No. 1 (Providence, 1883), p. 35.
[47] *OR*, ser. 1, 48, pt. 2:230.
[48] Dyer, "Scrapbook" clipping from *Galveston Daily News*, 26 November 1922; Price, "Ships," 11:276.
[49] Price, "Ships," 11:276; C.S. Treasury ledger accounts, p. 7, Sampson MSS.
[50] Bradlee, *Blockade Running*, pp. 128-130; the Treasury ledger accounts in the Sampson MSS shows that the *Owl* ran out on 3 May with 478 bales of Treasury cotton; also see *OR*, ser. 1, 48, pt. 2:517.
[51] *Galveston Daily News*, 24 February 1865; Dyer "Scrapbook," clipping from *Galveston Daily News*, 26 November 1922; Taylor, *Running the Blockade*, pp. 101-110; Treasury ledger accounts, Sampson MSS.
[52] Bradlee, *Blockade Running*, p. 119; Stuart "Scrapbook," clipping from *Galveston Daily News*, 16 July 1911.
[53] Watson, *Blockade Running*, pp. 149-150.
[54] Typescript entitled, "Galveston City Guide," p. 72, Rosenburg Public Library; Taylor, *Running the Blockade*, pp. 168-169. J. B. A.: See an example herein in Part II, Chapter 3-5.
[55] Hutchinson, "Life on the Texan Blockade," pp. 6-16.
[56] Black to Sampson, 15 December 1864, Sampson MSS.
[57] *Ibid.*
[58] Price, "Ships," 11:277.
[59] *Ibid.*; Treasury ledger accounts, Sampson MSS.
[60] Black to Sampson, 26 April 1865, Sampson MSS.
[61] Magruder to Kirby-Smith, 28 April 1865, *OR*, ser. 1, 48, pt. 2:1288-1291. Magruder favored the overland supply line to Matamoros, a plan which he claimed Gray refused to endorse.
[62] Compilation of Treasury ledger accounts, Sampson MSS.
[63] "Abstract of Henry Sampson, Agent, Produce Loan for Texas on the 22nd day of June 1865." Sampson MSS.
[64] Hayes typescript, "The Island and City of Galveston," pp. 635-636.

7

Blockade-Running: The Economics of the Trade

By whatever route cotton left Gulf ports the majority of the staple found its way, at least on the first leg of its journey, to Havana, Cuba. The Confederacy had dispatched Major C. J. Helm as its purchasing and shipping agent to that port to handle the details of intermediate Confederate import-export. Helm, the former United States Consul to Cuba, found a sympathetic populace and immediate markets for Gulf coast cotton. Consignments of Enfield rifles began reaching him in November 1861 from Caleb Huse, Confederate purchasing agent in England. The sympathies of the British were evidenced by the efforts of their Consul-General in Havana, James T. Crawford, in helping Helm find British warehouses to store Confederate contraband.[1]

Crawford's correspondence to the Foreign Office is replete with reports of federal maritime outrages on British shipping between Havana and Matamoros. He stated, however, that "it is a fact that no British merchantman is molested when a British man-of-war is nearby."[2] As for Confederate maritime commerce, he dutifully recorded the number of blockade-runner arrivals in Cuban ports and added that cotton found a ready market. Some indication of the amount of cotton arriving in Cuba was made by the reporter for the *New York Herald*. This correspondent noted that between 18 and 28 April 1862, ten vessels arrived in Havana, and there was already an estimated 10,000 bales of cotton on hand.[3]

Havana, like Matamoros, experienced boom conditions, which intensified as blockade-running activities shifted increasingly from the Atlantic to the Gulf coasts. As the Cuban economy inflated, residents on fixed incomes found themselves hard pressed to make ends meet. An excellent commentary on Havana conditions is supplied by Robert Bunch, Crawford's successor:

> The labors of the Consular office is extreme, owing to the scale upon which blockade running is conducted. Here the port is thoroughly demoralized. Desertions are universal, the change also of vessels to English flags from American is a matter of almost daily occurrence . . . . [It] has driven prices beyond what is reasonable for a Consul's salary. In Havana you cannot open your mouth under a dollar and shut it under a doubloon.[4]

Profits as well as risks were causing a general price increase of cotton as well as consumer goods, throughout the blockade-running trade. The English shipyards at Merseyside and on the Clyde enjoyed peak production to supply the demand for fast steamers to run the blockade. The price of steamers rose rapidly. The *London Times* of 25 November 1863 noted that three fine steamers had recently been sold to run the blockade: the *Caledonia*, *Iona*, and *Fairy*. The latter vessel had recently been used by Prince Alfred for royal touring. All vessels were second hand but brought more than their original cost.[5]

Freight and passenger rates on the vessels engaged in the trade were equally inflated. From $300 to $500 in gold was charged in advance for passage for one person, while $2,500 in gold was paid as freight from Bermuda to Wilmington on a box of medicines that could be stowed in one cabin.[6] The salaries of masters commanding runners in the Galveston trade ranged from $500 to $1,000 in gold per round trip to Havana. Even mess boys on these same runners were paid $60 a month in gold. British men-of-war on the West Indian station were plagued with desertions, as Consul Bunch intimated, by the temptations of good pay on runners. A typical table of expenses for one steamer for one month is as follows:

Expenses:

Captain	$ 5,000
First Officer, $600, second officer, $250, third officer, $170	$ 1,020
One boatswain	$ 160
One carpenter	$ 160
One purser	$ 1,000
One steward, $150, three assistants, $180	$ 330
One cook, $150, two assistants, $120	$ 270
One Engineer and three assistants	$ 3,500
Twelve firemen and coal heavers	$ 2,400
Coal, 240 tons @ $20 ton	$ 4,800
rations for crew	$ 2,700
tallow and packing	$ 1,000
Stevedores	$ 5,000
Pilotage out and in	$ 3,000
Sea Insurance	$ 3,500
Wear and tear	$ 4,250
Incidental expenses	$ 1,000
Interest	$ 875
Risks, 25 per cent	$ 37,000
Provisions for passengers	$ 3,000
total costs	$ 80,265

Revenue:

800 bales cotton for C.S. Government	$ 40,000
800 bales cotton for owners	$ 40,000
Return freight for owners	$ 40,000
Return freight for C.S. Government	$ 40,000
Passenger fare, round trip	$ 12,000
total revenue[7]	$ 172,000

While the wages of captains might run $5,000 a month and that of pilots $2,000 to $5,000 occasionally some would do even better. The *Times* quoted the case of a captain who made $5,000 per round trip from Nassau to Wilmington. Additionally, he had the privilege of purchasing for himself twelve bales of cotton for fifteen pounds, sterling, per bale which yielded seventy-five pounds in Liverpool.[8]

So lucrative was the contraband trade that blockade-running stock companies sprang up on both sides of the Atlantic. On 13 November 1863, the *Charleston Mercury* announced the payment of handsome dividends by three such companies. One of these paid a dividend of $500 a share. In eight successful round trips the blockade-runner *Banshee* paid her stockholders 700 percent profit. On another occasion four Southerners purchased a schooner and a cargo of cotton at a total outlay of $30,000. The vessel successfully ran to Nassau, where ship and cargo were disposed of at three times the initial cost.[9] One veteran captain who had made sixteen successful trips, claimed profits of 800 per cent. Occasionally net profits of $300,000 on one round trip were made.[10]

While blockade-running was the chief means of supplying the Confederacy with the sinews of war it was also draining the South of both raw materials and wealth. The case is best summed up by a resident of Mobile, who viewed unrestricted blockade-running as a definite evil:

... the running of the blockade, unless restrained, will ruin us forever. The fast sending of all our cotton, or a large part–some little [cotton] is captured–[but] that which passes had better be; the most of it laid out in brandies, wine and flimsy gewgaws which bring exorbitant prices, but little in articles which produce substantial goods.

The writer went on to declare that high prices were corrupting the people by the lure of speculation, all of which was undermining the morale of service personnel. He concluded that Confederate currency was the worst victim of the system.[11]

The unbelievably high prices commanded by "gewgaws" and luxury items in the Confederacy was exemplified by the report of General John Dix, who noted that 1,200 boxes of matches costing seventy-five dollars in Baltimore would bring $9,072 in Richmond. Confederate authorities were not unaware of this economic rapine. Larkin Smith, the Assistant Quartermaster General, reminded the Secretary of War that the South had to rely almost totally on imports of all manufactured goods. "The prices of all goods . . . from abroad [are] bringing readily in open market 300 or 400 percent," he stated.[12] Smith added that depending on the private contract system was to trust in an unreliable source of goods as well as to pay enormous profits to importers.

The effect of high prices and cotton exportation on the deficit balance of Confederate payments was pointed out by General W. H. C. Whiting. He explained that foreign-owned vessels arrived in Southern ports with articles that brought $1,000,000 in Confederate currency. With only fifty thousand of this currency the owners bought as much cotton as they could take out. In addition to the fantastic markup the cotton brought in Liverpool, the excess Confederate currency was converted into exchange certificates or gold, yielding, even at a lopsided ratio, a handsome profit. He cited the simple case of a sailor who had brought in six demijohns of gin, which had cost him twenty-four dollars in Nassau. He sold them in Wilmington at $900, and then exchanged his profits for gold at a ratio of nine dollars for one. The trouble was, Whiting pointed out, that these profits were not spent in the Confederacy but were invested in gold or in foreign debentures. "Every single bale of cotton that goes abroad on other than Government account to establish Government Credit abroad," wrote Whiting, "does us injury at home."[13]

The truth of Whiting's forecast was evidenced by a contract entered into between Confederate authorities in the Trans-Mississippi and an Arkansas speculator which was to be made on government account. The contractor proposed to buy at personal expense 2,000 bales of cotton which the Confederate government would assist in getting to foreign markets. The contractor was to pay, in Confederate script, $100 per bale. The contractor proposed to sell the cotton at either New York, where it would bring $400 a bale, in greenbacks, or in Europe where it would bring $300 in gold. He was to make on his investment of $200,000 Confederate $600,000 or $800,000 in stable currency. He then proposed to invest the proceeds in army supplies to be delivered within the Confederate lines in the Trans-Mississippi Department. It was agreed that his profit would be sixty-five percent of the amount expended. The Confederacy would owe the contractor $800,000 plus his commission of $520,000 which would be paid in cotton at the usual rate of $100 a bale. The contractor would receive from this transaction 13,000 bales of cotton worth over five million dollars if sold in New York.[14]

The Confederate Quartermaster Major who reported the affair to the departmental commander refused to endorse the contract, which had been approved by his immediate superiors. In his report he explained that the contract would not only involve some collaboration with the enemy at Confederate expenses, but also the South would lose over 15,000 bales of cotton for which they would receive credit on only 2,000 bales. The notorious contract was reported to the Secretary of the Treasury by Colonel Broadwell of the Cotton Bureau. He urged that new laws be enacted either raising the export duty on cotton, or placing closer restrictions on the contract system. He added that the aforementioned contract was based on the assumption that the contractor would uphold his end of the agreement, ". . . but experience has proven," he stated, "that contractors rarely ever comply with their engagements if permitted to export our staple in advance of the introduction of the supplies." In this case, he noted, the only benefit to the government was the miserable pittance paid the people in Confederate money for the initial lot of cotton.[15]

Even G. A. Trenholm, successor to Memminger as Confederate Secretary of the Treasury, had to admit that the contract system was demanding too heavy a burden on Southern cotton exports. In spite of the fact that it was his Liverpool firm that benefitted immensely by the system, he had to request the Secretary of War to reevaluate his foreign supply estimates. To meet War Department requests of foreign supplies purchased with cotton would entail cotton exports of 6,000 bales per month for the next six months. The tonnage requirements, noted Trenholm, would employ fifteen steamers a week at the current capacity. This was clearly impossible, he stated, and Confederate purchasing operations abroad would have to be coordinated within the capabilities of agent McRae.[16]

Interestingly enough, throughout all this correspondence there is not the slightest intimation that Confederate imports and exports might be interdicted by the federal blockade. McRae's report of 7 October 1863 stated that successful blockade-running was a foregone conclusion. This fact was proven by Huse's experiment of purchasing four blockade-runners which had in two months made twenty-two successful trips through the blockade without a single loss.

The decision of the Confederate government to enter the blockade-running business evolved from the exorbitant freight rates charged by commercial shippers as well as their propensity to import luxuries. The first ship to run the blockade on government account alone was the *Fingal*. It reached the South safely in November, 1861, and shortly thereafter three more steamers were purchased for government account.[17] In addition to the four vessels purchased for use by the Ordnance Department, three-fourths interest were purchased in five steamers operated by Crenshaw and Company. Later in the war, as McRae's proposals for governmental regulations of shipping were implemented, plans were made to obtain twenty-seven steamers for government use. Only six of these reached the Confederate coast in time to take one or more trips before the war ended.[18]

A great deal of credit is due McRae for the part he played in perfecting the "New Plan" of Confederate purchasing operations abroad. As mentioned heretofore, the plan evolved in conjunction with the Confederate shipping regulations of 6 February and 5 March 1864. As general agent over all agents in Europe McRae steadily perfected the purchasing system and Confederate finances. The agents in Europe gave him their hearty cooperation, enabling his plans to proceed smoothly. He could report by 4 July 1864, that the credit of the Confederacy was rapidly improving and through proper management would meet all the government's needs by the end of the year.[19]

Despite the stringent Confederate regulations controlling imports and exports, by January 1865 there were more vessels engaged in blockade-running than ever before. For a while commercial houses had sought to embarrass the government by restricting running activities, but relented as the price of cotton escalated. In fact, many new steamers were understood to be on the way to engage in the business.[20] Trenholm agreed that the regulations had benefitted the government and were still highly profitable to commercial interests. President Davis in his message to the Senate in December, 1864, cited instances to prove the commercial profitability under the New Plan. He stated that the shares of one company which had originally sold for $1,000 each, were, by July, 1864, selling for $20,000 each and currently brought $30,000. Those of another firm increased from $2,500 to $6,000, and, in fact, shares in all blockade stocks commanded an enormous price on the market.[21]

The plan, continued Davis, had ended the wasteful and ruinous contract system. Now, instead of giving contractors large cost plus profits on cheap cotton, the Confederacy could purchase abroad at cost and pay in cotton overseas, which brought 24 pence a pound. According to Davis' estimate, 100 bales of cotton exported by the government would purchase the same amount that 600 bales would under the private contract system.[22] Indeed, the plan appears to have improved with the passage of time and was perhaps one of the most intelligent strategic moves of the Confederacy. One can only speculate on the results of the war had the plan been inaugurated earlier. It is ironic that in its death throes the Confederacy was, from the standpoint of its balance of payments, more solvent than at any time previous.[23]

The constant addition of steamers to Farragut's squadron caused a tightening of the blockade of Gulf ports that lasted through 1864. In 1862 the percentage of successful runs into and out of these ports was only sixty-five per cent as compared to sixty-eight per cent in the Carolina trade. In 1863 successes in the Gulf trade were sixty-two percent and in 1864, sixty-five per cent. In each case the odds were not as favorable as those of the Carolina trade, most likely because of the continued activity of a large number of sailing vessels in the Gulf trade. However, by 1864 the chances of steamers succeeding in the Gulf stood at eighty-seven per cent which was two points higher than those steamers engaged in the Carolina trade. In 1864 Gulf steamers were completing ninety-four runs out of a hundred, whereas Carolina steamers completed only sixty-six per cent.[24]

The sharp increase in the number of successful runs in the Gulf was due in no small measure to the growing sophistication of the trade. Blockade-running became a profession, and a profitable one at that, for all those engaged in it. There were participating in the Gulf trade Mexicans, Spaniards, Cubans, Americans, French, and especially, the British. To the enterprising English is due the credit for building and running the first ships ever deliberately designed to break someone's national laws.[25]

The hazards of the trade caused English ship designers to construct hulls of steel that could enter shallow inlets, be narrower in beam for speed, carry power plants adequate for ships four times their size, and still have cargo space. Some of the Clyde-built runners broke nautical speed records, attaining speeds of twenty-four nautical miles per hour, and Rear Admiral Porter declared that the new runners just "played around" federal blockaders.[26]

Spars on these new ships were reduced to two masts without cross-arms and, in some cases, hinged to fold flat against the deck. The hull sat low in the water, showing only six feet above waterline, and was painted a dull gray, described as lead color. Lifeboats were lowered square with the gunwales and the funnels were, in some cases, telescoping. Everything was done to reduce the silhouette of the vessel and when running fast on hazy days, burning smokeless coal, and wearing camouflage color, runners gave the illusion of something ghostly passing over the water. Occasionally they succeeded in running through the blockading squadrons undetected in daylight. Steam valves were vented to blow off underwater to reduce noise, and even roosters were omitted from the cook's stores of live fowl shipped aboard.[27]

The first blockade-running steamer built entirely of steel was the *Banshee I*. Even though captured on her ninth trip, her previous successes earned 700 percent profit for her owners. *Banshee II* followed and entered the Gulf trade, barely escaping the federals while running into Galveston. One of the swiftest steel vessels running in the Gulf was the *Virgin*, which made several successful trips between Mobile and Havana. Because of their revolutionary design, and the haste to enter the trade, some steel runners were constructed in flimsy fashion, sacrificing hull strength for engine size. According to a British ship engineer of that period, the powerful engines almost forced the framework through the delicate steel hull, starting rivets and buckling plates. The *Bat* was so badly sprung that she leaked twenty inches in twenty-four hours.[28]

The new fast steel runners were not required for entering the Gulf trade; they just promised better chances of success. While most were owned by the large stock companies, some were one-owner vessels. Typical of the lesser entrepreneurs of the Gulf trade was William A. Swann of Fernandina, Florida. Swann became a blockade broker, managing shipments, collecting goods, and raising capital. He accompanied shipments, buying and selling profitably, taking his own profit in the opportunity to share in the investment, plus five percent commission on sales. On the first trip out a federal blockader chased them, and Swann was forced to destroy the vessel and escape to shore. It had cost the investors $7,615.72.[29]

Swann was more fortunate with the second vessel which he purchased for $400 and then loaded with forty-two bales of cotton. In Havana, Swann sold both vessel and cargo, realizing $650 profit on the latter. With better capital he then purchased a larger vessel for $2,100. As was often the case Swann's vessel had been captured by the federals, tried and condemned by an admiralty court in Florida and sold at auction. Naturally the sales attracted those interested in blockade-running, and it often happened that former owners of a condemned ship bid high and later resumed blockade-running. With an enlarged enterprise Swann traded in luxuries, arms, and medicines. He put in at Mobile where ship and cargo were disposed of at handsome profits during the running frenzy of the spring of 1863.[30]

Swann and his associates were small operators whose investments ranged from a few hundred up to a thousand dollars. At that, their profits were good, some clients taking their gains in supplies rather than cash. One backer who had invested $1,453 more than doubled his money in two months. Rates of returns for the other investors were similar. How many of these small investors and brokers were engaged in the Gulf trade is impossible to determine, but the evidence indicates they were numerous, especially in the more isolated regions of the coast.[31]

Interestingly, the most successful blockade-runner in the Gulf trade was not specifically built for that purpose. The *Denbigh* was built in 1860 by the famous English shipbuilding firm of Laird and Company of Birkenhead, presumably for normal commerce. She was described as an iron paddle steamer of one hundred horsepower, and her registered owner was Fenton Magnall of Manchester, England. Her length was 182 feet, her beam was twenty-two feet, and she drew eight and seven-tenths feet of water. (JBA: actually the depth of hold not draft.) Her departure from Liverpool to enter the running trade on 19 October 1863, was promptly reported by the United States Consul in that port.[32]

Information forwarded to Seward on the vessel described her as a schooner-rigged side-wheeler, with artificial quarter galleries, an elliptic stern, and straight stem. A house with a binnacle on top was located mid-ships between the paddle houses. Consulate information on the *Denbigh's* appearance was quite detailed so that she might be readily identified by Farragut's blockaders. Her crew was reported as consisting of a captain, two mates, two engineers, six seamen, seven firemen, a cook and a stevedore. Her speed was variously reported as eight and fourteen and one-half knots. She ran to Havana from Mobile until that Confederate port was closed, at which time she shifted her base to Galveston. Her arrivals in Havana became so routine that she was called "the packet," and her success was a source of considerable annoyance to Seward and Farragut. Between the time of her entry into the Gulf trade in early 1864 and her grounding and destruction by federals on 24 May 1865, she successfully completed twenty-six round trips.[33]

As in the case of the *Denbigh*, the longer a vessel stayed in the illegal trade, the greater her chances of being captured or destroyed. If captured, she was taken to the nearest admiralty court–in the Gulf, either New Orleans or Key West–where the case was tried. If found guilty, she was condemned and sold at public auction. The proceeds of the sale were distributed as prize money, one-half to the officers and men on the vessel making the capture and, later in the war, one-half to the Naval Pension Fund. The dream of all federal sailors was for their ship to capture, unaided, a rich prize like the *Memphis*. When the money from this prize was distributed, the Lieutenant of the capturing blockader, *Magnolia*, received $38,318.55 and each ordinary seaman received over $1,700.[34]

The difficulty of the federal government in securing justice in prize cases was due to the legal expertise of those retained to defend British ship owners. Their object, some federals were convinced, was to delay the sale which would run up costs of condemnation, thus reducing the amount of prize awarded. A congressional investigation disclosed that costs ranged from 5.83 percent at Boston to 15.39 percent at New York. Legal delay was especially critical in cases where cargoes consisted of perishables. In one such case delaying tactics caused a cargo of fish to be sold for $105, when a few days before the value was estimated at $5,000.[35]

When Secretary Welles made his annual report to Congress in December, 1864, he could state that there had been captured to that point, 1,379 vessels of all descriptions. The gross proceeds derived from the condemnation of these vessels, the report continued, was $14,396,250.51 of which costs accounted for $1,237,153.96. Ten years later Welles estimated that thirty millions of dollars in illicit property had been captured by blockaders. A list comprised in 1912 by Congress shows that between 1,400 and 1,500 vessels were captured or destroyed. Of these, it appears that 495 were accounted for in the Gulf trade.[36]

Lest these attrition figures appear high it should be borne in mind that the blockade was estimated to have been violated 8,250 times during the war. Price's statistics indicate that there were a total of 1,143 vessels engaged in the Gulf trade during the period 1861–1865, who collectively violated the blockade, roughly, 3,000 times. Their wartime percentage of successes was: for steamers, ninety-one percent; others, eighty-one percent, and for all types, eighty-three percent. For the last few months of the war, when the West Gulf Squadron had only the Texas ports to guard, there was a seventy percent success rate for all types of runners.[37]

That blockade-running in the Gulf was successful to a surprising degree is evident. Indeed, Thomas Taylor, speaking as a schooner captain engaged in the Gulf trade, contended that running the blockade was surprisingly easy for experienced captains who displayed boldness and who commanded vessels adequate to the task. Other statistics bear out his contention. One authority estimated that not only did the South export 1,000,000 bales of cotton to Europe in the last three years of the war, but also an additional 450,000 bales entered U.S. ports. Virtually all of this came from the blockaded South, and more than half of it was estimated to have cleared through Matamoros.[38]

Confederate Secretary of War Seddon, in evaluating the importance of blockade-running, stated that it was of inestimable value to the armies of the Confederacy. "The importations of blankets, shoes, arms and supplies of every description, have been of the utmost service and it is difficult to say how we should have done without the material aid thus rendered," he said.[39] One pro-Union observer of blockade-running contended that the value was even greater. According to him, the prolongation of the war from one to two years resulted from the success of the running trade. Additionally, running provided the communications link with the outside world, he contended. Most importantly, perhaps, blockade-running was of tremendous morale value to an otherwise crumbling Confederacy, because, he noted: "It cheered and encouraged the Southern heart that would otherwise have felt ostracized from the family of nations."[40]

As one historian has noted, blockade-running was perhaps the most successful large-scale campaign attempted by the Confederacy. Its cessation, he contends, was not due to any inherent weakness in the system but to the general collapse of the Confederacy itself.[41] Indeed, the importation of goods at Galveston reached its wartime peak during the last four months of the war.

Chapter 7 Notes

[1] Richardson, *Messages and Papers of the Confederacy*, 2:114-117, 132-133; John Wilkinson, *The Narrative of a Blockade Runner* (New York: Sheldon and Company, 1877), p. 93.

[2] Crawford to Russell, 13 April 1863, Foreign Office, Record Group 72, Spain, 1071. Complaints against Federal action are found in Despatches number 5, 11, 15, 16, 1862, and number 11, 15, 37, 1863, *ibid.*

[3] Cited in Scharf, *History of Confederate Navy*, p. 471. For a typical list of arrivals of blockade-runners in Cuba.

[4] Bunch to Russell, 5 December 1864, Foreign Office, Record Group 72, Spain, 1088.

[5] Hooper, "Blockade Running," p. 105.

[6] Wait, "Blockade," pp. 918-919.

[7] Ben C. Stuart, "Some True Stories of the Blockade," *Galveston Daily News*, 16 July 1911. These figures are quoted for the period subsequent to the Act of 6 Feb. 1864, reserving half cargo space for Confederate use. Cotton quotations also appear low, computing a 300-pound bale at $50; the cotton sold on C.S. account made nothing for the owners.

[8] Hooper, "Blockade Running," p. 107; Wait, "Blockade," p. 919.

[9] Cowley, *Leaves from a Lawyer's Diary*, p. 112; Merrill, *The Rebel Shore*, p. 63.

[10] Wait, "Blockade," p. 919.

[11] E. S. Dargan to James A. Seddon, 11 June 1863, *OR*, ser. 4, 2:585.

[12] Smith to G. W. Randolph, 31 July 1862, *ibid.*, ser. 4, 2:30.

[13] Whiting to James A. Seddon, 23 March 1863, *OR*, ser. 1, 18:939.

[14] Maj. John B. Burton to Gen. Theophelous Holmes, 4 Jan. 1864, *OR*, ser. 1, 34, pt. 2:823.

[15] Broadwell to Maj. J. P. Johnson, 19 Jan. 1864, *OR*, ser. 1, 26, pt. 2:577-578.

[16] Trenholm to Seddon, 12 August 1864, *OR*, ser. 4, 3:588.

[17] James D. Bulloch, *The Secret Service of the Confederate States in Europe or How the Confederate Cruisers Were Equipped*, 2 vols. (New York: Thomas Yoseloff Co., 1959), 1:111-127.

[18] Thompson, *Confederate Purchasing Operations*, p. 94; Memminger to Davis, 26 Nov. 1863, *ORN*, ser. 4, 2:1013-1016; *ORN*, ser. 2, 2:686, 720, 765-767; Vandiver, *Ploughshares into Swords*, p. 90.

[19] McRae to Seddon, *OR*, ser. 4, 3:525-529; Thompson, *Confederate Purchasing*, pp. 94-95; Vandiver, *Blockade Running*, p. 36.

[20] Scharf, *History of Confederate Navy*, p. 487; Seddon to Davis, 10 December 1864, *OR*, ser. 4, 3:948-930.

[21] Thompson, *Confederate Purchasing*, pp. 96-97; *OR*, ser. 4, 3;948-953.

[22] *OR*, ser. 4, 3:948-953.

[23] Vandiver, *Blockade Running*, p. xli; Thompson, *Confederate Purchasing*, pp. 96-102.

[24] Price, "Ships," pp. 270-271.

[25] Hanna, "Incidents," pp. 218-219.

[26] Cited in Hooper, "Blockade Running," p. 106; Wait, "Blockade," p. 917.

[27] Stuart, "Some True Stories," *Galveston Daily News*, 17 July 1911.

[28] Taylor, *Running the Blockade*, pp. 148-157; Bradlee, *Blockade Running*, pp. 114-115, 118-119.

[29] Hanna, "Incidents," pp. 219-221.

[30] *Ibid.*

[31] *Ibid.*

[32] Arthur C. Wardle, "Blockade Runners Built on Merseyside and Registered at the Port of Liverpool during the American Civil War, 1861–1865," typescript dated 23 Nov. 1941, Liverpool Public Library; Price, "Ships," p. 271.

[33] Price, "Ships," p. 271.

[34] Hooper, "Blockade Running," p. 109.

[35] *Ibid.*, 106.

[36] U.S. Congress, House, *Report of the Secretary of the Navy*, 38th Cong., 2d sess., House Exec. Doc. No. 1; Welles, *Lincoln and Seward*, p. 151; Owsley, *King Cotton Diplomacy*, pp. 260-261; Price, "Ships," 12:237.

[37] Bernath, *Squall*, p. 11; Price, "Ships," 12:236.

[38] Taylor, *Running the Blockade*, p. xvi; Owsley, *King Cotton Diplomacy*, pp. 264-266.

[39] *OR*, ser. 1, 35, pt. 3:638.

[40] Hooper, "Blockade Running," p. 108.

[41] Vandiver, *Blockade Running*, p. xli.

8

Conclusions

Any evaluation of the performance of the West Gulf Blockading Squadron must be measured within the perspective of the magnitude of the task, international as well as national restrictions, and, finally, results. Considering the extent of the area of operations assigned the squadron, and its relatively low logistical support priority, it is remarkable that the squadron did as well as it did. To have done an outstanding job of blockading the Gulf coast would have required hundreds of more vessels, not only immediately picketing the coast, but, in addition, forming an outer cruising perimeter. It also seems apparent that a new class of blockader–one to match the capabilities of the best runners–was needed in large numbers to have effectively blockaded the Gulf coast.

Even then effective enforcement would have to have been confined to a domestic zone exclusively which was possibly only within the territorial waters off the southern ports. The Navy Department operated almost as much in international waters and near neutral ports as it did within the limits of its own ports. In so doing, federal officers, who for the most part were novices at international law, discovered that the rules of the game had been established by other countries through long usage. This was the first American blockade in history and its enforcement was, at times, vaguely understood by the State and Navy Departments. The secretaries of these two departments disagreed, vehemently at times, among themselves over the interpretation of the rules of blockading, and the State Department was constantly embroiled with foreign powers over the interpretations. Understandably, this pressure reflected on the navy's ability to blockade effectively. It was particularly awkward for a nation which had completely reversed its policy concerning maritime rights.

Quite possibly, the judicial branch of the federal government played a more significant role in tightening the blockade than is generally credited it. District, or admiralty, courts tended to readily condemn cargoes and ships. Only a small percentage of these cases were appealed, and, when they were, the high courts tended to delay decisions. This produced a lingering state of uncertainty among shippers engaged in the nefarious running trade. Yet, the American judicial system, on the whole, was considered by some foreign powers to be a fair arbiter of delicate international cases. Often, under pressure from the Executive and State Departments, who, realistically or not, feared war with foreign powers over maritime infringements, American courts almost invariably found in favor of the defendants.

Those cases arising from maritime incidents at or near the mouth of the Rio Grande particularly weakened the effectiveness of the West Gulf Squadron. The "Matamoros cases," which resolved some of the most serious maritime incidents of the war in England's favor, resulted in that port being left virtually unguarded for the last half of the war. The federal navy, warned off the mouth of the river by a gun-shy State Department, had to be content with periodic cruising in the vicinity. Matamoros literally became the open back door of the Confederacy through which was funneled enormous amounts of supplies for the South and by which equally large amounts of cotton left. Yet, it is also apparent from this study that the supply line through Matamoros mostly aided the Trans-Mississippi Confederacy, especially after the closure of the Mississippi River. But it can equally be said that the Trans-Mississippi Department was predominantly sustained by the Matamoros trade. At least from a quartermaster standpoint the prospects of having adequate stocks of supplies in that department were never brighter than in the closing weeks of the war. The one critical drawback to the Matamoros-Confederacy supply route was the overland transportation shortage.

Because of this logistical shortage, as well as other considerations, blockade-running through other Gulf ports increased in volume throughout the war. As more and more federal warships joined the West Gulf Squadron, the escalating prices of cotton drew an even greater proportion of shippers into the running trade. Fast, sophisticated steamers designed specifically to break the blockade entered the Gulf, and their success rate attested to the inefficiency of the squadron in stopping this traffic. The blockade was efficient enough to inflate the price of everything involved in the trade, from consumer prices within the Confederacy to cotton prices in Europe. But the blockade was never efficient enough to seriously curtail the ever-increasing amount of blockade-running. This study has shown that blockade-running was more voluminous, as well as most successful, in the last months of the war.

The West Gulf Squadron operated under additional handicaps in addition to logistical shortages and low priority allocations. One of the most frustrating hindrances came from the federal government. By act of Congress and Presidential proclamation, various federal agencies were armed with the authority to purchase Southern cotton. The promulgation of this program was carried out in a gray zone of loosely controlled regulations. Even a bureau staffed with perfectly honest personnel would have had difficulty purchasing an exclusive enemy commodity without aiding, directly or indirectly, the enemy. Whether through graft, inefficiency, or legitimate intercourse, the United States Treasury Department was of inestimable value to the Confederate economy. No one realized this more painfully than the officers of the blockading squadron.

If the only mission of the squadron had been blockade duty, their record might be more favorable. As it was, major land campaigns drew heavily on the squadron's ships and personnel; in fact, the land campaigns had priority. Because of this dual responsibility, Farragut, as squadron commander was overextended. The assignment of a vigorous vice-admiral to command the blockading division, while Farragut was preoccupied with the New Orleans, Vicksburg, and Mobile campaigns, would have relieved the aging Admiral considerably, and made for a more efficient blockade.

Yet the experience at Galveston proved that under the most favorable blockading conditions blockade-running was highly successful. At a time when this was the last major open Gulf port, and there were no other priorities to prohibit full concentration on guarding this port, the maritime traffic was at its heaviest. So successful was blockade-running that some Confederate authorities felt confident that only four government steamers could meet the supply needs of the Trans-Mississippi as well as ship out cotton to meet obligations in Europe. This study has shown that Confederate and private cotton was getting to Europe through the Galveston blockade in significant amounts. The last six months export as shown by Confederate Treasury records, was in excess of 10,000 bales. Assuming that an average bale ran 450 pounds, this represents a Liverpool value of between two and three million dollars.

While wartime exports from Texas fall short of peacetime figures, the cotton trade was at least keeping alive a limited civilian economy. Confederate currency had already ceased to be of much value and was not widely used by the Treasury Department in the Trans-Mississippi. The greatest value of the cotton trade may have been to the general economy of south Texas in providing, at least minimally, a solid medium of exchange. All parties involved in the trade, with the not too surprising exception of the planters, had demanded and received payment in specie, cotton, or exchange certificates. Since there is no evidence to indicate that the certificates were not honored, then all three mediums retained exchange value after the close of the war as they had before. Thus, the collapse of the Confederacy, this report suggests, brought no cataclysmic change in either the lives of the cotton shippers or in the economy of the Houston and Galveston area.

In summary, the overall performance of the West Gulf Blockading Squadron can only be rated as mediocre. Mostly for reasons that were not of their own doing. That they did the best they could under the circumstances is conceded. The record of this naval command tends to bear out the maxim that blockades do not win wars. At the most blockades stifle, and at the least they inconvenience. Somewhere between the two is found the true evaluation of the effectiveness of the West Gulf Blockading Squadron.

Bibliography

PRIMARY SOURCES

Manuscripts

Austin. University of Texas Library. William Pitt Ballanger Papers.

Austin. University of Texas Library. Thomas William House Papers.

Galveston. Rosenburg Public Library. George W. Grover Papers.

Galveston. Rosenburg Public Library. Clinton G. Wells Papers.

Galveston. Rosenburg Public Library. Henry Sampson Papers

Galveston. Rosenburg Public Library. Catalogue item no. 71-0043, entitled "Report of the Cotton Bureau."

Great Britain. Public Record Office MSS. Foreign Office, America, Record Group 5, vols. 848, 908, 909, 970, 1009, and 1029.

Great Britain. Public Record Office MSS. Foreign Office, Spain, Record Group 72, vols. 1041, 1071, 1088, and 1108.

Great Britain Public Record Office MSS. Foreign Office, Mexico, Record Group 50, vols. 368, 377, 382, and 384.

Government Publications

Great Britain. Public Record Office. *List of Indexes*, vol. 52, America, 38th series.

Great Britain. Parliament. *Parliamentary Papers* (House of Lords), 1862–1863, vol. 25. "Papers Relating to the Blockade."

U.S. Congress. House. *Report of the Secretary of the Navy*. 37th Cong., 3rd sess., vol. 3, 1862; 38th Cong., 2nd sess., vol. 6.

U.S. Congress. House. *House Committee Reports*. 38th Cong., 1st sess., Report no. 111.

U.S. Congress. House. *Report of the Committee on Public Expenditures*. 38th Cong., 2nd sess., H. Doc. no. 25.

U.S. Congress. Senate. *Cotton Sold to the Confederate States*. 62nd Cong., 3rd sess., vol. 8, 1912.

U.S. Department of State. Consular Despatches, Matamoros, Record Group 56, National Archives.

U.S. Department of State. "Diplomatic Instructions, Great Britain, 1801–1906." Record Group 59, National Archives.

U.S. Department of War. *The War of the Rebellion: A Compilation of the Official Records of the Union and Confederate Armies*. Ser. 1, vols. 4, 9, 26, 41, pt. 4; 48, pts. 1-3; ser. 2, vol. 1. 128 vols. Washington, D.C.: Government Printing Office, 1880–1901.

U.S. Department of War. *The War of the Rebellion: A Compilation of the Union and Confederate Navies*. Ser. 1, vols. 16-22. 22 vols. Washington, D.C.: Government Printing Office.

Books

Beale, Howard K., ed. *Diary of Gideon Welles: Secretary of the Navy under Lincoln and Johnson.* 3 vols. New York: W. W. Norton and Co., 1960.

Bulloch, James D. *The Secret Service of the Confederate States in Europe, or How the Confederate Cruisers Were Equipped.* 2 vols. Reprint. New York: Thomas Yeseloff Co., 1959.

Cowley, Charles. *Leaves from a Lawyer's Life Afloat and Ashore.* Lowell, Massachusetts: Penhallow Printing Co., 1879.

Fremantle, Arthur James Lyon. *The Fremantle Diary Being the Journal of Lt. Colonel James Lyon Fremantle, Coldstream Guards on His Three Months in the Southern States.* Edited by Walter Lord. New York: Capricorn Books, 1960.

Hill, Frederick Stanhope. *Twenty Years at Sea.* Boston: Knopf Publishing Co., 1893.

Holton, William C. *The Cruise of the United States Flag-Ship* Hartford *1862–1863.* New York: L. W. Paine Printer, 1863.

Huse, Caleb. *The Supplies for the Confederate Army, How They Were Obtained in Europe and How Paid for.* Reprint. Houston: Deep River Armory, Inc., 1970.

Hutchinson, William F. *Life on the Texan Blockade.* Providence: Rhode Island Soldiers and Sailors Historical Society, 1883.

Johnson, Robert Underwood, and Buell, Clarence Clough, eds. *Battles and Leaders of the Civil War.* 4 vols. Grant-Lee edition. New York: The Century Co., 1884.

Lubbock, Francis R. *Six Decades in Texas or the Memoirs of Francis Richard Lubbock.* Edited by C. W. Raines. Austin: Ben C. Jones and Co., Printers, 1900.

Mahan, Alfred Thayer. *The Gulf and Inland Waters.* vol. 3. 3 vols. New York: Charles Scribner's Sons, 1883.

Moore, John Bassett. *A Digest of International Law As Embodied in Diplomatic Discussions, Treaties and Other International Awards: The Decisions of Municipal Courts, and the Writings of Jurists and Especially in Documents, Published and Unpublished, Issued by Presidents and Secretaries of State of the United States, the Opinions of the Attorneys-General, and the Decisions of Courts, Federal and State.* Washington: Government Printing Office, 1906.

North, Thomas. *Five Years in Texas.* Cincinnati: Worley Brothers, 1870.

Osbon, Bradley S. *Handbook of the U.S. Navy; Being A Compilation of the Main Events of Every U.S. Naval Vessel,* April 1861–May 1864. New York: Bradley and Co., 1864.

Richardson, James D. *Compilation of Messages and Papers of the Confederacy.* 2 vols., Nashville: United States Publishing Co., 1906.

_____. *Messages and Papers of the Presidents.* 10 vols. Washington: Bureau of National Literature and Art, 1907.

Scharf, J. Thomas. *History of the Confederate States Navy from its Organization to the Surrender of its Last Vessel.* New York: Rogers and Sherwood, 1887.

Scheliha, Viktor Ernst Karl Rudolf von. *A Treatise on Coast-Defense Based on the Experience Gained by Officers of the Corps of Engineers of the Army of the Confederate States and Compiled from the Official Reports of the Officers of the Navy of the United States.* Westport, Connecticut: Greenwood Press, Publishers, 1968.

Semmes, Admiral Raphael. *Service Afloat: Or, the Remarkable Career of the Confederate Cruisers* Sumter *and* Alabama, *During the War between the States.* Baltimore Publishing Co., 1887.

Soley, James Russell. *The Navy in the Civil War.* New York: Charles Scribner's Sons, 1885.

Taylor, Thomas. *Running the Blockade*. London: John Murry, Printer, 1896.

Thompson, Robert Means and Wainwright, Richard, eds. *Confidential Correspondence of Gustavus Vasa Fox Assistant Secretary of the Navy 1861–1865*. 2 vols. New York: DeVinne Press, 1918.

Welles, Gideon. *Mr. Lincoln and Mr. Seward*. New York: Sheldon and Company, 1874.

Wilkinson, John. *The Narrative of a Blockade Runner*. New York: Sheldon and Company, 1877.

Articles

Anonymous. "Blockade." *United States Naval Institute Proceedings* 11:423-463.

Hooper, W. R. "Blockade Running." *Harpers New Monthly Magazine* 42:105-108.

Wait, Horatio L. "The Blockade of the Confederacy." *Century Magazine* 61:914-928.

Newspapers

Corpus Christi Ranchero. 1 July 1861–1 July 1865.

Galveston Weekly News. 1 January 1861–1 July 1865.

The Houston Daily Telegraph. 9 January 1865.

Houston Tri-Weekly Telegraph. 1 January 1861–1 July 1865.

London Herald. 16 September 1962.

New Orleans Daily Picayune. 1 January 1861–1 January 1865.

New York Herald. I January 1862–1 July 1865.

Texas Republican (Marshall). 20 January 1865.

Texas State Gazette (Austin). 1 July 1861–1 July 1865.

The Times (London). 19 April 1862, 10 February 1862.

SECONDARY ACCOUNTS

Books

Adams, Ephraim Douglas. *Great Britain and the American Civil War*. 2 vols. New York: Longmans, Green and Co., 1925.

Barnes, James. *Midshipman Farragut*. New York: D. Appleton and Co., 1896.

Bernath, Stuart L. *Squall Across the Atlantic: American Civil War Prize Cases*. Berkeley: University of
California Press, 1970.

Bradlee, Francis B. C. *Blockade Running During the Civil War and the Effect of Land and Water Transportation on the Confederacy*. Salem, Massachusetts: The Essex Institute, 1925.

Briggs, Herbert Whittaker. *The Doctrine of Continuous Voyage*. The Johns Hopkins University Studies in History and Political Science. Baltimore: Johns Hopkins Press, 1926.

Callahan, James M. *Diplomatic History of the Southern Confederacy*. Baltimore; Johns Hopkins Press, 1901.

Clark, Joseph L. *The Texas Gulf Coast: Its History and Development*. New York: Lewis Historical Publishing Co., 1955.

Cochran, Hamilton. *Blockade Runners of the Confederacy*. New York: Bobbs-Merrill Co., 1958.

Coulter, E. Merton. *The Confederate States of America. A History of the South Series*. Vol. 7. Baton Rouge: Louisiana State University Press, 1950.

Davis, Charles S. *Collin J. McRae: Confederate Financial Agent*. Confederate Centennial Studies No. 17. Tuscaloosa: Confederate Publishing Co., Inc., 1961.

Gallaway, B. P., ed. *The Dark Corner of the Confederacy*. Dubuque, Iowa: William C. Brown Book Co., 1968.

Horner, Dave. *The Blockade Runners*. New York: Dodd, Mead and Co., 1968.

Hyde, Charles C. *International Law, Chiefly As Interpreted and Applied by the United States*. 3 vols. Boston: Little Brown and Co., 1945.

Johnson, Ludwell. *The Red River Campaign: Politics and Cotton in the Civil War*. Baltimore: Johns Hopkins Press, 1958.

Lea, Tom. *The King Ranch*. 2 vols. Boston: Little Brown and Company, 1957.

Lewis, Charles Lee. *David Glasgow Farragut*. Annapolis: United States Naval Institute, 1941–1943.

McCartney, David Edward. *Mr. Lincoln's Admirals*. New York: Funk and Wagnalls Publishing Co., 1956.

Merrill, James M. *The Rebel Shore: The Story of Union Sea Power in the Civil War*. Toronto and New York: Little Brown and Co., 1957.

Miller, Hunter. *Treaties and Other International Acts of the United States of America*. 8 vols. Washington: Government Printing Office, 1937.

Nichols, James L. *The Confederate Quartermaster in the Trans-Mississippi*. Austin: University of Texas Press, 1964.

Ormerod, Leonard. *The Curving Shore*. New York: Harper and Rowe Publishing Co., 1955.

Owsley, Frank L. "America and Freedom of the Seas, 1861–1865." In *Essays in Honor of William E. Dodd*, edited by Avery O. Craven. Chicago: University of Chicago Press, 1935.

_____. *King Cotton Diplomacy*. Chicago: University of Chicago Press, 1959.

_____. *The C.S.S. Florida*. Philadelphia: University of Pennsylvania Press, 1965.

Randall, James G. *Constitutional Problems Under Lincoln*. Urbana: University of Illinois Press, 1951.

Randall, James G., and Donald, David. *The Civil War and Reconstruction*. 2[d] ed. Lexington, Massachusetts: D. C. Heath and Co., 1969.

Robinson, William M. *The Confederate Privateers*. New Haven: Yale University Press, 1928.

Savage, Carolton J. *Policy of the United States Toward Maritime Commerce in War*. 2 vols. Washington: Government Printing Office, 1934.

Spears, John Randolph. *David Glasgow Farragut*. Philadelphia: G. W. Jacobi and Company, 1905.

Thompson, Samuel Bernard. *Confederate Purchasing Operations Abroad*. Chapel Hill: University of North Carolina Press, 1955.

Vandiver, Frank. *Confederate Blockade Running Through Bermuda, 1861–1865*. Austin: University of Texas Press, 1957.

Ziegler, Jesse A. *A Wave of the Gulf*. San Antonio: Naylor Publishing Co., 1938.

General References

Editorial Staff. *American Jurisprudence: War*. Vol. 56. 58 vols. Rochester, N.Y.: Lawyers' Cooperative Publishing Co., 1948.

Editorial Staff. *Words and Phrases*. Vol. 5. 46 Vols. St. Paul, Minnesota: West Publishing Co., 1962.

Lucas, Francis J., and Gilbert, Harold J., eds. *Corpus Juris Secundum*. Vol. 93. 101 vols. Brooklyn, N.Y.: American Law Book Co., 1956.

Webb, Walter Prescott, ed. *The Handbook of Texas*. 2 vols. Austin: University of Texas Press, 1964.

Articles

Barr, Alwyn. "Texas Coastal Defense, 1861–1865." *Southwestern Historical Quarterly* 45(1961):1-30.

Baxter, James P. "The British Government and Neutral Rights." *American Historical Review* 34(1928):1-29.

Blume, August N. "An Escape From Galveston." *National Republic* 54(1935):22-25.

Coulter, E. Merton. "Commercial Intercourse with the Confederacy in the Mississippi Valley, 1861–1865." *Mississippi Valley Historical Review* 5(1918–1919):377-395.

Delaney, Robert W. "Matamoros, Port for Texas During the Civil War." *Southwestern Historical Quarterly* 63(1955): 473-487.

Diamond, William. "Imports of the Confederate Government from Europe and Mexico." *Journal of Southern History* 6(1940):502-520.

Fitzhugh, Lester N. "Saluria, Ft. Esperanza and Military Operations on the Texas Coast, 1861–1864." *The Southwestern Historical Quarterly* 61(1957):66-100.

Gregory, Charles N. "The Doctrine of Continuous Voyage," *Harvard Law Review* 24(1911):170-195.

Hanna, Katherine A. "Incidents of the Confederate Blockade." *Journal of Southern History* 11(1945):253-275.

Hendren, Paul. "The Confederate Blockade Runners." *United States Naval Institute Proceedings* 59(1933):506-512.

O'Flaherty, David. "The Blockade that Failed." *American Heritage Magazine* 6(1964):30-41.

Paullin, C. O. "A Half Century of Naval Administration in America." *United States Naval Institute Proceedings* 38(1912):13-9-1336; 39(1913):165-195.

Price, Marcus W. "Ships that Tested the Blockade of the Gulf Ports." *The American Neptune* 11(1951):262-290; 12(1952):229-238; 13(1953):154-161; 15(1955):97-131.

Ramsdell, Charles W. "The Texas State Military Board, 1862–1865." *Southwestern Historical Quarterly* 27(1924): 260-275.

Roberts, A. Sellow. "High Prices and the Blockade in the Confederacy." *South Atlantic Quarterly* 24(1925):154-163.

_____. "The Federal Government and Confederate Cotton." *American Historical Review* 34(1927):262-275.

Trexler, H. A. "The *Harriet Lane* and the Blockade of Galveston." *Southwestern Historical Quarterly* 35(1931):112-126.

Ulmer, Grace. "Economic and Social Development of Calcasieu Parish, 1840–1912." *Louisiana Historical Quarterly* 32(1955):351-374.

Windham, William T. "The Problem of Supply in the Trans-Mississippi Confederacy." *Journal of Southern History* 27(1961):159-171.

Dissertations and Theses

Boozier, Jesse Beryl. "The History of Indianola, Texas." M.A. Thesis, University of Texas, 1926.

Cowling, Annie. "The Civil War Trade of the Lower Rio Grande Valley." M.A. thesis, University of Texas, 1926.

Dickeson, Sherrill L. "The Texas Cotton Trade During the Civil War." M.A. thesis, North Texas State University, 1967.

Felgar, Robert Pattison. "Texas in the War for Southern Independence." Ph.D. dissertation, University of Texas, 1935.

Ferguson, Stuart Alfred. "The History of Lake Charles, Louisiana." M.A. thesis, Louisiana State University, 1931.

Garner, Ruby. "Galveston During the Civil War." M.A. thesis, University of Texas, 1927.

Holkstra, S. Robert. "A Historical Study of Texas Ports During the Civil War." M.A. thesis, Texas College of Arts and Industry, 1951.

Holladay, Florence Elizabeth. "The Extraordinary Powers and Functions of the General Commanding the Trans-Mississippi Department." M.A. thesis, University of Texas, 1914.

Hoovestol, Paeder Joel. "Galveston During the Civil War." M.A. thesis, University of Houston, 1950.

Jackson, Vivian Gladys. "A History of Sabine Pass." M.A. thesis, University of Texas, 1930.

Jager, Ronald B. "Houston Texas During the Civil War." M.A. thesis, University of Houston, 1964.

Lamble, Agnes Louise. "Confederate Control of Cotton in the Trans-Mississippi." M.A. thesis, University of Texas, 1915.

Ogden, Brother August Raymond. "A Blockaded Seaport, Galveston, Texas C.S.A." M.A. thesis, St. Mary's University, 1939.

Settles, Thomas Michael. "The Port of Galveston During the Civil War." M.A. thesis, Trinity University, San Antonio, 1968.

Sydner, Laura. "The Blockade of the Texas Coast During the Civil War." M.A. thesis, Texas Technological College, 1938.

Webb, Jesse Owens. "The History of Galveston to 1865." M.A. thesis, University of Texas, 1924.

Unpublished Duplicated Works

Hayes, Robert. "The Island and City of Galveston." Carbon typescript. Galveston, Texas: Rosenburg Public Library, 1879.

Stuart, Ben C. "Stuart Book: A Series of Articles of Historic Interest Relating to Galveston and Texas, Published in the *Galveston News* During the Years 1906 to 1911." Scrapbook. Galveston: Rosenburg Public Library, 1913.

Tucker, Philip C. "The Royal Yacht." Carbon typescript. Galveston, Texas, Rosenburg Public Library, no date.

Unknown. "Blockades and Battles." Carbon typescript. Galveston: Rosenburg Public Library, no date.

Wardle, Arthur C. "Blockade Runners Built on Merseyside and Registered at the Port of Liverpool During the American Civil War, 1861–1865." Typescript. Liverpool, England. Liverpool Public Library, 1941.

Part II

Archival Material and Notes

by

J. Barto Arnold III
Denbigh Project Director
Institute of Nautical Archaeology

Introduction to Part II

by

J. Barto Arnold III

The present author has made a practice of including in *Denbigh* Project publications contemporary accounts and archival documents relevant to Civil War blockade activities in the Gulf where the *Denbigh* operated as a runner. In this book, we concentrated on the question, "What were the Union blockaders up to?" The subject was addressed by showing examples of memoirs, official reports, and the files generated by the requirements of prize law when a capture was made. It was helpful to gather these materials together in the project reports for the convenience of interested readers and researchers. Even more important were the insights such documents often provided. In addition to details relevant to the *Denbigh* Project, researchers may find different gems of detail relevant to their own inquiries. The present author herein used footnotes and commentary to highlight aspects of interest in the study of the *Denbigh*.

Chapter 1 contains two memoirs of sailors of the blockading fleet. Chapter 2 shows official Union reports. The records of these two chapters were available online, but it was desirable to gather them together and print them here to call out points of interest and for ease of access.

Even more interesting were archival documents generated by the capture of prizes. The litigation to condemn and confiscate a captured ship and cargo conformed to international law and custom dating back centuries. The documentation was very specific, formal, even ritualized. The court files yielded great stories and fascinating details. The Union navy officers and crews were highly motivated to capture the runners and collect the prize money that was the resulting reward. The court documents showed how the system worked in detail. Far from dry, the files often revealed the whys and wherefores of behavior entailed in the operations. One set of prize records generated by the *Denbigh* derived from the ship going aground while departing Galveston one night in April 1865. Over two hundred bales of cotton were thrown overboard to refloat the ship. The next morning this floating treasure was retrieved by and became a prize of two of the blockading ships, the U.S.S. *Gertrude* and U.S.S. *Cornubia*.

Chapter 3 presents the documents of the two prize cases dealing with the jettisoned cotton, one set for each ship. Chapter 4 shows the deck logs of the Union ships involved with the *Denbigh* on this occasion and from the following month when the *Denbigh* finally grounded and wrecked. Both sets of deck logs were presented here as a logical group. Chapter 5 includes the records generated after the court cases, when the prize money for the cotton bales of April 1865 was paid out. Documents concerning how the money was paid and how much for each man turned out to be quite interesting. Altogether the prize documents demonstrated not only how the system worked, but also why things were done in a particular way.

Some authorities identified the merchant steamer *Alabama* as a sister ship of the *Denbigh*, i.e., owned by the same blockade-running company, the European Trading Company (E.T.C.). The prize files from the capture of the *Alabama* were examined in hopes of determining which of the several ships of that name might be the one relevant to the E.T.C. The *Alabama*, a 510-ton steamer of Mobile, was almost certainly the one in question. We hoped that the ship's papers, often included in the prize case documents, might reveal the owners. The results on ownership were not clear-cut, being easily obfuscated by such mechanisms as frontmen or leasing of ships, a common but sometimes undocumented practice (see below, Chapter 6).

What the court file did reveal was a conflicting set of claims by the several Union ships involved in the capture. Because of the prize money at stake, the maneuvering of the Union commanders became frenzied and reflected poorly on the naval service. One does not have to read very far between the lines to see what was going on, and it did not speak well of the participating officers. Most of the other prize case files examined by the author might consist of a few dozen documents. The *Alabama* case file ran to over three hundred. Most of the *Alabama* documents are shown in Chapter 6.

Chapter 7 is the conclusion. Reading an historical overview and analysis like Glover's in Part I of the present book was of undoubted utility. Even better would be to combine that analysis with reading actual archival documents, such as those contained in Part II tracing activities of the blockading ships and the litigation entailed in several captures. The present author believes the combination presented herein is well worthwhile.

The document transcripts in Part II were not edited and corrected but rather illustrate the documents as written. When the original handwriting was illegible, question marks were inserted in the transcript.

1

Union Navy Blockade Memoirs

This chapter presents two examples of memoirs concerning U.S. Navy blockade activities on the Texas coast and vicinity. For getting the flavor of daily life, these sources reporting and recounting events in the participants' voices bring the reader close to the action. Particularly amusing were four extracts from *On a Man-of-War* told in colloquial voices.

Davenport, the author, had an excellent store of sea stories. He did not, however, simply tell tall tales. For example, consider the scale of shipping off the Rio Grande. Davenport observed seventy ships at anchor, a bottleneck that resulted from heavy Confederate trade across the Texas border to circumvent the Union blockade. There were also ships with supplies for the war between Maximilian and the insurgents in Mexico. All of these ships had to be unloaded and loaded by liter, a very slow process. In the shipping business time is money, and this kind of delay was a major problem.

Davenport shared in the prize money resulting from two small captures. In discussing prize money, he made a point of exaggerating how long it took to be paid, that being ten or twelve years. The usual time observed by the present researcher was more on the order of half a year, although a few years might have been a common experience. (See Part II, Chapter 5.)

Another interesting memoir concerned the sinking of the U.S.S. *Hatteras* by the C.S.S. *Alabama*. An officer on one of the other blockading ships at Galveston recounted his observations of the battle, including his thoughts and feelings on the incident. Just one item of interest in this tale was an explanation of how the news of the action spread through his ship when Union survivors arrived the next day.

While on blockade duty off Galveston one night, Davenport's ship grounded near Fort Point while scouting the channel. Running aground was a predicament experienced by many a blockade-runner, including the *Denbigh*, and there were anxious hours and a detailed description of getting the Union ship off the bar just before daylight. Davenport was quite open in describing feelings of stress and trepidation when on the eve of battle or under fire from the batteries.

Perhaps the most valuable parts of Davenport's stories concern detailed tactics and practices aboard a Union ship on blockade duty. At least in part he addressed the questions, "How did they do that?" and "Why did they do things that way?" concerning the blockade.

Excerpt 1 from *On a Man-of-War*[5]

LETTER VIII

ORDERED TO A BLOCKADING STATION—GETTING UNDER WEIGH IN A HURRY—CAPTURING A CONFEDERATE—THE RIO GRANDE—A VERY INCREDULOUS CAPTAIN—HOW A YOUNG NAVIGATOR SAILED A SHIP ALONG THE CREST OF THE ANDES—HOW THE CAPTAIN WAS CONVINCED—THE PECULIAR HABITS OF SAILORS OF DIFFERENT NATIONS—CUTTING OUT A SLOOP—TWO SAILORS' YARNS.

---

[5] Davenport, F. O. *On a Man-of-War: A Series of Naval Sketches.* Detroit: E. B. Smith & Co., 1878, pp. 76-84. The referenced ship, the U.S.S. *Portsmouth*, was a wooden-hulled, sail-powered sloop-of-war launched in 1843. Later he transferred to the U.S.S. *Scioto*, a ninety-day gunboat, in which he served off Galveston.

Shortly after our arrival at Ship Island, Gulf of Mexico, we were ordered by the flag officer, D. G. Farragut, to blockade the mouth of the Rio Grande River. So, one morning bright and early the pipes of the boatswain's mates were heard, followed by the repeated cry of,

"All hands,"
"All hands,"
"Up all hammocks,"
"Now tumble up there,"
"Show a leg,"
"Get out of there, you idlers,"
"Clear the berth-deck, master-at-arms,"
"No one excused,"
"It's up anchor this morning."

In ten minutes the hammocks were all up and, stowed in the nettings, and the captains of the different parts of the ship were quickly moving to and fro, quietly directing their men so as to get as ready as they dared before the order "up anchor" was actually given. In a few moments after the anchor was hove short, and in obedience to the order of the executive officer,

"Aloft, sail-loosers,"

the men swarmed aloft, and in less time than it takes to write it the gallant little *Portsmouth* was under way with all plain sail, standing to the southward and westward, bound for "La Boca del Rio del Norte," as the mouth of the Rio Grande is called by our Mexican friends on the south side.

THE FIRST PRIZE.

We sailed along pleasantly without incident until almost within sight of the coast, when we discovered a schooner standing to the southward, which we quickly overhauled. Running up the French flag we were much pleased to be answered by the display of the rebel stars and bars. Amid considerable excitement a boat was lowered, and with twelve men armed to the teeth I pulled off for the schooner, the *Portsmouth* at the same time hauling down the French flag and displaying the stars and stripes at the peak. We pulled alongside, and clambering up the side of the schooner (I with my sword in my teeth, being armed to the teeth), we sprang on board, prepared to cut down almost anything-excepting, of course (I cannot tell a lie), any cherry trees. There was one poor devil on deck that was quietly steering the schooner. After lowering the sails I boldly, yet cautiously, advanced upon this man and sternly asked,

"Where is the captain, sir?"

"Oh! we're all captains here," he answered *nonchalantly*, "but Captain B—— is below," he added, "if you want *him*."

We persuaded the captain and all hands to come on deck, and found that our prize was the schooner *Wave*, from New Orleans, bound to the Rio Grande. We transferred his cargo of sugar to one of the supply steamers shortly afterward, and used the schooner itself as a target for exercise at great guns.

I think I got some $43 prize money about twelve years afterward from the sale of the *Wave's* cargo.

The next day we anchored off

THE MOUTH OF THE RIO GRANDE,

where some seventy merchant vessels lay at anchor. I was surprised to find that the mouth of the river appeared to be only a couple of hundreds of yards in width instead of a mile, as I had always imagined.

I found that the commander shared my opinion, and decidedly refused to believe me, as navigator of the ship, declining to accept any such dirty little river as the great river of the north. In vain I pleaded that the sun could not lie, that figures were figures. He declined to accept the situation, until I returned from boarding most of the vessels at anchor and reported that they were all under the same impression.

SAILING OVER THE ANDES.

I could not blame him. He knew the story of a young officer who was attached to a ship bound for Rio, as navigator. The captain distrusted his ability and secured all the charts of the ship so the navigator could not see them, requiring him to send in the latitude and longitude every day as usual, ascertained by observations of the sun. On the arrival of the ship in Rio the captain showed the young mathematician that by his reckoning, as sent in from day to day, the ship had skirted Andes Mountains all the way down; being impartially sometimes on one side and sometimes on the other of the lofty range, congratulating him on the feat of balancing this sloop-of-war successfully on a mountain peak, 23,000 feet in height, without knowing it.

Of course, the navigator not having access to the charts, could not see where his latitude and longitude would place the ship, and supposed all the time that he was the equal, if not the superior of Mr. C. Columbus, as a navigator of the seas. Well,

THE CAPTAIN NEVER LET UP

on me; he would not believe that that little stream was the Rio Grande River. I took sights every day and required the midshipmen to take sights with me, showing, accurately, our position by the sun; he would not believe it, and, had it not been that an American, who had lived in Matamoras for ten years, and came armed with a permit from Hon. W. H. Seward, Secretary of State, to trade, assured him on his word of honor, that the opening in question, *was* the Rio Grande, he never would have believed it, and the baleful stigma of deception, want of accuracy, and general untrustworthiness, would have been equally shared between the sun of our solar system and the son of my father.

The really good old gentleman has, however, long since passed away, and I hope that he reached the port for which he was bound, his course being marked out for him by a skillful and unerring navigator in whom he could have implicit trust.

I had a great deal of amusement in

BOARDING THE DIFFERENT VESSELS

lying at anchor off the port. The Englishmen would hospitably offer a "drop a sherry," proffering a brimming tumbler to carry out the idea; the Frenchmen showed their friendly feeling in cognac, and the German kindness overflowed in the shape of Schiedam schnapps, and Holland gin. One old German had a little old galiot, that looked as if it had served as a tender to old Von Tromp, when he carried the broom at his masthead in the English Channel; he had his wife, nurse and four children on board, cooped up in the little cabin, 10x12 feet, and they all seemed as fat and contented as if they occupied a *Schloss Unter den Linden*, Berlin. He was very hospitable, and was profuse in his offers of "yust a ledle more cherry cordial."

The appearance of a little German-American, shortly after, showed that the climate of Mexico—Texas, tempered by the cool breeze of the Gulf of Mexico, must have assimilated greatly with that of their own Fatherland.

We remained on the coast of Texas some two or three months, part of the time at anchor, and part of the time cruising off and on. Now and then the monotony would be varied by a terrible gale of wind called a "Norther," which put our ground-tackle or our canvas to a severe test, according as we were at anchor or under sail.

THE OTHER PRIZE.

One noon we saw a sloop standing in shore toward the mouth of the river. I started in pursuit with a twelve-oared cutter, and we had a neck and neck pull to cut him off before getting within range of the guns at the mouth of river. Fortunately for us there was not much wind, and we captured the sloop within a mile and a half of her destination. It proved to be from New Orleans, with a cargo for Matamoras. So, we transferred the cargo to our own ship, and sunk the sloop.

I believe I got some $25 prize money from the capture some ten years after.

Being short of fresh water, the commander resolved to go to the mouth of the Mississippi for a supply, and we accordingly got under weigh, and after a pleasant trip of a few days we anchored off South-west Pass in time to participate in the capture of Forts Jackson and St. Philip.[6]

Apropos of Mexico,

THEY TELL A STORY

of the captain of a brig at Vera Cruz who took a sailor who spoke Spanish on shore with him to interpret for him. The conversation was somewhat as follows:

Sailor—"*Habla usted Español Señor?*"

"*Si, Señor, perfectamente bien,*" replied the Mexican.

"*Bueno,*" said the sailor, "*in cuantas dias* can you make a new main yard for the brig?"

"*No entiendo*" (I don't understand), said the Mexican.

"*No en ten day?*" said the sailor.

"Oh, come on, captain, he says he can't do it in ten days."

Another linguist on shore, at the same port, came up excitedly to a native and asked:

"Look here, Señor, *ha visto usted* a *caballero* a cavorting down the streets on a derned big gray horse with a Mexican saddle on?"

"*No entiendo?*" said the native with a peculiar shrug of his shoulders pertaining to the race.

"*No entiendo?*" don't you understand your own lingo, you infernal Dago?

The boatswain of the U.S.S. *Portsmouth* was very profane, and showed a great deal of disgust at the agricultural aspect of many of the crew, really good men, but quite unused to a man-of-war. One day he apostrophized them on the foreyard somewhat as follows:

"Pick up that sail, will you?"

"No! pass in the leech first; that's no way to stow that bunt, Oh! you farmers!"

"Hold on with that bunt jig, will you?"

"Who in the d—l told you to pull up that bunt jig? My grandmother would make a better sailor than you."

"Look aloft; the devil would have been a sailor only he couldn't look aloft," etc.

One day hearing me hail the lookout aloft,

"Fore topmast crosstrees there,"

several times without reply, the boatswain who was standing on the forecastle said:

"That's a farmer up there, sir; he don't know that he's on the crosstrees, say haymow and he'll jump overboard."

---

[6] The battle dated 18-28 April 1862.

Excerpt 2 from *On a Man-of-War*[7]

LETTER XI.

THE *SCIOTO* OFF GALVESTON—WATCHING FOR A SAIL—ORDERED TO PASS THE BATTERIES—THE *HATTERAS* CHASES THE *ALABAMA*—AND THE *ALABAMA* SINKS HER—THE PAYMASTER'S OMELET—A STICKLER FOR RANK—HE FOUND THE MAN WHO FURNISHED THE CHEESE—A ROLLING GUNBOAT.

The *Scioto* (Figure 8) was for some six months on the blockade

OFF GALVESTON, TEXAS.

There was a good deal of rivalry between the gunboats as to which should report first that a sail was in sight. So in order to stimulate our mast-head lookouts to watchfulness I had given directions that if our lookout reported a sail *first,* he would be relieved at once, but if the lookout of a rival gunboat got ahead of him he was to be kept up there all day. Thanks to this competitive system, we were generally the first to report a sail.

ALWAYS READY.

The steam was kept low with the fires banked; the chain brought to the capstan, with the anchor just under foot, ready to trip at a moment's notice, and orders given, so that at the cry of,
"Sail ho!"
from the mast-head, the men sprang to the bars and commenced to heave round, the engineer spread his fires, the quartermaster bent on the signal number 1258, which telegraphed to the flagship,
"Strange sail to the eastward,"
and as soon as answered up went interrogatory 896, meaning,
"Can I give chase?"
The captain came up at once, and in *one minute* from the discovery, the ship was under weigh, standing out to sea.
One Saturday the commodore announced that on Monday, if pleasant, we would

ATTEMPT TO PASS THE BATTERIES

and enter Galveston; so we were all feeling correspondingly uncomfortable, not, of course, because we were afraid at all—in fact we were the original parties who were "longing for the fray," men of gore, whose trade was war and rapine, more particularly, perhaps, the latter. Still we were familiar with the reputation of the Texas riflemen as marksmen, and we knew there were a good many chances in favor of some losing the number of their mess, and leaving their families unprovided for; some of us were even not insured, so reckless had we become, inured to toil and danger as we were, bronzed by the tropic sun for, say, three or four months.
Well, we looked up our little matters, some of us had our hair cut; others hunted up their bibles, which were safely stowed away in their lockers, and all wore a pretty serious air, I

---

[7] Davenport, F. O. *On a Man-of-War: A Series of Naval Sketches.* Detroit: E. B. Smith & Co., 1878, pp. 107-137.

assure you. Sunday was a delightful day, and the prospect that Monday would be pleasant was very good, or rather, very bad, indeed.

About 4 P.M. we observed that

THE *HATTERAS* WAS UNDER WEIGH,[8]

with signal,

"Strange sail to the eastward,"

flying. I accordingly doomed our unfortunate lookout to stay up there until sunset for not seeing it first; as it turned out, I have forgotten whether he was ultimately rewarded for it or not. As the *Hatteras* steamed off in pursuit, having permission from the flagship so to do, she signaled,

"Strange sail suspicious,"

and later,

"Strange sail positively an enemy,"

soon after disappearing in pursuit. Let me say here that this last signal was not understood at the time. About dusk (say 7 P.M.), as the captain and myself were pacing the quarter deck, we noted flashes of heat lightning, as we thought, to the eastward, but soon a low thunder which followed led us to believe that it was firing of great guns. We timed the flash and report, and estimated the distance to be twenty miles, allowing 1,120 feet per second for the velocity of sound.

The captain at once took his gig and went on board the flagship, leaving me to get the ship under weigh. After some delay the flagship *Brooklyn*, the *Katahdin*, and the *Scioto* were standing out to sea, in pursuit. Well, we steamed until 1 A.M., and finding nothing the *Katahdin* and *Scioto* returned to their anchorage.

Figure 8. U.S.S. *Scioto*, off Galveston. The *Scioto* was an *Unadilla*-class gunboat, one of the "ninety-day gunboats" built for the blockade. *Naval Historical Center.*

---

[8] 11 Jan. 1863.

THE *HATTERAS* SUNK BY THE *ALABAMA*.

About 7 A.M. the next morning a white whale boat was seen approaching the ship from the shore. We watched it with considerable curiosity, as we supposed it to be another flag of truce from the "rebs." As it dashed up alongside it proved to be the *Hatteras's* gig. The acting master in charge came on board and touched his cap, and answered my question of,

"Where is the *Hatteras*, sir?"

with,

"The *Hatteras* was sunk at 7.30 last evening, sir, by the 290."

"Walk down in the cabin, sir, and report to the captain,"

I said, and giving directions to drop the boat astern, and let the crew come aboard to breakfast, I was obliged to await the result of his interview with the captain, anxious as I was to learn the particulars. The crew of the boat were the center of attraction forward, and our men soon knew as much as they did.

We at once got under weigh, and put to sea in pursuit of the *Brooklyn*. As soon as she saw us she came after us, under full steam, hoping that we were the *Alabama*, and seemed correspondingly disgusted upon discovering our true character. Upon comparing notes we found that the *Brooklyn* had picked up two of the *Hatteras's* boats, lashed together, with clothing in them stained with blood, and that she discovered the *Hatteras*, sunk in nine fathoms, with her mastheads sticking out of water, where she now is, I suppose.

A SHORT FIGHT, BUT A HOT ONE.

We afterwards learned that Capt. Blake was positive in his mind that the vessel he was approaching was the *Alabama*; he had his cutlass ground sharp, and determined to run down his enemy, far his superior in force, and carry him by boarding.

As he approached he hailed,

"What vessel is that?"

"Her Britannic Majesty's ship *Spitfire*," was the reply, at the same time running up the English flag.

"I'll send a boat aboard of you," said the *Hatteras's* captain.

"Aye, aye, sir,"

was the reply, and as the gig approached her the *Alabama* lowered her ladder, and showed a light over the side, it being just dusk. At this exciting moment the captain of the *Alabama* called out:

"This is the 290,"

and bang! went his broadside into the *Hatteras*, down came the English flag, and up went the Stars and Bars.

The *Hatteras* replied nobly and struck the *Alabama* twenty-two times with her shot and shell.

The shot of the *Alabama*, however, soon pierced her thin iron hull, and penetrating her boilers rendered her helpless, enveloped in a cloud of steam, the *Alabama* steaming round in a circle pouring in shot and shell.

Finding it impossible to do else, Capt. Blake surrendered and transferred his crew to the *Alabama* just in time to escape the going down of the *Hatteras*. The *Alabama* at once steamed off to the eastward, and plunging into the darkness was soon beyond pursuit.

Capt. Semmes remarked that he didn't want to fight any more men-of-war, as he suffered considerably with his fight with the *Hatteras*, but he did try it again off the coast of France, and was sunk by the United States steamer *Kearsarge* in a fight of two hours, escaping himself in the English yacht *Deerhound*, which by some peculiar notion of neutrality picked him up and ran off with him.

"It is an ill wind that blows nobody good," is a true proverb. By the loss of the *Hatteras* we were too weak to attack the Galveston forts, and the attack was postponed. Hence these tales.

AN EGG STORY.

One day our commander happened to be on board the flagship, when an officer came on board from shore with a flag of truce. While waiting to be shown into the cabin the officer recognized in him an old class and shipmate. They shook hands and gossiped a while, and, upon leaving the ship, the "reb" offered to send our skipper a couple of dozen of eggs. When the eggs arrived the commander sent a half dozen down into the wardroom to the paymaster. Well, Pay was delighted. Eggs was *eggs* just then, as we had lived on our rations for the last three months; so he bragged about his eggs until the rest of the mess were dissolved in envy. The following morning, at breakfast, an omelet was placed in front of the paymaster which certainly contained, at least, five eggs. The paymaster was furious.

"Steward! where's the steward?" he shouted.

While the boy went forward after the steward, Pay regarded the omelet gloomily, and coldly invited each member of the mess to take some. All declined, save two, who ate with great satisfaction the portion allotted them. Notwithstanding his evident annoyance, Pay commenced to eat some of the omelet, when the steward appeared with a covered dish in his hands.

"Steward!" shouted the paymaster, "what in the d—l do you mean by cooking all of those eggs at once? Besides, I told you I wanted 'em boiled."

"Here's your eggs, sir," said the unruffled steward, uncovering the dish and setting five eggs down on the table, "one of 'em was bad."

A smile broke over the face of the paymaster, and after finishing the omelet and offering the boiled eggs to each, he reached out and took one himself. He looked at it curiously, and with a muttered swear he dashed it down, and rising from the table, rushed on deck.

I did not understand, but managed to gather from the explanation furnished by one of the omelet eaters, interrupted by frequent laughter, that he and the other confederate had sat up half the night blowing out the contents of the eggs with straws through small orifices; the result of the blowing was made into omelet, and the shells being boiled filled with water, and for a moment deceived a person into taking one.

I was real glad that I did not take omelet with mine. The missing sixth egg was accounted for by the blower's stating that in his haste and fear of detection, he had dropped it on the wardroom floor. It might have appeared at breakfast, however, as a dropped egg, but didn't.

Every day or two after the paymaster would burst into the wardroom in a rage because some allusion to eggs had been dropped.

AN OFFICER OVERBOARD.

They tell the story of the eccentric old captain, now dead, formerly in command of the *North Carolina*. He was a martinet and very profane. On one occasion he fell overboard in crossing the gangway plank from the cobb dock to the ship. The sentry in the gangway promptly called out:

"Man overboard!"

"An officer, you blasted fool!" spluttered the captain, as he rose to the surface for the second time,

"An officer, sir."

NO CHANCE FOR THE WINNER.

The same captain bantered the executive officer of the ship into a wager to race with him, the former having, as he supposed, a crack boat and crew.

The race came off, but the irascible commander, seeing that he was being badly beaten, made the surrounding air blue with his sulphurous oaths, while executing a war dance in the stem of his boat, ordered the other to,

"Lay on his oars,"

and, upon their return to the ship, put the executive officer under arrest for disrespect to his superior and commanding officer in *daring* to pass the former without permission.

There is nothing to show that the captain ever paid the bet, but the ship's log records the fact that not long after the executive officer was transferred to a sea-going ship, where racing boats to win, could be more profitably engaged in with safety to the leading boat.

JUST THE MAN HE WANTED.

An old man-of-wars-man took his seat in a passenger car one day, attracting some considerable attention by his dress and manner. One of those meddlesome sort of people, described in that laughable book *On Wheels*, moved over, and took a seat alongside the sailor.

"In the navy, eh?"

The sailor nodded affirmatively.

"Well," said the interlocutor hesitatingly, "I am not *exactly in* the navy myself. I am a contractor—that is, I furnish cheese to the navy."

"Oh! you *are,* are ye?" said the sailor menacingly, "you are just the chap I've been looking for."

And accordingly he knocked the aspirant for naval honors over the car seat, and added, as he looked inquiringly up, and down the car,

"Now show me the son-of-a-gun that furnishes butter."

A ROLLING GUNBOAT.
"Twice ten tempestuous nights I rolled, resigned
To roaring billows and the warring wind."

The *Scioto* rolled terribly, when in the trough of the sea, to the great detriment of our crockery; though we always had sand bags lashed round the rim of the table to save the pieces. I have seen an officer vibrate between the table and the bulk-head holding a plate of soup in his hands, his chair slipping back and forth over the smooth oil cloth of the ward room floor.

One of our officers, returning from a visit, said, admiringly,

"That *New London* is a bully little steamer."

"Why?"

"Because she rolls so confounded fast that the dishes don't have time to slip off the table."

I have heard an old sailor yarn, where a schooner *he* was in, rolled *clear over* one night, and so easy, too, that they'd never have known it, but that every man had a *round turn in his hammock clews.*

Excerpt 3 from *On a Man-of-War*[9]

LETTER XII.

"Now on their coasts our conquering navy rides,
Waylays their merchants, and their land besets."

THE BLOCKADE OFF GALVESTON—A LITTLE BATTERY PRACTICE—"HE WHO FIGHTS AND," ETC.—IN A VERY SERIOUS PREDICAMENT—WHY THE CAPTAIN "SET 'EM UP FOR THE BOYS . . . ."

While on the blockade, off Galveston, the gunboats used to get under weigh at daylight, and run down to the flag-ship for company, returning to their stations just after dark. This enabled the officers to visit one another during the day, and tended to mislead the rebels as to where we lay during the night. Had we selected any particular anchorage, it would have been easy for blockade runners to have run in by a route far enough away from the gunboats to have escaped observation in the darkness; and again, a permanent anchorage might have enabled a ram to come out some pleasant, obscure evening, and sink a gunboat or two.

DRAWING THE ENEMY'S FIRE.

One afternoon we steamed slowly in toward Galveston, and threw some shell into the city, aimed at the captured steamer *Harriet Lane*, which lay at a wharf inside. We succeeded in having her towed away up the bay, and also succeeded in drawing the fire of the shore battery near by, as well as fire from Fort Point, some two and a half miles distant.

I remember looking through the glass, trying to see the battery, as located by the captain, when a shot came whistling just over us; and, do you know, that I could not get a focus on that glass to save me, and it was a good glass, too. The long shots from the batteries on the Point were

TRYING TO THE NERVES.

I assure you, on account of the time elapsing between the puff of smoke and the arrival of the shot; the time was probably only ten seconds, but if a fellow was dancing around you with a big club, and you were waiting for him to hit you most anywhere, you wasn't sure where, time would be time. A puff of white smoke would shoot out from the fort, and we knew that something was coming. After a while you would hear a murmuring sound, like the wind in a distant grove, growing deeper and fuller, until, like the blast of a hurricane, it rushed over and struck the water nearby, throwing a column fifty feet into the air, simultaneously relieving the suspended respiration of 150 sets of lungs, whose owners were earning their living literally by the sweat of their brows. Well, we just put our little helm a-starboard, and dusted out of that, a parting shot throwing the spray quite on board, and having the extraordinary effect of increasing the revolutions of the screw to a maximum.

After dark we got under weigh, to go to our anchorage for the night, and steamed off to the northward and eastward.

As we approached Fort Point, the captain thought he would explore the channel a little, and stood close in toward the fort. Suddenly, with an easy grating slide, the little steamer was

---

[9] Davenport, F. O. *On a Man-of-War: A Series of Naval Sketches*. Detroit: E. B. Smith & Co., 1878, pp. 119-123.

HARD AND FAST AGROUND.

As the fellow who asked for gape [sic] seed, in New York, would say, we backed her, and we backed her, and we backed her; and we rolled her, and rolled her, and rolled her; for two hours we worked to try and get her off, without success. We sent a boat to the flagship for assistance; got a heavy kedge out with a hawser to the capstan, and backed her and rolled her again, but to no purpose.

The captain then gave me orders to throw over the coal, and to lighten the ship as best I could. I asked for one more trial before throwing away coal that was worth, down there, $20 a ton in gold, and he consented to hold on a little longer. You can imagine that we were anxious to get out of there before daylight revealed our position to the batteries not a mile distant, and as the daylight would bring us certain demolition, we cast an anchor out of the stern of the ship, and *dreaded* the day.

OFF AT LAST.

I stationed every man in the ship along the starboard side, and made them a little speech, and at the order

"Rush,"

they rushed violently over to the port side; again the "rush" order, and back went every son of 'em, laughing, as if it were a good joke. Then we manned the capstan again and walked away with the hawser. The man in the chains quietly watched his lead, and reported no movement; the engines were backing all they knew how; the quartermaster reported quietly,

"The kedge is coming home, sir;" round went the capstan.

"Heave and walk him up, bullies," I said, "and we'll back the kedge and try again."

Just them the kedge tripped under the stern, the ship swung back to port and slowly moved; the imperturbable leadsman in the chains remarking quietly,

"She's going astern, sir;" and, sure enough, she was. As soon as I had her safely clear, I sent down word to the captain, and after we were safely anchored he embraced me warmly, invited us all down in the cabin and "set 'em up for the boys."

Excerpt 4 from *On a Man-of-War*[10]

LETTER XIII.

"He like a foolish pilot hath shipwreck'd
My vessel gloriously rigged."

A FRUITLESS CHASE OFF GALVESTON—MAN OVEBBOABD—A COLLISION ON THE MISSISSIPPI RIVER—HEAVE ROUND.

THE ROCKET.

One beautiful evening while lying at anchor on the blockade, off Galveston, a rocket was suddenly seen to seaward.

"Stand by to slip," called the officer of the deck;

"Boy, tell the engineer to spread fires,"

---

[10] Davenport, F. O. *On a Man-of-War: A Series of Naval Sketches*. Detroit: E. B. Smith & Co., 1878, pp. 128-137.

"Ship, go ahead, one bell," and in five seconds we were standing out to sea, all hands excited, and hoping that it would prove to be a blockade runner worth a *million* dollars. As we passed the *Itasca*, her commander called out:

"What is it, L——?"

"A rocket to seaward," replied the skipper, "come along."

"All right," was the reply, and soon the *Itasca* was steaming up alongside, and was shortly lost to view on our starboard bow.

The night was clear, with only a light breeze blowing, the sea was smooth, the moon shone brightly, and as I stood on the ship's rail, leaning on the boarding netting, I thought, well, there is a bright side to even blockading; what a lovely night, and how easily the little *Scioto* runs her ten knots an hour; no sea on, no motion; if we could only have it this way always. Suddenly the cry,

"MAN OVERBOARD,"

rudely dispelled my contemplation of the beauties of the sea, and the bright side of blockading.

"Stop her,"

"Three bells, a turn back;

"Away there, life-boat's crew, clear away the gig," I sang out as I jumped, and let go the life-buoy myself. The gig was lowered, the men sliding down the falls and tumbling into the boat.

"Pull right astern and keep the lights in range," I said, as, by the captain's orders, the quartermaster hoisted a light forward, and the gig plunged into the darkness, and was soon far astern. By keeping the ship headed as she was when the man fell overboard, and hoisting range lights forward and aft, the gig would, of course, be able to pull back exactly over the track made by the ship, and must pass close to the man, if afloat.

"Who was it?" was the question next asked.

"Old Rogers, sir, the gunner's mate," answered the captain of the forecastle; "he leaned against one of the pivot ports, sir, and it dropped down, sir, and he went overboard."

"Yes, sir," said the man in the chains, "and I hove the lead-line right between his hands, and he couldn't catch it, sir."

Pretty soon the gig returned with the life-buoy and the mournful report:

"We couldn't see nothing of him, sir."

"Hook your boat on. Lay aft to the gig's falls," was the order, and the men silently hoisted the boat, and off went the *Scioto* again in pursuit of the rocket.

True, she was short a hand, but then you know if the rocket proved to be an enemy, and showed fight, why we would probably be short more than one.

INSPIRATION FOR AN ARTIST.

I often think of that scene, and wish that a painter could have stood with me on the rail of that steamer—the painting of it would eclipse any marine view ever yet exhibited. The gig had just pulled off astern, the quartermasters standing on the stern rail of the ship burning Coston's signals, illuminating the sea for a mile round, and as the bright flame now green, now red, lighted up the eager faces of the crew, all crowding and gazing aft, alternately with a ghastly pallor and rosy light, the ship rising and falling easily on the long swell of the Gulf of Mexico, with every thread of rigging standing out bright on the dark background of the sky, it was a most beautiful picture.

HUNTING IN COUPLES.

The *Scioto* and *Itasca* were ninety-day gun-boats, or, as they were called, the 23's; they carried an eleven-inch pivot gun amidships, a twenty-pounder rifle Parrot on the topgallant forecastle, and four howitzers aft. By an arrangement with the *Itasca*, we were to hunt in couples, the *Scioto* fighting her starboard battery, and the *Itasca* her port battery. By this means we could pivot our heavy eleven-inch to starboard and carry it athwartships ready for use, instead of securing it fore and aft, as was the usual custom.

It was the duty of

THE GUNNER'S MATE

to see that the pivot ports were hauled up, and stopped only with a yarn, so that in case we had to use the gun suddenly, the ports could be instantly dropped and the gun fired at once. On this fatal evening, he had secured the ports as usual, reporting the same to the officer of the deck at eight P.M., and forgetting what he had done, leaned against one of them, which giving way, plunged him into the water; being an old man and encumbered in a heavy pea-coat, he was unable to keep up.

He had been in the navy all his life, and used to tell how he was on board the brig *Somers* when young Spencer was hanged for mutiny.

Well, we ran on until about midnight, and still saw nothing to explain the rocket. Up came the *Itasca*, and sneeringly asked:

"WHERE'S YOUR BLOCKADE RUNNER?"

so we concluded to just keep steerage way on and let the watch turn in, all hands having been on deck all the evening, and wait for daylight.

I lay down in my clothes, with my sword and revolver in my belt, and went to sleep. It seemed as if I had only about closed my eyes, when whir-r-r went the rattle, and I climbed on deck in a hurry, I tell you.

The captain met me at the head of the ladder, and with a stage whisper, led me forward by the pivot gun; I stooped down at his bidding and looked.

"What do you think of it?" he whispered anxiously.

"Think?" said I, "it's the *Great Eastern*."

Just then along came our old friend, the *Itasca*; we pointed out the steamer, and then both started for her, the *Itasca* on her starboard hand and we on the port. Well, the *Itasca* was the fastest and darted ahead. We approached the steamer cautiously, when, suddenly there was a rushing sound from forward, and the officer of the forecastle sang out:

"Hard a-port; quick's your play."

We jammed down our helm, when whiz went the *Itasca* past us, running about ten knots. She had made the circuit of the steamer, and coming back we narrowly escaped a collision that would have sunk one or both of us.

A DISAPPOINTMENT.

Well, our rocket was only a friend, another man-of-war, come to join the squadron, and he had concluded to anchor for the night when we discovered him. We all stood in for our anchorage, and passed close under the stern of the flagship about 7.30 in the morning. The commodore was on deck, and he hailed us with:

"Good morning, captain. What vessel is that?"

"The United States steamer *B—*," was the reply."

"How did you know she was out there, sir?"

"Saw her signals about 8.30 last evening; been out after her all night, sir," answered our captain.

As we steamed slowly over to our anchorage, I saw the commodore pulling some hair out of his chin whiskers and gesticulating to an unhappy quartermaster, and I thanked my stars that I didn't have the first watch last night aboard the flagship.

ON A SINKING GUN-BOAT.

One bright June day the *Scioto* entered the Southwest Pass, Mississippi River, and taking a pilot, steamed up the river, bound for New Orleans for repairs, and a hospital for me.

I was lying in a cot just under the ward-room hatch outside my state-room. All hands were pleased to hear the news of the fall of Vicksburg, and looked forward to a pleasant visit to New Orleans after several months' blockading.

As we steamed steadily along, I became aware that we were approaching or meeting a steamer coming down the river. I heard the contradictory orders of the pilot,

"Starboard,"

"Port,"

"Steady,"

some confusion, and then an easy grating sound and motion exactly like the gentle glide of a boat upon a sloping beach. The descending steamer struck us just abaft the fore chains, cutting into us clear to the keelson. Soon I heard the master-at-arms come aft, and report,

"Five feet of water on the berth deck, sir;"

someone else cried out,

"We are sinking."

The engines were started ahead again, and the ship was run ashore; the engineer came up and reported:

"The water is over the fire-room floor, sir,"

again,

"The fires are out, sir."

The engine pegged away a few minutes longer and then slowly stopped, and the impassible [*sic*] chief engineer came up and reported:

"The engines have stopped, sir."

(We had a fussy sort of officer at the Naval Academy, when I was a midshipman, and the cadets said he liked the evening gun because it always *reported* when it went off.)

The doctor then came down, and had me carried on deck in my cot, and put into one of the quarter boats. (They were about to lower the boat with the crew in her, but I knew that the eye bolts in the bow and stern would not hold, and I made the men get out, as I feared a plunge in the Mississippi would not help me any in my somewhat weak condition.)

I was put safely on board the steamer which ran into us, which turned out to be the *Antona*, a captured blockade runner, now a store-ship. About dark a tug came along, and I started for New Orleans where I arrived at daylight the next morning.

The *Scioto* sunk to her spar deck, and the men and officers spent the night on deck with the mosquitoes, being taken off and sent to New Orleans the next day. The mosquitoes are so large on the banks of the lower Mississippi, that they may be killed with a shot-gun, sometimes.

I had quite comfortable quarters in the Army Hospital at New Orleans, it being previously the St. James Hotel.

I received many calls, and presents of delicacies, from the ladies of New Orleans, though very Confederate in their sentiments, and spent three weeks there very comfortably. The Confederate ward was just above me, and was well filled with wounded "rebs" from Port Hudson.

As the visitors to the Confederates had to pass my door, I made many acquaintances among them, and as I have said received considerable attention.

IF YOU CAN ONLY KEEP IT.

While lying at the New York Navy Yard one of our men, a captain of the foretop, was returning from liberty and coming down the wharf, bound for the ship; the night being dark, he ran into a big man in a heavy coat, who was coming in the opposite direction.

"Do you know who you are running into?" said the stranger.

"No, I don't," said the inebriated son of the sea, "and, what's more, I don't care a continental."

"Well, sir, I am Admiral ——, and I am in command of this yard."

"Well, admiral," said the unabashed man-of-wars-man, while remembrances of the ups and downs in his own checkered career on board ship dashed through his mind, "that's a mighty good billet if you can only *keep it*."

HEAVE ROUND.

An old lady passing along the dock saw some sailors on board of one of the lake schooners heaving up anchor. The anchor was up to the hawse-hole, but the men not noticing it, continued hauling, with a,

"Yah *heave* oh."

"Well!" said she, "you may 'Yah *heave* oh' just as much as ye like, but if you pull that crooked iron through that little hole in a hurry, I'm mistaken."

Excerpt from *Forty-Five Years Under the Flag*[11]

CHAPTER III

WITH FARRAGUT IN THE GULF—ORDERED TO VERA CRUZ
1861–1862

The *Niagara* (Figure 9) was ordered to New York, where stores, coal, and ammunition were taken to her full capacity and a few needed repairs made to boilers, engines, and pumps. The orders received there at first contemplated the defense of the gateway to the capital by placing the *Niagara* off Annapolis. One or two members of the famous Seventh Regiment of New York, left behind when that excellent regiment had gone to the front, came on board for passage. But the necessity of establishing the blockade of Charleston under the President's proclamation being regarded as paramount, the *Niagara* proceeded off Charleston, and on May 12th established the blockade of that port.

In the old days of sail Charleston was a great cotton port, and the offing was filled with vessels bound in and out. Under the law of blockade, the *Niagara's* duty on arriving before the port was to board all vessels bound in and to endorse the fact of the blockade of the port on the ship's papers[12] and to warn them off the coast. A very busy day was spent in doing this duty

---

[11] Schley, W. S. *Forty-Five Years Under the Flag*. New York: D. Appleton and Co., 1904.
[12] With the notice of blockade entered by handwritten note on the ship's papers, the next time she was captured by the U.S. Navy in trying to run the blockade, she would be a lawful prize since her captain could not claim ignorance of the blockaded status of the Confederacy. See the interrogatories put to the *Alabama* officers below, Chapter 6.

on arrival, as the *Niagara* was the only ship on the station. The great desire of some of the masters bound inward to communicate by signal their arrival to agents in Charleston emboldened some to attempt to do this after having been boarded and warned off; or, when the *Niagara* had steamed some distance north or south to overhaul others bound in, occasional attempts were made by others to gain entrance, although they had been already warned. A shell across their bows generally reminded such masters that it was dangerous to continue.

One vessel, however, the *General Parkhill*, whose master was a skillful navigator, and a courageous Marylander, persisted, after being warned, in the attempt to run the blockade anyhow, and but for the fact that the wind was light he might have succeeded. One or two shots were fired across his bow and one over his vessel. It was realized that his ship was in great danger, he concluded to haul his wind and head off shore, but it cost him the capture of his ship. The writer, then a young midshipman, was detailed, with a prize crew of twelve men, to take the *General Parkhill* to Philadelphia. The master, the first and second mates, as well as the crew, were left on board. After taking charge of the *Parkhill* a thorough search of the ship was made for arms or explosives without result. The first search made had secured everything with which injury might be done to themselves or to others.

The fact was discovered, however, that the vessel was without American colors and that her master had determined to force the blockade if he found that to be possible on reaching the offing of the port. A large Confederate flag, found on board, indicated the master's sympathy with the Confederacy, and explained his determined action in attempting to gain the harbor entrance before his vessel should be overtaken.

\# NH 75895   USS Niagara off Boston, Massachusetts, in 1863, after modification

Figure 9. U.S.S. *Niagara*, steam frigate, 1863. *Naval Historical Center.*

It was decided from the outset of the voyage to Philadelphia to employ the prize crew, armed, on deck at the wheel and in handling the rigging and as guards and sentries, and to employ the ship's crew aloft and in handling the sails. This arrangement safeguarded the ship from recapture, which was further provided against by confining the officers to their rooms under armed guard. Against this latter expedient the master protested, though ineffectually, until after the ship had reached Delaware Bay and was taken in tow by the tug *America,* bound for Philadelphia, where she arrived about the last week in May and dropped anchor off the old Navy Yard, then commanded by Captain S. F. DuPont, to whom the orders given to the writer had directed him to report.[13]

This vessel was the first square-rigged prize captured during the Civil War. There was, at this early period of operations, some confusion and uncertainty about the forms of law governing prize cases and the legal methods of dealing with vessels so indicted. Feeling ran high with the tap of drums, the tramp of troops moving to the front, and the enthusiasm of loyal attachment to the Union, all of which suggested the thought, at times expressed in the press, that masters and crews of such vessels ought to be classed as pirates. This idea, however, soon vanished as the number of captures increased and the dockets of the courts of admiralty were filled with cases. It caused a number of uncomfortable hours to those whom the laws of war had brought before the courts, and some anxiety to those who had to search the records of our earlier wars for precedents. It ended, however, in releasing the crews after taking their evidence and examining the log-books or other papers of the ship, and in holding the ship and cargo only under condemnation and sale.

These forms having been observed by Marshal Millward of the District of Pennsylvania, to whom the ship had been delivered in due process of law and precedent, the writer was detached and ordered to the *Keystone State,* Captain Gustavus Scott commanding. Before she could be fitted out orders detaching the writer were received, and a short leave of absence was allowed. This time was passed in Frederick with his family and friends. In July following the writer was promoted to the grade of Acting Master and ordered to the frigate *Potomac* as navigating officer.[14] On August 31, 1861, he was promoted to master in the line of promotion.

---

[13] Herewith are the orders under which the writer had undertaken this work:
U.S. STEAM FRIGATE *Niagara,*
Off CHARLESTON, May 12, 1861.

MIDSHIPMAN W. SCOTT SCHLEY.

SIR: You will take charge of the ship, *General Parkhill,* this day captured as the property of the citizens of South Carolina, and proceed with her, under your command, to Philadelphia. You will be very vigilant to prevent any attempt at recapture, and to that end keep *your own men on deck* and employ the others aloft. On your arrival you will deliver the ship to the Marshal of the United States together with the papers and two Palmetto flags, found on board, which are now put in your charge for that purpose.

You will report your arrival to the commanding naval officer.

Wishing you a happy voyage and safe arrival, I am, Respectfully,
Your obedient servant,
WM. W. MCKEAN, *Captain.*

[14] Officers of the *Potomac,* 1861:

Captain, L. M. Powell.	Midshipman, D. D. Wemple.
Lieutenant, Saml. Marcy.	Master's Mate. Kane.
Lieutenant, L. A. Kimberly.	Master's Mate, Cressy.
Lieutenant, George Law.	Surgeon, T. D. Miller.
Lieutenant, A. V. Reid.	Asst. Surgeon, G. R. Brush.
Master, W. S. Schley.	A. A. Surgeon, A. O. Leavitt.
Master, W. T. Sampson.	Paymaster, J. D. Murray.
Acting Master, W. N. Wood.	Captain, G. W. Collier, USMC.
Acting Master, Jerry Smith.	Boatswain, C. A. Bragdon.
Acting Master, E. D. Bruner.	Gunner, W. H. French.
Acting Master, David Magune.	Carpenter, O. F. Stimson.
Acting Master, George Wiggins.	Sailmaker, Geo. Thomas.
Midshipman, Merrill Miller.	Captain's Clerk, Bradley.
Midshipman, C. H. Humphrey.	

The *Potomac* was fitting out in New York. The captain was an officer of high character and a valiant Virginian, Levin M. Powell. The needs occasioned by a large increase of ships, officers and men were met by purchasing everything in the open market that could turn a wheel or hoist a sail. The need of officers and men was supplied from the large number of those in the merchant service that were thrown out of employment.

As the white wings of our commerce in those days were seen on every sea, the nation availed itself of this resource from which to draw many skilled officers and men to its service; and it can he said that scores of these good sailors rendered incalculable service to the navy in its great work for the country during four long, weary years of blockading the coasts and reducing the fortified places on them, from Cape Hatteras to the Rio Grande. The history of this meritorious service, with its experiences, its hardships, its privations and its unceasing perils, will live forever in the song and verse of a grateful people.

A number of these gentlemen were appointed to the *Potomac*. They were good and experienced sailors and ready in learning the drills, discipline, and routine of the Navy. They served with merit and distinction until the war ended, when they returned to their former calling in the merchant service.

The *Potomac*, with an excellent crew of officers and men, sailed in August, 1861, bound for the Western Gulf blockading squadron, then commanded by Commodore McKean. Soon after her arrival Flag Officer D. G. Farragut was assigned to this important command. It was surprising in those days to observe how accurately the men knew and gauged their officers. It often happened in the long hours of a watch that the deck officer would consult with the quartermasters, always old and experienced seamen, about the weather or matters touching the qualities of the ship, etc. In one of these confidences James Barney, an old and competent quartermaster, said that "the men for'd had heard that the commodore (McKean) was ill and had to be sent home." Almost immediately he volunteered the suggestion that if he had anything to do about it, he "would pick out Cap'n Davy Farragut" to take his place. He added that if "Davy Farragut" came down there, "it wouldn't be long till the fur was a-flying." Captain Farragut did come down to relieve Commodore McKean, and Barney's predictions were verified in a short time afterward, for Farragut showed himself, in all that followed, to be one of the greatest and grandest of American captains.

The *Potomac*, on reaching the station, was assigned to the blockade of Pensacola, and was present on the occasion of an attack upon Colonel "Billy Wilson's" regiment in camp outside Fort Pickens by a force of Confederates landed during the night on Santa Rosa Island. A piece of artillery and a company of blue jackets from the *Potomac* were landed, under the writer's command, to the eastward of the Confederates to cut them off. This hastened the abandonment of the island by the Confederates, who had been roughly handled by Fort Pickens and Wilson's regiment combined.

When Flag Officer Farragut joined his command in the Gulf, the *Potomac* had been attached to the blockade off Mobile. Her position was in the main channel, about four miles from Sand Island Lighthouse, where she lay at anchor for several months. It was wondered often, as she lay helpless in calms, why the Confederate steamers did not venture out for a shot or two with their longer range artillery. It was a fact at that time that the *Potomac's* battery consisted of long 32s and 8-inch smooth-bore shell guns, with two 20-pound Parrott rifles. But as this fact was not known to the Confederates, the *Potomac* lay undisturbed or unchallenged, except by winds and waves, which now and then gave the old ship many an uncomfortable night during this long vigil.

From time to time the monotony of blockade duty was broken by some vessel attempting to elude the squadron's vigilance. The excitement of chasing or that of advancing to attack any unfortunate vessel that had run aground under the guns of Fort Morgan relieved some of the tedious hours of this wearing duty.

As the war grew apace, vessel after vessel arrived to reinforce those guarding the entrance to Mobile Bay until the fleet contained from twelve to fifteen vessels, the larger proportion being steamers. The custom was to mask all lights at night and to take up the night position after sundown nearer the beach or closer in to the channels and just before daylight to drop off shore out of range. Many unsuspecting blockade runners ran into the web thus woven.

On the night of December 26, 1861, a schooner bound out before a fine northerly wind was forced ashore under the guns of Fort Morgan and was discovered at daylight. Signal was made at once to the *Water Witch,* Commander A. K. Hughes, commanding, to go in and destroy her, if possible. The writer asked authority to go in the *Water Witch* with two boats for any duty Commander Hughes might require. As the *Water Witch* closed in on this schooner, using her rifled gun, the guns of Fort Morgan took up the gauge and returned the fire with a long range gun, assisted by another gun east of the fort. Though many projectiles struck near the boats and the *Water Witch,* many others passing over and beyond them, neither was struck. When a point had been reached where the shoal water prevented the *Water Witch* from approaching nearer to the schooner, the boats were manned and a dash made for the vessel, but before they could reach her neighborhood her crew set fire to and abandoned her near Fort Morgan. The *Water Witch,* as well as the boats, remained in the neighborhood of the schooner under fire for quite near an hour, until she was completely destroyed. A number of shells landed very near the boats, but none was accurately enough aimed to strike.

The object of the commanding officer having been fully accomplished, the *Water Witch* and boats withdrew. This was the first instance in which the writer had ever been under fire. The sound of projectiles whistling over the vessel or boats was an entirely novel sensation. The jets of water thrown into the air as the shells struck the surface might have been more beautiful under other circumstances, but the greater danger as they came nearer and nearer suggested that the boats turned bows on to the fort presented a smaller target, and thus minimized the chances of hits; and so this experience proved of value then.

In the month of January following (1862), a brig was discovered by the steamer *R. R. Cuyler,* Commander Francis Winslow, near the beach, some twelve miles to the eastward. Not long after the *Cuyler* had reached this brig's locality, heavy firing of guns was heard. The senior officer present in command called the *Huntsville,* within hail, to direct her to proceed to the *Cuyler's* assistance, as that vessel was some miles away and out of signal distance. The writer requested permission to accompany the *Huntsville* with two boats armed for near service. It proved later that these boats were of much service to the *Cuyler's* commander. On nearing the *Cuyler* she was found lying stern to the beach, distant some 250 yards, disabled by the parting of a hawser which her own boats had carried to the brig to pull her off the beach. In parting, the end of the hawser whipped back, was then taken up by her propeller, and a number of turns were taken around its hub, between the propeller and stern-post, completely bringing her engines up. At the same time her two boats were broadside on to the beach in a light surf, with all hands in them wounded save one or two by the rifle fire of the coast guard.

As the assisting boats approached the *Cuyler* to report for orders, her gallant commander hailed the officer in charge, stating that he would not order anyone in under such a fire, as all hands in his own boats were probably killed; but if the boat officer would secure the boats from the surf near the beach he would perform an eminently important public service. A dash was made at once through this fire from the coast guard defending the prize, though not without some loss, a sergeant of marines, the coxswain and one or two men in the assisting boats being wounded. The *Cuyler's* boats were rescued with their dead and wounded, and, though riddled with bullets, were towed back to the *Cuyler,* much to the gratification of her noble commander, who complimented the service highly.

Reaching the deck, Commander Winslow was found on the quarter-deck, and though under heavy musketry fire from the riflemen on the beach, received the report of the rescue of his boats. He remained until the officers of his ship had cleared the propeller by using a cutting spade from a boat under the *Cuyler's* stern, this boat being protected by the boats from the *Potomac* and the battery of the *Huntsville,* Commander Cicero Price. The second hawser from the *Cuyler* was carried to the prize by the *Potomac's* boats, and toward sundown the prize was pulled off the beach and captured. The assisting boats, with the *Cuyler* and *Huntsville,* returned to the blockade of Mobile about 8 P.M., and the wounded men of the *Potomac's* boats. were cared for most tenderly. The prize proved to be the brig *Wilder.*

In the decree of the prize court afterward, at New York, through some error made on board the *Potomac* or in transmitting the prize list, the writer's name and those of the assisting crews were omitted, so that although this capture had been made possible by the work of the *Potomac's* boats, they did not participate in the proceeds of the prize.

For some weeks this dreary and wearisome blockade was maintained. The lack of exercise and nutritious food was felt by officer and man, and scurvy, that dreaded pest of ship-life in the olden times, was only avoided by the occasional relief which came to them afterward from the steamers bringing supplies of fresh meats and vegetables in amounts about enough for two or three days. The diet for the rest of the month was composed mainly of salted meats, cheese, hard bread, bad butter, inferior coffee and positively bad tea. It is indeed a wonder that the efficiency of the personnel was maintained at all under such conditions.

In the early part of 1862 news reached the squadrons blockading our coasts that a large fleet, consisting of English, French and Spanish war vessels, with a division of the French army, had descended upon Vera Cruz, Mexico, with the ostensible purpose of collecting debts due to the subjects of each of those countries from Mexican merchants. The *Potomac* was selected to proceed to Vera Cruz to ascertain the purpose of this expedition, so far as that might be possible, on the spot.

On her way to Vera Cruz the *Potomac* looked in at Pass à l'Outre, one of the several mouths of the Mississippi River, to inquire from the *Vincennes* the circumstances of the death of Lieutenant Samuel Marcy, who had recently been transferred from the *Potomac* to the command of the *Vincennes.* Marcy had been killed some days before in a boat attack upon a vessel attempting to force the blockade. After this information had been obtained, the *Potomac* proceeded on to South-West Pass to fill her tanks with water. While doing this, she was attacked by a river steamer, which was driven off after firing about a half-hour.

Vera Cruz was reached early in February. A large fleet of English, French, and Spanish war vessels was found anchored in the harbor. Among them were such vessels as the line-of-battle ship *Donegal, Imperieuse, Guerriere,* and many others of smaller class. The combined forces were commanded by such distinguished officers as Commodore Dunlop, Admiral Graviere, Marshal Prim, and General Lorenzes. Preliminary to the inquiry into the purposes of such a fleet in adjacent waters, the customary salutes and courtesies were exchanged with the several commanders.

2

Union Navy Reports

The views of Union captains and squadron commanders about the general situation and individual incidents were reported up the chain of command. More formal that the memoirs of the last chapter, the reports were equally filled with details providing context. They were helpful in addressing the general question of "What was the Union navy up to off Galveston and the Texas coast?" The reports were available in the *ORN* but were gathered here for *Denbigh* context.

Report of Union Commander at Galveston
upon First Establishing the Blockade.[15]

The *South Carolina* was the Union ship that first arrived off Galveston to impose the blockade. Captain Alden, her commander, had a unique story to tell. His experiences off this port provided a baseline for the development of more effective practices later in the war. Galveston was at the far end of a lengthy supply line. Providing the blockaders with maintenance, personnel, coal, and fresh food and water was a continuing headache throughout the war.

Hon. SECRETARY OF THE NAVY,
Washington, D. C.

[Enclosure.]

U.S.S. *SOUTH CAROLINA*,
Off Galveston, July 8, 1861.

SIR: I have the honor to report that I arrived here on the 2^d instant, and immediately hoisted a signal for a pilot for the purpose of communicating with the shore. In a short time a pilot boat came off, bearing a flag of truce and a document, a copy of which is herewith sent. In reply to it I enclosed a copy of your declaration of blockade, with a single remark that I was sent here to enforce it, which I should do to the best of my ability. The next morning, the 3^d instant, at daylight, a small, high-pressure steamer evaded our vigilance, and under cover of a thick rain squall slipped in, but, on the following day, the 4th instant, we succeeded in capturing six schooners; on the 5th two, besides one ran ashore by the gunboat; on the 6th one, and on the 7th one, making in all eleven sail destroyed or captured. A descriptive list is herewith sent. It will be seen that they are all rather small, and, of course, with cargoes not very valuable. Out of the whole there are but three of them coppered, viz., one called the *Shark* and the two pilot boats. On the 5th instant it was represented to me that some of our prizes were worthless. I thereupon caused a survey of them to be held, when three–the *Venus, McCanfield,* and *Louisa*–were condemned as unseaworthy and unfit to lie even at anchor in any seaway. At this time, my prisoners being increased in number to 52 (an inconvenient crowd to handle with our small crew, which is very much reduced by manning the prizes), I determined to put them on board the condemned vessels and send them in, which was accordingly done. To-day (the

---

[15] *ORN*, ser. 1, 16:576-577.

8th instant), I have discharged the *Coralia,* another worthless one, into one of the other prizes, and shall send 33 more prisoners in her to Galveston this afternoon. This leaves me six vessels with four cargoes to dispose of, the latter being worth about $20,000.

The two pilot boats being empty, one of them I have made a cruiser, and the other I send to you with this dispatch. Among other things captured we have 13 mail bags and 31 bags containing express matter. Be pleased to inform me what I shall do with these last. Might I not let the private citizens have their letters after giving them a thorough overhauling, and extracting anything of a public nature? Am I right in turning over such poor craft as I have to their penniless owners by using them as cartels for prisoners, or ought I to destroy them and retain the captives?

This activity incident to chasing our prizes in detail has, of course, caused a greater consumption of coal than was contemplated, and we have not now more than seven days' fuel on board; this, however, I can economize very much by lying still and keeping the two gunboats constantly underway and only chase when they are unable to bring a vessel to. I would suggest that a steam frigate, or some of that class, even a sail vessel would do, to anchor off the entrance to Galveston, where she can ride safely all the year round and serve as a nucleus for the rest of us to center about. The vessel, with the two gunboats, the *Aid* and *Dart,* together with another of our prizes which I contemplate taking into service, called the *Shark,* a vessel of light draft and strongly built, if armed with a 24-pound howitzer, would be most efficient. These sent to cruise to the south as far as the Rio Grande and the east as far as the South West Pass of the Mississippi, would, I believe, most effectually close up every hole and corner along that line of coast.

If a vessel cannot be sent now, I shall require at least 25 men and 2 officers, which will be no more than sufficient to fill vacancies made by manning prizes, and if you approve the plan of arming the *Shark;* please send me in addition to that number 12 more men and a 24-pound howitzer. The *Dart* I shall arm with one of the 12-pound howitzers obtained from the *Powhatan* as soon as I can get a slide. I send you some stirring news from the North of a very late date, Washington, the 4th instant. I write in great haste and am compelled to omit many things which perhaps you ought to know.

But I am sure the bearer of this, our master, Mr. [Rodney] Baxter, will be able to fill up any omissions, and will, of course, give you in detail some interesting facts, which I have been barely able to touch upon. He is ordered to make the best of his way to Pensacola and report to you, and I will venture to hope that as he is my right bower you will be pleased to return him to me at the earliest practicable moment.

With great respect, I am truly, your most obedient servant.

JAMES ALDEN,
Commanding U.S.S. South Carolina.

Flag Officer WM. MERVINE,
Commander in Chief U.S. Gulf Squadron.

Report to Secretary of the Union Navy Regarding the Vessels and Disposition of the West Gulf Blockading Squadron

This section presents a series of reports that were particularly helpful in understanding the Union navy resources off Texas. The dates were Nov.–Dec., 1864, and coincided with the beginning of the *Denbigh's* runs from Havana to Galveston and back. The reports provided important context and first-hand details of U.S. Navy activities countering the blockade-runners. The runner traffic picked up many fold after the fall of Fort Morgan at Mobile Bay in early August 1864.

Stations of vessels composing the West Gulf Blockading Squadron, November 1, 1864.[16]

[No. 520.]

FLAGSHIP *HARTFORD*,
Mobile Bay, November 4, 1864.

SIR: I have the honor to report the following as the position of the vessels of this squadron on the 1st instant, viz.:

At the East Pass of Santa Rosa Island.–Steamer *Bloomer*, sailing vessel *Charlotte*.

In Pensacola Bay.–Steamers *Cayuga, Genesee, Kanawha, Kennebec, Octorara, Pinola, Sebago*; tugs *Jasmine* and *Pink*, and sailing vessels *Potomac, W. G. Anderson, Kittatinny*, and *J. C. Kuhn*. Most of these steamers are repairing.

In Mobile Bay.–Steamers *Hartford, Chickasaw* (ironclad), *Lackawanna, Manhattan* (ironclad), *Metacomet, Monongahela, Owasco, Port Royal, Richmond, Rodolph* (tinclad), *Selma, Stockdale* (tinclad), *Winnebago* (ironclad); sailing vessel *M. A. Wood;* tugs *Althea, Buckthorn, Cowslip, Narcissus,* and *Tritonia*.

In Mississippi Sound.–Steamer *J. P. Jackson* and tug *Rose*.

At Ship Island.–Sailing vessel *Vincennes*.

In Lake Pontchartrain.–Steamers *Elk* (tinclad), *Fort Gaines*, and sailing yacht *Corypheus*.

In the Mississippi River.–Steamers *Antona, Arizona, Arkansas, Conemaugh, Estrella, Meteor, Mobile, Milwaukee, Oneida, Princess Royal, Tennessee* (ironclad), and *Virginia*; sailing vessels *Fearnot* and *Portsmouth*. Most of these steamers are at the Quarantine station.

At South West Pass.–Sailing vessel *Pampero*.

In Berwick Bay.–Steamer *Carrabasset* (tinclad), *Glide* (tinclad), *and Nyanza* (tinclad).

At Calcasieu.–Steamers *Aroostook* and *Penguin*.

Off Sabine Pass.–Steamers *Gertrude, New London*, and *Pocahontas*.

Off Galveston.–Steamers *Ossipee, Bienville, Chocura, Cornubia, Katahdin, Kineo, Pembina, Sciota,* and *Seminole*.

Off the Rio Grande.–Steamer *Itasca*.

Off Brazos Santiago.–Steamer *Penobscot*.

The *Augusta Dinsmore* has been employed as a dispatch vessel between New Orleans and the coast of Texas, the *Glasgow* between Mobile Bay and New Orleans, and the tugs *Ida* and *Hollyhock* between New Orleans and the Passes [of the Mississippi].

Since my last report the *Galena* has been transferred from this squadron to the East Gulf, and the *Milwaukee* has joined this squadron from the Mississippi Squadron.

Very respectfully, your obedient servant,

D. G. FARRAGUT,
Rear-Admiral, Commanding West Gulf Blockading Squadron.

Hon. GIDEON WELLES,
Secretary of the Navy, Washington, D. C.

Memorandum of Commander Le Roy, U.S. Navy, giving, list of vessels belonging to Third Division, West Gulf Blockading Squadron, with stations, conditions, and remarks.

[Memorandum left for the guidance of my successor.][17]

---

[16] *ORN*, ser. 1, 21:712-715.

The blockading points at present are:

1st Off Calcasieu, two vessels being found necessary, and is of great importance.

2d Off Sabine Pass, two vessels being found necessary, and is of great importance, the *Granite City* and other vessels being there. It is a much more quiet anchorage than off Galveston, and I purposed making it the principal coaling station, keeping only one vessel off Galveston to supply immediate wants.

3d Off Galveston, the principal point, and requiring a number of vessels. When desiring to communicate by flag of truce, a white flag hoisted on board ship is responded to by a white flag hoisted on a large building in Galveston, near the west end of the town. It has been the custom to make the beacon by the main channel the place of meeting, our boats not going much inside, and the enemy's not coming much outside of it. Brigadier-General J. M. Hawes commands the defenses of Galveston, and in all my official intercourse with him and the flag-of-truce boat I have found the utmost courtesy. I have directed the vessels of the blockade not to approach within range of any of the batteries on the beach, as they are said to have a long-range gun, and the property in Galveston being owned by people at the North, or who are loyal, I did not wish to be compelled to return the fire and probably injure the property of those friendly to the Union.

4th San Luis Pass and Velasco. From the paucity of vessels, I have been compelled to place both of these passages under the surveillance of one gunboat. They are important, being used by the light-draft vessels.[18]

5th Velasco to Pass Cavallo. Could only be cared for by an occasional passer-by and the vessel stationed to guard San Luis Pass and Velasco on the one side; and,

6th Pass Cavallo, Espiritu Santo, and Aransas Pass. Pass Cavallo and Espiritu Santo are considered important points, and I have had a vessel stationed there when I have had a sufficient number at my command.

7th Brazos Santiago. A vessel is especially directed to be kept off Brazos Santiago, where we have a military station, the only one now on this coast. I believe the vessel has been ordered there on account of some irregularities on the part of some of the military, who had gone to work shipping cotton or something else other than their legitimate occupation. As far as assistance the army or troops might require, that could be easily given by the vessel stationed off the Rio Grande.

8th Off the Rio Grande. Owing to evacuation of Brownsville the blockade of the Rio Grande has been resumed; but our vessels accomplish nothing, Matamoras offering all the facilities for contraband goods the enemy may require and as an avenue for the exportation of cotton from Texas.

9th Rio Grande to Tampico. The *Princess Royal,* Commander Woolsey, as I was advised by Admiral Farragut in a communication received on the 1st of November, is to cruise between the Rio Grande and Tampico, and the admiral directs, when necessary, etc., that one of the vessels of this division relieve him. The order upon the subject is among the archives of the division.

Divisional.

The coal and oil reports are usually sent in before the end of the month to the divisional commander, who prepares an estimate of the quantity that will be required for the ensuing month, and a report is made to the admiral and one to the commanding naval officer at New Orleans.

---

[17] Enclosed with the above Farragut letter.

[18] It was important to note that there were insufficient Union ships efficiently to blockade a huge section of the Texas coast. Five entrances did not have full time coverage by a blockader on station. That left many opportunities for safe entrance and departure. Three of the entrances were suited to smaller vessels such as schooners. Two were suitable for steamers.

The *Arkansas* and *Augusta Dinsmore* are considered as regular supplies from New Orleans, but they are irregular in their coming.

All prizes are sent to New Orleans for adjudication.

The guard flag, with the meal (red) pennant underneath, I have been in the habit of hoisting when there was a chance to send letters, or letters were received for any coal vessel lying here, having an understanding, with the captains who would answer the signal by hoisting their ensigns.

Blockade runners, steamers.

Watson.

Triton.

Marie or *Maria.*

Fannie or *Fanny.* American built; walking beam; said to be fast.

Denbigh or *Danby.* An old Mobile blockade runner; said not to be very fast, but a most successful vessel; English build.

Susanna. A small, fast, light-draft steamer.

William Curry, formerly engineer of the *Isabel,* is said to be on board the *Watson.* He is reported to have violated his parole and escaped from guard ship at Philadelphia.

Captain Blakesly, formerly of the *Cumberland,* made his escape under similar circumstances and at the same time with Curry.

The American steamer *Ike Davis,* of and from New Orleans, was captured by her passengers a month or two since, in September or October, after leaving the Rio Grande and taken into Texan waters. It is supposed her capture did not occasion her owners much regret.

All trading vessels in this region are, I think, a fit subject for suspicion, there being too many of our rascally countrymen who care but little about our nation's welldoing so long as gold comes into their coffers.

A small steamer, the *Susanna,* is rather successful in passing in and out of Galveston. She is of very light draft, and I believe can pass in and out by San Luis Pass. She is fast.

Order for court-martial of persons on board the *Kineo* and *Bienville* could not be attended to, owing to the absence of those vessels, etc., and bad weather.

Report of Commander of Union Blockade of Texas, Nov. 1864, Concerning the Poor Condition of His Ships in the Early Days of the *Denbigh's* Runs.

Report of Commander Le Roy, U.S. Navy, commanding Third Division, regarding the impaired condition of the vessels of his command, and referring to the capture of British schooner Albert Edward.[19]

U.S.S. *OSSIPEE,*
Off Galveston, November 2, 1864.

ADMIRAL: On the 1st instant the *Fort Morgan* arrived early in the morning and I was enabled to dispatch her by noon to supply the vessels at the Rio Grande and intermediate points. On the morning of the 1st the *Augusta Dinsmore* left on her return to New Orleans. Fortunately, a few hours quiet enabled us to get our supplies from the *Fort Morgan* without unnecessary delay. In the last fortnight or more the weather has been so stormy or rough as to greatly interfere with communication between the vessels of the blockading fleet off the port, and that, with the absence of a number of the vessels, has prevented my organizing the court ordered sometime since upon a person on board the *Kineo,* that vessel also having been until

---

[19] *ORN,* ser. 1, 21:716-717.

within a few days stationed off the Rio Grande. As soon as it is practicable I will endeavor to have the several cases tried.

On the night of the 30th an alarm was given of a vessel running out from this port, and three of the blockade went in pursuit. The *Chocura* returned yesterday morning, having failed in capturing anything; but in the afternoon the *Katahdin* returned, towing the British schooner *Albert Edward,* loaded with 150 bales of cotton, captured on the 31st ultimo, having run out from Galveston the previous night. A steamer got out the same time and was lost in the thick weather the following day, after throwing overboard some of her deckload of cotton. The *Albert Edward* was dispatched yesterday afternoon to New Orleans with a prize officer and crew.

Early this morning the *Itasca* arrived from the Rio Grande, and this afternoon the *Seminole* arrived from her chase after blockade runners of the night of the 30th. Owing to a heavy blow that commenced last evening and continues at present, I have been unable to communicate with either of the two vessels.

The blockade-running seems to be very brisk, and with the promise of a much greater number of vessels soon to be engaged in that business, but our means at present are so limited I fear but little can be done to break this trade up. At present I have off here, in addition to this ship, the *Seminole* and *Bienville,* the only vessels that can do anything in chasing. The *Itasca,* just arrived, I know nothing of her condition; but the *Katahdin, Kineo,* and *Chocura* are in danger of breaking down at any moment, the former having but one boiler she can depend upon. The *Kineo's* hull and machinery represented in such condition her commander does not consider her seaworthy, and I am only awaiting an opportunity to send a survey on board to ascertain her condition. A few days since Lieutenant-Commander Gillis, stationed off Velasco, writes, requesting a survey upon the *Sciota,* reporting her disabled in hull and machinery. So constant are the complaints of commanding officers in this division of the bad condition of their vessels that I look upon every arrival with anxiety. The *Aroostook,* now off the Rio Grande, was delayed here some days repairing, her machinery being out of order.

I regret to trouble you with these complaints, but the condition of the vessels is too patent to permit me to pass them by, and I feel also it is due to the commanding officers to state [that] I believe as a body they are most anxious to be of service–to work–but the constant unceasing work their vessels have been called upon to perform has used them up.

*November 5.–*The *Seminole* reports having chased a steamer, but lost her at night.

*November 5.–*Lieutenant-Commander Skerrett reports through Lieutenant-Commander Brown, of the *Itasca,* that one of his boilers has given out.

The health of this division I am pleased to report as good.

I am, sir, very respectfully, your obedient servant,

WM. E. LE ROY,
Commanding Third Division West Gulf Blockading Squadron.

Rear-Admiral D. G. FARRAGUT, *U.S. Navy,*
Commanding West Gulf Blockading Squadron, Mobile Bay

Report of Rear-Admiral Farragut, U.S. Navy, regarding the impaired condition of the vessels of his command.[20]

[No. 523.]

FLAGSHIP *HARTFORD*,
Mobile Bay, November 6, 1864.

---

[20] *ORN*, ser. 1, 21:719-720.

SIR: I regret to inform the Department that the boilers of our gunboats are becoming so badly used up that it is almost impossible to keep a sufficient number on the coast of Texas to maintain an efficient blockade.

What we want are vessels that can run from 10 to 14 knots, if possible, with light batteries and capacity for much coal. I should suppose the captured and purchased vessels would be preferable for blockading; our large vessels are fast enough, perhaps, but they are easily seen at great distances and consume an immense quantity of coal, and coming on the coast is a serious matter.

The *Mobile* and *Bienville* are the only vessels that can vie in speed with the blockade runners, and although they are perhaps a little faster than the latter, yet not sufficiently so to overtake them before night closes in, and they have, therefore, in nearly every case, made their escape.

The *Mobile* is now broken down by a gale off the Rio Grande and must go North for extensive repairs. She was condemned fourteen months ago, but has done me good service since. She is the fastest vessel I have in this squadron, and I will be glad to have her back again as soon as repaired.

Very respectfully, your obedient servant,

D. G. FARRAGUT,
Rear-Admiral, Commanding West Gulf Blockading Squadron.

Hon. GIDEON WELLES,
Secretary of the Navy, Washington

Report of Rear-Admiral Farragut, U.S. Navy, regarding the impaired condition of the vessels of his command.[21]

[No. 532.]

FLAGSHIP *HARTFORD*,
Mobile Bay, November 13, 1864.

SIR: I consider it my duty to call the attention of the Department to the present condition of the force on this station.

The vessels are becoming used up, the boilers of most of them require several months repairs, and some of the officers consider their vessels unsafe to keep the sea.

The class denominated double-enders[22] I have had but little service out of compared with the screw gunboats. The latter have done pretty much all the blockading duty, on account of the general complaint of the commanders of the double-enders of their vessels being so weak as to work and twist in a gale and also to settle amidships, some of them to the extent of 4 inches when ready for sea, besides which they all steer so badly that I forbid them crossing the bows of any vessel, although I have had a third rudder put on three out of five of them. None of them that I can find out use the forward rudder; this arises, I believe, from two causes. First, the rudder is embedded in the woodwork, so that the water does not act sufficiently upon it; and, second, there is not sufficient space in the bow or stern to allow the helm to be put at a sufficient angle to the keel before striking the side, and from their great length and little depth they work so much that the commanding officers are generally afraid of them and are always anxious to avoid sea duty. As they are valuable vessels in smooth water, they are usually sent on inside duty when this can be done. The *Conemaugh,* now gone home for repairs, has only been out here eight or nine months, and during that time she has not been outside two months,

---

[21] *ORN*, ser. 1, 21:725-728. This and the three following documents are covered by this reference.
[22] The double-ender gunboats (many of which were converted ferryboats) received a discussion of their structural weaknesses that made them near useless outside of protected waters like bays and bayous.

but kept mostly in the Mississippi River and Sound, and she was sent North because, by the report of the board of survey, it would occupy all our mechanics too long a period to repair her (Figure 10).

Figure 10. U.S.S. *Conemaugh*, double-ender gunboat purpose built for the navy. *Naval Historical Center*.

The *Octorara* has recently been four months in New Orleans, in the hands of the machinists, and lately, one month in Pensacola, and has not done four months' sea duty since she has been on this station, but, like the others, kept in the Mississippi Sound when we could spare her.

The *Metacomet* and *Sebago* (Figure 11) are the best of them; the latter is much the strongest built of all.

Figure 11. U.S.S. *Sebago*, gunboat. *Naval Historical Center*.

The double-enders that I have named are well commanded, and yet, with the exception of the two last mentioned, are unfit to go to sea, according to the report of their commanding officers, without taking off some of their guns, which I shall do if I find myself compelled to send them to keep up the blockade off Texas.

The *Genesee* I do not know the extent of the weakness of. She has now been upward of two months in Pensacola repairing, but her commander is one of those nervous, complaining officers, that it is not easy to form a very just idea of his vessel; he has always said that "she is good for nothing."

The *Port Royal's* boilers are also said to require extensive repairs, as well as new bottoms; the *Sebago's* also. The screw gunboats nearly all require new bottoms and extensive repairs to boilers, so that although with these vessels I manage to keep up the blockade of the Dog River Bar, it is only by changing them every few days for repairs, and I am being very hard pushed to maintain the blockade of Texas with vessels that can not catch anything but sailing vessels and schooners. Some of these screw gunboats are weakly built. The *Owasco* was condemned by survey the other day and sent up here. I find her lodge and hanging knees rent and split so badly that the officers thought she would roll herself to pieces in a gale off Texas.

With all these lame vessels I can only keep three, and at times four gunboats and one ironclad (the *Chickasaw*) to blockade the enemy, who has three ironclad casemated rams, two iron casemated floating batteries, and two gunboats, also protected as to their machinery, with iron, all lying under a line of forts and the whole protected on the outside by a line of obstructions, piles, and torpedoes. I had a beautiful view of everything yesterday when up the bay. The day was remarkably clear and we could count every gun, and I could not but regret the loss of time through the want of troops to render aid to our forces in Georgia.

I have made the above statement of the condition of the gunboats in order that the Department may not be surprised if we do not catch all the blockade runners on the coast of Texas.[23]

Very respectfully, your obedient servant,

D. G. FARRAGUT,
Rear-Admiral, Commanding West Gulf Blockading Squadron.

Hon. GIDEON WELLES,
Secretary of the Navy, Washington.

List of vessels comprising the Third Division, under command of Captain Marchand, U.S. Navy.

BLOCKADE OF GALVESTON,
Monday, November 14, 1864.

The very small and crippled force with which I am expected to blockade She coast from Calcasieu to the Rio Grande is composed as follows:

At Calcasieu.–Steam gunboat *Penobscot*, Lieutenant-Commander A. E. K. Benham; steamer *New London*, Acting Master Lyman Wells.

At Sabine Pass.–Steam gunboat *Pocahontas*, Lieutenant-Commander M. P. Jones; steam gunboat *Itasca*, Lieutenant-Commander George Brown; steam gunboat *Pembina*, Lieutenant-Commander J. G. Maxwell; steamer *Gertrude*, Acting Master H. C. Wade.

At Galveston.–Steam sloop *Lackawanna*, Captain J. B. Marchand, commanding; steam gunboat *Chocura*, Lieutenant-Commander R. W. Meade, Jr.; steamer *Bienville*, Lieutenant H. L. Howison; steamer *Cornubia*, Acting Volunteer Lieutenant John A. Johnstone.

Off Velasco and San Luis Pass.–Steam gunboat *Katahdin*, Lieutenant-Commander John Irwin.

Off the Rio Grande.–Steam sloop *Seminole*, Commander A. G. Clary; steam gunboat *Aroostook*, Lieutenant-Commander J. S. Skerrett.

Letter from the Secretary of the Navy to Rear-Admiral Farragut, U.S. Navy, transmitting copy of the President's order revoking permission granted to Andrew J. Hamilton for the shipment of cotton from Texas.

---

[23] The Texas blockade must have been quite porous, and Farragut pointed out that the poor condition of his ships was a primary reason for the lack of success.

NAVY DEPARTMENT, *November 14, 1864.*

SIR: The Department has received your No. 511, dated the 30th ultimo, in reference to Andrew J. Hamilton and his agents.[24]

The Executive order of the 9th of August last, directing that if Mr. Hamilton or his authorized agents should come out of Galveston or Sabine Pass with any vessel or vessels freighted with cotton shipped to the agent of the Treasury Department at New Orleans, such vessels should be permitted to pass, has been revoked, as you will perceive by a copy of an order of the President, dated 11th instant, herewith enclosed. It is not unlikely that you may have already received the order of revocation.

Very respectfully, etc.,

GIDEON WELLES,
Secretary of the Navy.

Rear-Admiral D. G. FARRAGUT,
Comdg. Western Gulf Blockading Squadron, New Orleans.

[Enclosure.]

EXECUTIVE MANSION,
Washington, November 11, 1864.

An Executive order to Rear-Admiral David G. Farragut having been issued on the 9th of August last, directing that, if Andrew J. Hamilton, or any person authorized in writing by him, should come out of either of the ports of Galveston or Sabine Pass with any vessel or vessels freighted with cotton shipped to the agent of the Treasury Department at New Orleans, the passage of such person, vessels, and cargoes should not be molested or hindered, but should be permitted to pass to the hands of such consignee, the said order is from this date to be considered as revoked.

ABRAHAM LINCOLN.

Report of Commander Woolsey, U.S. Navy, commanding U.S.S. Princess Royal, *regarding the capture of the schooner* Alabama.[25]

U.S.S. *PRINCESS ROYAL,*
Off Galveston, December 7, 1864.

SIR: I have the honor to make the following report of the capture by this vessel of the schooner *Alabama*.[26]

At daylight this morning, whilst cruising off San Luis Pass, a schooner was discovered inshore, standing in. This vessel immediately gave chase. A shot was fired to bring her to, when she rounded to, being at [the] time in the breakers and aground. She then lowered a boat on the inshore side, in which all her people immediately deserted the vessel and pulled for the

---

[24] Hamilton, a Texan who stayed loyal to the Union, had the shocking privilege to ship cotton out of Texas through the blockade bound for New Orleans. Clearly a very well connected man in Washington, he lead Union troops and became a Reconstruction governor of Texas.

[25] *ORN*, ser. 1, 21:751-752.

[26] This report of capturing a blockade-runner was interesting in several aspects. The ship's papers provided many details such as the merchant in Havana who sent the ship, Rafael Perez de St. M. His name also appeared on many cargo invoices of the *Denbigh* (see Arnold, J. B. *The* Denbigh's *Civilian Imports*. College Station, Texas: Institute of Nautical Archaeology, *Denbigh* Shipwreck Project Publication 5, 2011.). The cargo was consigned to T. W. House, Houston merchant and sometime mayor, who was active in blockade-running in a major way. House's name also appeared in connection with the *Denbigh* when she was on the Galveston run fall 1864 through May 1865.

beach, keeping the schooner between us and themselves. This ship could go in no farther for want of water.

Two boats were immediately sent to board her, when she was worked off into deep water, a prize crew kept on board, and she was ordered to Galveston in order that I might more thoroughly search her hold.

She is a very sharp, clipper-built schooner of about 52 tons and looks like a yacht, and, being of light draft, was probably built for a blockade runner.

The following articles were found on board, viz.:

1st One English ensign.

2d Logbook of the schooner *Alabama,* from Havana to Matamoras, commanded by Captain Davis, commencing 30th November 1864; kept by Alexander Johnson, mate. Said log is not written up beyond the 4th instant.

3d Bill of health, signed by the visiting medical officer at Havana and endorsed by the Mexican consul-general at that place.

4th Letter dated Havana, 29th November 1864, addressed to T. W. House, Galveston, and notifying said House of invoices, etc., of the cargo of "*my* British schooner *Alabama,* Captain Davis," and requesting him to invest the proceeds in cotton, "*nothing* lower than middling class." The letter is signed by Rafael Perez de St. M., and goes on to state that he intends to keep the *Alabama* running regularly to Galveston.

5th Shipping articles, registered at Nassau, J. H. Davis, master, signed at Havana, November 29th, 1864, by Alex. Johnson, mate; Edward McGrath, second mate; John Miller, cook; Robert Burns, William Gallagher, Edward Geary, and C. Saxtorph, seamen.

6th Certificate of British register, dated Nassau, 23d March 1864. Register tonnage 52 84/100.

7th Invoices and bills of lading showing her cargo to consist of 15 kegs bicarbonate soda, 15 bales India bagging, 20 coils manila rope,[27] 100 barrels flour, 102 iron bars, assorted; 42 iron bars, square; 36 bundles iron bars, square; consigned to Mr. T. W. House, Matamoras.

8th Letters of introduction to persons in Tampico, Vera Cruz, and Matamoras.

9th A pocketbook containing a two-dollar bill of the Confederate States, and some private papers not in any way connected with or relating to the vessel.

The *Alabama* will be sent to New Orleans for adjudication; to sail this evening.

I have the honor to be, most respectfully, sir, your obedient servant,

M. B. WOOLSEY,
Commander.

Hon. GIDEON WELLES,
Commander. Secretary of U.S. Navy, Washington.

Report of Commander Woolsey, U.S. Navy, commanding U.S.S. Princess Royal, *regarding the capture of the schooner* Cora.[28]

U.S.S. *PRINCESS ROYAL,*
Off Galveston, December 20, 1864.

SIR: I respectfully report the particulars of the capture of the schooner *Cora* last night, as follows:

At 7:50 p.m., while at my night station, a vessel was discovered inshore of us, standing to the southward. I immediately made the signal of alarm to the squadron, slipped the chain, and chased, running about 4 miles to the southward and westward of my station and 6 miles from

---

[27] The rope and bagging were materials used to bale cotton and were constantly in short supply in Texas.
[28] *ORN,* ser. 1, 21:763-764.

Galveston, where I came up with the vessel and caused her to lower her headsails and anchor. She proved to be the schooner *Cora,* hailing from Belize, [British Honduras], and had just run out of Galveston with a cargo of 175 bales of cotton. I immediately took possession of her. Her chronometer book shows that her chronometer was rated at Houston on the 27th ultimo. On board of her were found the following papers, etc.:

1st Certificate of British registry, dated Belize, 16th October 1863, 47 50/100 tons, owned by William Ardill, of Manchester, England.

2^{d} Articles of agreement, signed by seven persons, registered at Belize, endorsed by British consul at Vera Cruz, April 20, 1864, and also endorsed by a certificate of James Sorley, collector of customs, port of Galveston, dated 11th November 1864, to the effect that "John Greenough is master British schooner *Cora* in place of Henry Sherffins, late master."

3^{d} Pass signed by Captain Stephen D. Yancey, Houston, August 16, 1864, by order of Major-General J. B. Magruder, permitting Mr. T. W. House to send to sea, from Galveston Bay, the schooner *Cora,* loaded with cotton.

4th Manifest of cargo (175 bales of cotton).

5th Bill of health, dated Galveston, 27th August 1864.

6th Letter to Messrs. Choppinger & Co., Havana, written in Spanish, dated Houston, December 31, 1863, introducing Mr. Taylor, former master of schooner *Cora,* signed by A. Maseras.

7th Memorandum of articles to be purchased at Havana, one of which was "50 to 100 kegs rifle powder," besides a number of minor papers, mostly private, which I have enclosed to the judge of the district.

8th One English ensign.

She will be sent to New Orleans to-day for adjudication. Captain George F. Emmons was in command of this division at the time. There was a thick fog, and none of the vessels of the squadron were in sight, nor within signal distance, this vessel being about 6 miles distant from the fleet off Galveston at the time of capture.

Her crew consists of J. Greenough, master, and 6 men.

I have the honor to be, most respectfully, sir, your obedient servant,

M. B. WOOLSEY,
Commander.

Hon. GIDEON WELLES,
Secretary of U.S. Navy, Washington.

Report of Captain Emmons, U.S. Navy, commanding Second Division, giving list of captures by the vessels of his command.[29]

U.S.S. *LACKAWANNA,*
Off Galveston, Tex., December 29, 1864.

SIR: I have the honor to report the following captures made by the Second Division of the West Gulf Blockading Squadron, under my command, during the last month ending the year 1864, which have already been furnished to the commander in chief of this squadron.

---

[29] *ORN,* ser. 1, 21:777-778.

Date 1864. Dec.	Vessel	Rig	Cargo	Captured by	Commander	Disposed of
4	*Hilligonda*	Brig	Assorted	*Pembina*	Lt. Comdr. Maxwell	Sent to New Orleans for adjudication
4	*Lowood*	Schnr.	221 bales of cotton	*Chocura*	Lt. Comdr. R. W. Meade	Do
5	*Julia*	do	Iron, paper, medicines	do	do	Do
6	*Lady Hurley*	do	Steel and medicines	do	do	Do
7	*Alabama*	do	Cotton, etc.	*Princess Royal*	Comdr. Woolsey	Do
8	*Mary Ann*	Sloop	do	*Itasca*	Lt. Comdr. Brown	Chased on shore and destroyed.
19	*Cora*	Schnr.	175 bales of cotton	*Princess Royal*	Comdr. Woolsey	Sent to New Orleans for adjudication.
26	*Belle*	do	95 bales of cotton	*Virginia*	A. V. Lt. Brown	Do

December 28, small sailboat with 5 deserters from Galveston, including F. W. Stephens, a surgeon of the rebel Marine Department; John Wern, carpenter of the *Bayou City*, steamer, and 3 privates of the rebel army, which have been sent to New Orleans.

I have the honor to be, with great respect, your obedient servant,

GEORGE F. EMMONS,
Captain, Comdg. 2ᵈ Div. West Gulf Blockdg. Squadron.

Hon. GIDEON WELLES,
Secretary of the Navy, Washington, D.C.

Extract from Journal of Captain Marchand, U.S. Navy, Commanding U.S.S. *Lackawanna*.[30]

March 13, 1864.–Blockade of Galveston. At about 5 in the afternoon the *Ossipee*, with Captain Gillis, left for the blockading force off Mobile and I assumed command of the western division of the blockading squadron, which now becomes the Third Division. It embraces the whole coast of Texas and the western part of Louisiana and is composed of the following vessels: At Sabine Pass, steamer *Princess Royal*, Commander M. B. Woolsey; steam gunboat *Chocura*, Lieutenant-Commander B. Gherardi; steamer *Antona*, Acting Master A. L. B. Zerega. At Galveston, steam sloop *Lackawanna*, Captain J. B. Marchand (chief); steam gunboat *Owasco*, Lieutenant-Commander E. W. Henry; steam gunboat *Kanawha*, Lieutenant-Commander B. B. Taylor; steam gunboat *Cayuga*, Lieutenant-Commander W. H. Dana. At San Luis Pass, steamer *Virginia*, Acting Volunteer Lieutenant C. H. Brown. At Brazos River, steam gunboat *Penobscot*, Lieutenant-Commander A. E. K. Benham; steam gunboat *Aroostook*, Lieutenant-Commander C. Hatfield. At Pass Cavallo, steamer *Estrella*, Lieutenant-

---

[30] *ORN*, ser. 1, 21:809-823.

Commander A. P. Cooke. From Pass Cavallo to the Rio Grande, steam gunboat *Sciota*,[31] Lieutenant-Commander G. H. Perkins. Patrol of whole coast of Texas, etc., steamer *Arkansas,* Acting Volunteer Lieutenant D. Cate.

*March 14.–*A few minutes before the past midnight the *Aroostook* [arrived] from the Brazos River and anchored. The little schooner which arrived yesterday was a prize to her, being the *Marian.* Early this morning the latter, at anchor half a mile from us, made signal of distress, and the *Aroostook* going, found her in a foundering condition. Her cargo, consisting of a few sacks of salt and some bars of iron, was taken out and she was sunk in 4½ fathoms water.

*March 15.–*A little before 4 in the morning one of our blockading vessels, the *Aroostook,* which was to have come near at daylight, was reported to me as being underway inside of us and anchored outside. At daylight it was discovered to be the U.S. gunboat *Katahdin,* Lieutenant-Commander John Irwin, from New Orleans, to join the blockade of this place. Toward 8 in the morning he called upon me and reported.

*March 18.–*About 1 o'clock in the morning, whilst the *Katahdin* and *Kanawha* were chasing the coal schooner, we saw with perfect distinctness a small steamer, which we supposed to be one of the rebel picket boats, inside of the bar and running up the channel toward Galveston, and, as they are nightly seen, we entertained no uneasiness as to its being a blockade runner; but still it seemed singular she should come so far down when the night was so perfectly clear and moonlight, for we could see distinctly the city of Galveston, the land, forts, and even the *Harriet Lane* above Pelican Spit. Shortly after the supposed rebel picket steamer had passed up, the *Sciota,* stationed at Cylinder Channel, made signal that a steamer was running out and stood to the S. and W. By that time the *Katahdin* and *Kanawha* were well back to their stations and could see anything going out, so the *Lackawanna* was not moved, and I soon found, I afterwards learned, that it was an object on shore that was seen which caused the signal to be made. Three o'clock in the morning passed before I lay down to rest. On going on deck after 8 in the morning discovered that a small steamer was near Fort Point, a stranger to us, with English colors, and she must have run the blockade the past night, and probably have been the one which we supposed to be the rebel picket boat, although subsequently to-day I was told by the pilot of the *Sciota* that last night he saw four rebel picket boats between the beacon and Pelican Spit. If it was the blockade runner that we saw running in the past night, taking her to be a picket steamer, she must have followed close along the land from the southward and westward and skirting along the outer edge of the—, entered the Main Ship Channel unperceived whilst the *Kanawha* and *Katahdin* were chasing the coal schooner. Another supposition is that she followed close along the coast, entered by the Southwest Channel, and that the unusual number of the enemy's steam pickets were in the outer channels to attract our attention from her. At all events, it is mortifying that a vessel has gotten in. In the forenoon I took passage in the *Sciota,* skirting along the reef to Bolivar Island to have a close inspection of the blockade runner, which had shifted her berth to Pelican Spit; also to examine the rebel forts, as well as to get a more perfect knowledge of the localities, and on returning along the edge of the shoal the *Sciota* grounded on a spit, but was gotten off in an hour by carrying out a hawser and kedge, assisted by the *Kanawha.*

*March 26.–*They brought information that the *Clifton,* which formerly was one of the gunboats and captured by the enemy on the attempted invasion at Sabine several months ago, had attempted to run out of Sabine River, laden with cotton, and grounded on the bar, and to prevent her falling into the hands of our blockading boats was set on fire and burned. It was the cotton from the *Clifton* that drifted by the ship, as noted in my remarks of the 23[d] instant.

*April 4.–*Shortly before the past midnight the *Kanawha,* stationed a little to the west of the Main Ship Channel, made signal that a vessel was coming out. She was gotten underway, as was the *Sciota,* anchored off the Southwest Channel. Subsequently I learned that the

---

[31] Spelling varies in official sources: *Scioto* or *Sciota.*

Kanawha was delayed in slipping her cable and lost sight of the blockade runner in the darkness of the night, whilst the *Sciota,* keeping her in sight with the aid of sail and steam, overtook her after running two and one-half hours and 25 miles on a S. by W. course, capturing her. This afternoon she towed the prize near this ship and anchored her. The blockade runner is the rebel schooner *Mary Sorley,* formerly the U.S. revenue cutter *Dodge,* seized by the rebels at Galveston at the commencement of the rebellion in Texas. By her register she appears to be owned by Thomas W. House, of Houston, Tex., and belongs to Galveston. She is 75½ tons burden, laden with 257 bales of cotton; was bound to Havana, commanded by Charles Diericks, with a crew of 7 persons and 2 passengers. During the chase last night, at 1 a.m. we saw the flash and heard the report of a cannon fired at the blockade runner by the *Sciota;* but the capture was made beyond signal distance of the blockading squadron. In the early part of the afternoon the U.S. gunboat *Kineo,* Lieutenant-Commander John Watters, arrived from New Orleans to join the blockade of the Texan coast.

April 5.–At the break of day a boat came off from Galveston to the *Kineo,* at anchor to the westward of the Main Ship Channel, having on board 7 persons, viz., Antonio Orluffs, Peter Berg, Frederick Rudolph, Martin Ricke, deserters from the rebel First Texas Artillery (Colonel Cook's), and Henry Axtram and Frederick Hahn, of the rebel army (butchers), and Frank Teichman, carpenter, of Galveston, a minor. They gave the information that the British blockade-running steamer *Isabella* was to run out laden with cotton tomorrow night; that the only two schooners laden and ready to run out were those in sight from the blockading vessels now at anchor off Pelican Spit, the names of which were *Tip Top* and *Higher;* that the rebels had abandoned the idea of loading with cotton and sending to sea the bark *Cavallo,* having discharged her crew, or, rather, sent her to Houston; that several cotton-laden schooners had gone through West Bay to try and run out at San Luis Pass and were accompanied by one or more schooners armed with 12 or 24 pounders and each a crew of about 40 soldiers, with the deck covered with bales of cotton to make them resemble blockade runners to prevent the schooners from being cut out by the boats of the blockading vessel, as was done on a former occasion, or to capture our blockading steamer either by stratagem or boarding. They further represent that the guns of the *Harriet Lane* were taken out and are mounted on the fort at Bolivar Point, and that within the last few weeks cotton has been taken on board of her, and that she will be loaded and run out some time this week under command of a man named Maynard, who successfully ran a schooner through the blockading squadron in daytime a short period before the arrival of the *Lackawanna* here. The *Harriet Lane* is to be piloted out by a man named Davidson, who some time ago was captured in a blockade runner and released from Fort Lafayette on taking the oath of allegiance to the United States.

April 14.–In the forenoon the *Katahdin* came from her station at the N. E. end of the blockading line, bringing 7 deserters from the rebel army, namely: Eugene Picard, Lewis Bechet, Alphonso F. Rogers, Felix Loving, Francis Souffloit, John B. Nivard, and Francis Bouton, all Frenchmen, who had some eleven months ago come from Texas and settled in Brownsville, from which they were conscripted into the rebel army a short time before our army occupied that place. These deserters started in a canoe from the rebel fort, Mannahassett, 7 miles southwest of the Sabine Pass, for our blockading vessels off that place, but bad weather setting in they were compelled to run along the coast 46 miles to the blockading vessels here.

April 15.–In the early part of the afternoon two deserters from the rebel army came off to the squadron. Their names were Richard Allen and Matthew Dullaghan, belonging to the Eighth Texas Regiment, stationed on Bolivar Island. They are Irishmen by birth, and came off in an old boat which had on two former occasions brought off deserters and set adrift on each occasion. They report that our blockading vessel off San Luis Pass captured a schooner having liquor on board, and the prize crew getting drunk, the old crew rose and took her into San Luis Pass.

April 16.–The admiral brought the report from Acting Volunteer Lieutenant [C. H.] Brown, of the *Virginia,* that, through drunkenness, a schooner which he had captured was run ashore a short distance from San Luis Pass, and that 1 acting ensign and 5 men forming the crew were captured by the rebels.

April 20.–Acting Volunteer Lieutenant Brown, of the *Virginia,* reported to me that on yesterday he captured the Mexican schooner *Alma,* off San Luis Pass, and laden with gunny bags, iron hoops, army blankets, etc., and sent her to New Orleans; the vessel and cargo supposed to be worth $2,500.

April 22.–The *Kineo* came from her station to say that the *Owasco* was nearly out of coal, and she returned to San Luis Pass. To-morrow the latter vessel will be relieved. These two vessels captured a schooner yesterday.

April 24.–I mentioned yesterday that the *Owasco* arrived from the Brazos River, and owing to the gale no intercourse was had with her; but the sea being smooth, her commander, Lieutenant-Commander Henry, who is confined to his bed by rheumatism, wrote me that he had made three captures, viz.: British schooner *Lily,* on the 17th instant; the *Fanny,* on the 19th; and the English schooner *Laura,* on the 21st, all bound into the Brazos River from Havana.

April 29.–The following are the vessels now under my charge in various places along the coast: Steam sloop *Lackawanna,* Captain J. B. Marchand; steamer *Princess Royal,* Commander M. B. Woolsey; steam gunboat *Owasco,* Lieutenant-Commander E. W. Henry; steam gunboat *Chocura,* Lieutenant-Commander B. Gherardi; steam gunboat *Katahdin,* Lieutenant-Commander John Irwin; steam gunboat *Kineo,* Lieutenant-Commander John Watters; steam gunboat *Kanawha,* Lieutenant-Commander B. B. Taylor; steam gunboat *Cayuga,* Lieutenant-Commander W. H. Dana; steamer *Estrella,* Lieutenant-Commander A. P. Cooke; steamer *Arizona,* Acting Master H. Tibbits; steamer *New London,* Acting Master Lyman Wells; steamer *Virginia,* Acting Volunteer Lieutenant C. H. Brown; steamer *Augusta Dinsmore,* Acting Volunteer Lieutenant William Hamilton; steamer *Gertrude,* Acting Master H. C. Wade.

April 30.–This morning at daylight the *Katahdin* was not in sight, and at breakfast time the *Kineo* came near and the captain called, reporting that a schooner had succeeded in running in last night from the southward and westward, through the Main Ship Channel; that he slipped and ran in as far as he could without firing, for fear of drawing the other vessels from their stations, but the schooner was too far in and he sent a boat to capture her, but which ineffectually chased until near Fort Point, and that in attempting to regain his cable the slip rope parted and the chain dropped to the bottom. Soon after the *Kineo* stood to the southward and westward and after a while returned with the *Katahdin.* Lieutenant-Commander Watters, of the *Kineo,* then reported that the *Katahdin* had chased a steamer which came out of Galveston at 10 last night, kept her in sight for two hours, and, losing her, cruised about 40 miles off to the S. S. W. I really do not know what steamer could have come out last night, and fancy that it must have been one attempting to run in, as the three steamers we are watching, viz.: *Harriet Lane, Isabel,* and *Alice,* seem still to be lying near Pelican Spit, and those are the only sea steamers known to be in Galveston.

May 1.–A day of intense mental suffering to me, as the *Harriet Lane* and the two other steamers, *Alice* and *Isabel,* escaped last night from Galveston. The escape was unknown to me until daylight this morning, and then only by their not being seen at their usual place. At the same time the *Katahdin* was absent from her station, having in all probability gone in chase during the night. I should have mentioned that last night, until midnight, it was very dark, cloudy, and somewhat squally, with rain. The wind, which at 9 p.m. last night, was moderately fresh from the eastward, after veering to the southward, came out fresh from the westward at 11 at night and hauled to the north. It was just such a night as suited for running the blockade. The *Owasco* was stationed off Bolivar Passage; the *Arizona* off the north part Cylinder Channel, the *New London* off the south part of the same channel, the *Lackawanna* on eastern side of Main Ship Channel, the *Kineo,* underway, apparently, till nearly midnight, at entrance of Main Ship Channel, and the *Katahdin's* regular night station at South West Channel. The *Arizona, New London,* and *Kineo* were in sight from the *Lackawanna* all night. The darkest part of the night, and the time those vessels could have run out without being seen by the *Arizona, New London, Lackawanna,* and *Kineo,* was between 7 and 11:15 o'clock the past night. At about the latter hour the discharge of a couple or three heavy guns was heard in a direction about S. S. W., and the flash also seen, but that was doubtful, as frequent lightning flashed in all directions. It is supposed that between 9 and 10 o'clock the *Alice, Isabel,* and *Harriet Lane* followed each other through the South West Channel (which was supposed not to have sufficiently deep water for the *Harriet Lane*); that the *Katahdin* gave chase to the first comer and, leaving the channel unguarded, the others passed unobserved. This, however, is but a surmise, as the *Katahdin* has not yet (9 p.m.) returned. I am in ignorance of the time the *Katahdin* left, as she did not, in conformity with the rules of the blockade, burn the blue Coston light indicative that a vessel was coming out of that channel, nor did she hoist a red light to show that she was in chase. I should have stationed the *New London* on her arrival on the 26th ultimo to the southward and westward to assist the *Katahdin* and *Kineo,* although the distance to be watched is little more than half that assigned to the *Arizona* and *Owasco,* as the probabilities were greater that the attempt to run out would be made through the Main Ship and South West channels, but a few days previous the rebels had planted a large buoy, apparently near Bird Key, and I was apprehensive that an attempt would be made through Cylinder Channel. From all information I was led to believe that the *Harriet Lane* drew 9 or 9½ feet water, and as the Main Ship Channel could only afford that draft, it was watched with the greatest intensity, and the *Lackawanna* was kept near to give chase, as she possessed the greatest speed.

May 2, 1864.–At daylight 8 deserters from the rebel army, viz.: Bernard Aysen, William Koch, Oscar Meano, Julius Ludwig, Anton Hess, August Hammon, John G. Rear, and Henry Cook, came off near the South West Channel. They brought no news of interest except that the *Harriet Lane, Alice,* and *Isabel* succeeded in running out on Saturday night last, confirming our belief. Had I been certain yesterday morning that the *Harriet Lane* had gotten out, I would have sent another vessel after her. The *Katahdin* is still absent, having left some time on Saturday night, doubtless in pursuit of the other steamers which then ran out. The deserters say that a schooner attempted to go out the same night, but grounded near Fort Magruder.

May 4.–In the morning a schooner came from the southward and westward, which proved to be the British schooner *Agnes,* captured by the *Chocura* in running out of the Brazos River on the morning of the 3[d] instant, laden with 155 bales of cotton. She proceeded onward toward New Orleans.

May 5.–The army transport steamer *Sophia* arrived from Pass Cavallo and the Brazos River, bringing dispatches for me from the vessels below, amongst others, information that the *Chocura* had captured the Prussian schooner *Frederic II* on the 4th instant, having come out of the Brazos River with 114 bales of cotton. The *Virginia* also came up for a few minutes from San Luis Pass and reported her having captured the schooner *Experiment,* which had run the blockade to Galveston with 31 bales of cotton, and that the vessel being unseaworthy, the cargo was taken out and the schooner burned.

May 6.–I neglected to mention that yesterday at daylight a strange steamer *Coroica,* or else *Mail,* with English colors flying, was seen in the bay, which had run the blockade unseen during the preceding night.

May 11.–Besides the foregoing, there went, much to my convenience, Mr. A. M. Hobbs and Mrs. Ball, who came to this ship on the 20th ultimo. They both occupied my small cabin.

They came under permission of the admiral,[32] to try and arrange with a Mr. J. D. Champlin, of Houston, to run out the *Harriet Lane* to us, and they to have the cargo of cotton. Mrs. Ball was to personate Mrs. Champlin, who is sick in Tennessee, as an excuse for Mr. Champlin to come out and make the necessary arrangements. But the letter by the flag of truce remained unanswered. Should Mr. Champlin come out, I am to get a letter from him on the subject of the purchase of the *Harriet Lane,* now supposed to be in Havana, and about running out the barks *Cavallo* and *Elias Pike* here and the *Sachem* at Sabine Pass, and send the letter to A. M. Hobbs, care of *True Delta* Office, New Orleans. The name signed by Mr. Hobbs to the letter sent to Mr. Champlin to induce him to come out to see his sick wife was Samuel French.

*May 12.–*Acting Master Lyman Wells, commanding the *New London,* on the following morning, the sea then being smooth, sent in a flag of truce, which was met by a boat, when the information was obtained that on the 6th instant an attack was made from the land by rebel soldiers and both the *Granite City* and the tinclad (*Wave*) were captured. Nothing could be learned as to the mode of capture or loss of life. Information was further obtained that when the armed boat from the *New London* ran into the Calcasieu and approached the *New London* the rebel flag was flying at the peak, and that the officer of the boat, as well as the boat's crew, [thought] that those on board the *Granite City* were making fun of them, and that the officer of the boat, Mr. Jackson, fired a musket, upon which musketry was opened upon the boat, and the first shot struck Mr. Jackson in the head, killing him instantly; none others were injured. Subsequently, to-day, I learned from a rebel army deserter that the capture was made by a detachment from Colonel Griffin's command from the Sabine Pass of about 200 men, having no cannon, whilst the crews of the *Granite City* and *Wave* were fishing and amusing themselves ashore. During the day I also received by a flag of truce newspapers of yesterday printed in Galveston, wherein it was mentioned, under head of "*Houston,* May 10, 1864, 107 privates, 23 officers, and 3 n------s (Yankees), captured by Colonel Griffin on gunboats at Calcasieu, arrived here to-day from Beaumont by rail." I immediately ordered the *New London* to proceed to the blockade of Calcasieu, reporting to Commander Woolsey on the way, and she started back for that place soon after 12 o'clock the past night. This morning I also sent the *Owasco* for the same purpose. Neither the *Granite City* nor *Wave* had ever reported to me as belonging to this division, and I was almost ignorant of the location of the *Granite City* when I sent the *New London,* and knew nothing about the tinclad being in Calcasieu.

*May 14.–*A small rowboat, with flag of truce flying, came to the beacon, and a boat from the *Sciota* communicated with her, delivering the letter which I tried to send in yesterday and bringing late Galveston and Houston newspapers and letters from Acting Master C. W. Lamson, late commanding the U.S.S. *Granite City;* Acting Master Charles Cameron, late of the *Wave,* and M. F. Rogers and M. F. Fitzpatrick, and one signed with the Christian name of Arthur (directed to Mrs. Laurence Lemhan, New Orleans), formerly belonging to those vessels, all to the friends or relations. Mr. Lamson says nothing in relation to the capture of these two vessels. Mr. Cameron says that 2 were killed and 8 wounded on board the *Granite City* and 7 wounded on board the *Wave* at the time of the capture. Fitzpatrick wrote that none of the officers was hurt, except one. The above-mentioned Michael F. Rogers was second assistant engineer of the *Wave,* and writes that that vessel was surprised. The letters were all dated 12th instant, and the writers unite in saying that they were on the eve of starting as prisoners for confinement at Hempstead, in Texas. The Galveston and Houston newspapers brought off by the flag of truce give some accounts of the capture of the *Granite City* and *Wave.* Colonel

---

[32] An interesting and complex plot, seemingly endorsed by Admiral Farragut, to steal back several Union vessels that had been captured by the Confederate forces in Texas. The spies who planned to execute the recapture were to be compensated by the cotton cargo loaded on board in preparation to run the blockade, a pretty penny indeed. The plan hinged upon the fact that occasionally people resident in the Confederacy were allowed to pass the blockade for humanitarian reasons.

Commanding W. H. Griffin dates a dispatch, "Headquarters, Sabine Pass, on board C.S. (late U.S.) steamer *Granite City,* Calcasieu, May 6, 1864. Dear General (P. O. Hébert): I attacked the enemy this morning and captured two gunboats, the *Granite City* and *Wave,* 16 guns, and about 80 prisoners." A correspondent of the *Galveston News* writes that "the attack on the gunboats was made on Friday, 6th instant, a little after 5 a.m., Lieutenant-Colonel Griffin, commanding. Our force consisted of parts of Griffin's, Spaight's, and Daly's battalions and of Captain Creuzbaur's battery of 4 guns. The Federals had no steam up and were caught napping, but fought for about one hour, when they surrendered. Our casualties are 10 killed and 16 wounded. The Federals had only 1 man wounded, since dead; 28 of them were absent from the boats, but surrendered afterwards. The *Wave* is a stern-wheel steamboat, mounting 8 guns, badly used up by Captain Creuzbaur's guns. The *Granite City* is a side-wheel steamer, two-decked, an iron ram, having six guns below and two on the upper deck. Captain Loring was in command of the Federals." The rebel officer who came out in the flag of truce boat would give no information in relation to the capture of the *Granite City* and *Wave*.[33]

May 17.–Received orders from the admiral to turn over command of the Third Division to Commander M. B. Woolsey, of the *Princess Royal,* and repair with the *Lackawanna* off Mobile, as a threatened onslaught is making on our fleet there by the ironclad ram from Mobile Bay. About 9:30 p.m. the U.S.S. *Princess Royal* arrived from Sabine Pass. Commander Woolsey came on board, and to him turned over command, with all the papers.

May 20.–The Mobile blockade is composed as follows: Steam sloop *Richmond,* Captain Thornton A. Jenkins; steam sloop *Lackawanna,* Captain J. B. Marchand; steam sloop *Ossipee,* Commander William E. Le Roy; steam sloop *Seminole,* Commander Edward Donaldson; steam sloop *Oneida,* Lieutenant-Commander W. W. Low; steam double-ender *Metacomet,* Lieutenant-Commander James E. Jouett; steam gunboat *Genesee,* Lieutenant-Commander Edward C. Grafton; steam gunboat *Pembina,* Lieutenant-Commander L. H. Newman.

May 30.–The following vessels are here: Steam sloop *Hartford,* Captain P. Drayton, with Admiral Farragut; steam sloop *Richmond,* Captain T. A. Jenkins; steam sloop *Lackawanna,* Captain J. B. Marchand; steam sloop *Ossipee,* Commander W. E. Le Roy; steam sloop *Seminole,* Commander Donaldson; steam sloop *Galena,* Lieutenant-Commander C. H. Wells; steam sloop *Oneida,* Lieutenant-Commander W. W. Low; steam sloop *Monongahela,* Lieutenant-Commander J. H. Gillis; steam double-ender *Metacomet,* Lieutenant-Commander J. E. Jouett; steam gunboat *Genesee,* Lieutenant-Commander E. C. Grafton; steam gunboat *Pembina,* Lieutenant-Commander L. H. Newman; steam gunboat *Kennebec,* Lieutenant-Commander W. P. McCann; steamer *Penguin,* Acting Volunteer Lieutenant J. R. Beers; steamer *Tennessee,* Acting Volunteer Lieutenant Pierre Giraud.

June 1.–Blockade of Mobile. The fleet here is the largest I have ever been in whilst maneuvering. Composing it are the steam sloop *Hartford,* Captain P. Drayton, with the admiral; steam sloop *Richmond*, Captain T. A. Jenkins; steam sloop *Lackawanna,* Captain J. B. Marchand; steam sloop *Brooklyn,* Captain James Alden; steam sloop *Ossipee,* Commander W. E. Le Roy; steam sloop *Seminole,* Commander E. Donaldson; steam sloop *Galena,* Lieutenant-Commander C. H. Wells; steam sloop *Oneida,* Lieutenant-Commander W. W. Low; steam sloop *Monongahela,* Lieutenant-Commander J. H. Gillis; steam double-ender *Metacomet,* Lieutenant-Commander J. E. Jouett; steam double-ender *Port Royal;* steam gunboat *Genesee,* Lieutenant-Commander E. C. Grafton; steam gunboat *Pembina,* Lieutenant-Commander L. H. Newman; steam gunboat *Kennebec,* Lieutenant-Commander W. P. McCann; steam gunboat *Pinola,* Lieutenant-Commander O. F. Stanton; steamer *Penguin,* Acting Volunteer Lieutenant

---

[33] It seemed the practice of communication between Union blockaders and Confederate defenders ashore allowed for exchange of newspapers, but the Confederate officer who came out in the flag of truce boat would not discuss the fighting ashore. The story of the capture of the *Granit City* and the *Wave* was in the newspapers though. The inconsistency regarding intelligence security seemed striking when read today.

J. R. Beers; steamer *Tennessee,* Acting Volunteer Lieutenant Pierre Giraud; gunboat *Conemaugh,* Lieutenant-Commander J. C. P. de Krafft, commanding second division.

June 6.–About 2:30 this morning signals were seen to the N. and E. of an attempt to violate the blockade, and at daylight the *Metacomet* was missing from her station. In the forenoon she returned, bringing as a prize the steamer *Donegal.* It seems the prize attempted to run in, but, being seen and headed off, was pursued and captured by the *Metacomet.* The cargo consisted of 40,000 pounds of powder, medicines, and various contraband of war shipped at Havana.

June 11.–The blockade running is brisk. One instance was mentioned as having occurred at 11 last night, another took place toward 2 this morning, toward the N. E. end of the line, which was seen by one or two of the blockading vessels, who gave unsuccessful chase. The prize steamer *Donegal,* captured yesterday morning by the *Metacomet,* left for the North.

July 1.–During the past night signal was made of a blockade runner, and guns were fired. At daylight a blockade-running steamer[34] of large size was seen ashore apparently a mile to the eastward of Fort Morgan, and throughout the day the gunboats went near and kept up an irregular fire upon her. The admiral himself went pretty near in the *Glasgow,*[35] and, after taking a look, communicated with the firing vessels, after which a little more firing took place. I did not learn the reason that the firing ceased, as at dark the blockade runner was aground in the same place.

July 4.–Early in the afternoon several of the fleet, and this ship among the rest, ran in for a couple of hours, fired at the blockade-running steamer aground east of Fort Gaines (actually Fort Morgan). We were at too great range, and do not think that a shot struck her. Fort Morgan and various field pieces on shore fired in return without effect. The unknown shoalness of the water prevented our going closer in.

July 6.–About the past midnight an expedition consisting of three boats from the *Hartford* and one from the *Brooklyn,* towed in by a gunboat, boarded the blockade-running steamer aground to the east of Fort Morgan, found that her cargo and every portable thing had been removed on shore, and that several shot holes had been made in her hull by our fleet, causing her to sink in 6 feet water. The boarding party found no persons on board, but were fired at by sentinels on the beach 15 or 20 yards off without effect. They set her on fire in the forecastle and cabin and left. She burned for a few hours; but as she is an iron vessel, the conflagration was not very magnificent.

July 10.–About the past midnight signal was made that a blockade runner was going to the northward and eastward, but no firing by our vessels. Yet it seems that about daylight one succeeded in passing through our lines, when seven steamers were watching, and in attempting to go through the Swash Channel grounded. Toward 7 in the morning signal was made to some three or four of our vessels to go in, which did so and for a couple of hours shelled the vessel, and, as she was aground near Fort Morgan, that fort replied. After breakfast I called on the admiral and obtained permission to go in. We took a position somewhat nearer and threw shell around and far over her, making good shots, and from the fact of the rebels getting her afloat while we were firing and running her still farther ashore, the admiral thought we had inflicted a serious injury, particularly from a 150-pound shell striking her forward. We sustained no injury from the guns of Fort Morgan, although one shot struck the water under our bows.

July 12.–At daylight it was noticed that the blockade-running steamer which had been aground yesterday close to Fort Morgan had floated off and was at anchor inside of the bay, with English and rebel flags flying. The army transport *Clyde* arrived from New Orleans, hav-

---

[34] This vessel was the blockade-runner *Ivanhoe.* See McLean, Cecil W. and George R. Fischer. Investigation of the Civil War Blockade Runner *Ivanhoe.* Florida State University Department of Anthropology, 1991.
[35] Former blockade-runner *Eugenie.* This ship figured prominently below in Chapter 6.

ing on board the chief of staff of General Canby to make arrangements with the admiral about an attack on Mobile. It is said that 20,000 soldiers are ready at New Orleans for the expedition, and that two ironclads are at New Orleans and two at Key West, which, with the one at Pensacola, are to reinforce the fleet upon the contemplated attack.

July 20.–The following vessels are here: Steam sloop *Hartford*, Captain P. Drayton, with the admiral; steam sloop *Richmond*, Captain T. A. Jenkins; steam sloop *Lackawanna*, Captain J. B. Marchand; steam sloop *Brooklyn*, Captain J. Alden; steam sloop *Monongahela*, Commander J. H. Strong; steam sloop *Ossipee*, Commander W. E. Le Roy; steamer *Bienville*, Commander J. R. M. Mullany; steam sloop *Seminole*, Commander E. Donaldson; steam sloop *Oneida*, Commander T. H. Stevens; steam monitor *Manhattan*, Commander J. W. A. Nicholson; steam sloop *Galena*, Lieutenant-Commander C. H. Wells; steam double-ender *Metacomet*, J. E. Jouett; steam double-ender *Port Royal*, B. Gherardi; steam gunboat *Genesee*, Lieutenant-Commander E. C. Grafton; steam gunboat *Pembina*, Lieutenant-Commander L. H. Newman; steam gunboat *Penobscot*, Lieutenant-Commander J. G. Maxwell; steam gunboat *Kennebec*, Lieutenant-Commander W. P. McCann; steam gunboat *Sebago*, Lieutenant-Commander W. E. Fitzhugh; steam gunboat *Pinola*, Lieutenant-Commander O. F. Stanton; steam gunboat *Itasca*, Lieutenant-Commander George Brown; steamer *Tennessee*, Acting Volunteer Lieutenant Pierre Giraud; steam tender *Glasgow;* steam tender *Philippi.*

August 3.–We are still preparing ship for the great struggle in passing the forts, which event will come off in a day or two. It would have been done to-day had the *Tecumseh* been here from Pensacola. The *Sebago* and *Galena* arrived from Pensacola.

August 4.–To-morrow is contemplated the great battle for the occupation of Mobile Bay and the forts at its entrance. We were employed in the afternoon finishing preparations for the fight to-morrow. I have the conscientious feeling that every preparation has been made for the safety of the ship and the crew that was within my power. In the evening the *Richmond, Tecumseh,* and *Port Royal* arrived from Pensacola. In the morning all the commanding officers met on board the flagship by signal, and after consultation with the admiral we rejoined our ship with the following programme for the attack to-morrow: The gunboats *Genesee, Pembina, Penobscot, Sebago,* and *Pinola* and steamers *Bienville* and *Tennessee,* together with the tugboats, will remain outside. In going in it is not expected that we shall take the forts, but get possession of the bay, after which the forts must fall as a matter of course. Yesterday a detachment of our army landed on the west end of Dauphin Island and was fired upon by the enemy at Fort Powell. They will simultaneously attack Fort Gaines with our passage into Mobile Bay. What torpedoes or obstructions are in the ship channel we are ignorant. An effort on our part to pass in will be made, but the result is in the hands of the Almighty, and we pray that He may favor us.

August 5.–I thank the Almighty for the victory He has permitted us to gain this day and His special protection of us. The day was pleasant with a moderate westerly wind. Before its dawn all hands were called, an early and hurried breakfast eaten, and as the sun rose the vessels took positions alongside and lashed to each other, as required in the diagram mentioned yesterday. The *Lackawanna,* with the *Seminole* alongside, was in the center of the line of battle, and at 6 o'clock all had advanced up the Main Ship Channel, so that the leading vessels were enabled to open fire upon Fort Morgan, and each succeeding one did the same as soon as her guns would bear upon the fort. The fort opened fire upon us first, and the rebel boats *Tennessee, Morgan, Gaines,* and *Selma,* inside of the bay, raked our vessels with shot and shell. It was a magnificent sight. Every vessel with ensigns at her masthead and peak, the shot and shell flying through the air with their hissing sound. The dense volumes of smoke from the guns sometimes hiding the nearest ship, then floating away toward the fort, and the loud cheers of all hands. Although shot and shell were flying around, none struck the *Lackawanna's* hull doing serious injury till we were within 400 or 500 yards of Fort Morgan, when a heavy elongated shot from the fort passed through the ship's side, killing and wounding 16 men at the 150-pounder rifle, where it carried away two stanchions of the fife rail, passed through the foremast, carried away the head of the sheet cable bits, and passed through the other side of the ship, falling into the water. Blood and mangled human remains for a time impeded the working of the 150-pounder. The firing of shells from our fleet was so continuous that the enemy were driven from their guns, and when we had to bear away for the bay to clear the Middle Ground, leaving our stern exposed, few shots were fired by them, and those doing no injury. But the rebel ironclad ram *Tennessee* and gunboats *Morgan, Gaines,* and *Selma* still kept up their discharges. The *Morgan* and *Gaines,* however, soon after ran to the eastward under the guns of the fort, while the *Selma* stood up the bay [followed] by the *Metacomet* and another one of our gunboats, and was captured. The rebel ram *Tennessee* kept hovering off the fort, unsuccessfully endeavoring to strike our vessels in passing. One of our ironclad monitors, the *Tecumseh,* Commander T. A. Craven, being in his station in advance and abreast the leading ship of the fleet, about 400 yards from the shore at Fort Morgan, struck upon a torpedo, which, exploding, sunk her with all the officers. Only about a dozen of the crew and the pilot escaped from her and were picked up by the boats of the *Metacomet* and another vessel. At 8:30 o'clock in the morning our fleet had passed the Middle Ground, beyond range of the guns of Fort Morgan, and many of them had anchored, when the rebel ram *Tennessee* was seen approaching. The admiral made signal to the *Monogahela, as* being nearest, to run her down, and instantly afterwards the same was made to me. The *Monongahela* struck her angularly near the stern and glanced away. I was more fortunate and struck her at right angles to her keel. The concussion was tremendous and we rebounded, but soon after drifted against her, broadside to broadside, head and stern, when our marines and some of the crew with muskets and revolvers opened fire into her ports, preventing the reloading of her guns, which had been fired into our bows when almost touching, exploding two shells and sending one solid shot into our berth deck, killing and wounding many of the powder division and the already wounded under the surgeon's care. In running against the *Tennessee* we did her no perceptible injury except demoralizing the crew but our stem was cut and crushed far back of the plank ends. All the time I was standing on the bridge, and while alongside, looking into the ports of the *Tennessee,* one of the crew looking out, but standing at a distance from the port, hallooed out to me, "You d----d Yankee son of a b---h," which, being heard by the crew of the *Lackawanna,* redoubled their discharges of small arms into the rebel ports, and as some of them had not small arms in their possession, one of them threw a spitbox and another a hand holystone at the fellow. Our guns had been pivoted on the opposite side in anticipation of swinging head and head, so that but one IX-inch gun could be sufficiently depressed to bear upon the *Tennessee,* which was fired nearly into one of her ports, causing the port shutter to jam, becoming useless during the remainder of the engagement. We then separated in different directions by her going ahead, and we having nothing to hold on by, I ordered the helm hard over, to bring the ship around, to make another attempt at ramming the *Tennessee,* but our great length and the shoalness of the water, which sometimes was not more than a foot under the keel, prevented our turning rapidly and in going round, we collided with the *Hartford,* knocking two of her quarter-deck ports into one, although every effort was made on my part by backing the engine to prevent the occurrence. We sustained no injury by the collision. As soon as we cleared the *Hartford* I again started to run down the *Tennessee,* but before reaching her she had hauled down the flag, hoisted a white one, and surrendered to the fleet, which had by that time gotten round her and opened an incessant discharge of solid shot against her casemate. Our fleet then anchored, having added to it the *Selma* and *Tennessee,* both in good fighting order. It was three hours and four minutes from the time that the fleet opened fire on Fort Morgan until the surrender of the *Tennessee,* during which time the loss of officers and men was great in the fleet. Our loss alone was 4 killed instantly and 35 wounded. One of the latter died in the evening. Among the prisoners was Admiral Buchanan, who was severely wounded in the leg. In the evening the severely wounded, among them the captured admiral, were sent on board the *Metacomet* to leave, tomorrow morning under a flag of truce for the Pensacola hospital. After night the rebels abandoned and set fire to the quarters in Fort Powell, leaving us an open communication through Grant's Pass into Mississippi Sound. During the engagements of the day we fired 105 cannon shot and shell.

In my remarks of the 5th of August I neglected to mention the names of the officers of the *Lackawanna*, they being on board during the action: John B. Marchand, captain; Thomas S. Spencer and Stephen A. McCarty, lieutenants; James Fulton, paymaster; Thomas W. Leach, surgeon; W. F. Hutchinson, acting assistant surgeon; Felix McCurley and John H. Allen, acting masters; George H. Wadleigh and Frank Wildes, ensigns; Clarence Rathbone, acting ensign (regular): James W. Whittaker, first assistant engineer; E. J. Whitaker, second assistant engineer; George W. Roche and Isaac B. Fort, third assistant engineers; David Hennessy, George W. Russell and Thomas W. Sillman, acting third assistant engineers; Thomas Kelly, acting boatswain; John G. Foster, gunner; W. J. Lewis, C. H. Foster, and John C. Palmer, acting master's mates; Thomas Chessan, paymaster's clerk.

In my remarks of the 5th I did not mention the names of vessels comprising the fleet which safely passed Forts Morgan and Gaines and entered in triumph Mobile Bay, but with the loss of many men. They were: Steam sloop *Hartford*, Captain P. Drayton, with the admiral; steam sloop *Richmond*, Captain Thornton A. Jenkins; steam sloop *Lackawanna*, Captain J. B. Marchand; steam sloop *Brooklyn*, Captain James Alden; steam sloop *Monongahela*, Commander J. H. Strong; steam sloop *Ossipee*, Commander W. E. Le Roy; steam sloop *Oneida*, Commander J. R. M. Mullany.

August 8.–Mobile Bay. Another pleasant day until evening, when rain commenced falling. At 10 o'clock in the morning Fort Gaines surrendered and was occupied by our army. The number of prisoners obtained by the surrender was 850, who were taken away by army transports. The rebel officers had the option of surrendering their swords to the army or navy, and they all delivered them to the navy. Upon occasion of hoisting the Union flag on Fort Gaines three cheers were given by the fleet upon signal to that effect having been made by the admiral. In the forenoon called upon the admiral agreeably to a signal, and from there, in company with Captains Jenkins and Le Roy, went and held a survey on the captured steamer *Selma*. In the evening dined on board the *Ossipee* with Captain Le Roy. The other guests were Captain Strong, of the *Monongahela*, and Commander James D. Johnston, late commanding the rebel ironclad ram *Tennessee*, now a prisoner, captured on the 5th and in custody of Captain Le Roy on board the *Ossipee*.

August 9.–At an early hour in the morning our army, which had assembled at Fort Gaines, was transshipped in one tinclad and some tugboats across the bay and landed on the peninsula in Navy Cove, about 3½ miles east of Fort Morgan. The boats of the fleet were sent to put the soldiers ashore, but the wharf was perfect and the boats were unnecessary. Some of our ironclads and gunboats covered the landing of the troops, but the enemy made no opposition. About 9:30 in the forenoon I received orders from the admiral to go and shell Fort Morgan, and immediately a similar order was given to the *Monongahela*. The steam in both vessels was low. We started first, but the *Monongahela* having more steam, passed us and opened fire first upon the fort. Not long after we got to the Middle Ground and also opened on the fort. At that time the engine was stopped that the aim of the guns would be more perfect and it could not be started again on account of the lowness of the steam. For forty-eight minutes we lay waiting for steam to raise in the boiler, and all that time firing the pivot guns, none of the shells, however, seeming to do the enemy injury. While lying there the rebels fired but once at us, the shot passing directly over our ship and striking the water about 300 yards off. It was just noon when our engine started and we were making a circuit in turning to continue the fire, when the admiral made signal to cease firing and sent a flag-of-truce to demand the surrender of Fort Morgan. The reply was a refusal, and as it was late nothing more was done by the fleet except some firing on the fort by the captured ironclad *Tennessee*, which our fleet had officered and manned and had been towed for the occasion near, when she steamed by herself. The captured *Tennessee* is a great acquisition to our effective fleet, as is the *Selma,* which also fired several times at the fort. The ironclad *Tennessee* appearing nearer the fort than she really was, and the admiral fearing that she might be in some danger of grounding, sent the *Manhattan* to her assistance, but none being required, both were coming out when the latter grounded on the Middle Ground and was some time in getting off.

August 11.–The fleet doing little else than planting buoys preparatory to joining in the attack on Fort Morgan when the army, which is now preparing their batteries, are ready to open fire.

August 12.–Frequently during the day there was skirmishing between the advance posts of our army and the rebels at Fort Morgan. Several cannon shots from the fort were fired at them. The army people are working like beavers among the sand hills in their approaches toward Fort Morgan; they seem to be about 600 yards from it.

August 13.–Yesterday the captured ironclad *Tennessee* was fitted with a smokestack brought from Pensacola (hers having been shot away in the action), and this afternoon she steamed near Fort Morgan and for an hour fired upon the fort, after which the ironclads *Manhattan* and *Chickasaw,* from their anchorage east of the fort, also opened fire and kept it up until after 8 o'clock in the evening. The fort returned a severe fire, but did our vessels no injury.

August 14.–Firing was kept up all last night at intervals of about twenty minutes by our ironclad *Chickasaw* on the fort and ceased at daylight.

August 22.–At 4 in the morning all hands were called, and as daylight appeared we were underway and felt by the soundings, going slowly, the way in the bay to the north edge of the Middle Ground, so as to bring Fort Morgan to bear south, and anchored with 8 inches water under the keel on hard, sandy bottom. We had not reached our berth before the sun rose and at that moment the army and navy batteries on shore opened the bombardment on Fort Morgan. We were followed to our anchorage by the vessels designated yesterday, which opened fire with their rifle guns, making our ship their beacon for avoiding the Middle Ground. The tide set so strong that I could not wind the ship to bear upon the fort until after the signal was made by the admiral to cease firing, as it endangered the army sharpshooters that had approached near the glacis of the fort. During the subsequent part of the day the batteries on shore kept up an incessant firing, so that a rebel could not show himself on the fort. Consequently not a gun was fired from the fort after the bombardment commenced. Our ironclads also kept up a perpetual fire, at times being close to the wharf by the fort. After dark the firing continued brisk, and soon after 8 o'clock in the evening the shells thrown into the fort set fire to the rebel quarters. The shot and shell fell into the fort at the rate of about three a minute, preventing the extinguishing the flames. Throughout the night the conflagration continued, and it was magnificent. The sight of the flames, the burning course of the shells from the mortars, and the bursting shells from other guns combined with the most dense volumes of smoke. About midnight I went tired to my bed, and I slept with all the noise of whirring shot and exploding shells around, satisfied that the rebels could not long hold out in the burning fort.

August 23.–The bombardment of Fort Morgan continuing all night, was kept up until 6:30 o'clock this morning, when a white flag was hoisted on the fort in token of submission of the rebels. All firing ceased, and subsequent interview between the navy and army on our part, and General Richard [L.] Page, formerly of the Navy, now commanding here, eventuated in an unconditional surrender of Fort Morgan and its garrison of 800 men. At 3 p.m. the surrender was made, and by signal from the admiral I was invited to attend on the occasion. The rebel troops were marched out of the fort and, confronting about an equal number of our soldiers, stacked their arms and were marched on board steamer at the wharf, which came in for the purpose to carry them to New Orleans. The rebel officers were mustered between the two lines of soldiers and their swords delivered. Very few had any, and it was generally believed that they had destroyed them and prevaricated as each was interrogated. Even General Page had no sword. Having in former years been intimate with General Page, I had two or three private talks with him and offered him every hospitality, purse, and mess, etc., but as he was to leave immediately for New Orleans it could not be accepted.

September 12.–For an hour or two *yesterday* evening heavy cannonading was heard toward the northward and eastward. This morning it was explained by our gunboats at Fish River, on the eastern side, as the shelling of the woods where rebel sharpshooters were concealed firing at our vessels.

October 9.–Pensacola Navy Yard. The following vessels are here: Steam sloop *Lackawanna*, Captain J. B. Marchand; sailing frigate *Potomac*, Commander Alex. Gibson; double-ender *Metacomet*, Lieutenant-Commander J. E. Jouett; steam gunboat *Kanawha*, Lieutenant-Commander B. B. Taylor; steam gunboat *Cayuga*, Lieutenant-Commander H. Wilson; steam gunboat *Genesee*, Acting Master C. H. Baxter; sailing bark *Arthur*, Acting Master J. E. Stannard; sailing bark *J. C. Kuhn*, Acting Ensign S. H. Newman; schooner *Kittatinny*, Acting Ensign N. J. Blasdell; bark *W. G. Anderson*, Acting Ensign R. H. Carey; supply steamer *Bermuda*, Acting Volunteer Lieutenant J. W. Smith.

3

Cotton Cargo Jettisoned by the *Denbigh*, 1865

The *Denbigh* departed Galveston with a load of cotton on the night of 19-20 April 1865. The runner promptly ran aground on the shallow sand bars choking the bay mouth. The ship likely was using a shallow inshore channel and must have become securely stuck. In order quickly to refloat the ship in this emergency, the deckload of the cotton was thrown overboard, and the *Denbigh* returned to Galveston unobserved by the blockading fleet. Some bales of the *Denbigh's* cotton drifted ashore on Bolivar Peninsula, generating an interesting entry (Figure 12) in the orders book of the C.S. Army District of Texas, shown here.[36] Other bales drifted offshore near the Union blockading vessels *Cornubia* and *Gertrude*, which salvaged eighty-nine and fifty bales, respectively. This capture was a fine irony, as both ships were former runners taken into the U.S. Navy after capture.

The prize case documents for the *Cornubia* and the eighty-nine bales of cotton[37] appeared in Powell et al.[38] The current volume presents the litigation documents for the *Gertrude* and the fifty bales of cotton.[39] The salvaged cotton was treated as a prize of war and sent to New Orleans for adjudication, just as would have been a captured ship. The court documents and the Treasury Department documents for payment to the crews were particularly interesting for the *Denbigh* Project. Therefore, they were discussed in an abbreviated form in Powell et al. and shown here in a complete form. Also included were the after-action reports to the West Gulf Blockading Squadron headquarters and the Navy Department, plus the deck logs of the blockading fleet showing the activities of the Union ships the day of the incident. It was informative to examine the loss and recovery of the cotton bales from the varying points of view provided by the archival documents. The sets of documents supported speculation concerning what the opposing Union and Confederate groups might have been thinking. A careful reading of the documents enabled picking out interesting details of behavior not available elsewhere. Footnotes herein discuss such factors and are particularly commended to the reader's attention.

Confederate Orders Book Entries

The whole page was here transcribed in addition to paragraph 25 about recovering the jettisoned bales of cotton on Bolivar Peninsula. Administrative orders revealed pertinent detals of military life in Galveston. (Parts of paragraph 25 shown in bold type for emphasis.)

Concerning the note on increased rations for the steamboat crews, see the daily reports from the observatory atop the Hendley Building in Galveston.[40] Three subjects comprised by far the greatest number of entries in the observatory's logbooks: movements and activities of the Union blockading

---

[36] National Archives and Records Administration (NARA) Record Group (RG) 24, Chapt. 2:103-128.
[37] The United States Steamer *Cornubia*[,] John A. Johnston, U.S.N. Commander[,] The Officers and crew thereof vs. Eighty-nine Bales of Cotton, Cause #8036. NARA.
[38] Powell, G. R., M. S. Cordon, and J. B. Arnold III. *Civil War Blockade-Runners: Prize Claims and the Historical Record, Including the* Denbigh's *Court Documents*. College Station, Texas: Institute of Nautical Archaeology, *Denbigh* Shipwreck Project Publication 6, 2012.
[39] The United States Steamer *Gertrude*[,] Benjamin C. Dean, U.S. Commanding[,] the Officers and crew thereof vs. Fifty Bales of Cotton, Cause #8307. NARA.
[40] JOLO Log Book, 1861, Rosenberg Library, Galveston, Texas and Chief Signal Reports from the Houston Observatory etc., 1863–1864, RG 109, NARA. Online at https://research.archives.gov/id/7682655. The NARA entry mistakenly said the lookout was located in Houston when the observatory was actually in Galveston.

fleet, weather conditions, and comings and goings of several steamboats on government service in the bay. The steamboats were constantly in use going back and forth from the Galveston wharves to Fort Jackson on Pelican Spit, Bolivar Peninsula, Houston, and West Galveston Bay. The busy crews indeed rated extra rations.

S.O. (Special Order) No. 111
Galveston, April 21, 1865

23. On account of the nature and severity of the service, hereafter, one and a half rations will be issued to all the Ship Carpenters and steamboat men. The Chief C.S. will issue the necessary instructions.   E.P.S. (?)

24. Mr. Alexander Campbell, Master's Mate, C.S. Navy, is hereby permitted to proceed on board the C.S. Steamer *Lark* with the view of obtaining a permanent assignment as Purser on board of her.   E.P.S.

25. Capt. McGary, Comdg. **Str. *Diana*** now under orders to take the C.S. Str. *Owl* (Figure 13) in tow to Lynchburg (near Houston) on tomorrow morning after transporting ordnance stores to Fort Jackson on Pelican Spit **will proceed to Bolivar and collect as much as possible of the cotton necessarily cast overboard by the *Denbigh* which cotton had drifted ashore on Bolivar.** As soon as this is accomplished she will return to this place and take in tow the Str. *Owl* as previously ordered. **This order is given for the reason that most of the cotton referred to is Govt. cotton** and in consideration of the fact that little time will be lost in getting the *Owl* (now located) on Red Fish Bar (part way up Galveston Bay on the way to Lynchburg).
      E.P.S.

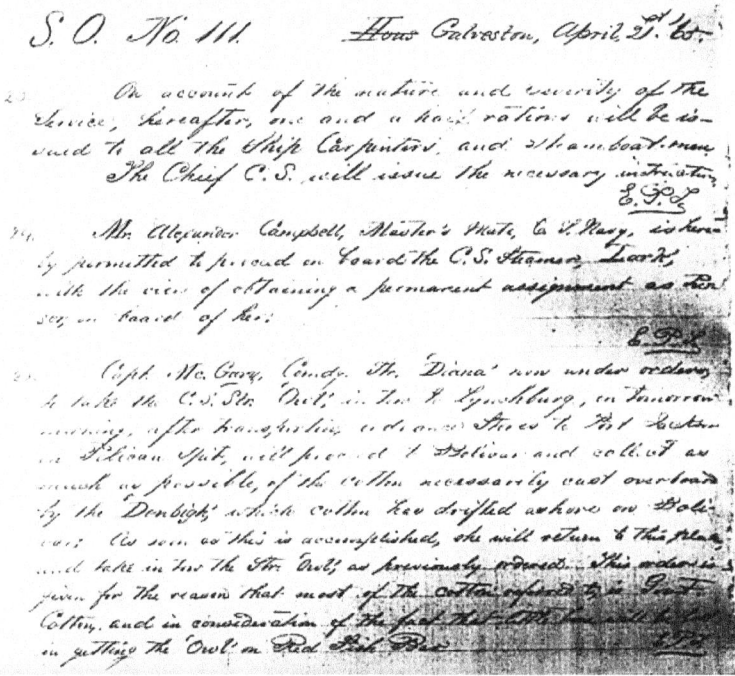

Figure 12. Special Order 111 issued at Galveston concerning recovering part of the *Denbigh's* jettisoned cotton bales and other interesting details of Confederate naval life in the blockaded port. *NARA.*

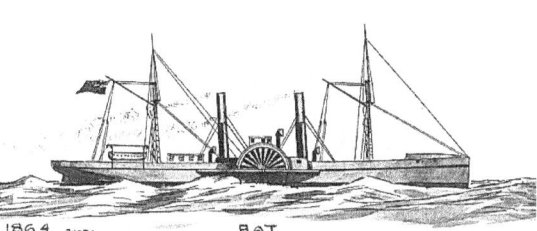

Figure 13. The blockade-runner *Owl* was similar to the *Bat*, her sister ship. *Naval Historical Center.*

Reports and Prize Documents
for U.S.S. *Gertrude* vs. Fifty Bales of Cotton

Navy Reports

This section begins with summary reports on the jettisoned cotton recovery that were sent up the chain of command from the Union squadron blockading Galveston.[41]

Capture of cotton by the U.S. steamers Cornubia *and* Gertrude, *April 20, 1865.*

Report of Captain Sands, U.S. Navy, transmitting reports of commanding officers.

U.S.S. *FORT JACKSON,*
Off Galveston, Tex., April 21, 1865.

The steamer was seen next under Fort Point and returned to the city. The bar was very rough and the night hazy.

I have the honor to be, very respectfully, your obedient servant,

B. F. SANDS,
Captain, U.S. Navy, Comdg. Third Div. West Gulf Squadron.

The steamer that [we] tried and failed to get was the *Denbigh*, and lost 200 bales in the attempt.

(To) Acting Rear-Admiral H. K. THATCHER,[42]
Comdg. West Gulf Squadron, Hdqrs., New Orleans.

Report of Acting Volunteer Lieutenant Dean, U.S. Navy, commanding U.S.S. *Gertrude*.

U.S.S. *GERTRUDE,*
Off Galveston Bar, Texas, April 21, 1865.

SIR: I have the honor to report that on the morning of the 20th instant, whilst at anchor off Bolivar Channel, at 6 a.m., discovered a number of bales of cotton that had been thrown overboard from a blockade runner which had got aground on the bar the night preceding while attempting to run out, and finding he was closely watched and would undoubtedly have been

---
[41] *ORN*, ser. 1, 22:125-137.
[42] The letter format in the Civil War period placed the addressee at the bottom.

destroyed, he threw overboard said cotton and succeeded in getting off safe and ran back to the town.

We picked up 50 bales, which I shall send to New Orleans by the first supply vessel, for adjudication, as per order of senior officer, Capt. Benjamin F. Sands, Commanding Third Division West Gulf Squadron.

Hoping this may meet your approval, sir, I remain,
Your obedient servant,

BENJ. C. DEAN,
Acting Volunteer Lieutenant, Commanding.

Hon. GIDEON WELLES,
Secretary of the Navy, Washington, D. C.

Report of Acting Volunteer Lieutenant Johnstone, U.S. Navy,
commanding U.S.S. *Cornubia.*

U.S.S. *CORNUBIA,*
Off Galveston, Tex., April 22, 1865.

SIR: I hereby respectfully inform the Department that on the 20th instant, off Galveston Bar, this ship, under my command, picked up 89 bales of cotton, part of the cargo of a blockade runner that got ashore on the bar the night previous in attempting to come out. The cotton will be sent to New Orleans for adjudication by the earliest opportunity.

Enclosed is a list of the officers and crew attached to this ship and borne upon her books at the time the cotton was picked up.[43]

Very respectfully, your obedient servant,

JOHN A. JOHNSTONE,
Acting Volunteer Lieutenant.

Hon. GIDEON WELLES,
Secretary of the Navy.

Report of Acting Volunteer Lieutenant Johnstone, U.S. Navy, commanding U.S.S. Cornubia, *regarding the capture of the schooner* Chaos, *with a cargo of cotton.*

U.S.S. *CORNUBIA,*
Off Galveston, Tex., April 22, 1865.

SIR: I have the honor to report the capture of the English schooner *Chaos*, with a cargo of 170 bales of cotton, by this ship, under my command, about 10 p.m. on the 21st instant, while attempting to run the blockade from the port of Galveston.

The U.S.S. *Fort Jackson* came up and spoke us a few minutes after the capture was made.

As the schooner was brought to within 3 miles of the bar, I have no doubt that other blockading vessels were within signal distance at the time, but I do not know how many, or their names.

The vessel will be sent to New Orleans for adjudication with all the necessary witnesses and documents.

Enclosed is a list of the officers and crew attached to this ship and borne on her books at the time of the capture.

Very respectfully, your obedient servant,

---

[43] The most up to date list of officers and crew was required in order to determine who would receive prize money for the capture of the cotton. The pay rate of each man also was reported on that list and used in making the calculations to divide up the prize money.

JOHN A. JOHNSTONE,
Acting Volunteer Lieutenant.

Hon. GIDEON WELLES,
Secretary of the Navy.

Report of the U.S. prize commissioner at New Orleans of the receipt of schooner Chaos[44] *and captured cotton.*

U.S. PRIZE COMMISSIONER'S OFFICE,
Custom-house, New Orleans, May 12, 1865.

SIR: As prize commissioner of the U.S. District Court, Eastern District of Louisiana, I have the honor to report the arrival of the prize schooner *Chaos*, of 63 tons burden, with a cargo of 170 bales of cotton, captured on 21st day of April, 1865, by the U.S.S. *Cornubia*, John A. Johnstone, U.S. Navy, commanding, off Galveston Bar, coast of Texas.

Also of 89 bales of cotton picked up at sea by the said steamer *Cornubia*, on the 20th April, 1865.

Also 50 bales of cotton picked up at sea by the U.S.S. *Gertrude*, B.C. Dean, U.S. Navy, commanding, on the 20th April, 1865.

Very respectfully, your obedient servant,

U.S. Prize Commissioner.

Hon. GIDEON WELLES,
Secretary of the Navy.

Prize Case Documents from the U.S.S. *Gertrude*

Upon arriving in New Orleans where the prize court was located, the *Gertrude's* prize master filed a form (Figure 14) with the prize commissioner's office to initiate the case.[45]

District Court of the United Sates for the Eastern District of Louisiana.

In the Matter of
50 Bales of Cotton[46] R. R. Brawley Prize
And her cargo   Master of fifty bales of cotton
A Prize.

Assistant Acting Paymaster U.S.N. detached from the U.S.S. *Gertrude* of which B. C. Dean is commander, being duly sworn deposes and says, that the fifty bales were picked up at sea off Galveston Harbor Texas, on the 20th of April 1865.
(Remainder of standard form crossed out as the verbiage did not apply.)
(signed) R. R. Brawley
A. (Acting) A. (Assistant) Paymaster

Sworn to and signed before me United States Prize Commissioner, this the 9th day of May 1865, New Orleans, La.

---

[44] The capture of the runner *Chaos* was reported at the same time as that of the floating cotton.
[45] The United States Steamer *Gertrude*[,] Benjamin C. Dean, U.S.N. Commanding[,] the Officers and crew thereof vs. Fifty Bales of Cotton, Cause #8307. NARA.
[46] Underlined sections indicate blanks in the printed form that were filled in manually. The blanks were filled in by hand written content.

(signed) T. B. Thorpe
Prize Commissioner

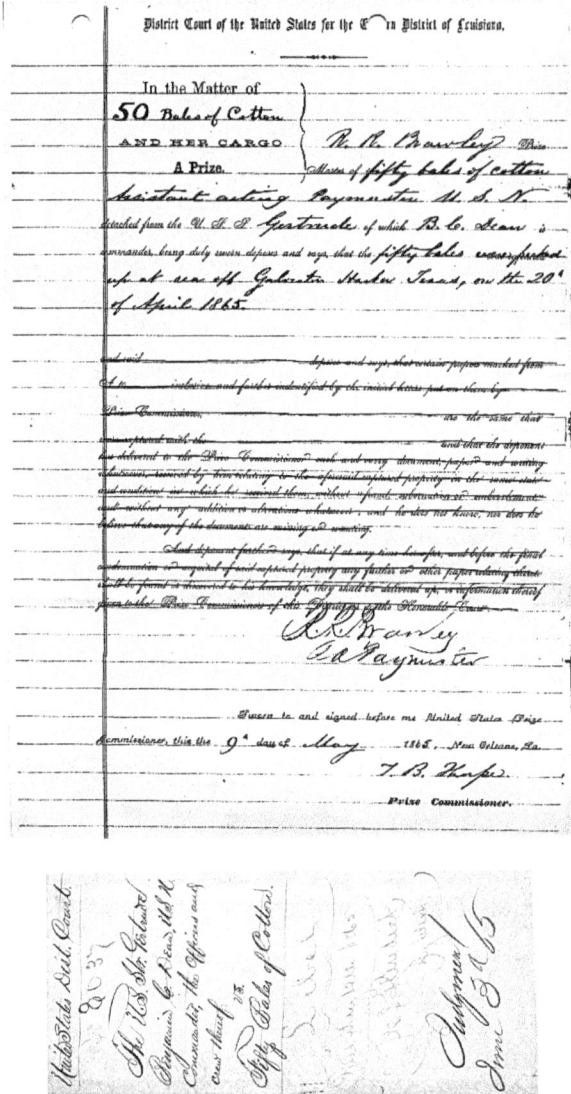

Figure 14. (Above) When the prize master from the *Gertrude*, R. R. Brawley, first arrived in port accompanying the prize, he reported to the U.S. prize commissioner, an officer of the court. There he filed this form stating the basic facts of a new prize case. Dated 9 May 1865. (Below) A note on the back of the document showed that the court issued a judgment in favor of the *Gertrude's* claim on 3 June 1865. *NARA*.

The *Gertrude's* Monition regarding the cotton

The monition was the substantial document filed with the court initiating the legal action against the prize and providing more complete details of the capture (Figure 15). In some ways, the case was like a combination prize and salvage litigation. The "dangerous service" involved "to save from total loss" were terms and conditions of a valid salvage case rather than a prize of war.

In the United States District Court
Eastern District of Louisiana.

The United States Steamer *Gertrude*
Benjamin C. Dean, U.S. Commanding
The Officers and crew thereof   In Admiralty
        vs.
Fifty Bales of Cotton.

      To the Honorable Edward H. Durell, Judge of the District Court of the United States for the Eastern District of Louisiana:

      Benjamin C. Dean, Commander of the U.S. Steamer *Gertrude*, for himself as well as for the officers and the several persons composing the crew of said Steamer *Gertrude*, exhibit this their libel against Fifty Bales of Cotton, (names of the owners of said cotton to said Libellants unknown,) now in the Port of New Orleans in the District aforesaid, and within the admiralty and maritime jurisdiction of this Hon. Court, and against all persons intervening for their interests therein.
      And thereupon the said Libellants, Benjamin C. Dean, Commander &c. for himself and Co-Libellants, alleges and articulately propounds as follows to wit:
      First: That on the 20th day of April 1865, the aforesaid Steamer *Gertrude*, whereof the aforesaid, Benjamin C. Dean was commander, with her crew was cruising off the coast of Texas in the service of the United States.
      Second: That while proceeding on her said cruise, to wit: on the day and year aforesaid the said Steamer *Gertrude*, being at sea, descried a number of bales of cotton floating derelict—that orders were thereupon given on board the said steamer *Gertrude* to bear down upon the scattered and floating cotton and pick up the same, which was immediately done. That the said libellants [*sic*], on the said 20th day of April 1865, picked up and saved Fifty Bales of Cotton. That the same were saved by a great deal of hard and dangerous service, and was attended with the risk of their lives. That the said cotton was brought to this port.
      Third: That but for the assistance and dangerous service so rendered by said libellants the said fifty bales of cotton would have been totally lost. That said libellants, by reason of the dangerous services and perils necessarily incurred, and the great importance and nature of the services rendered by them in saving the said cotton from total loss, reasonably deserved to have, and they therefore claim a commensurate reward by way of salvage, therefor. That all and singular the premises are true as above stated.
      Wherefore the said libellants pray that process in due form of law may issue against the said Fifty Bales of Cotton, and that this Hon. Court will pronounce for the demand of the libellants, and will decree for them such compensation and reward, by reason of the premises, as shall appear to be just and reasonable, together with their costs and expenses, and such other and further relief as to right and justice may appertain, and as the Court is competent to give in the premises.

                                                           (signed) Rupert Waples
                                                                 U.S. Atty.

In the United States District Court
Eastern District of Louisiana.

The United States Steamer Gertrude
Benjamin C. Dean, U.S. Commanding
the Officers and crew thereof } In Admiralty.
vs.
Fifty Bales of Cotton.

To the Honorable Edward H. Durell, Judge
of the District Court of the United States for the
Eastern District of Louisiana:

Benjamin C. Dean, Commander of
the U.S. Steamer Gertrude, for himself as well as
for the officers and several persons composing the crew of said
Steamer Gertrude, exhibit this their libel against
Fifty Bales of Cotton, (names of the owners of said
cotton to said Libellants unknown,) now in the Port
of New Orleans in the District aforesaid, and within
the Admiralty and maritime jurisdiction of this
Hon. Court, and against all persons intervening for
their interests therein.

And thereupon the said
Libellants, Benjamin C. Dean, Commander &c. for
himself and Co-Libellants, alleges and articulately
propounds as follows, to wit:

First: That on the 20th day of April 1865,
the aforesaid Steamer Gertrude, whereof the

aforesaid Benjamin F. Sands was Commander, with her crew was cruising off the Coast of Texas in the service of the United States.

Second: That while proceeding on her said cruise, to wit: on the day and year aforesaid the said Steamer "Gertrude", being at sea, descried a number of bales of cotton floating derelict — That orders were thereupon given on board the said Steamer Gertrude to bear down upon the scattered and floating cotton and pick up the same, which was immediately done; That the said libellants, on the said 20th day of April 1865, picked up and saved fifty bales of cotton; That the same were saved by a great deal of hard and dangerous service, and was attended with the risk of their lives; That the said cotton was brought to this port.

Third: That but for the assistance and dangerous service so rendered by said libellants the said fifty bales of cotton would have been totally lost; That said libellants, by reason of the dangerous services and perils necessarily incurred, and the great importance and nature of the services rendered by them in saving the said cotton from total loss, reasonably deserve to have, and they therefore claim a commensurate reward by way of salvage, therefor; That all and singular the premises are true as above stated.

Wherefore the said libellants pray that process in due form of law may issue against the said Fifty bales of cotton, and that this Hon.

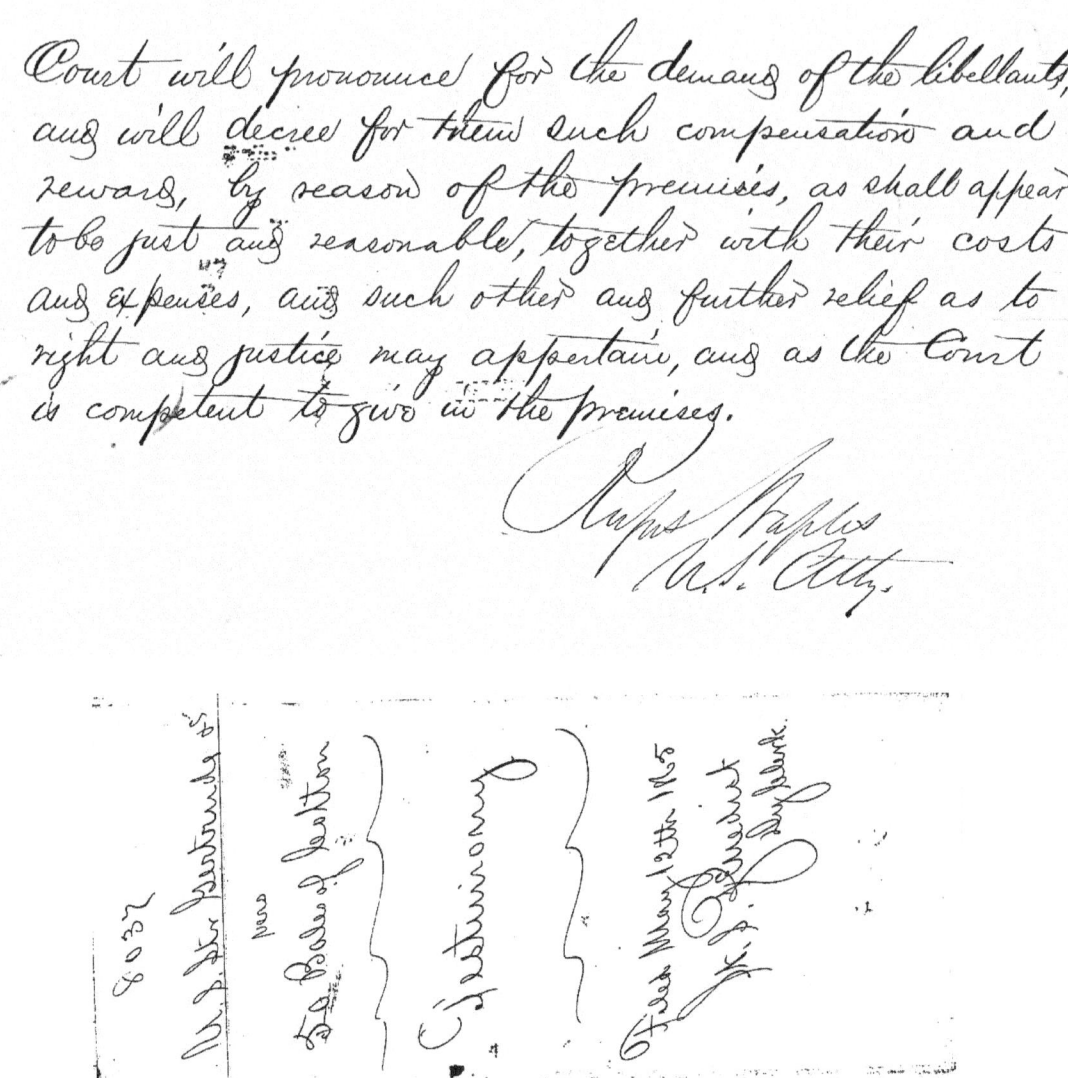

Figure 15. The monition, a document that initiated the prize case of the *Gertrude* vs. Fifty Bales of Cotton jettisoned by the *Denbigh* when she ran aground leaving Galveston on the night of 19-20 April 1865. Three-page document (above) and the inscription on the document cover (below). Filed 12 May 1865 in New Orleans. *NARA*.

Gertrude crew testimony

Several members of the *Gertrude's* crew sailed to New Orleans with the prize cotton. It was important for eyewitnesses from the capturing navy vessel to be present to testify as to the details of the capture (Figure 16). This category of accounts, from the researcher's viewpoint, provided some of the most interesting stories. Most of the documents in the case were in stiff legal language; the crew testimony was more personal in tone.

In the matter of ~~the~~
50 bales of Cotton
and Cargo, a Prize

                      District Court of the United States,
                      For the Eastern District of Louisiana

Deposition of Morris Garvin, Seaman on board of the U.S.S. *Gertrude* a witness produced, sworn and examined *in preparatorio* on the 10th day of May in the year eighteen hundred and sixty five at the U.S. Prize Commissioner,

On the standing interrogatories established by the District Court of the United States, for the Eastern District of Louisiana, the said witness having been produced for the purpose of such examination by R. R. Brawley Acting Paymaster U.S.N. in behalf of the captors of a certain 50 bales of cotton.

To the first interrogatory the witness answers that he is a native of Portsmouth, N.H., owes allegiance to the United States, no family, by profession a seaman, now a coal heaver on the U.S.S. *Gertrude*, of which B. C. Dean is commander. Between six and 7 o'clock in the morning of the 20th of April 1865, I was forward on the U.S.S. *Gertrude*, when I heard that there were goods floating on the sea thrown over from a blockade runner. I looked over the bow and saw the U.S.S. *Cornubia* about two miles and a half off, with her boats out picking up what proved to be bales of cotton. The executive officer of the *Gertrude* now came forward, and ordered the anchor to be taken up, when we steamed up near the *Cornubia*, when all the boats of the *Gertrude* were lowered and manned, and the crews of the boats went to work picking up this cotton. The amount of bales I understand was fifty bales. This cotton was picked up at sea off the harbor of Galveston, Texas. After it was picked up, the cotton, with that picked up by the boats of the *Cornubia*, was put on board of the U.S.S. *Denismore*, and traveled to New Orleans.

                      (signed) Maurice Garvin

Sworn to and signed before me this the 10th day of May 1865, New Orleans La.
                (signed) T. B. Thorpe
                United States Prize Comm.

Robert Duffey, first class boy, on board of the U.S.S. *Gertrude*, a witness produced, sworn, and examined *in preparatorio*, on the 19th of May 1865, before me, T. B. Thorpe, U.S. Prize Commissioner, says, that he was on board of the *Gertrude* on the 20th of April, and that he was in the boats belonging to the *Gertrude* and assisted in gathering up the cotton in amount about fifty bales. That this cotton was picked up on the open sea off Galveston, Texas. The boats of the *Cornubia* U.S.S. were also engaged in picking up cotton at the same time, the *Cornubia* and the *Gertrude* at the time we were picking up the cotton being near each other. I am a native of Louisiana, loyal to the United States. The fifty bales of cotton picked up by the *Gertrude*, and that picked up by the *Cornubia* were transported to New Orleans on board of the U.S.S. *Denismore*.

                      (signed) Robert Duffey

Sworn to and signed before me this the 10th day of May 1865, New Orleans, La.
                (signed) T. B. Thorpe
                U.S.P.C.

In the matter of the
50 bales of Cotton
and Cargo, a Prize

District Court of the United States,
For the Eastern District of Louisiana.

Deposition of Morris Garvin Seaman on board of the U.S.S. Gertrude a witness produced, sworn and examined in preparatorio on the 10th day of May in the year eighteen hundred and sixty-five at the U.S. Prize Commissioners,

on the standing interrogatories established by the District Court of the United States for the Eastern District of Louisiana; the said witness having been produced for the purpose of such examination, by R. R. Bawley Acting Paymaster U.S.N. in behalf of the captors of a certain 50 bales of cotton.

To the first interrogatory the witness answers that he is a native of Portsmouth N.H. owes allegiance to the United States, exclusively, by profession a seaman, now a coal heaver on the U.S.S. Gertrude, of which B. C. Dean is commander. Between six and 7 o'clock in the morning of the 20th of April 1865, I was forward on the U.S.S. Gertrude, when I heard that there were goods floating at sea thrown over from a blockade runner. I looked over the bow and saw the U.S.S. Cornelia about two miles and a half off, with her boats out picking up what proved to be bales of cotton. The executive officer of the Gertrude now came forward, and ordered the anchor to be hove up, when we steamed up near the Cornubia, when all the boats of the Gertrude were lowered and manned, and the crews of the boats went to work picking up this cotton. The amount of bales I understood was fifty bales. This cotton was picked up at sea, off the harbor of Galveston, Texas. After it was picked up, the cotton, with that picked up by the boats of the Cornubia, was put on board of the U.S.S. Dinsmore, and brought to New Orleans.

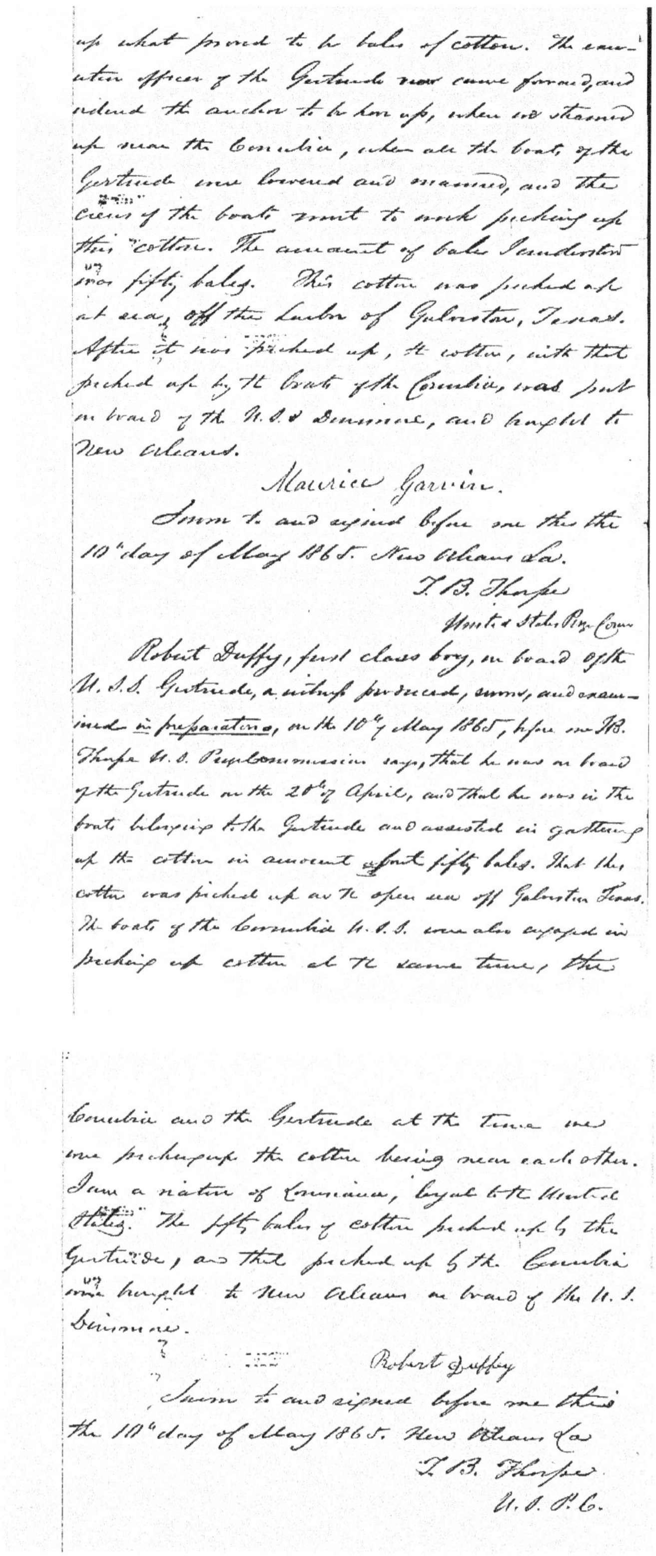

Maurice Garvin.

Sworn to and signed before me the the 10th day of May 1865. New Orleans, La.

T. B. Thorpe
United States Prize Comr

Robert Duffy, first class boy, on board of the U.S.S. Gertrude, a witness produced, sworn, and examined in preparation, on the 10th of May 1865, before me T. B. Thorpe U. S. Prize Commission says, that he was on board of the Gertrude on the 20th of April, and that he was in the boats belonging to the Gertrude and assisted in gathering up the cotton in amount about fifty bales. That this cotton was picked up at the open sea off Galveston Texas. The boats of the Cornubia U.S.S. were also engaged in picking up cotton at the same time, the

Cornubia and the Gertrude at the time we were picking up the cotton being near each other. I am a native of Louisiana, loyal to the United States. The fifty bales of cotton picked up by the Gertrude, and that picked up by the Cornubia were brought to New Orleans on board of the U. S. Dinsmore.

Robert Duffey

Sworn to and signed before me this the 10th day of May 1865. New Orleans, La.

T. B. Thorpe
U. S. P. C.

Figure 16. Depositions of two crew members of the *Gertrude*, Morris Garvin, seaman (coal heaver) and Robert Duffey, first class boy. They provided firsthand, detailed accounts of recovering the cotton bales. (This page and the prior page.) Filed 12 May 1865. *NARA*.

Admiralty Warrant for the Arrest of the Cotton

The court issued an order for the U.S. marshal to arrest (or take into the court's possession pending adjudication) the cotton bales brought in by the crew of the *Gertrude* as a prize (Figures 17-19). The printed form gave official notice to the owners to appear at the hearing if they wished to file an argument against the seizure.

The President of the United States,

To the Marshal of the Eastern District of Louisiana,

or to his lawful deputy, greeting:

You are hereby commanded, forthwith to seize, and into your possession take, the

Fifty Bales of Cotton

now libeled by the U.S. Str. *Gertrude*, B. C. Dean, U.S.N. Comdg. & the Officers & crew thereof

for the causes set forth in the Libel now pending in the District Court of the United States for the Eastern District of Louisiana; that you do cite and admonish the owner or owners and all and every other person or persons having, or pretending to have, any right, title or interest in or to the same, to be and appear before a District Court of the United States, for the District aforesaid, to be holden at the City of New Orleans, on or before the third Monday from the service hereof, to show cause, if any they have or can, why this said

Fifty Bales of Cotton

should not be condemned and sold agreeably to the prayer of the libellant: and how you have executed this warrant, that you make return according to law.

Witness, the Honorable Edward H. Durell, Judge
of the said Court, at New Orleans this 12th
day of May 186_5_, and
the 89th year of the Independence of
the United States.
(signed) M. G. Benedict
Dty. Clerk.

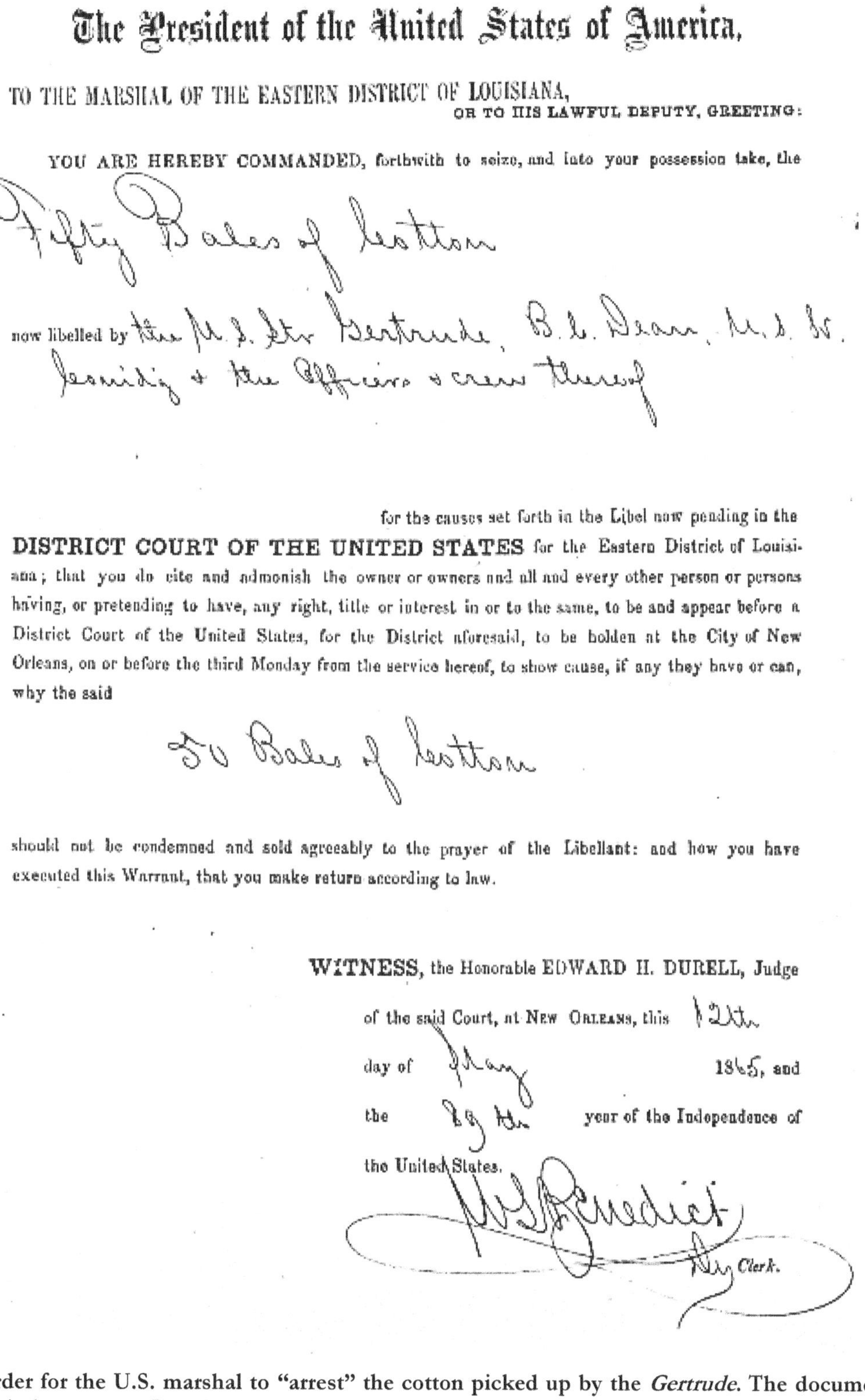

The President of the United States of America,

TO THE MARSHAL OF THE EASTERN DISTRICT OF LOUISIANA,
OR TO HIS LAWFUL DEPUTY, GREETING:

YOU ARE HEREBY COMMANDED, forthwith to seize, and into your possession take, the

Fifty Bales of Cotton

now libelled by *the U.S. Str Gertrude, B.L. Dean, U.S.N. Jennings & the Officers & crew thereof*

for the causes set forth in the Libel now pending in the **DISTRICT COURT OF THE UNITED STATES** for the Eastern District of Louisiana; that you do cite and admonish the owner or owners and all and every other person or persons having, or pretending to have, any right, title or interest in or to the same, to be and appear before a District Court of the United States, for the District aforesaid, to be holden at the City of New Orleans, on or before the third Monday from the service hereof, to show cause, if any they have or can, why the said

50 Bales of Cotton

should not be condemned and sold agreeably to the prayer of the Libellant: and how you have executed this Warrant, that you make return according to law.

WITNESS, the Honorable EDWARD H. DURELL, Judge of the said Court, at NEW ORLEANS, this *12th* day of *May* 1865, and the *89th* year of the Independence of the United States.

W. S. Benedict, Dy Clerk.

Figure 17. Order for the U.S. marshal to "arrest" the cotton picked up by the *Gertrude*. The document was called an admiralty warrant. Dated 12 May 1865. *NARA*.

Picayune

United States District Court.
No. 8037
U.S. Str. *Gertrude* &c.
vs.   Ad. Warrant.
Fifty Bales of Cotton

Marshal's Return

29th May

Received May 12th 1865, and in obedience to this precept I seized and took into my possession the property within described, posted a copy of this Warrant, Libel, and Judge's order on the door of the court house, and published the Monition in the *Picayune*, on the 13th, 18th, 23rd & 27th May 1865.

Recd. New Orleans   (signed) M. Egan
May 31, 1865   Dpy. U.S. Marshal

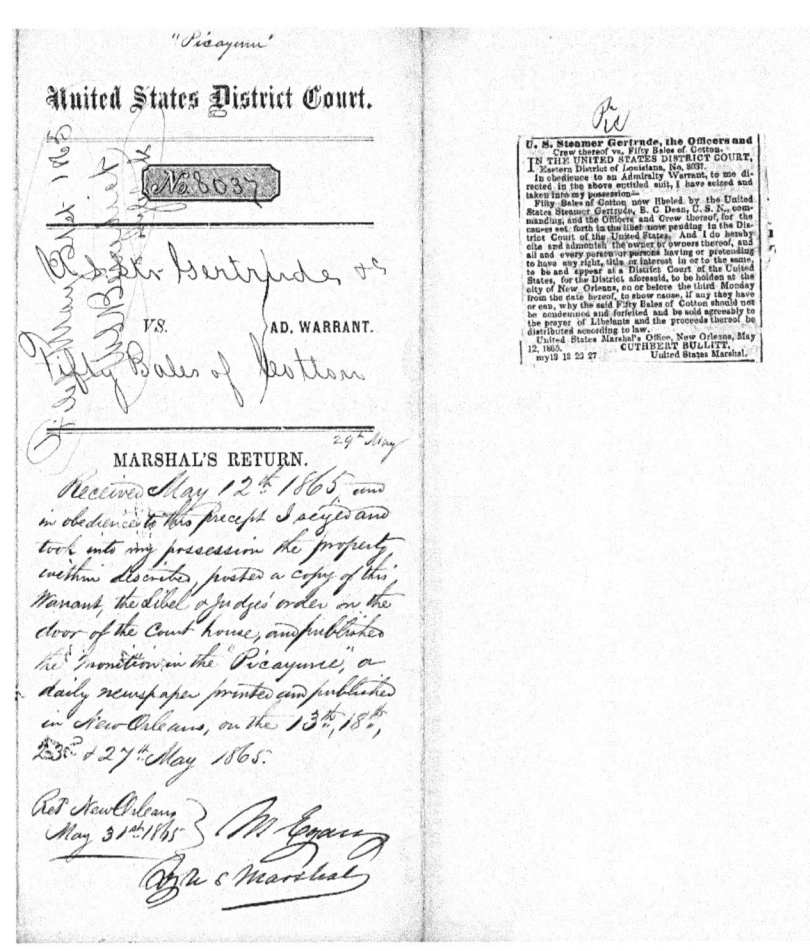

Figure 18. The back of the admiralty warrant provided a form for the marshal's return or proof of execution of the arrest and advertisement calling for the owners to come forward to dispute the seizure. The form included a copy of the newspaper ad that provided the public notice. Dated 29 May 1865. *NARA*.

> U. S. Steamer Gertrude, the Officers and Crew thereof vs. Fifty Bales of Cotton.
> IN THE UNITED STATES DISTRICT COURT, Eastern District of Louisiana, No. 8037.
> In obedience to an Admiralty Warrant, to me directed in the above entitled suit, I have seized and taken into my possession—
> Fifty Bales of Cotton now libeled by the United States Steamer Gertrude, B. C. Dean, U. S. N., commanding, and the Officers and Crew thereof, for the causes set forth in the libel now pending in the District Court of the United States. And I do hereby cite and admonish the owner or owners thereof, and all and every person or persons having or pretending to have any right, title, or interest in or to the same, to be and appear at a District Court of the United States, for the District aforesaid, to be holden at the city of New Orleans, on or before the third Monday from the date hereof, to show cause, if any they have or can, why the said Fifty Bales of Cotton should not be condemned and forfeited and be sold agreeably to the prayer of Libelants and the proceeds thereof be distributed according to law.
> United States Marshal's Office, New Orleans, May 12, 1865. CUTHBERT BULLITT,
> my 13 18 23 27   United States Marshal.

Figure 19. Advertisement (enlarged) from the marshal's return. Note the dates published at the lower left. *NARA.*

U.S. Marshal's Report of Cotton's Condition

The U.S. marshal, an officer of the court, assumed control of the cotton bales when the *Gertrude's* men presented themselves to the U.S. prize commissioner with the captured cotton. In this case, the marshal reported to the court that the bales needed a quick sale as they were damaged by salt water (Figure 20). The damp posed a continuing danger of spoilage. Presumably the purchaser would dry or process the cotton before shipping it on to its final destination.

U.S. Str. *Gertrude*
B. C. Dean, U.S.N. Comdg.
the Officers & Crew thereof
       vs.
Fifty Bales of Cotton

To the Hon. E. H. Durell, Judge of the Dist. Court of the United States, East. Dist. of La.

The undersigned U.S. Marshal, respectfully reports that the cotton libeled herein is seriously damaged ~~and that~~ by salt water and is now in a perishable condition and every day deteriorates in value, and in justice to the parties interested would recommend that the said cotton be sold and the proceeds deposited according to law.

                           (signed) Cuthbert Bullett
                                   U.S. Marshal

On Motion of R. Waples, U.S. Atty.

Ordered that the above cotton, after due advertisement, be sold, and the proceeds be deposited according to law.

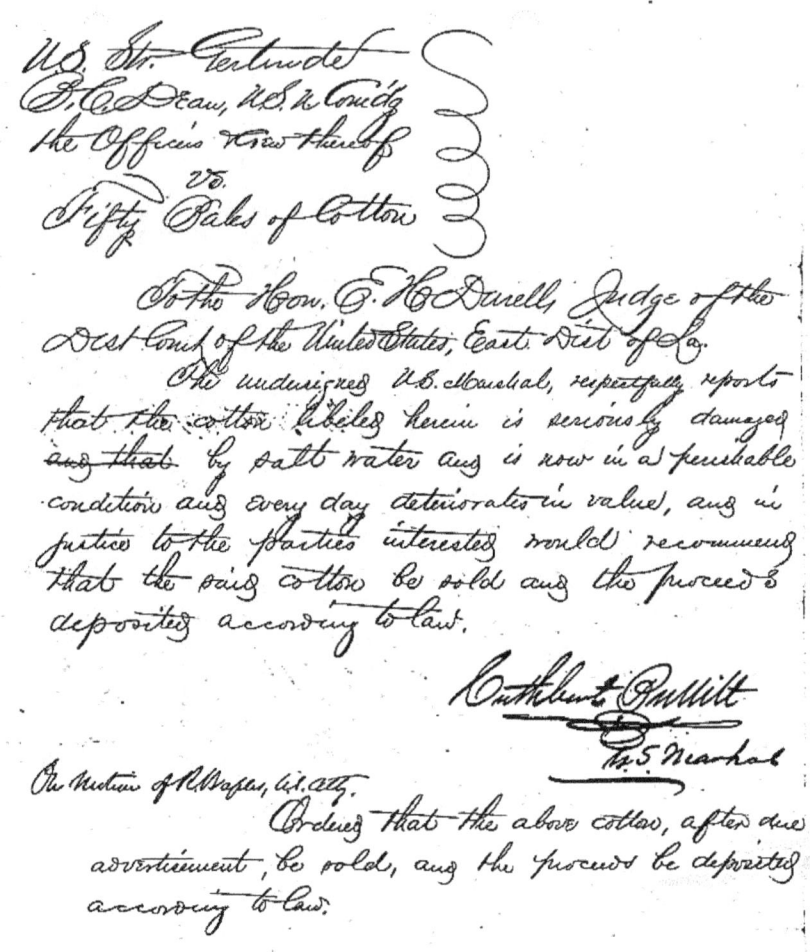

Figure 20. The U.S. marshal reported to the court that the cotton was damaged and deteriorating due to salt water and should promptly be sold instead of awaiting the conclusion of adjudication. The court agreed and ordered a quick sale after due advertisement. Filed 15 May 1865. *NARA*.

U.S. Prize Appraisers' Report of Cargo

The court had official appraisers to carry out the valuations required by the prize cases before it (Figures 21-22). The appraisal prior to auction sale was good business practice and clearly necessary for an honest process. In court, the original owners of the cotton conceivably might have prevailed against the capturing ship. In that case, the proceeds of the auction sale would have eventually been remitted to the original owners. An honest prize proceeding was required to maintain the integrity of the judicial process, bearing in mind that this was a process of international law, and U.S. citizens might someday be at the mercy of foreign prize courts with the shoe on the other foot.

No. 8037 Office of the United States Prize Appraisers.

Appraisers' Report of Cargo.

United States, Str. *Gertrude*
Officers & Crew thereof New Orleans, May 24, 1865
vs.
52 Bales Cotton

We the undersigned Prize Appraisers, of this Honorable Court, duly appointed and sworn, have the honor to report that we have this day carefully examined the ~~cargo of the Prize~~ 52 Bales of Cotton and are of opinion that the actual market value or wholesale prices of the said merchandise are this day and we do therefore appraise the same as follows:

Marks	No.	Description of Merchandise or Vessel	Value
Small 1 /@15	15	Bales Good Ordinary Reduced (?) pr. lb.	34¢
Small 2 /@20	20	do Ordinary do " "	30¢
Small D /@17	17	do Pickings " "	18¢
	52	Bales	

(Signature block omitted from transcript.)

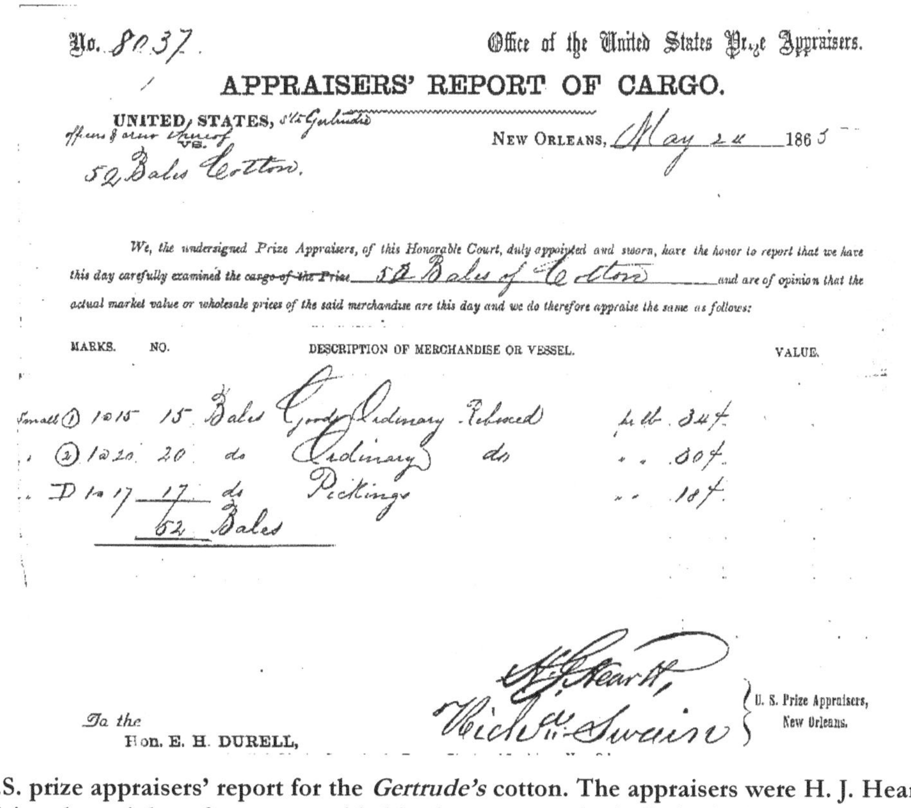
No. 8037. Office of the United States Prize Appraisers.

APPRAISERS' REPORT OF CARGO.

UNITED STATES, Str Gertrude
Officers & crew thereof
vs.
52 Bales Cotton.
New Orleans, May 24 1865

We, the undersigned Prize Appraisers, of this Honorable Court, duly appointed and sworn, have the honor to report that we have this day carefully examined the ~~cargo of the Prize~~ 52 Bales of Cotton and are of opinion that the actual market value or wholesale prices of the said merchandise are this day and we do therefore appraise the same as follows:

MARKS.	NO.	DESCRIPTION OF MERCHANDISE OR VESSEL.	VALUE.
Small ① 1@15	15	Bales Good Ordinary Reduced	pr lb. 34¢
② 1@20	20	do Ordinary do	" " 30¢
D 1@17	17	do Pickings	" " 18¢
	52	Bales	

H. J. Heartt,
Richd. Swain
U. S. Prize Appraisers, New Orleans.

To the Hon. E. H. DURELL,

Figure 21. U.S. prize appraisers' report for the *Gertrude's* cotton. The appraisers were H. J. Heartt and Richard Swain. Using the weights of cotton provided in the account of sale (below), the total estimated value was $5,473.26 before deducting costs. The per pound price was the market or wholesale value of the cotton in its damaged condition. Dated and filed 24 May 1865. *NARA*.

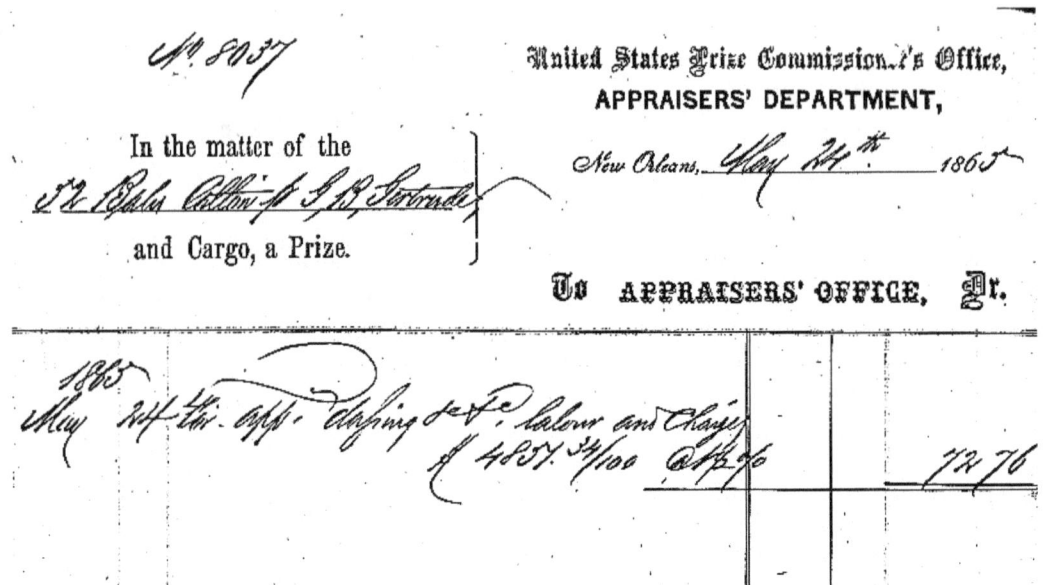

Figure 22. Invoice for services of the appraisers based on ½% of total appraised value of cotton after deduction of costs, $4,857.34, equaling $72.76. Dated 24 May 1865 and filed 27 May 1865. (Sufficiently legible, no transcript.) *NARA*.

Court Order for Sale of Cotton Bales

The court order for the sale of the cotton appears below (Figure 23).

District Court of the United States of America.
for the
Eastern District of Louisiana,
Holding Sessions in the City of New Orleans.
<u>May</u> Term, A. D., 186<u>5</u>
New Orleans, <u>Monday</u> the <u>15</u>th day of <u>May</u> 186<u>5</u>
Court met pursuant to Adjournment. Present, the Honorable
Edward H. Durell, Judge

<u>U.S. Str. *Gertrude*, her</u>
<u>Officers & crew</u>
vs.   No. <u>8037</u>
<u>50 Bales of Cotton</u>

On motion of the District Attorney, and on referring the Court to the report of the U.S. Marshal on file.
It is ordered that the above cotton be sold after due advertisement and that the proceeds be deposited according to law.

True copy from the Minutes
(signed) H. S. Benedict
Dy. Clerk

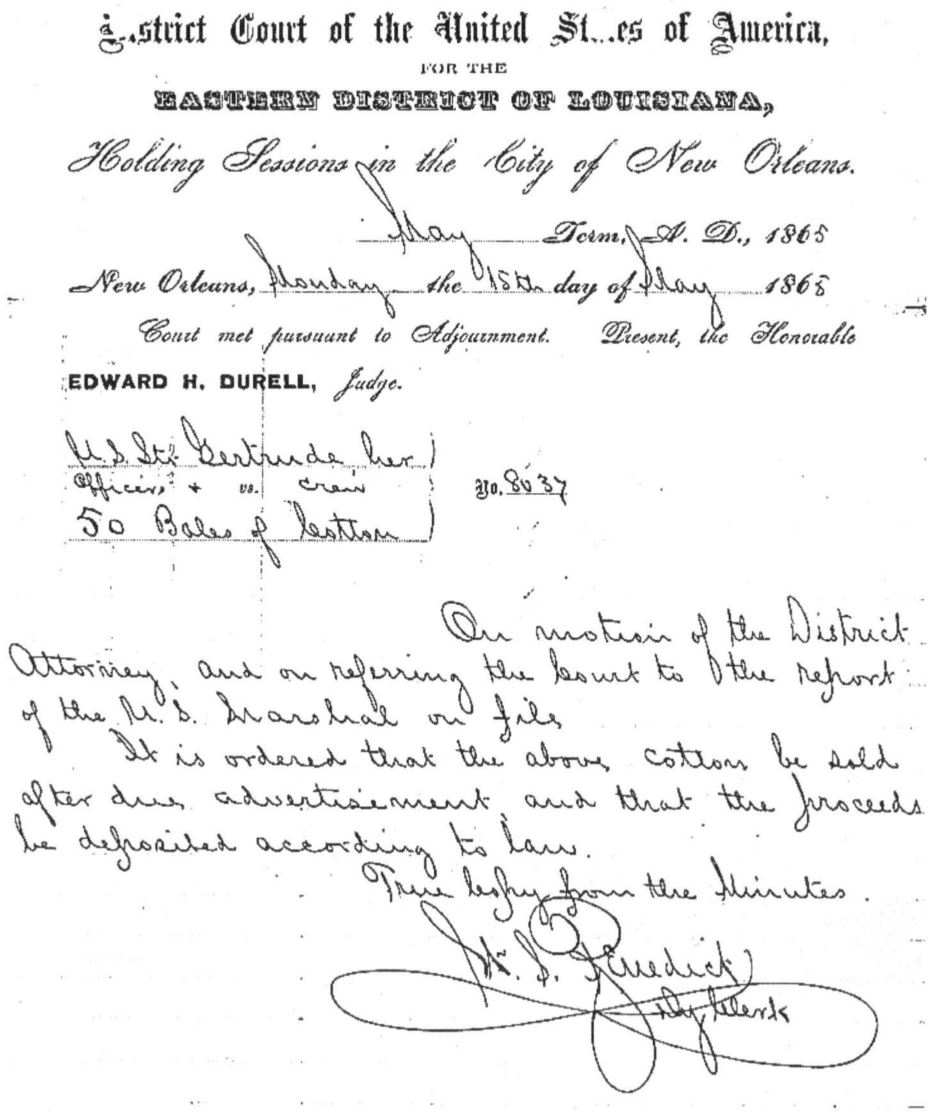

> **District Court of the United States of America,**
> FOR THE
> **EASTERN DISTRICT OF LOUISIANA,**
> Holding Sessions in the City of New Orleans.
>
> May Term, A. D., 1865
> New Orleans, Monday the 15th day of May 1865
> Court met pursuant to Adjournment. Present, the Honorable
> EDWARD H. DURELL, Judge.
>
> U.S. Str. Gertrude her ┐
> Officers + vs. crew   │ No. 8637
> 50 Bales of Cotton   ┘
>
> On motion of the District Attorney, and on referring the Court to the report of the U.S. Marshal on file
>
> It is ordered that the above cotton be sold after due advertisement and that the proceeds be deposited according to law.
>
> True copy from the Minutes.
>
> W. S. Benedict
> Dy Clerk

Figure 23. Following the marshal's return, the court ordered the sale of the *Gertrude's* cotton to be completed. Dated 15 May 1865. *NARA*.

Gertrude Advertisement of Sale and Note of Receipt for Deposit of Proceeds

This was a two-sided form with the court order of sale on one side (Figure 23) and an example of the newspaper advertisement of the auction with a report of the sale results on the other (Figure 24).

United States District Court
Eastern Dist. of Louisiana
New Orleans June 2nd 1865.

In obedience to the within order, I offered for sale at public auction to the highest bidder for cash, the property named in the within order, at the Merchant's Press Yard No. 5, in the first district of this city, after having advertised the same ten days in the *True Delta*, one of the newspapers printed in this city, in which the day, place & hour, as well as terms of said sale were fully set forth, and by the crier at the time of sale, and on Friday the 26th day of May 1865, at the hour of ten (10) o'clock A.M., at which said sale the same was sold in the lots at the prices and to the parties as set forth in the account of sale of G. A. H. U.S. Prize Auctioneer here to annexed.

By which said sale the gross proceeds amounted to four thousand eight hundred & forty dollars, 24 cents, which I pay over to C. C. Claiborne Esq., the Clerk of the Court aforesaid, to be disposed of as the Court directs.
          Attest   (signed) H. S. Benedict, Dy. Clerk

Received from Cuthbert Bullitt, U.S. Marshal, four thousand eight hundred and forty dollars, 24 cents, being as he saith the gross proceeds of the foregoing sale, the money to be disposed of as the Court directs, and for which I have signed duplicate receipts.

New Orleans June 2nd 1865   (signed) H. S. Benedict
                                                                                        Dy. Clerk

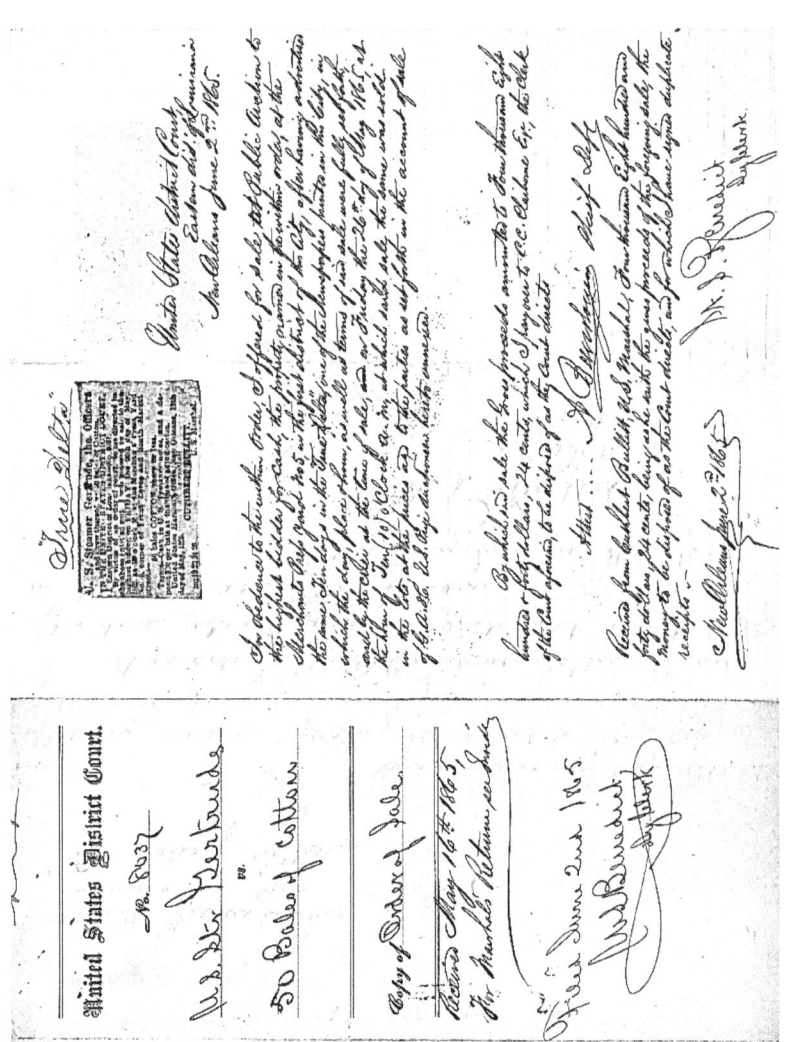

Figure 24. Order for and results of the sale. Sale date 26 May 1865. This information was presented on the reverse side of the form shown above in Figure 23. Dated 16 May 1865. Filed 2 June 1865. *NARA*.

Account of Sale of Fifty-Two Bales Cotton

This document presented the detailed results of the auction sale including the names of the buyers (Figure 25).

8037
The U.S. Steamer *Gertrude*
    vs.   Suit No. 8037
Fifty Bales of Cotton

Account Sales of Fifty-two Bales Cotton sold by order of Hon. Cuthbert Bullitt U.S. Marshal for the Eastern District of La. on Friday, May 26th 1865 at 10 o'clock A.M. at the Merchants Press Yard No. 5, Cor. South Market and Front Levee Sts.

Purchaser	Articles	Amt.	Total
H. McCall	15 Bales Cotton Wg. 5,631 " @ 32¢	1,801.92	
As$^{n.}$ Jones	20 Bales Cotton Wg. 7,834 " @ 24¢	1,880.16	
A. Conroy	17 Bales Cotton Wg. 6,714 " @ 17¼	1,158.16	
E.&O.E.			$ 4,840.24
New Orleans May 26th 1865			
To The Hon. Cuthbert Bullitt			
	U.S. Marshal		G. A. Hall
	E. D. La.		U.S. Prize Auct$^{r.}$

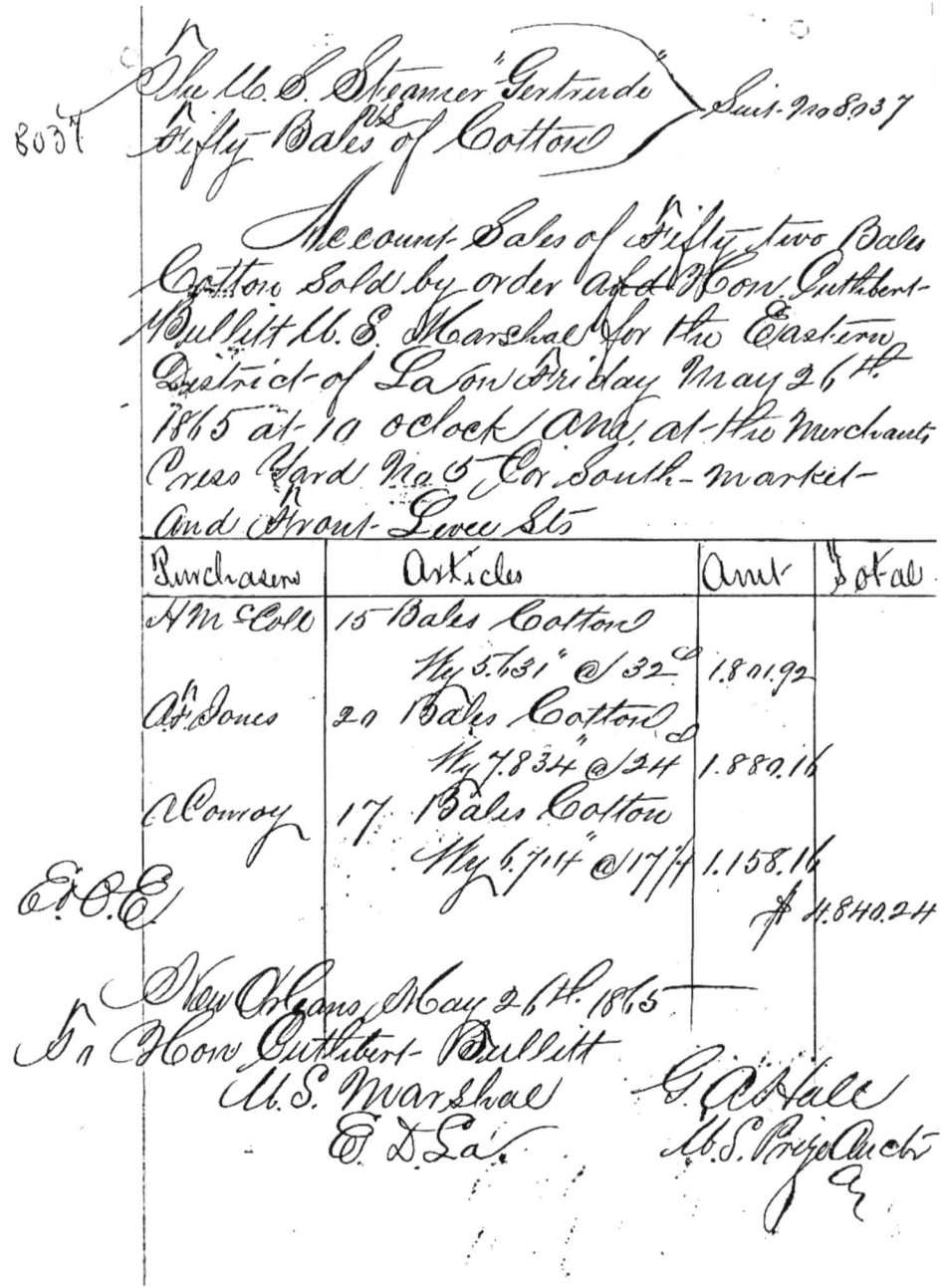

Figure 25. Account of the sale of the *Gertrude's* cotton that yielded $4,840.24 as opposed to the appraised total $5,473.26 a difference of $633.02. The cotton brought at auction less than estimated: grade good ordinary short by 2¢ per lb., ordinary by 6¢, and pickings by ¾¢. Dated 26 May 1865. *NARA*.

U.S. Marshal's Office Expenses

The listing of the marshal's expenses gave an excellent idea of the activities required in the preparations and the auction sale of the *Gertrude's* fifty bales of cotton (Figure 26).

By far the largest expense was the bill from the Merchant's Press for "Putting in Order." Perhaps this suggests drying and rebaling as there was such concern about the damage being caused by salt water. Had the *Denbigh* remained stuck on the sand bar, the Union blockaders would have destroyed the ship when daylight revealed her predicament. So in the emergency, the deck cargo was dumped overboard. The cotton bales floated with the tide for some hours before pickup by the U.S.S. *Gertrude* and the U.S.S. *Cornubia*. "Weighing and Reweighing" might also suggest that the bales were weighed upon arrival in New Orleans (wet) and then a second time after drying and rebaling. Logically the buyer would be interested in the final dried out and reprocessed weight.

United States District Court, Eastern Dist. of La.:
United States Marshal's Office. New Orleans June 7/65.
U.S. Str. *Gertrude* &c. &c.
vs   No. 8037
50 Bales of Cotton

1865			
May	12	Admiralty Warrant, Libel & Orders $6 Posting $1 Copy $1	$ 8.00
"	"	Attendance of Deputy to receive from U.S. Str. *Dinsmore*	5.00
"	"	Monition in *Picayune* $13.50 Drayage & Labor to Press $35.00	48.50
"	16	Order to Sell and Copy $2.50 Discharging Cl'k. $3 Publishing Sale $12	19.50
"	"	Bill of Stevedore Disch'g. $30 Weighing and Reweighing $25.50	55.50
"	26	Attendance at Sale $5 Commission on Sale $24.20 3 Proclamations .90	30.10
"	"	Merchant's Press Bill for Putting in Order	358.08
"	"	Bill of Storage and Labor in Merchant's Press	15.00
"	"	Marshal's Custody 15 days $37.50 Commission on disbursements $12.51	50.00
"	"	Recording Process Verbal $1 Return on Vends. Exp. $1	2.00
			$ 591.69

(signed) Cuthbert Bullitt
U.S. Marshal

Cuthbert Bullitt, U.S. Marshal being duly sworn doth depose & say that the above bill is correct & true and the charges made were actually incurred for the purposes set forth.

Sworn to and subscribed before me this 9th day of June 1865.
(signed) K. Loew
Clk. (Clerk)

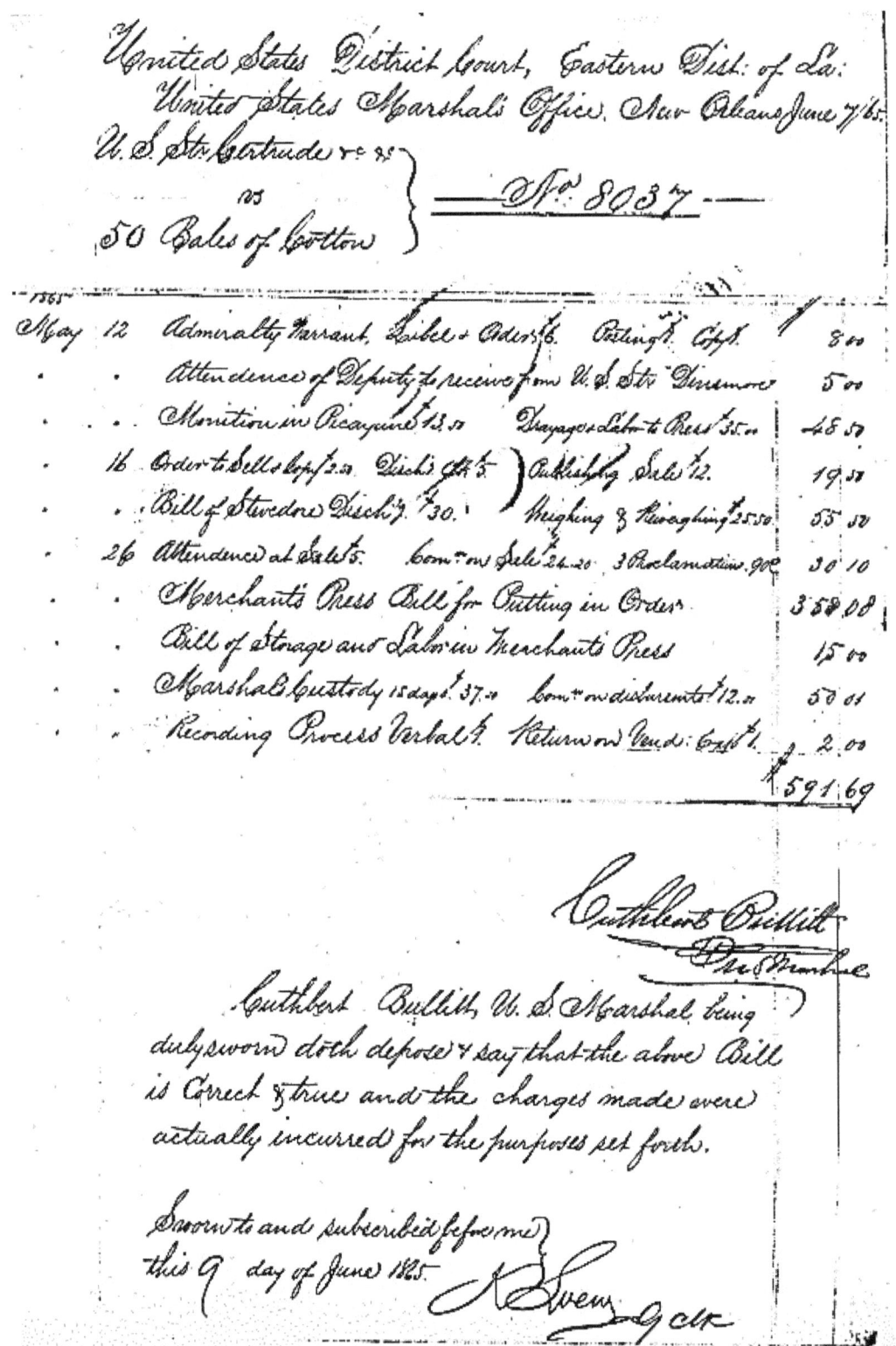

Figure 26. Marshal's bill showing the expenses in preparing and selling the *Gertrude's* captured prize cotton. Dated and filed 9 June 1865. *NARA.*

Auctioneer's Newspaper Bills

G. A. Hall, U.S. Prize Auctioneer, arranged notice in several newspapers advertising the sale of the prize cotton captured by the *Gertrude* (Figures 27-29). Proper legal proceedings required public notice. The newspaper invoices were printed with interesting graphics, but with (nowadays) confusing terminology. In spite of the antique form of expression,

"G. A. Hall (is)
        To the *New Orleans Times*, Dr. (Debtor)."

was in this context to be understood as "G. A. Hall owes to the *New Orleans Times*" thus and such an amount for advertising fifty bales of cotton. In modern accounting "Dr." translated as "debit," but notice there was no "r" in "debit." The "r" was leftover from the older form of expression wherein "Dr." stood for "debtor."

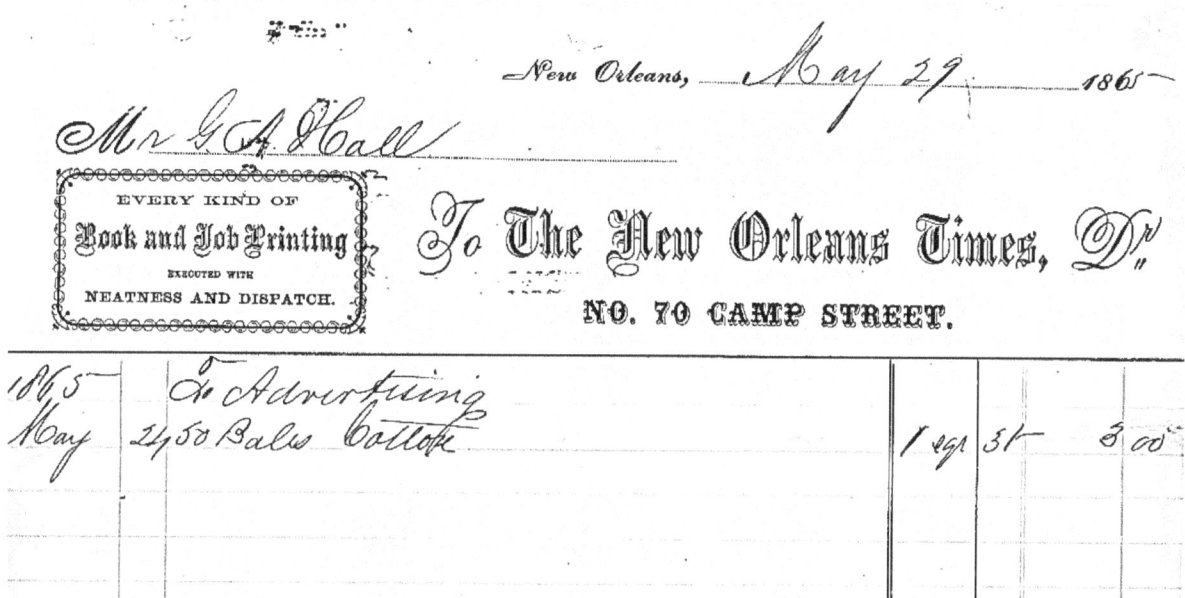

Figure 27. *New Orleans Times* bill for advertising sale of the *Gertrude's* cotton. Dated 24 May 1865. *NARA*.

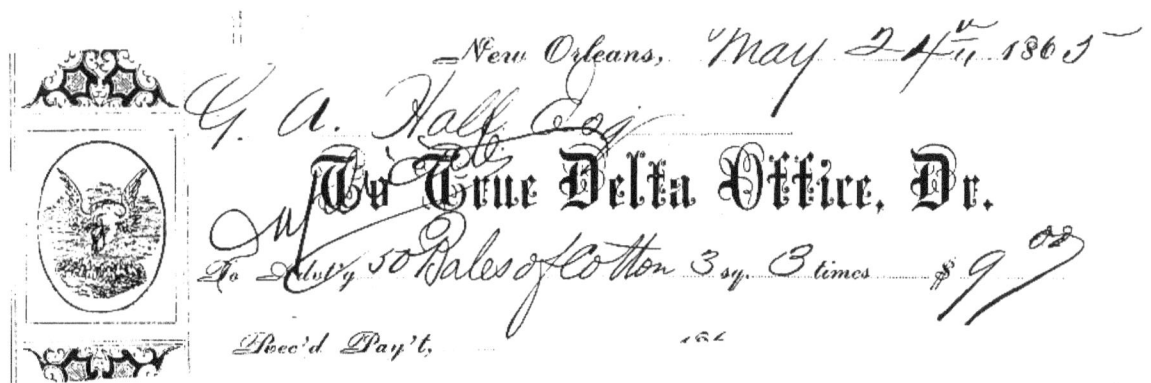

Figure 28. *True Delta* bill for advertising the sale of the *Gertrude's* cotton. Dated 24 May 1865. *NARA*.

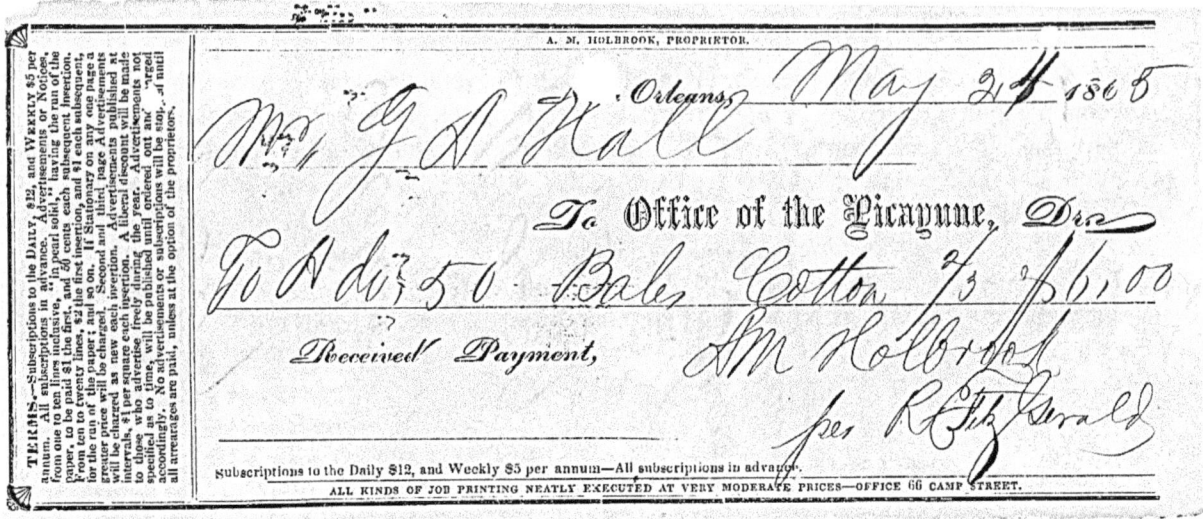

Figure 29. *Picayune* bill for advertising the sale of the *Gertrude's* cotton. Dated 24 May 1865. *NARA*.

Auctioneer's Bill for Sale of Cotton

The list of expenses for the auctioneer included advertising plus a fee based on 1% of the $4,840 proceeds from the sale or $48.40 in this case (Figure 30).

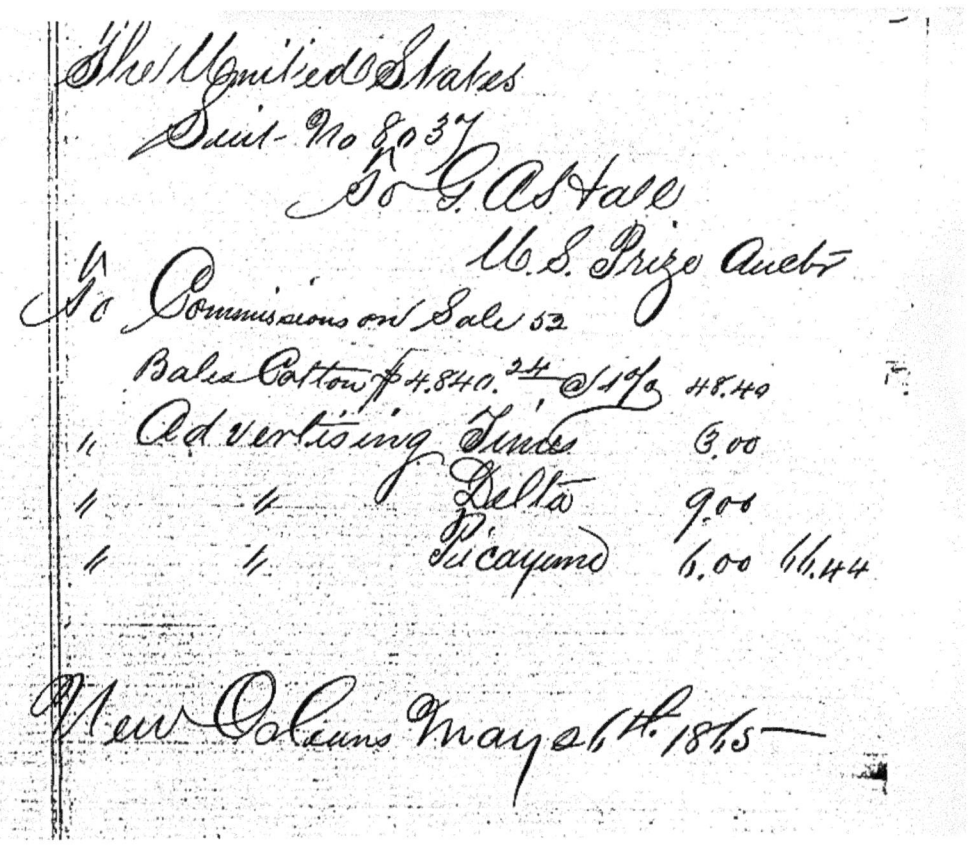

Figure 30. Bill of expenses of the U.S. prize auctioneer. Dated 26 May 1865. (Legible, no transcript.) *NARA*.

The Court's Docket Sheet

The docket presented a calendar and listing of documents and steps in the *Gertrude's* cotton prize case (Figure 31). After deducting the expenses involved, two equal payments were made from the court to the U.S. Treasury. One was the government's share, the other for the naval personnel.

8037 U.S. Str. *Gertrude* Benj. C. Dean U.S.N. Commander
& the Officers & Crew thereof
vs.
Fifty Bales of Cotton

1865			Salvage
May	12	Libel & order	
"	"	Henry Durrel etc. Cert. & S.	
"	"	" Libel etc. "	
"	"	Testimony before P. C. (U.S. Prize Commissioner)	
"	15	Marshal's Report	
"	"	Order etc. to Sell Cotton	
"	24	Appraisement	
"	29	Appraiser's Bill 72.76	
June	2	Marshal's Return proceeds in Registry	$ 4,840.24
"	3	Decree of Condemnation default	
"	9	Marshal's Bill $591.69	
"	"	Auctioneers Bill 66.44	
"	"	Taxation of Costs	
"	12	Judgment of distribution	
"	17	Paid amt. Treasurer $1,939.37[47]	
'66 June	30	Order to pay the balance into the Treasury	
July	11	Paid amt. Treas. $1,939.37 less $5.50 Costs	

---

[47] About five months after 17 June 1865, the court transferred to the Treasury Department the Navy Department's half of the proceeds. On 19 Nov. 1865, *The New York Times* published notice that the *Gertrude's* prize money regarding the fifty bales of cotton was ready for payment by navy paymasters. (See below pp. 229-230.) That was prompt considering the processing and calculating required. The court did not give over the government's half until 11 July 1866. Why did the court wish to carry those funds on its books for so long, nearly 13 months longer than the navy's half?

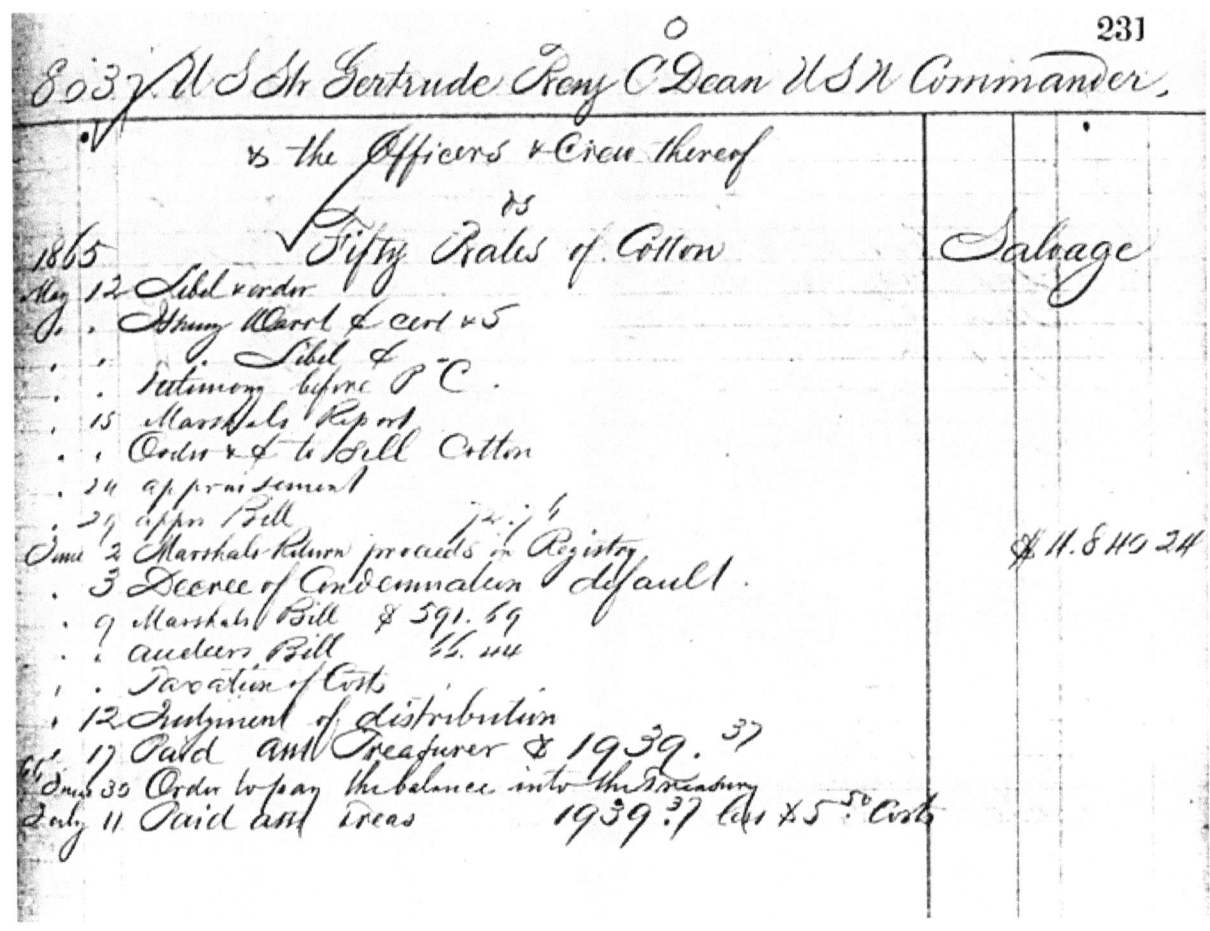

Figure 31. Docket sheet for the *Gertrude* claim on the *Denbigh's* jettisoned cotton. *NARA*.

4

Galveston Blockading Squadron Deck Logs Regarding the *Denbigh*

The deck logs of the Union navy[48] ships stationed off Galveston provided another, more immediate, perspective on dealing with the *Denbigh* and runners in general. In the present section the author selected the log entries for two *Denbigh* incidents: the jettisoned cotton bales from the grounding of 20 April 1865 (Figures 32-34) and the log entries for the final loss of the *Denbigh* on 24 May 1865 (Figures 35-48). Of particular interest in the latter case was determining which ships fired their guns at the stranded blockade-runner. The artillery engagement totaling at least 95 shots left no damage apparent during INA's excavation of the wreck. Perhaps this was another example of the poor gunnery about which Admiral Farragut often complained when his ships were in action against blockade-runners. The log entries recorded in detail which guns fired and the ammunition expended.

Vessels Firing	No. of Shots Fired
Cornubia	47
Princess Royal	20
Albatross	15
New London	13
Penguin	shots fired but not recorded

The log entries for the *Denbigh's* destruction included two small arms injuries. One sailor died by the accidental discharge of his musket while getting into a boarding party's boat. A second sailor was shot through the hand and wrist while unloading small arms from a boarding party's boat. When INA excavated the site, we found alcoholic beverages on board. The U.S. Navy had, during the Civil War, eliminated daily grog (alcohol rations) for its men. Could the sailors have gotten into the blockade-runner's booze? That might well have been the cause for the gunshot injuries. Alcohol was involved for the two sailors, likely from the boarding parties, who were put in irons for drunk and disorderly conduct.

The logs demonstrated the somewhat limited viewpoint of the officer of the deck who recorded the entries at the conclusion of each watch. In real time, one could not be expected to have the complete story that became clearer after some time elapsed as was the case for the summarizing reports sent to higher command. On the other hand, the deck logs did provide a wealth of detail, from weather conditions to which crewmen were assigned to which tasks.

---

[48] NARA RG 24 Navy Deck Logs.

U.S.S. *Cornubia* Log Concerning Capture of Eighty-nine Bales of Floating Cotton

This entry contained a couple of examples of interesting naval parlance. "At 3 made a vessel bearing…" meant "made out" or sighted a vessel. Under way (also underway) was proper usage for getting a ship moving. "Under weigh" was not proper English, and that variation came about through confusion with the expression "weigh anchor."[49]

Narrative comments from the remarks column of the *Cornubia* log, Acting Volunteer Lieutenant John A. Johnstone, commanding (Figures 32-33):

Remarks on this 20th day of April 1865.
Off Galveston, Texas

From Midnight to 4. At 3 made a vessel bearing E by N. Slipped anchor[50] & stood for her. Proved to be the U.S. Steamer *Princess Royal*. Put about and returned to night station. Anchored in 4½ faths. (fathoms) water veering 30 faths. chain.
(signed) William H. Wood[51]

From 4 to 8. At daylight saw large quantities of cotton floating on the water. Got underway and stood towards them. At 7 lowered all boats and commenced to pick them up.
(signed) Geo. H. Russell[52]

From 8 to Meridian. All hands engaged in picking up and hoisting cotton on board. At 11 anchored off Bolivar Point. (signed) Frank Millett[53]

From Meridian to 4. At 2 all boats returned to the ship having picked up eighty nine (89) bales cotton. Got underway and stood out to station. (signed) George F. Braley[54]

From 4 to 6. As per cols. (i.e., columns to the left showing conditions).
(signed) William H. Wood

From 6 to 8. As per cols. (signed) Geo. H. Russell

From 8 to Midnight. As per cols. (signed) Frank Millett

---

[49] *Webster's New International Dictionary of the English Language*, Second Edition, Unabridged, 1943.
[50] To drop the anchor chain in order to quickly give chase rather than haul up the anchor, a time consuming process. With a buoy attached, the end of the anchor chain could later be easily retrieved. The blockaders were prepared promptly to pursue a runner.
[51] Acting Master's Mate. Officers' ranks found by cross-referencing crew listing from the *Cornubia* vs. Eighty-Nine Bales of Cotton prize documents.
[52] Acting Master's Mate.
[53] Acting Ensign.
[54] Acting Ensign.

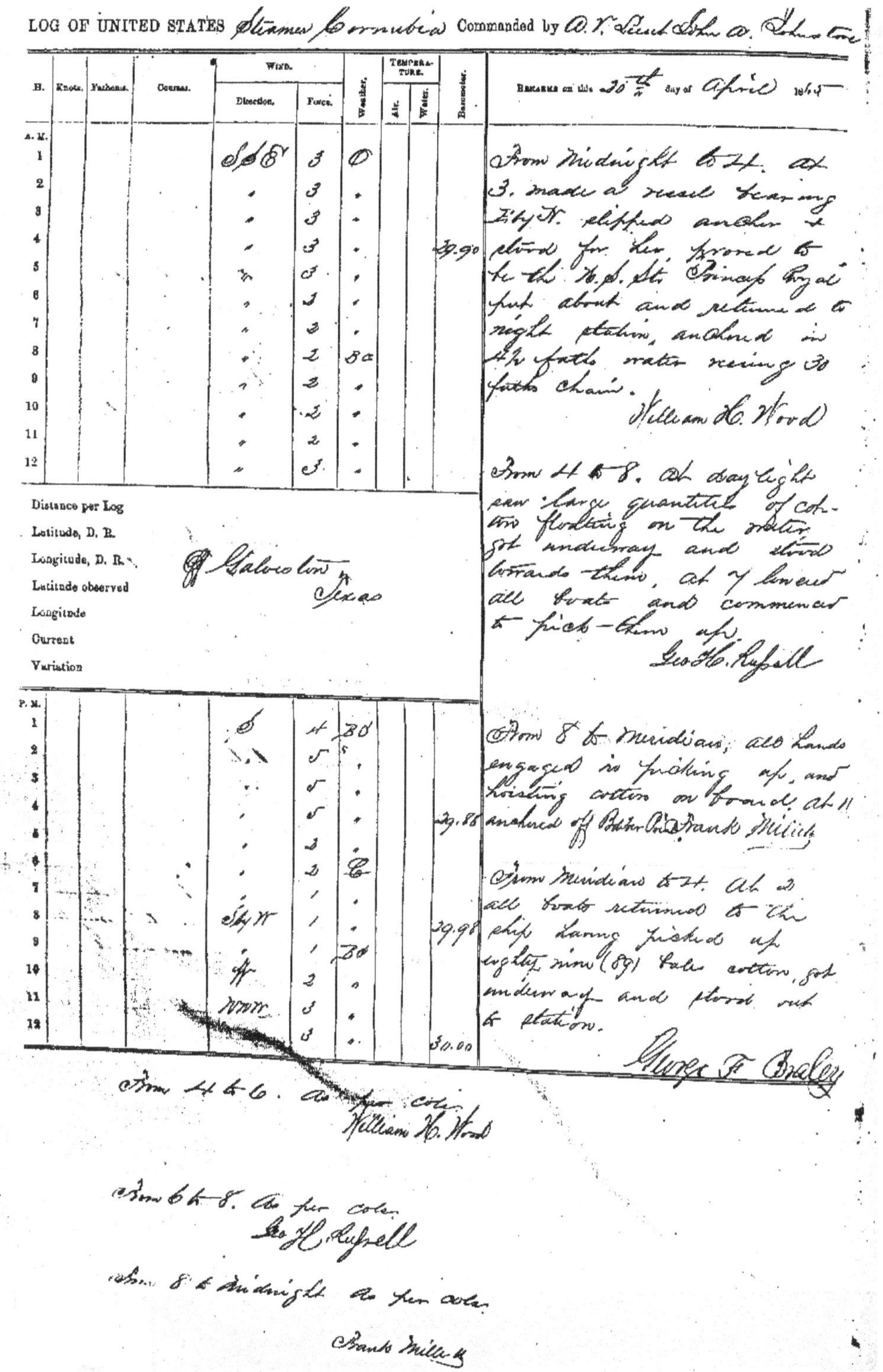

Figure 32. The U.S.S. *Cornubia's* deck log including the recovery of the floating cotton. Dated 20 April 1865. Wind force was reported hourly based on the Beaufort Scale.[55] Winds here reported were Force 1=1-3 kph (knots/hour), light air; Force 2=4-6 kph, light breeze; Force 3=7-10 kph, gentle breeze; Force 4=11-16 kph, moderate breeze; Force 5=17-21 kph, fresh breeze producing moderate waves 4-8 ft. with many whitecaps and some spray. *NARA*.

---

[55] NOAA web site: http://www.spc.noaa.gov/faq/tornado/beaufort.html.

LOG OF UNITED STATES Steamer Cornubia Commanded by A.T. Snell Ash. a. Volns. &c.   Remarks on this 26th day of April 1865—

H.	Knots.	Fathoms.	Courses.	Wind Direction.	Wind Force.	Weather.	Temperature Air.	Temperature Water.	Barometer.
A.M.									
1				SbE	3	O			
2				"	3	"			
3				"	3	"			
4				"	3	"			
5				"	2	"			
6				"	2	"			29.90
7				"	2	Bc			
8				"	2	"			
9				"	2	"			
10				"	2	"			
11				"	3	"			
12									

From midnight to 4. at 3. made a vessel steering EbyN. slipped anchor & stood for her, found to be the U.S.S. Princess Royal just about and returned to night station, anchored in 5½ faths. water veering 30 faths chain. William H. Wood

From 4 to 8. At daylight saw large quantity of cotton floating on the water. Got underway and stood towards them, at 7 lowered all boats and commenced to pick them up. Geo. H. Ryfall

Distance per Log
Latitude, D. R.
Longitude, D. R. — Off Galveston Bar Texas
Latitude observed
Longitude
Current
Variation

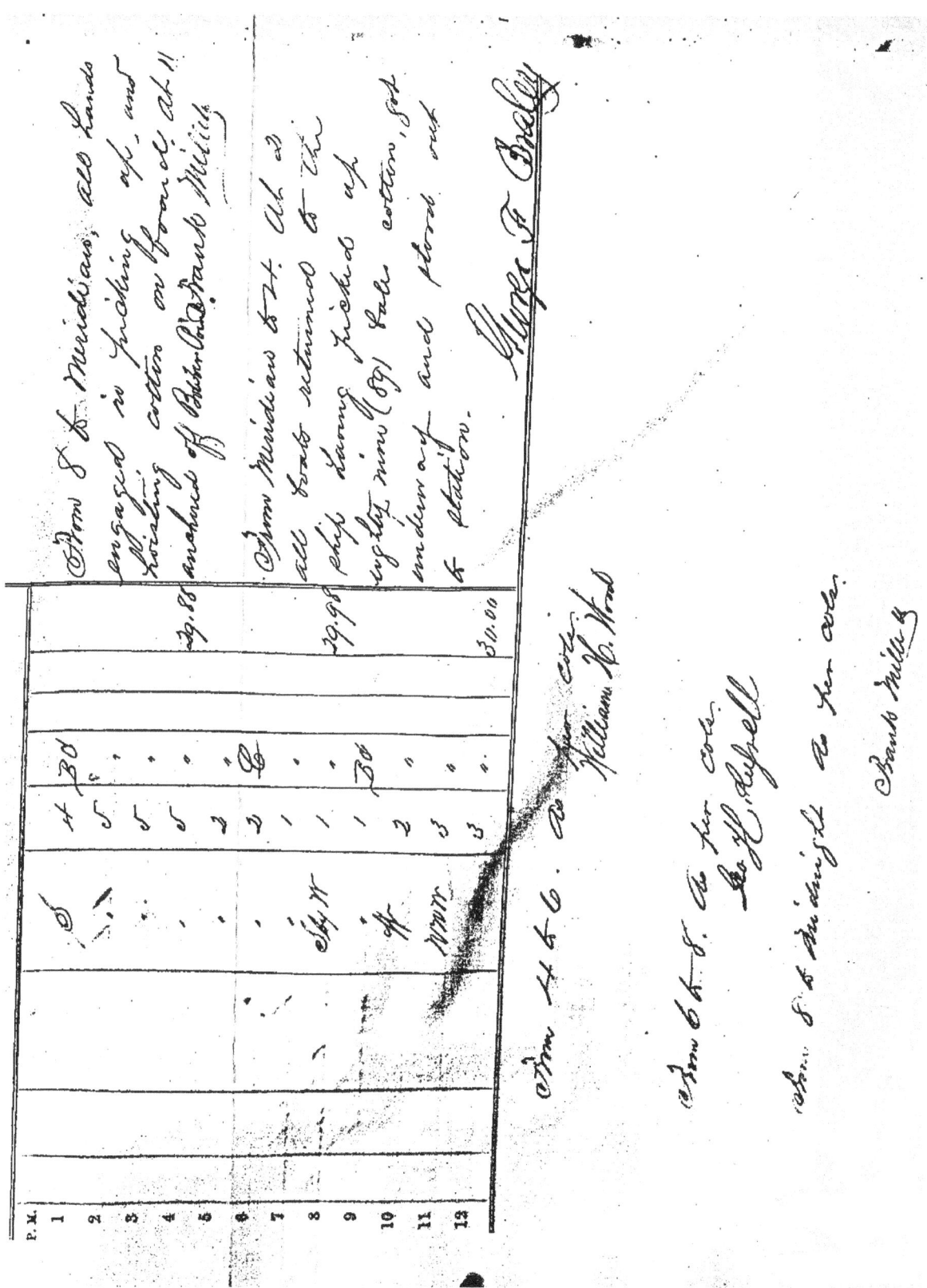

Figure 33. *Cornubia's* deck log for 20 April 1865 enlarged for legibility. (Two pages.) *NARA*.

U.S.S. *Gertrude* Log Concerning Salvage of Fifty Bales of Floating Cotton

The example below mentioned using the ship's boats on nighttime picket duty. The practice involved a small craft patrolling or anchoring in the shallow inshore channels that were a favorite entry for the blockade-runners. The low profile, dark painted runners were well camouflaged against the loom of the land at night. The rockets mentioned were a picket boat's signal about sighting a blockade-runner coming in or going out of Galveston.

The location of the *Denbigh's* grounding near or "under Fort Magruder" indicated that the runner's departure was planned close inshore along Galveston Island rather than along Bolivar Peninsula.

The following were the narrative comments from the remarks column of the *Gertrude* log, Acting Volunteer Lieutenant. B. C. Dean, commanding (Figure 34).

Remarks on this 20th day of April 1865.
Off Galveston, Texas

Midnight to 4 A.M. As per Cols. At 3 saw rockets sent up from P. (picket) boat in Easterly direction. (signed) Chas. (Charles) A. Osborne[56]

4 to 8 A.M. As per Cols. At 4 discovered a steamer aground under Fort Magruder (located near the east end of Galveston Island fronting the Gulf). At 5.30 discovered that the *Cornubia* was standing inshore picking up cotton. At 6.30 got under way and ran inshore. At 6.40 lowered all boats and commenced picking up cotton.

8 to Meridian. As per Cols. All hands employed picking up cotton.
  (signed) I. (Isaac) W. Goodrich[57]

Meridian to 4 P.M. As per Cols. At 1.10 all boats returned. Hoisted the cotton on board. At 1.45 got underway and stood off shore. Anchoring at 2.30 in 5½ fms. (fathoms)[58] water, with port anchor veered 45 fms. chain. (signed) Chas. A. Osborne

4 to 6 P.M. As per Cols. All hands employed stowing cotton. At 4.45 got it all stowed.
  (signed) J. W. Monroe

6 to 8 P.M. As per Cols. (signed) I. W. Goodrich

8 to Midnight. As per Cols. (signed) Chas. A. Osborne

Note. Picked up 50 bales of cotton.

---

[56] Acting Mate. Officers' ranks found by cross-referencing crew listing from *Gertrude* vs. Fifty Bales of Cotton prize documents.
[57] Acting Ensign.
[58] 1 fathom = 1.8 meters or 6 feet so 5½ fathoms = 9.9 meters or 33 feet.

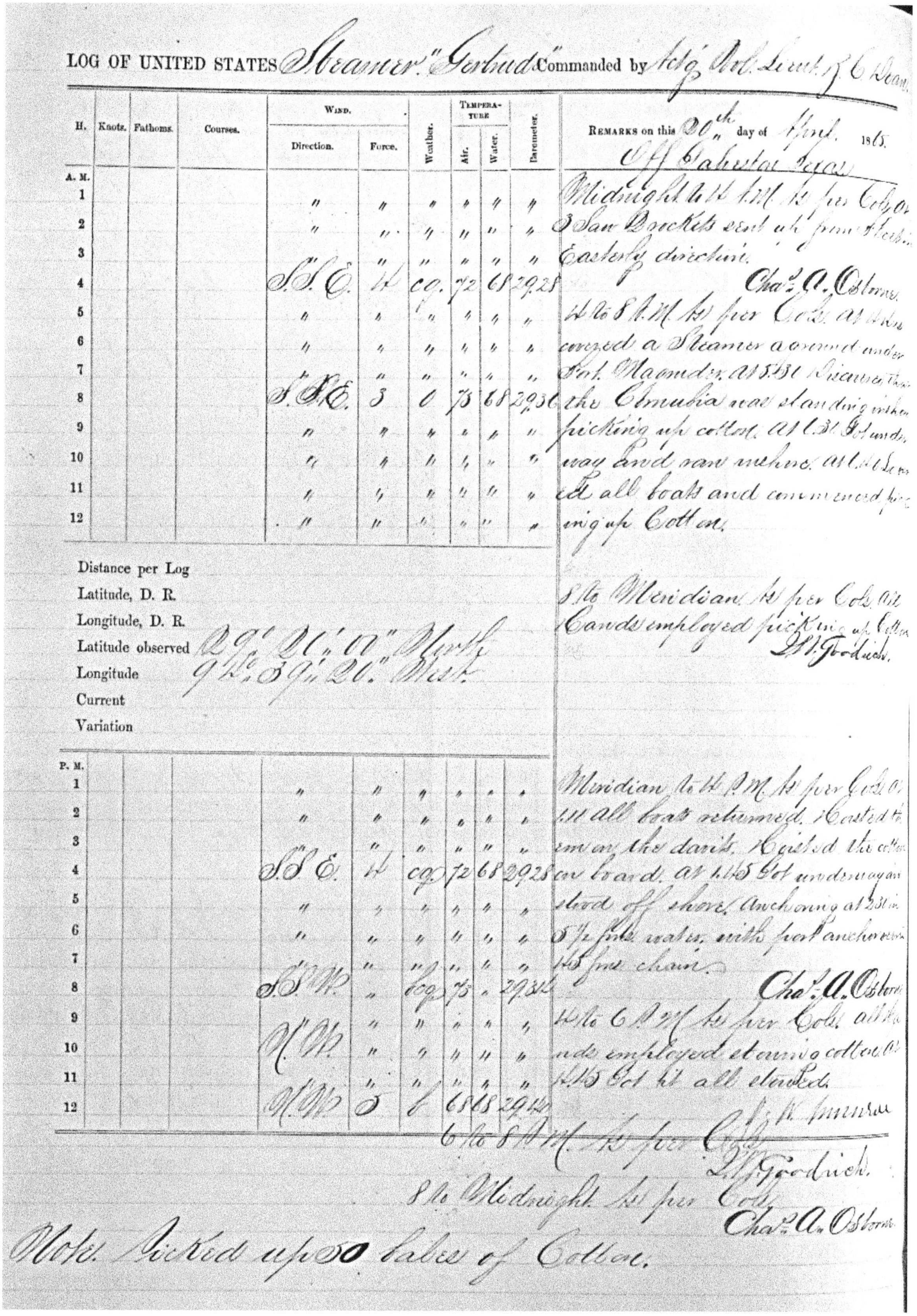

Figure 34. The *Gertrude's* deck log covering the recovery of floating cotton. Dated 20 April 1865. *NARA*.

Vessel Log Book Entries Regarding the *Denbigh's* Grounding and Wreck

Note that Confederate forces mutinied and abandoned Galveston on 22 May. The commanders hoped in vain to keep the army together while negotiating surrender terms. Probably the *Denbigh* wrecked attempting to enter Galveston due to a simple fact. When it got dark on 23 May, there was no one left to set the range lights safely to guide runners through the shallow inshore channel near Bolivar Point.

The logbooks for the day of the *Denbigh's* destruction were important in understanding Union navy operations at Galveston. The present author hoped to determine which vessels actively participated in firing on the stranded runner. A careful reading of individual logs beginning with the main actors mentioned in Captain Sands' report led from the log of one ship to another in quite an interesting series. In no document were the ships firing all listed at once.

U.S. Steamer *Cornubia* 24 May 1865

Narrative comments from the remarks column of the *Cornubia* log, Acting Volunteer Lieutenant John A. Johnstone, commanding (Figure 35-36):

Remarks on this 24th day of May 1865.
Off Galveston, Texas

From Midnight to 4. As per cols.

From 4 to 8. At 4.15 got underway and stood out from night station. Discovered a steamer ashore under north beach.[59] Put for her, and beat to quarters. Came to anchor in 2 faths. of water and commenced firing on the Str. (Steamer). At 7 a boat from the *Seminole* boarded the Str. and set her on fire. It was the Str. *Denbigh* from Havana attempting to run in. At 7.10 sent a boat with officer in charge to assist in firing. At 8 all boats returned to the ship. 7.50 beat the retreat. Expended the following ammunition. 2 shot (solid), 13 Per. (percussion?) shells, 1. 15" (second) timer and 7. 28" timer shell 30 pdr. (pounder) Parrott, and 15 percussion and 9 shots (solid) at (from) the 12 pdr. rifled howitzer.[60]   (signed) Frank Millett

From 8 to Meridian. At 8.20 got under weigh with all boats in tow and stood out to the flag ship. At 9 *Fort Jackson* made signal B code 152. At 9.30 anchored near the *Fort Jackson*. At 10.30 quarters for inspection, confined Robert Freeman, Lds. (Landsman)[61] in double irons with bread and water for 5 days for neglect of duty (while on lookout) at the mast head.[62]
                                        (signed) George F. Braley

From Meridian to 4. At 3.15 got under weigh and stood in towards Bolivar Point. At 3.45 anchored in 2 fath. Water veering 10 fath. chain.   (signed) Geo. H. Russell

---

[59] Bolivar Peninsula.
[60] This passage unclear and open to interpretation. The author attempted to aid understanding by adding commas.
[61] A raw recruit on a ship, a rank indicating an inexperienced sailor as opposed to the higher naval rank of seaman.
[62] Probably indicated he was asleep, a very serious infraction.

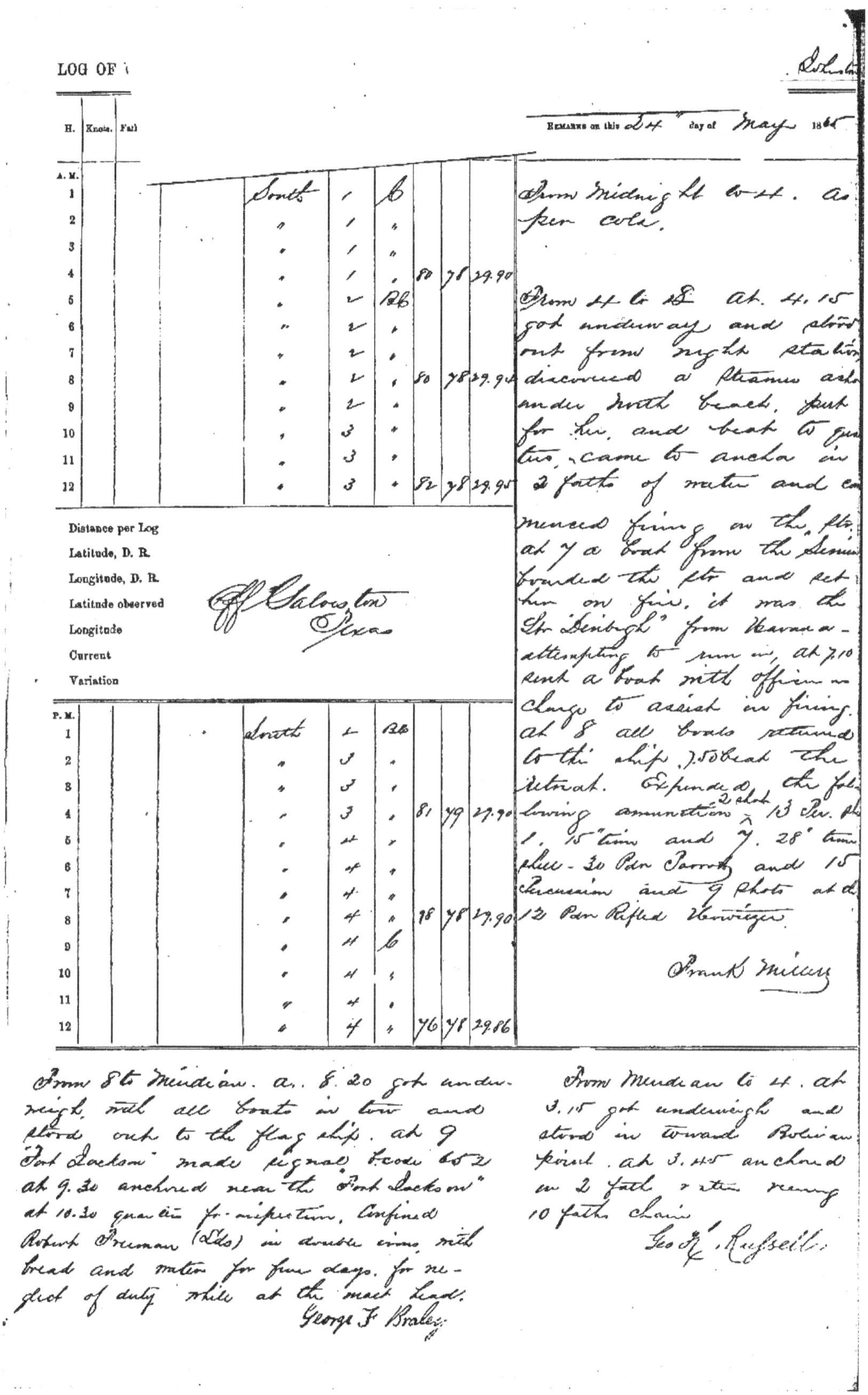

Figure 35. The *Cornubia's* deck log covering the destruction of the *Denbigh*. Dated 24 May 1865. *NARA*.

Continuation (page) of Remarks of May 24th

From 4 to 6. At 4.10 called no. 1 gun's crew and powder divisions at quarters and fired on a schooner out from Galveston wrecking (salvaging) the Steamer *Denbigh*. Which (schooner) immediately got underway and stood in towards the city when a shell from our gun caused her to heave too [sic] and lower all sail the crew taking to their boat and landing on Bolivar Point. An armed boat with the men in charge of Acting Ensign Frank Willey was sent from this vessel to capture the schr. who boarded her and captured four Confederate soldiers. The schr. proved to be the Rebel guard boat *Lecompt* armed with a 24 pdr. Dahlgren howitzer and several small arms. On boarding it was found that the schr. was fast aground, bilged,[63] and half full of water and rudder and forefoot[64] gone. After fruitless attempt to get her off we threw the howitzer overboard, took the small arms, a Confederate flag, and prisoners in the boat, abandoned her, and returned to the ship. We received the information from the prisoners of the evacuation of Galveston on the 22nd. At 6 the boat returned on board the vessel. Chas. De Vassy[65] (sea.)[66] was accidentally shot through the hand and wrist while passing small arms out of the boat.

Expended three percussion shell of the 30 pdr. rifle.

Names of the men captured on the schr. John Gallaher, Robert Hilliard, Patrick Clair, John Murphy. (signed) George F. Braley

From 6 to 8. At 6.10 made signal B code Geog. 404. 2760. 3469. Answered by the *Fort Jackson*. Got underway and stood out to the fleet and communicated with the flagship. At 8 anchored in our night station in 3¼ fath. veering 30 faths. chain. (signed) Frank Millett

From 8 to Midnight. At 10.45 sighted strange sail running in,[67] slipped anchor, and made for her. Buoy rope caught in the wheel, stopped engine and cleared it. At 11 anchored with port anchor veering 30 faths. chain. (signed) Geo. H. Russell

---

[63] Indicated hole or holes in the bottom.
[64] Forward part of hull where the keel and stem joined. This hull was damaged past saving.
[65] Last name was illegible. Crosschecked against the crew list for prize claim regarding *Cornubia* vs. Eighty-nine Bales of Cotton.
[66] Seaman.
[67] Probably the blockade-runner *Lark*, the last runner clearing a Confederate port.

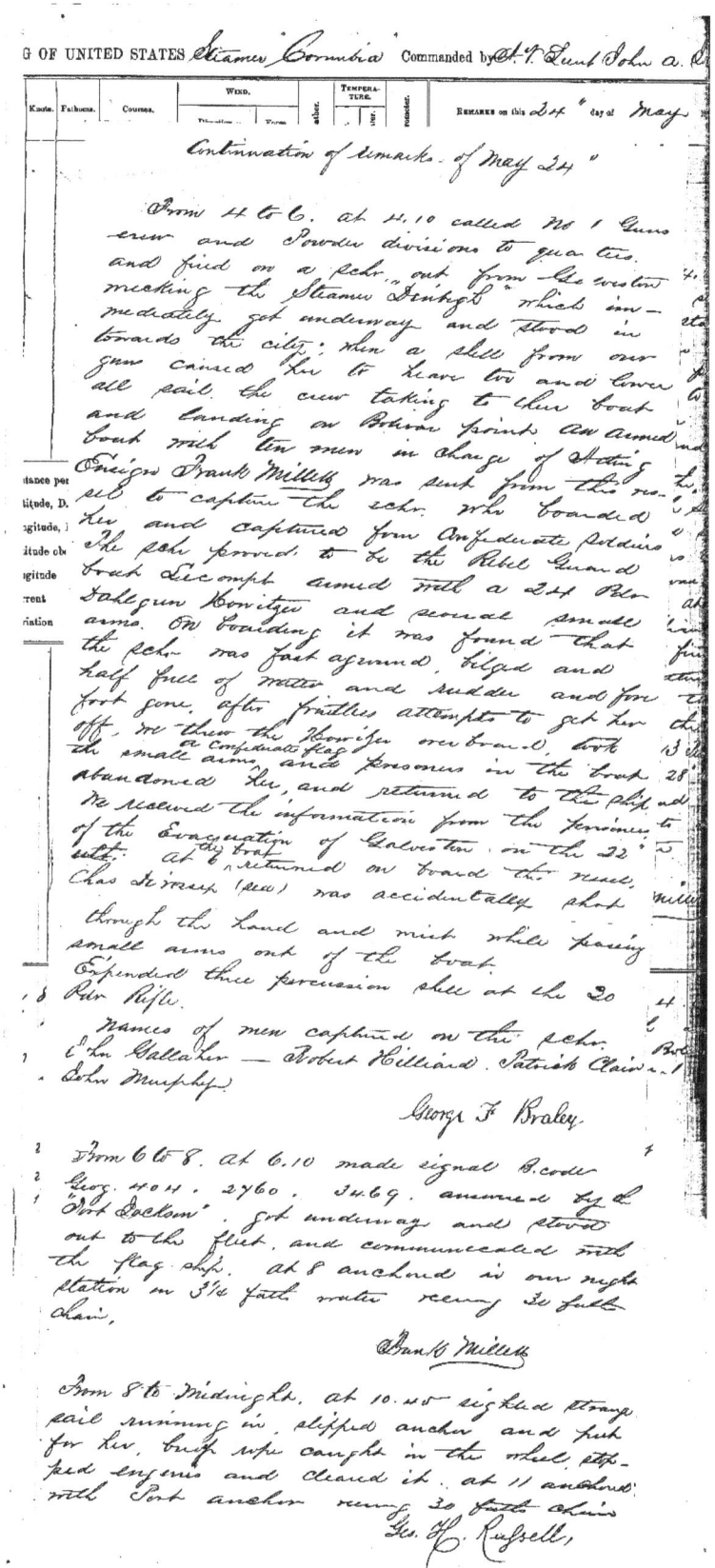

Figure 36. Continuation sheet of the *Cornubia's* deck log. Dated 24 May 1865. *NARA.*

U.S. Steamer *Kennebec* 24 May 1865

Narrative comments from the remarks column of the *Kennebec* log, Lieutenant Commander T. (Trevett) Abbott, commanding (Figure 37):

Remarks on this 24th day of May 1865.

From Midnight to 4 A.M. As per cols. (signed) J. D. Ellis

From 4 to 8 A.M. As per Columns. At 5.30 discerned a blockade runner (Str.) ashore bearing N. by E. (?). Got under weigh and stood towards the buoy. At 6 sent two cutters and gig in charge of Actg. (Acting) Ensign Ellis and Mate Mann, all under the command of Acting Master A. R. Emerson, to board her. At 7 *Fort Jackson* made (BC) sig. (signal) Nos. 454. U.S.S. *Penguin*, *Princess Royal*, (and) *Cornubia* firing on her. At 7.30 boats boarded her and set her on fire. At 7.30 made (BC) Sig. Nos. 5737-8438-599. (signed) ??? ???

From 8 A.M. to Meridian. As per cols. At 9.45 all boasts returned. (signed) A. A. Hann

From Meridian to 4 P.M. As per cols. Coal Brig *Moses Rogers* anchored in the fleet. At 12.30 discovered a strange sail to the North$^{d.}$ and East$^{d.}$ (signed) J. J. Butler

From 4 to 6 P.M. As per columns. (signed) A. A. Hann

From 6 to 8 P.M. As per columns. At 6 inspected crew at quarters. Coal brig arrived at the fleet. At 6.40 got under weigh and stood from station. At 7.50 came to with port anchor in 4½ fathoms water. Veered to 45 fathoms cable. (signed) J. D. Ellis

From 8 P.M. to Midnight. As per columns. At 9 saw a rocket bearing S. West. At 11 U.S.S. *Cornubia* made Coston Sig. No. 3.[68] Saw steamer running out.[69] (signed) A. L. Emerson

---

[68] Coston night signal flares invented by Martha J. Coston in 1862.
[69] The last blockade-runner from Galveston, the *Lark*.

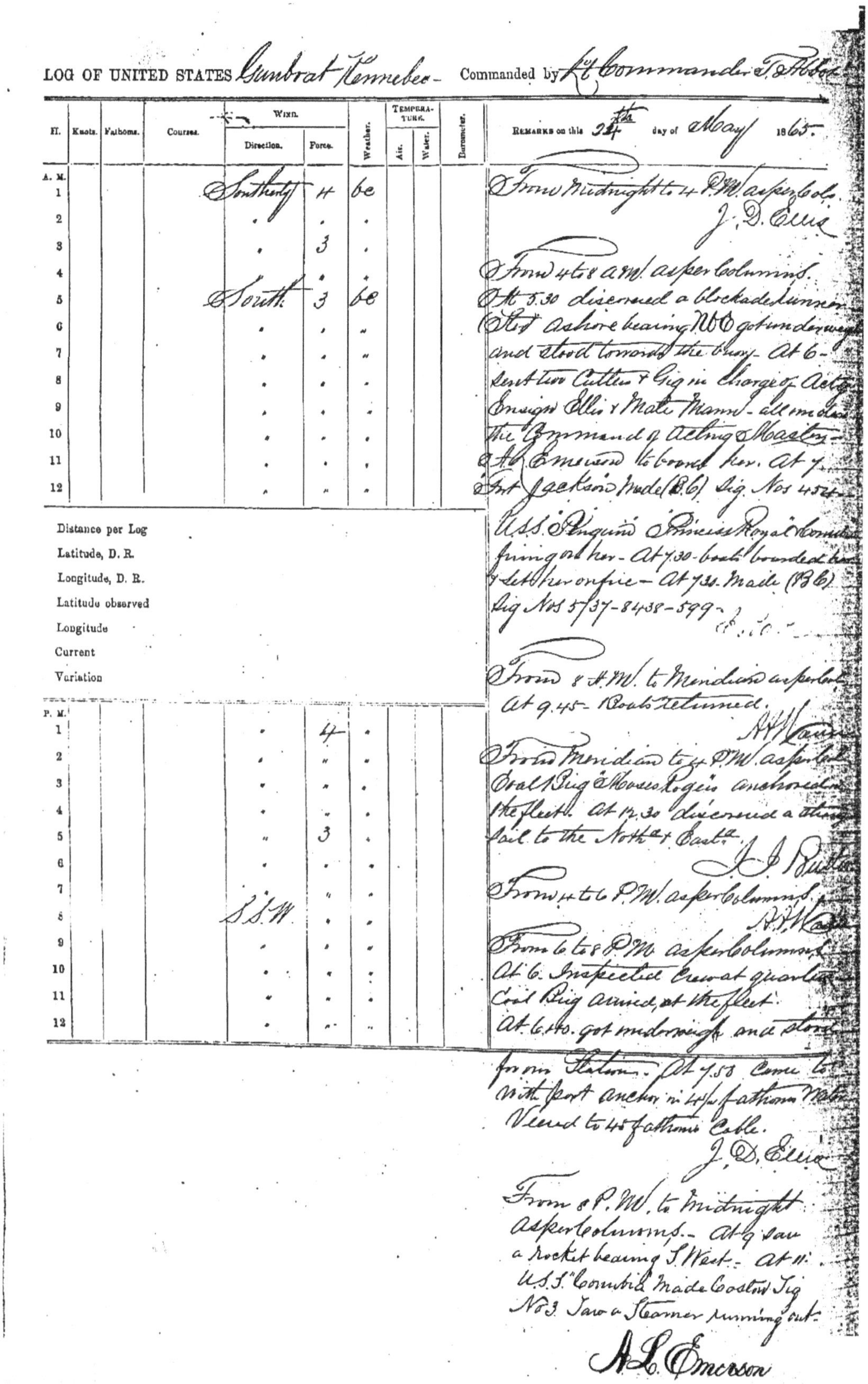

Figure 37. The *Kennebec's* log covering the destruction of the *Denbigh*. Dated 24 May 1865. *NARA*.

U.S. Steam Sloop *Seminole* 24 May 1865

Narrative comments from the remarks column of the *Seminole* log, A. (Albert) G. Clary, commanding (Figure 38):

Remarks on this 24th day of May 1865.
Off Galveston Texas

Mid. (Midnight) to 4. As per cols. Steam 15 lbs.   (signed) J. A. Bennett Act. Ensign

4 to 8. As per cols. At 5.30 discovered a blockade runner str. ashore near Bolivar Point. Sent 1st cutter with an armed boat's crew in to her. Several of the fleet in shelling her. Also discovered a brigantine (rigged ship) bearing S.E. standing towards the fleet. Communicated with Senior Officer by signals. Steam 15 lbs.   (signed) F. Kempton, Act. Ensign

8 to Mer. (Meridian). As per cols. At 10.30 1st cutter returned to ship having boarded and set on fire the blockade runner *Denbigh* the crew having escaped to the shore. Luke Robbins was killed by the accidental discharge of his own musket while getting into the boat (to return to the *Seminole*). Wm. Flagg, Sea. and Samuel Dow, Lds. confined in irons by order of the Comdg. (Commanding Officer) for drunkenness and disorderly conduct.[70]
    (signed) W. S. Church, Act. Ensign

Mer. to 4. As per cols. *New London* in shore shelling blockade runner at intervals. Steam 15 lbs.   (signed) D. K. Perkins, Act. Ensign

4 to 6. As per cols. As 4.30 got under weigh and stood to the southward. At 6 committed the remains of Luke Robbins, Sea. (deceased) to the deep. Steam 15 lbs.
    (signed) J. A. Bennett, Act. Ensign

6 to 8 P.M. As per cols. At 7 anchored at station with port anchor veered to 45 fathoms chain in 5½ fathoms water. Steam 15 lbs.   (signed) F. Kempton, Act. Ensign

From 8 to Mid. Steam 15 lbs.   (signed) W. S. Church, Act. Ensign

---

[70] Likely the two men found and consumed alcoholic beverages on board the *Denbigh*.

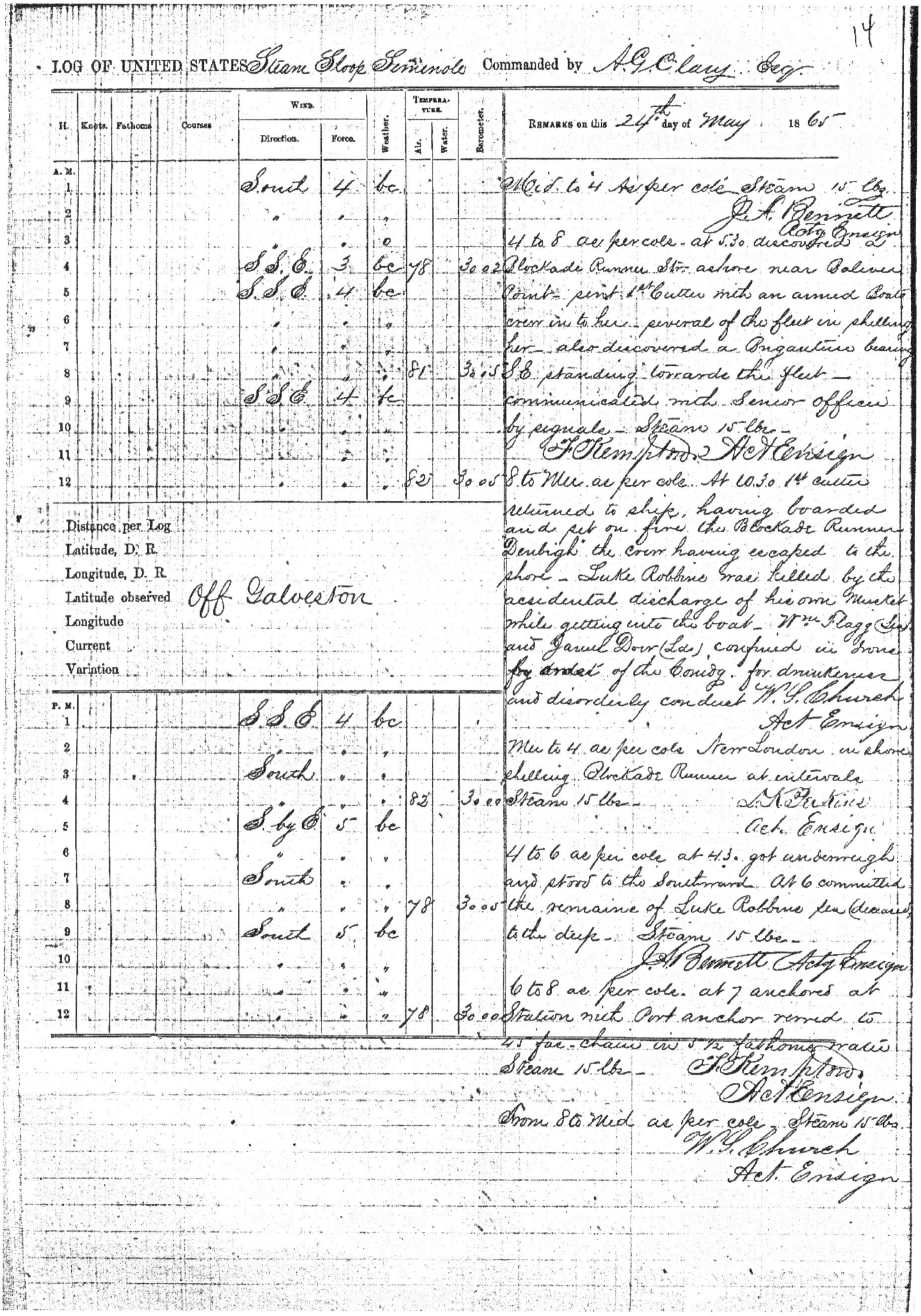

AND THE UNION NAVY

LOG OF UNITED STATES Steam Sloop Seminole Commanded by A. G. Clary Esq.

Remarks on this 24th day of May 1865

H.	Knots	Fathoms	Courses	Wind Direction	Force	Weather	Air	Water	Barometer	Remarks
A.M. 1				South	4	bc				Mid. to 4 As per cole. Steam 15 lbs.
2				"	"	"				J.F. Bennett Acty Ensign
3				"	"	"				4 to 8 as per cole. at 5.30 discovered a
4				S.S.E.	3	bc	78		30.02	Blockade Runner Str. ashore near Bolivar
5				S.S.E.	4	bc				Point – sent 1st Cutter with an armed Boats
6				"	"	"				crew in to her – several of the fleet in shelling
7				"	"	"				her – also discovered a Brigantine bearing
8				"	"	"	81		30.05	S.E. standing towards the fleet –
9				S.S.E.	4	bc				communicated with Senior Officer
10										by signals – Steam 15 lbs –
11										F. Kempton Act Ensign
12							82		30.05	8 to Mer. as per cole. At 10.30 1st cutter

Distance per Log
Latitude, D.R.
Longitude, D.R.
Latitude observed Off Galveston
Longitude
Current
Variation

returned to ship, having boarded and set on fire the Blockade Runner Denbigh – the crew having escaped to the shore – Luke Robbins was killed by the accidental discharge of his own Musket while getting into the boat. Wm Flagg (Sea) and Jamel Dow (Lds) confined in Irons by order of the Comdg. for drunkenness and disorderly conduct. W.S. Church Act Ensign

P.M.										
1				S.S.E.	4	bc				Mer to 4 as per cole New London in shore
2				"	"	"				shelling Blockade Runner at intervals
3				South	"	"				
4				"	"	"	82		30.00	Steam 15 lbs – L.K. Perkins
5				S by E	5	bc				Act Ensign
6				"	"	"				4 to 6 as per cole at 4.3. got underweigh
7				South	"	"				and stood to the Southward. At 6 committed
8				"	"	"	78		30.05	the remains of Luke Robbins Sea (deceased)
9				South	5	bc				to the deep – Steam 15 lbs –
10				"	"	"				J.F. Bennett Acty Ensign
11				"	"	"				6 to 8 as per cole. at 7 anchored at
12				"	"	"	78		30.00	Station with Port anchor veered to

45 fas. chain in 5½ fathoms water
Steam 15 lbs – F. Kempton
Act Ensign
From 8 to Mid as per cole – Steam 15 lbs
W.S. Church
Act Ensign

Figure 38. The *Seminole's* deck log for the destruction of the *Denbigh*. Dated 24 May 1865. **NARA**.

U.S. Steamer *Princess Royal* 24 May 1865

Narrative comments from the remarks column of the *Princess Royal* log, Lieutenant Commander C. ? W. Bohm, U.S.N., commanding (Figures 39-41):

Remarks on this 24th day of May 1865.
Off Galveston

Commences (?) and till 4. As per cols. ("signed"[71]) Cyrus K. Porter

From 4 till 8. At 5.00 discovered a strange sail bearing S.S.E. (pc). At 5.05 got under way and stood out to day station. At 5.20 Flagship made BC Sig. 652 ans'd. Flagship spoke us. We wore around and stood to the North'd. to destroy blockade runner ashore near Bolivar Point. At 6.10 beat to Genl. Quarters, came to anchor, 2½ fathoms water, 10 fathoms chain, stbd. anchor, and commenced shelling the steamer with forward and after rifles at 6.50. Ceased firing at 7. Steamer was boarded by a boat's crew from the *Seminole* and fired. 9.30 secured battery and piped down. (signed) William E. Cannon

From 8 till Mer(idian). *Cornubia* passed under our stern and reported the blockade runner to be Steamer *Denbigh*. At 8.30 got under way & stood out to day station. At 9.30 came to anchor 6 fathoms water, stbd. anchor 20 fathoms chain. At 11 discovered strange sail from the mast head bearing (pc) E.N.E. (signed) Thomas A. Witham

From Mer. Till 4. 1.15 Coal Brig anchored near the fleet. At 2 made BC Sig. interrog. 2296. 8076. Flag ship ans'd. making BC sig. 1157 ans'd. (signed) Lewis Johnson

From 4 till 6. 6 muster at Quarters. Loaded rifles with solid shot. (signed) Cyrus K. Porter

From 6 till 8. At 6.40 got under way. Spoke the flag ship and then stood to night station. At 7.25 came to anchor in 4 fathoms water, 30 fathoms chain, stb'd. anchor ready for slipping.
(signed) William E. Cannon

From 8 till Midnight. As per cols. (signed) Thomas A. Witham

Expended
    (20) Twenty cartridges 30 Pdr. Parrott
    (18) Eighteen Shell   30 Pdr. Parrott
    (2) Two shot   30 Pdr. Parrott
    (5) Five fuses 15" (second) Navy Gun
    (13) Thirteen fuses paper case. (signed) Thos. A. Witham, Act. Ensign

Figure 39. U.S.S. *Princess Royal*. Naval Historical Center.

---

[71] An exception to most of the other logbooks examined herein, this whole page was written in one person's hand. This in spite of the claim that each officer of the watch signed his entry.

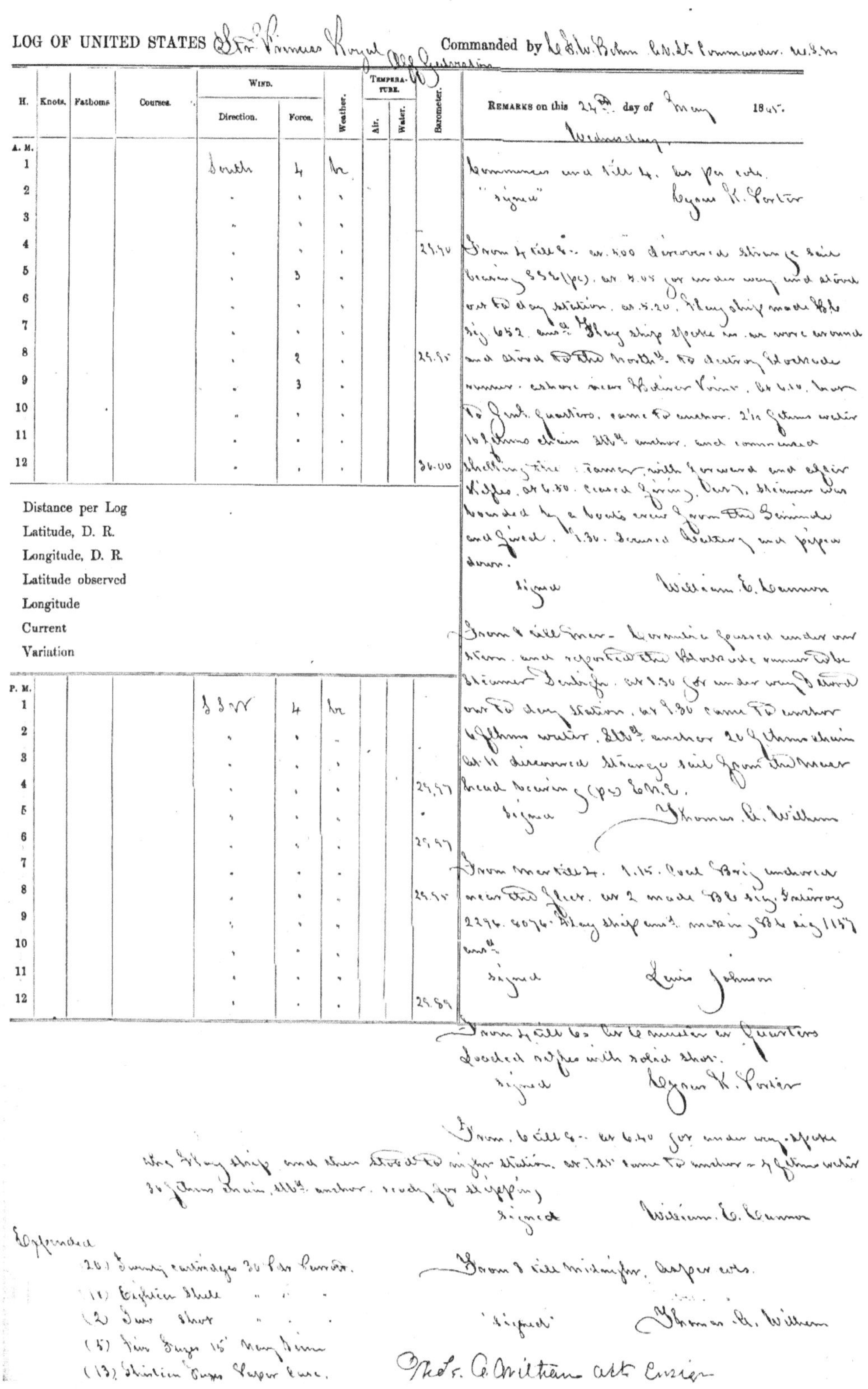

Figure 40. The *Princess Royal's* deck log for the destruction of the *Denbigh*. Dated 24 May 1865. NARA.

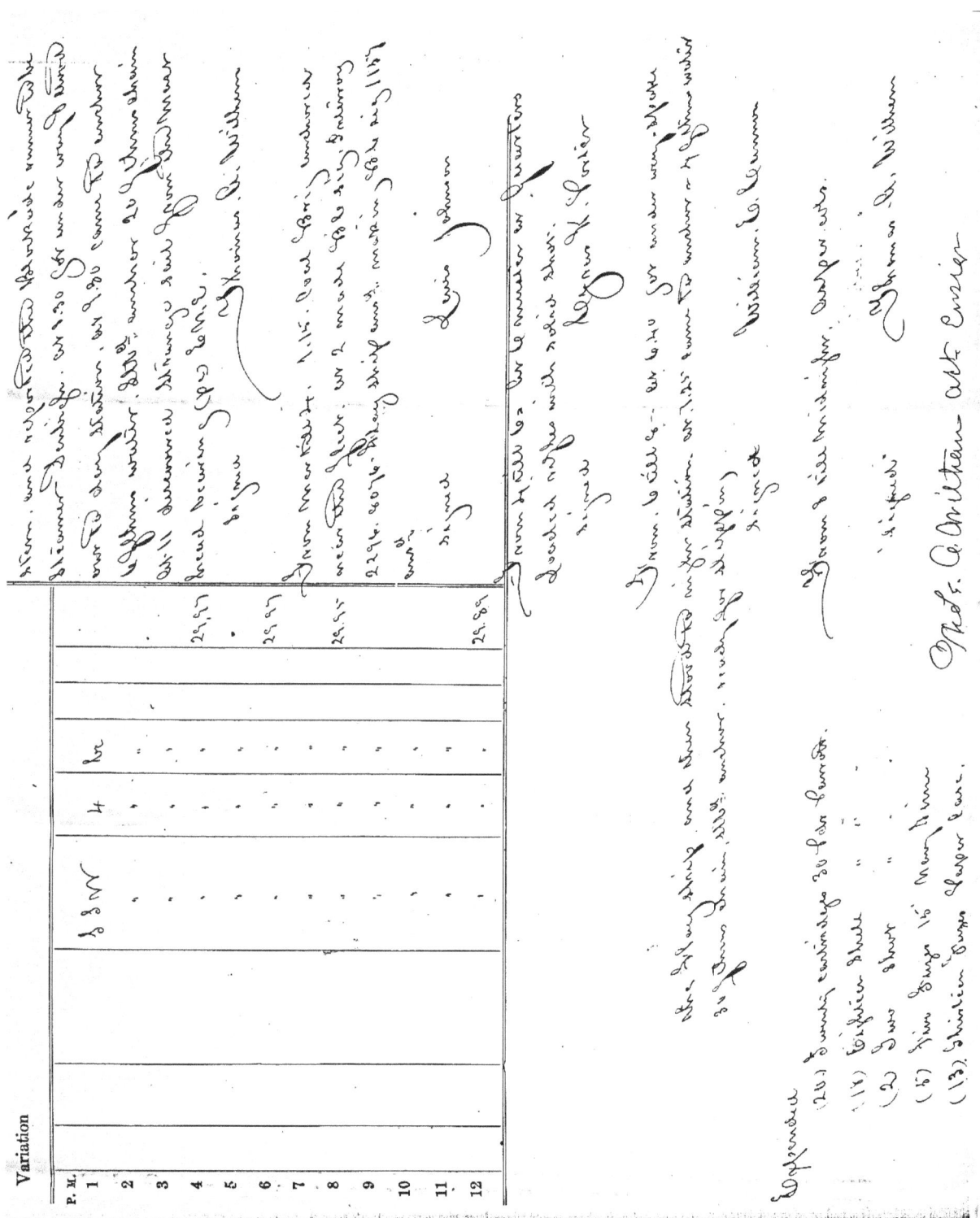

Figure 41. The U.S. Str. *Princess Royal's* deck log for the destruction of the *Denbigh* enlarged for legibility. Dated 24 May 1865. (Two pages.) *NARA*.

U.S. Steamer *Penguin*, 24 May 1865

Narrative comments from the remarks column of the *Penguin* log, Acting Lieutenant J. (James) R. Beers, commanding (Figure 42):

Remarks on this 24th day of May 1865.
Off Galveston.

From Midnight to 4 A.M. Steam 20 lbs.   (signed) Thos. G. Watson

From 4 to 8 A.M. At 5.30 saw a strange sail bearing S.S.E. which proved to be a coal brig. At 6 saw a blockade runner ashore off Bolivar Point with several of our gunboats going in for her. At 6.10 got under weigh for her. At 7.30 stopped engine on seeing the blockade runner on fire and the boats from our gunboats returning. Steam 23 lbs.   (signed) F. C. Almy

From 8 A.M. to Meridian. At 8.30 started for our station. At 9.15 anchored in 5 fath. water with starboard anchor & 20 fath. chain. At 9.30 saw a strange sail bearing N.E. Sig. No. 1116.361. The U.S.S. *Grand Gulf* repeated them to the Flagship. At 10.45 weighed anchor & steamed towards the strange sail which prove to be a coal brig. Changed course & steered for our station. Steam 23 lbs. Rev'd. 73.5.   (signed) C. ? Smith

From Meridian to 4 P.M. At 1 signalized to the U.S.S. *Fort Jackson* 7.66 234. At 1.25 came to in 6¼ fath. water with starboard anchor & 25 fath. chain. Fleet bearing S.W. Banked fires. Drilled 1st, Masters & howitzers with small arms & target practice. Steam 24 lbs.
                    (signed) N. A. Hannah

From 4 to 6 P.M. Steam 21 lbs.   (signed) Benj'n. Caullet

From 6 to 8 P.M. As per columns. At 6.40 went to Quarters. At 6.55 veered to 30 fath. chain ready for slipping. Steam 23 lbs.   (signed) Thos. G. Watson

From 8 P.M. to Midnight. Steam 21 lbs.   (signed) F. C. Almy

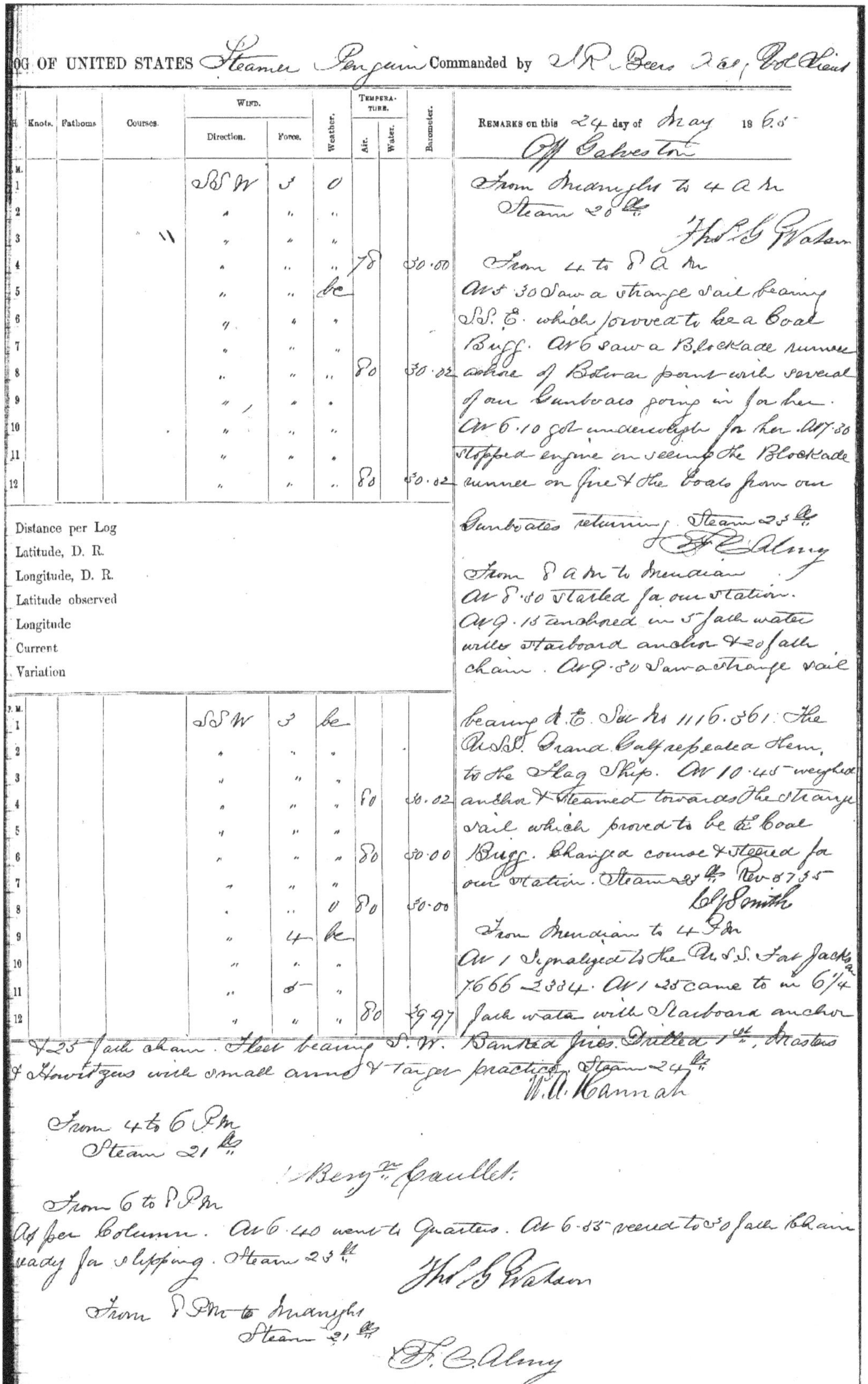

Figure 42. The *Penguin's* deck log for the destruction of the *Denbigh*. Dated 24 May 1865. **NARA**.

U.S. Str. *Albatross* 24 May 1865

Narrative comments from the remarks column of the *Albatross* log, Lieutenant Commander C. (Charles) S. Norton, commanding (Figures 43-44):

Remarks on this 24<u>th</u> day of <u>May</u> 18<u>65</u>.

From Midnight to 4 A.M. As per cols.   (signed) William Evens[72]

4 to 8 A.M. At daylight saw a blockade running steamer ashore on Bolivar Point bearing N. by W. (?). 6 A.M. Flagship signaled 1916-7017-1662-1662-4429-B. C. Got underway and went to General Quarters. At 6.30 commenced firing. At the same time, the U.S. Strs. *Cornubia*, *New London*, *Penguin*, and *Princess Royal* commenced firing. At 7.20 ceased firing. At 7.30 boats from *Seminole* and *Cornubia* boarded the steamer and set her on fire. She proved to be the English Steamer *Denbigh* bound in (to Galveston). At 7.40 came to anchor. At 8 secured battery and got under way and proceeded towards flagship. Expended four 13 sec. shell, 2 solid shot 32 pdrs., six distant firing charges for 32 pdr., 9 charges for 30 pdr. Parrott.
                    (signed) J. T. Chace

8 to Meridian. At 8.30 anchored near flagship, Galveston City bearing W. Bolivar Point N.W.
                    (signed) J. L. Brown

12 to 4. At 12.30 brig *Moses Rogers* of New York arrived with coal for the fleet. James Woodley, Lds. to have 5 days extra duty by day and confined in double irons by night for refusing duty by order of Comdg. Officer.   (signed) H. G. Martin

4 to 6. As per cols.   (signed) William Evens

6 to 8. At 7.10 got under way and at 7.30 came to with port anchor and 30 fathoms chain.
                    (signed) J. T. Chace

8 to Midnight. As per cols.   (signed) J. L. Brown

Coal remaining 240. T. (tons) 45 lbs.

Figure 43. U.S. Gunboat *Albatross*. *Naval Historical Center*.

---

[72] The handwriting style in this document was difficult to read. The signatures were particularly unclear.

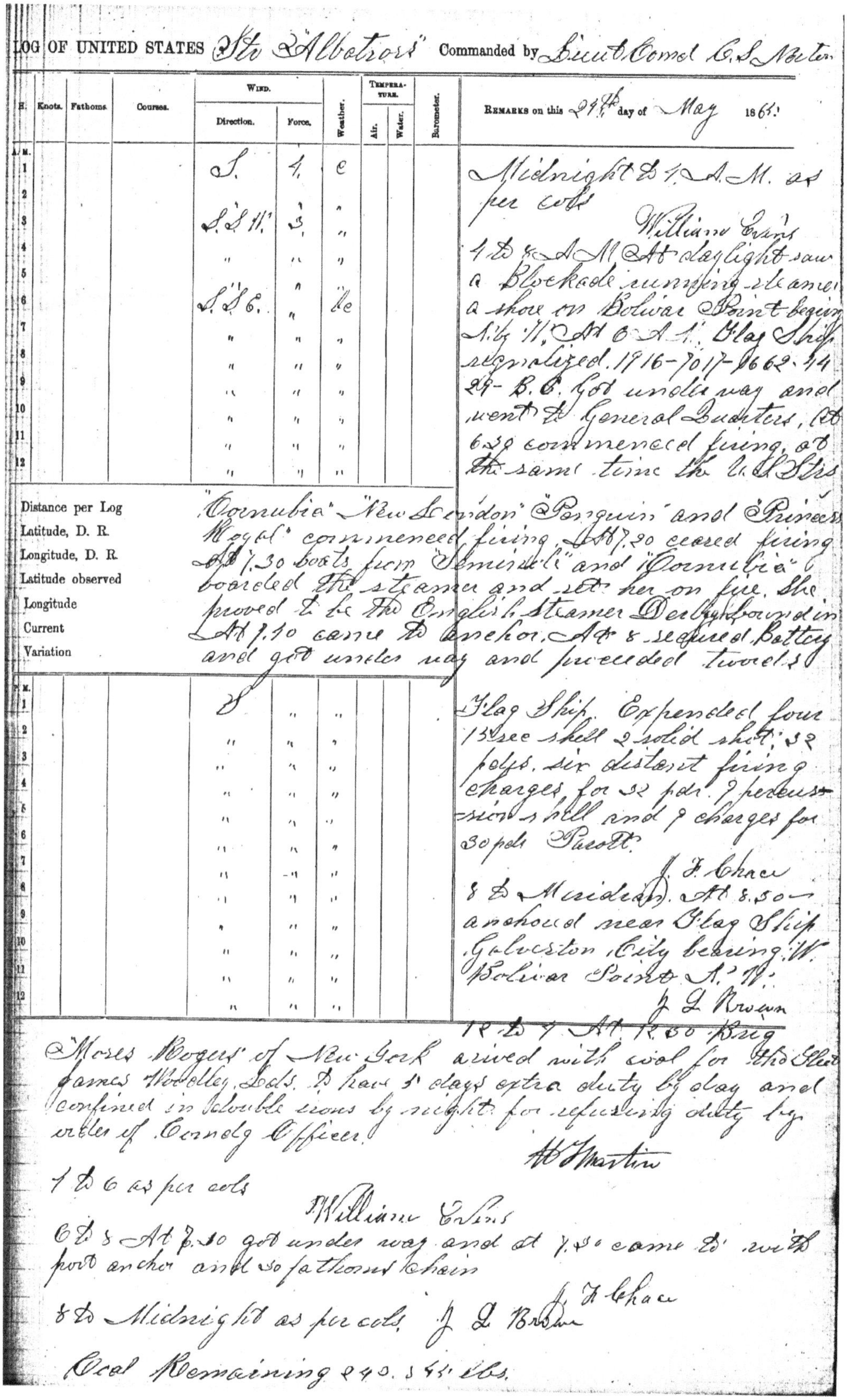

Figure 44. The *Albatross's* deck log for the destruction of the *Denbigh*. Dated 24 May 1865. **NARA**.

U.S. Steamer *New London*, 24 May 1865

Narrative comments from the remarks column of the *New London* log, Acting Volunteer Lieutenant W. Godfrey, commanding (Figure 45):

Remarks on this 24th day of May 1865.
Off Galveston.

From Mid. to 4 A.M. As per cols. (signed) V. W. Jones

From 4 to 8. At 4.30 discovered a steamer ashore on Bird Key. At 5 the U.S. Stmrs. *Cornubia, Princess Royal, Penguin,* and *Albatross* stood into Bolivar Channel and opened fire on her. At 5.20 got our anchor, stood in and anchored in 2¼ fathoms water and opened fire on the steamer with the starboard battery. Expended 6 shells and one shot. The stmr. was boarded by boats from the *Princess Royal* and *Cornubia* and set on fire. (signed) John H. Gregory

From 8 to Merid. At 9.30 called the Second and Powder divisions to quarters and fired two shells at the blockade runner. At 11 a strange sail was reported, bearing E.N.E.
(signed) W$^{m.}$ A. Prescott

From Merid. to 4 P.M. A herm (hermaphrodite) brig came in and anchored near the flagship. At 2.55 fired four 15 sec (second fused) shells at the blockade runner to drive away the wreckers. At 3.30 the U.S.S. *Cornubia* got underway and stood to the northward.
(signed) V. W. Jones

From 4 to 6. At 4.15 the *Cornubia* anchored off our port bow and opened fire on a schooner and sent a boat in. At 5.15 the boat boarded her. At 5.45 the boat returned bringing some prisoners and a white flag. At 5.45 the *Cornubia* fired again and made the following signal Geographical (BC) 404.2760.3769. (signed) John H. Gregory

From 6 to 8. Four refugees came off and brought the news that the enemy had evacuated the City of Galveston. (signed) W$^{m.}$ A. Prescott

From 8 to Mid. As per columns. (signed) V. W. Jones

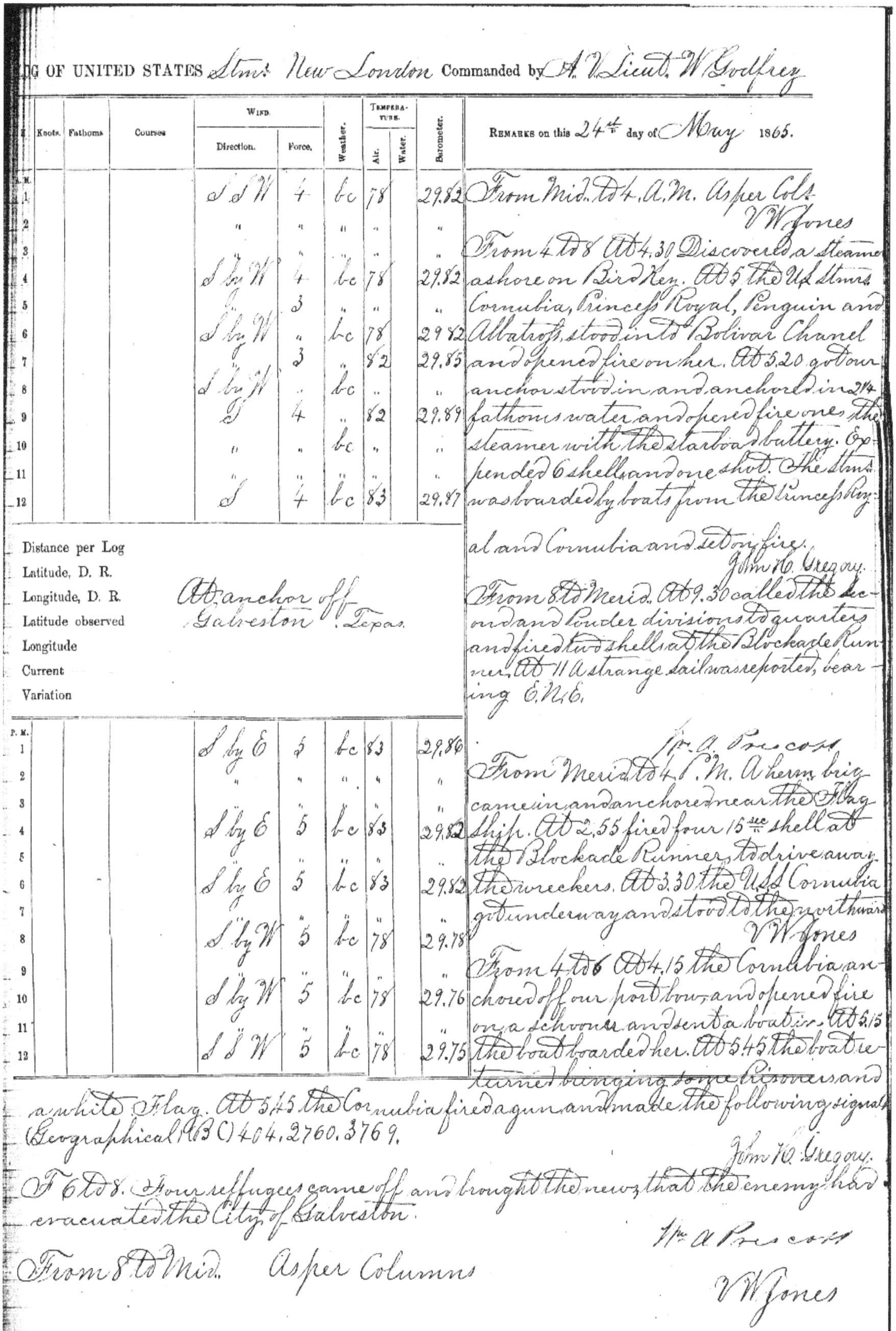

Figure 45. The *New London's* deck log for the destruction of the *Denbigh*. Dated 24 May 1865. *NARA*.

U.S. Steamer *Fort Jackson*, 24 May 1865

The *Fort Jackson's* logbook made a stark contrast with the other examples herein shown. There was a great deal more detail of description and context. The rest of the logs were by comparison quite bare bones. The *Fort Jackson*, flagship of the squadron, was commanded by a full U.S.N. captain of the old school who was also the squadron commander. Captain Sands had the logbook kept to a much higher standard. The information in the flagship's log provided a more complete picture of the day's activities.

Narrative comments from the remarks column of the *Fort Jackson* log, Captain B. F. Sands, commanding (Figures 46-48):

Remarks on this <u>Wed.</u> <u>24</u>th day of <u>May</u> 18<u>65</u>.
Off Galveston Bar, Texas.

Commences and until 4. Light S.S.W. wind, cloudy with hazy horizons.
(signed) John J. Reagean, Acting Ensign

From 4 to 8. Moderate breeze from the S$^{d.73}$ blue sky and detached clouds. At 4.50 saw a blockade runner ashore near Bolivar Point. Sent the *Cornubia*, *Princess Royal*, & *Albatross* in to destroy her. At 5.55 the *Cornubia* opened fire. At 6 saw a coal brig standing in from the E$^{d.74}$ At 6.15 made signal "destroy blockade runner." At 7 a boat from the *Seminole* went alongside & at 7.10 the runner was discovered to be on fire. At 7.50 the boat left the burning blockade runner which proved to be the Steamer *Denbigh*.
(signed) S. K. Harkins (?), Acting Ensign

From 8 to M. Moderate S. breeze, blue sky, cirrus clouds. At 9 shore boats took blockade runner's crew from the beach & proceeded towards town. Some of our ships firing occasionally at blockade runner. Isaac Morris, Lds., to have extra lookout for 5 nights for neglect of duty. Wm. Scott, Boy, to be kept under sentry's charge in watch below for 5 days for disobedience of orders, by order of Ex. (Executive) Officer. Lewis Frank, Lds., for skulking to be kept in double irons on bread & water for 5 days, by order of Capt. Sands. Rich$^{d.}$ Harris, Lds., to have mid watch on wheel house for 1 night for neglect of duty. At 10.50 *Penguin* got under way & stood E$^{d.}$ At 11 heavy gun fired from Pelican Island. Boat ret'd. from coal brig which proved to be the *Moses Rogers* from New Orleans. At 11.10 a sail was reported from aloft bearing N.E. by E. *New London* bearing N. by W. ¾ W., blockade runner N.W. ½ N., *Antonia* S.W., balance of the fleet anchored nearby.
(signed) Joseph S. Carey, Act'g. Master

---

[73] Southward.
[74] Eastward.

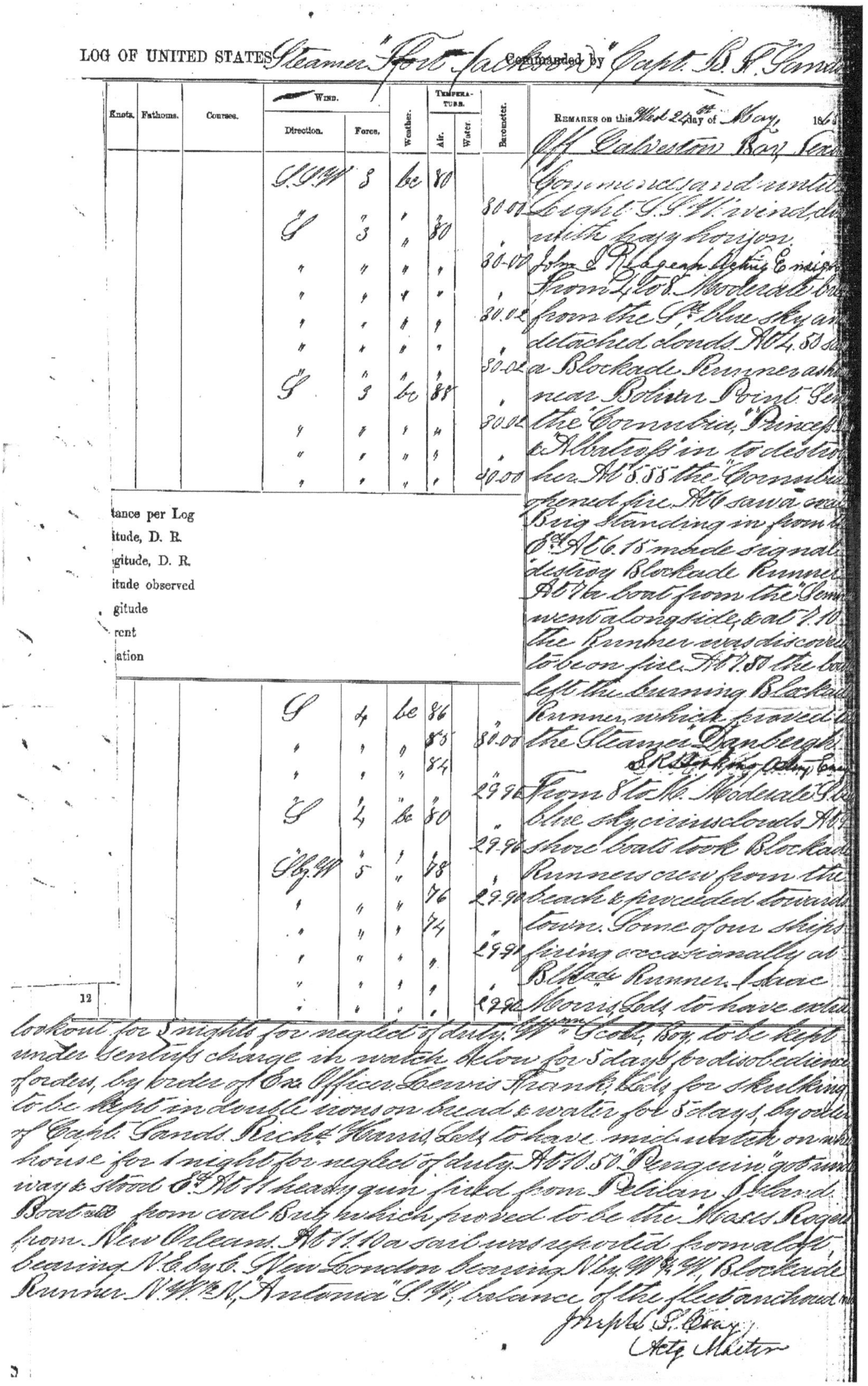

Figure 46. The *Fort Jackson's* deck log for the destruction of the *Denbigh*. Dated 24 May 1865. *NARA*.

Continuation (page) of Log of May 24th 1865

From M. to 3. Moderate breeze from the E^d, blue sky & detached clouds. At 12.10 the brig *Moses Rogers* anchored on our starb^d bow. Saw the smoke of several explosions from South Battery. John Austin to have extra lookout for five nights for insolence to petty officers. At 2 the *Penguin* made signal that the strange sail bearing N.E. was a collier.

(signed) H. F. Moffat, Act'g. Master

From 3 to 6. Light S. wind and cloudy. At 4.30 the *Cornubia* sent a boat to the wreck of the bl'kade. runner. At 6 the boat returned.

(signed) John J. Reagean, Acting Ensign

From 6 to 8. Moderate breeze from the S^d Blue sky & detached clouds. At 6.30 the *Cornubia* stood out towards the fleet. (signed) S. K. Harkins, Acting Ensign

From 8 to Mid. Fresh Southerly breeze with considerable sea. At 8.30 saw two rockets W. by S. Observed several lights on shore. At 10.25 Coston No. 3 was burned by a vessel bearing N.W. ½ W. shortly after saw a night challenge and answer near same direction.

(signed) Joseph S. Carey

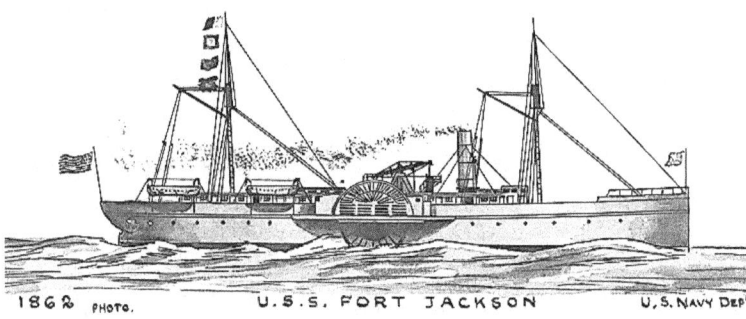

Photo # NH 63873   USS Fort Jackson. Artwork by Erik Heyl

Figure 47. U.S.S. *Fort Jackson*, flagship of the Galveston blockading squadron. *Naval Historical Center.*

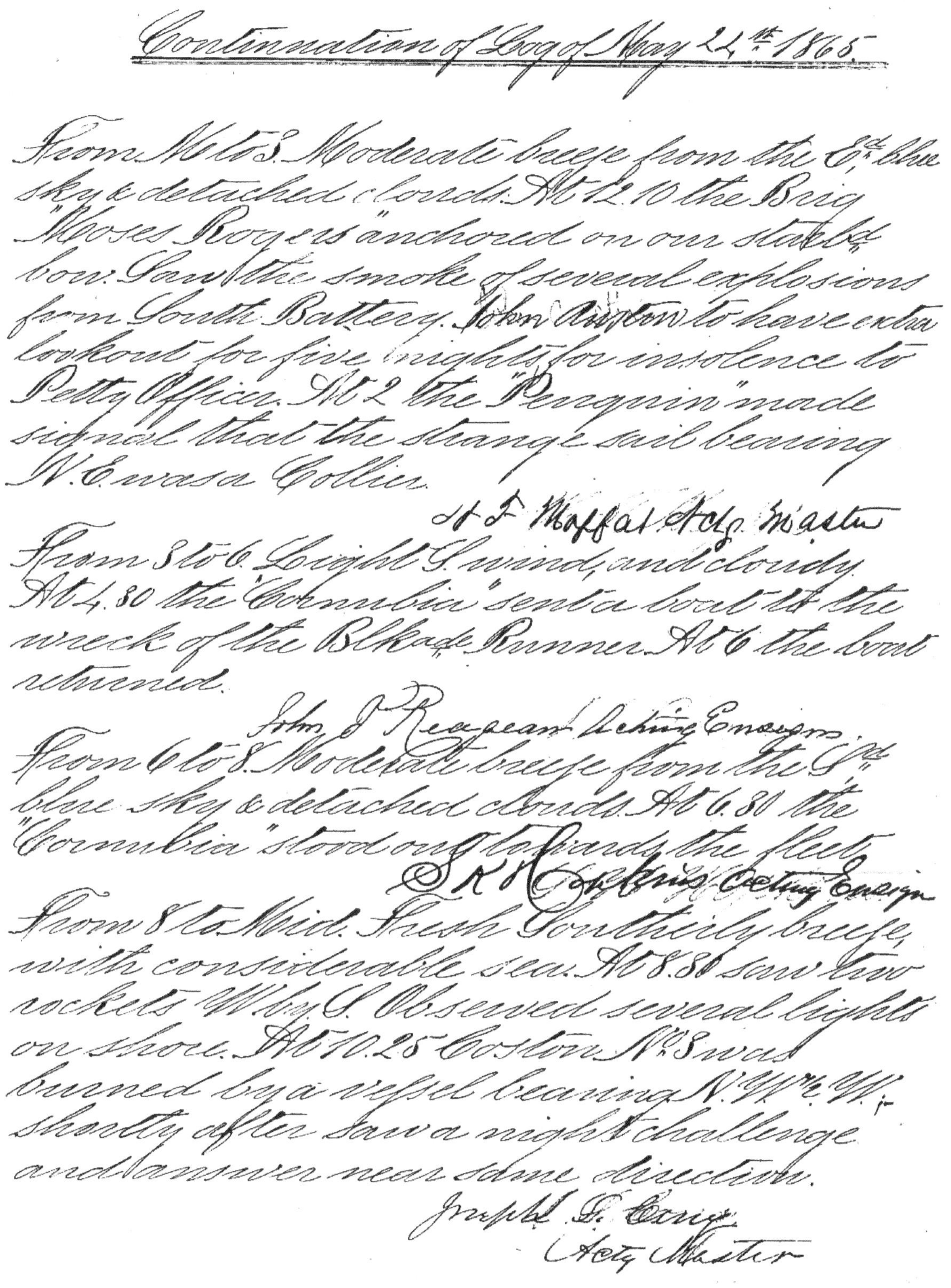

Continuation of Log of May 24th 1865

From Mid to 3. Moderate breeze from the 8th, blue sky & detached clouds. At 12.10 the Brig "Moses Rogers" anchored on our starb'd bow. Saw the smoke of several explosions from South Battery. John Ayton to have extra lookout for five nights for insolence to Petty Officer. At 2 the "Penguin" made signal that the strange sail bearing N.E. was a Collier.

                H. F. Moffat Act'g. Master

From 3 to 6. Light S. wind, and cloudy. At 4.30 the "Cornubia" sent a boat to the wreck of the Blk'de Runner. At 6 the boat returned.

                John J. Reagan Acting Ensign

From 6 to 8. Moderate breeze from the S'd, blue sky & detached clouds. At 6.30 the "Cornubia" stood on towards the fleet.

                S. R. Hopkins Acting Ensign

From 8 to Mid. Fresh Southerly breeze, with considerable sea. At 8.30 saw two rockets W by S. Observed several lights on shore. At 10.25 Coston N°3 was burned by a vessel bearing N.W. ½ W.; shortly after saw a night challenge and answer near same direction.

                Joseph F. Craig
                Act'g Master

Figure 48. The *Fort Jackson's* deck log continuation page. Dated 24 May 1865. **NARA**.

Other Log Entries

U.S. Steamer *Kennebec* 19th April 1865

Note the arrival of the news of Lee's surrender at Appomattox on 9 April, just 10 days after the fact. At this time the *Kennebec* was cruising off the Texas coast, not on station at Galveston. This seemed to the present author a rather quick transmission of news to a remote location.

Narrative comments from the remarks column of the *Kennebec* log, Lieutenant Commander T. (Trevett) Abbott, commanding:

Remarks on this 19th day of April 1865.

From Midnight to 4 A.M. As per columns. (signed) C. A. Thorne

From 4 to 8 A.M. As per columns. (signed) A. L. Emerson

From 8 A.M. to Mer. (Meridian). As per columns. At 9.30 inspected crew at quarters and drilled with small arms. At 9.45 discovered a strange sail to the East^d. Got underway and stood for her. At 10.30 went to genl. (general) quarters and cleared battery for action. Strange sail proved to be the Mexican Steamer *Fanny Fisk* from New Orleans bound to Matamoras. Found her papers correct and allowed her to proceed. Received the glorious news of the surrender of Genl. Lee's whole army to Genl. Grant. Dressed ship in commemoration of this cheering news. Secured battery and stood back for our station. (signed) J. J. Butler

From Mer. to 4 P.M. As per column. At 1 P.M. came to with port anchor in three faths. water veered to 45 faths cable. Salt works bearing N.E. by N. ½ E., Buoy N.E. ½ E. At 1.30 fired a gun and hoisted flag of truce and sent gig in charge of A. V. Lt. (Acting Volunteer Lieutenant) E. Baker on shore to communicate. (signed) J. S. Ellis

From 4 to 6 P.M. As per columns. At 5 gig returned from shore. (signed) C. A. Thorne

From 6 to 8 P.M. As per columns. (signed) A. L. Emerson

From 8 P.M. to Mid^t (Midnight). At 10.30 got underway and stood off into 4½ faths. water and anchored with port anchor. Veered to 60 faths. cable. (signed) J. J. Butler

9 March 1864 Schooner *Velocity* ran into Galveston in broad daylight

The following came from the logbook of the Confederate lookout station in Galveston. Sometimes known as the JOLO lookout for reasons lost with the passage of time, the watch was kept atop the Hendley Building, which provided a good view of the Gulf, bay entrance, and harbor. The sections below reported an all too common example of poor gunnery by the blockading fleet. In broad daylight, the Schooner *Velocity* ran in, seemingly unscathed, through a storm of 54 shots (Figure 49).

Observatory Wednesday March 9th 1864.[75]

7 A.M.	Morning cloudy and foggy with appearance of rain. Wind fresh from the S.S.W. Blackade is seen as follows Flagship and two supply vessels off Main (Channel). Two gunboats off Swash Channel. One off N(orth) Breakers and one off Cylinder Channel. Total seven.
10.30 A.M.	S. B. (Steam Boat) *Ruthven* leaves Central Wharf for upper bay.
11.15 A.M.	Gunboat from Swash makes way cruising to southward. After cruising in this fashion some five miles or more returned to the fleet.
11.35 A.M.	Gunboat No. 2 from Swash moves up to the fleet.

Observatory March 9th Continued

12 Meridian	Still cloudy and foggy. Fleet is same seven vessels. Wind S.S.W.
1.35 P.M.	Strange sail in sight from the southward coming up the Swash Channel proves to be a schooner with Texas flag at main.
1.50 P.M.	Gunboat from Swash (Channel) made way standing for stranger.
2.35 P.M.	Two gunboats made way from Main (Channel) & N(orth) Breakers. Gunboat No. 1 opens fire on schooner. Schooner nearly opposite Catholic College (in Galveston). Fires in all 21 shots.
2.50 P.M.	Gunboat No. 2 opens fire. Fires in all 16 shots. Gunboat No. 3 opens fire. Fires in all 13 shots.
3 P.M.	Flagship opens fire. Fires in all 4 shots. Total no. of shots fired 54.
3.05 P.M.	Fort Magruder opens fire, fires 4 shots.
3.15 P.M.	Enemy cease firing. Schooner arrives safe and anchors off Fort Point and enemy's boats take usual positions.
3.00 P.M.[76]	Schooner proves to be the *Velocity* now underway for Houston. . . .

---

[75] Chief Signal Reports from Houston Observatory Relating to Meteorological Conditions and Enemy Vessels, District of Texas, New Mexico & Arizona, 1863–1864, pp. 221-224. NARA, Record Group 109: War Department Collection of Confederate Records, 1825–1927. Series: Record Books of Executive, Legislative, and Judicial Offices of the Confederate Government, 1874–1899. On line at http://research.archives.gov/description/768265.

[76] Likely meant at 4:00 P.M. the schooner got under way for Houston after checking in at the fort.

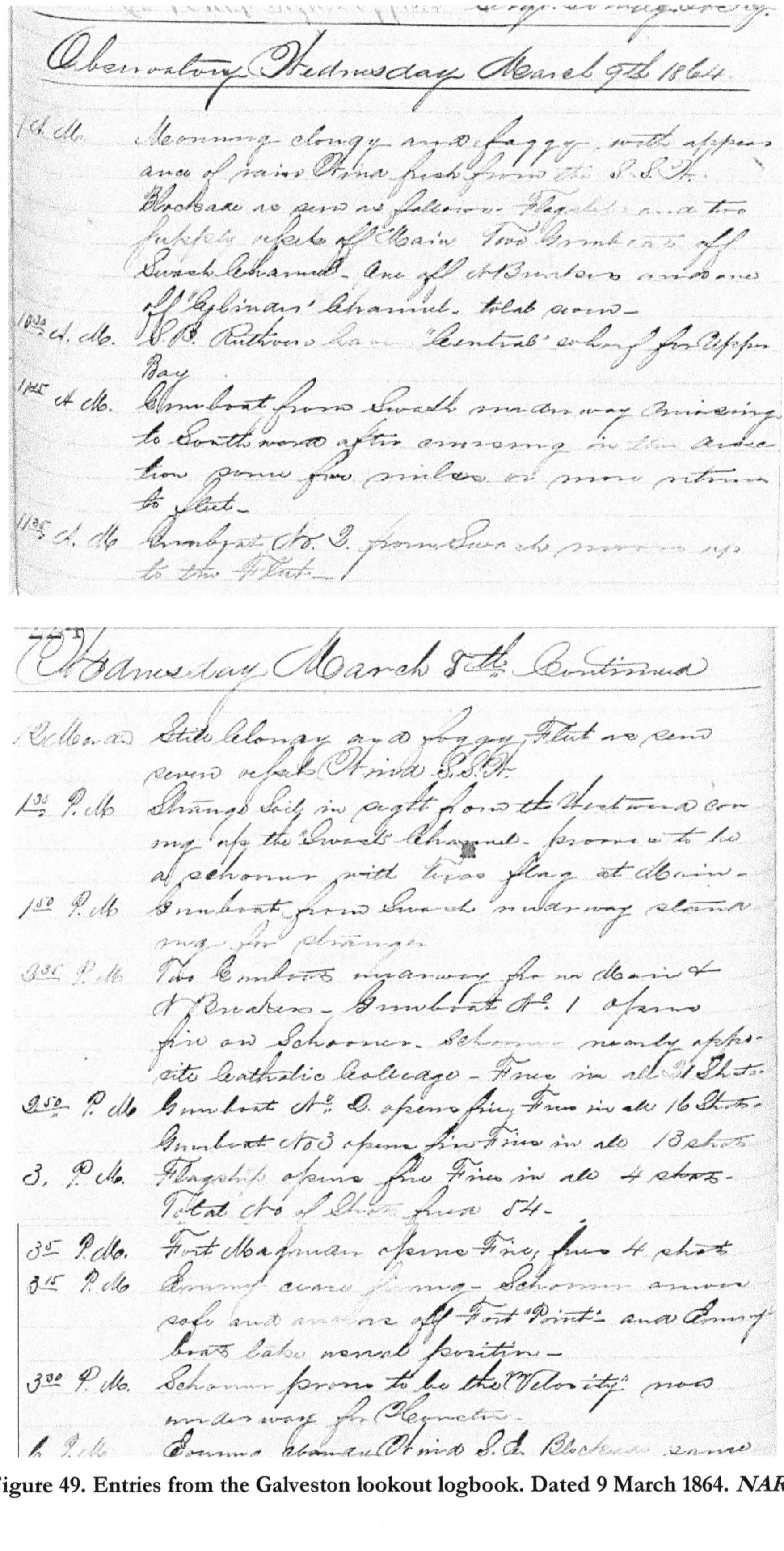

Figure 49. Entries from the Galveston lookout logbook. Dated 9 March 1864. *NARA*.

5
Prize Money Payout for
the *Denbigh's* Jettisoned Cotton

This chapter covers the multiple ledger entries generated in the prize money payment process. The specific examples related to the cotton bales thrown overboard when the *Denbigh* ran aground on 20 April 1865. The questions to be addressed were how much the crews individually realized in prize money and how the payout worked.

The NARA in Washington, D.C., has a set of several kinds of accounting records having to do with prize money. The ledgers detailed the appropriation and payout of prize money to the crews of each Union navy vessel that made a successful capture approved by the courts. Examples of the U.S.S. *Gertrude* and U.S.S. *Cornubia* entries in each type of ledger book were traced and herein displayed. Altogether the ledgers offered detailed and efficient ways to look up information on prizes. There were separate ledgers indexed by prize name, by capturing vessel name, by the names of officers and men of each crew entitled to payment, and by the individual appropriation with acknowledgement of payment for each man, one by one.

The navy was rapidly decreased in size when the Civil War ended. How did the scattered veterans find out their prize money was ready for payment? The prize money award depended on completion of the court cases, transfer of funds, and complicated calculations that frequently reached fruition some months or even years after the capture. Notices were probably posted at navy yards and other facilities. Public notice was published in newspapers like the *New York Times* herein shown.

The *New York Times* Notice

The *New York Times* published an announcement, the "List of Prize-Money to be Distributed" (see below pp. 229-230) on 19 Nov. 1865. Therefore, the 17 June payment to the U.S. Treasurer listed on the court docket for the *Gertrude* vs. Fifty Bales of Cotton must have been the half of the prize money allocated to the officers and crew. It took only three months June-Sept.) for the funds to go from the Treasury Department to the Navy Department for calculation of individual payouts and five months (June-Nov.) for the public notice to be published in the newspaper. The actual calculation results table of payments was certified and dated 18 Sept. for the *Gertrude* and 19 Sept. for the *Cornubia*. Then the word went finally to the paymasters at the various naval facilities who received notice of approval to pay and presumably the exact amounts payable to the individual people. The earliest payouts recorded took place 30 Sept. for the *Gertrude* and 6 Oct. for the *Cornubia*. All dates in this paragraph were in 1865.

The three-month processing period did not seem excessive in spite of complaints of long periods before getting paid. (See the Davenport's memoir above, Part II, Chapter 1.) However, note that it was the individual sailor's responsibility to apply to a navy paymaster for actual payment. Some recipients had agents located near the navy yards to secure the funds on their behalf. The relevant ledger book showed that some people never did collect their prize money. Some did collect many years after the fact. One sailor's record was marked "deserted" and the prize money not paid.

The *Denbigh's* Cotton Jettisoned at Galveston: Research in RG 217

The idea in this section was to discuss each of the relevant prize money ledger books and show examples of the information therein contained.

Ledger Book Listing Prizes

This book had a listing of prizes indexed alphabetically by the vessel name of the captured prize. It had sections with page tabs for each letter of the alphabet. In the present case, the *Denbigh's* lost cotton bales were captured cargo, but the blockade-running ship escaped. So there was no vessel name for the prize as the cargo was captured but not the ship. There was, however, a large section under letter "C" for "Cotton" as such captures were commonplace. That was where the author found the information on the incident of interest (NARA RG 217, Entry 821, Vol. 1 of 1.). The data entries were as follows:

Year	Prize	Court	No.	(notes in red ink)
1865	Cotton 50 Bales	New Orleans	822	(blank)
"	Cotton 89 Bales	New Orleans	833	See 822 – C

The column headings were often self-explanatory as follows:
1. Year: The year the prize was taken.
2. Prize: Usually the captured vessel name, but, in this case, the name of the items, that being "cotton bales" jettisoned by the *Denbigh* when she ran aground leaving Galveston.
3. Court: Location of the U.S. district court that adjudicated the admiralty case for the prize. This was important information for researchers (if not already known) since the court's file included many documents describing the prize, it's sale, and often supporting documents captured on board the prize ship. The court files were available in the NARA regional repository, in this example in Fort Worth, Texas, and well worth the study.
4. No.: The meaning and logic behind this reference number was unclear to the present author.
5. Blank column for notes: "See 822 – C." Perhaps this was a note to cross reference with the related case as both derived from the same incident.

Ledger Book Listing Capturing Union Navy Vessel

Again the book was divided into sections with page tabs for each letter of the alphabet. In this case, we consulted sections "C" for the U.S.S. *Cornubia* and "G" for the U.S.S *Gertrude* (NARA RG 217, Entry 825, 3 Vols.). The entries for those two ships showed how many prizes each navy vessel captured as follows:

Sed.	Captor	Prize	Amount	Decree of Court	Remarks
8/459	*Cornubia*	89 Bales Cotton	3,510.38	Apr. 20, 1865	Sept. 26, 1865
9/380	"	*Chaos*	7,307.78	Apr. 21, 1865	Feb. 9, 1866

Sed.	Captor		Prize	Amount	Decree of Court	Remarks
5/324	*Gertrude*	16.14	*Ellen*	2,293.32	Feb. 15, 1864	Feb. 3, 1865
7/227	Do.	21.84	*Wenona*	43,122.05	Nov. 29, 1863	" " "
5/496	Do.	56.66	*Warrior*	12,843.07	Aug. 16, 1863	Dec. 8, 1864
8/463	Do.		50 Bales Cotton	1,939.37	Apr. 20, 1865	Sept. 26, 1865
9/186	Do.	4	*Eco*	1,620.93	Feb. 19, 1865	Nov. 21, 1865

The column headings were as follows:
1. Sed.: a cross reference number? Meaning was not clear to the present author.
2. Captor: name of the U.S. Navy vessel that captured the prize. All the prizes of each capturing ship were listed in a group.
3. Prize: name of the prize vessel or, in this case, the number of the bales of cotton captured.
4. Amount: money due to the officers and crew of the capturing ship, being one-half of the total net proceeds from the sale of the prize. The other half, referred to as a moiety, went to the U.S. government.
5. Decree of Court: one of the final steps in the admiralty prize case was issuing the court's final decision. The date of the court's decree was a key cross reference from the court case to the prize money payment.
6. Remarks: this column held the date prize money was certified and became available for payment to the individual officers and crew. The date was a key cross reference for locating the table with the breakdown of payment to each man. Those tables were in a separate set of oversized ledgers with pages ordered by the date the prize money was certified (NARA RG 217, E 823).

Public Notice of Availability

And how did the individual crew members know when the prize money was approved for payment? Often this happened many months or even years after the capture. For examples like the *Denbigh's* cotton captured near the end of the war, the ship likely had been decommissioned, and the crew mustered out of the service. Official notice probably was posted at navy yards and shore installations, particularly where there were paymasters available for actually paying out the funds. More helpful for the men who had returned to civilian life were prize lists published in newspapers. The following was extracted in part from a lengthy list and showed the two prizes here relevant (bold type added) along with a few listings for other ships to illustrate the nature of the overall article.

The New York Times

Archive

TO NAVAL OFFICERS AND SAILORS. A List of Prize-Money to be Distributed. The Brevets in the Regular Army. Interesting Old Document.

Published: November 19, 1865
The following list of additional prizes payable is published by the Fourth Auditor of the Treasury:

Name of Captor.	Name of Prize.
Adolph Hugel	Cargo of schnr. Kate.
Antona	Betsy.
Ariel	Boat Buckshot.
Antona	Cecelia "D."
Argosy	Property, mules, &c.
Augusta	Cumbria cargo.
Augusta	Island Belle.
. . .	
Connenburgh	Queen of the Waves.
Corypheus	Sailboats and cargoes.
Commodore	4 boats, &c.
Choctaw	Volunteer.
Champion	Volunteer.
Chocura	Lote Harley.
Chocura	31 bales cotton.
Cornnum	**89 bales cotton.**
(Misprint of *Cornubia*.)	
Connecticut	90 bales cotton.
Clyde	42 bales and bags of cotton.
Cumberland	Young America.
Cayuga	Cotton.
De Soto	Bright and Mississippian.
De Soto	Jane, Adelie and Major Prim.
De Soto	Rapid.
. . .	
Gettysburg	Little Ada.
Gen. Price	Cotton, &c.
Gazelle	Cotton, &c.
Gettysburg	Armstrong.
Gettysburg	Lillian.
Gettysburg	Cotton.
Gettysburg	Blenheim.
Gertrude	**Cotton.**
Glide	Malta.
Gov. Buckingham	Cotton.
G. W. Blunt	Prince Royal.
Harriet Lane	Henry C. Brooks.
. . . .	

Payout of Prize Money to the *Gertrude's* Officers, Crew, and Commanders

Presented in this section were the lists and tables relating directly to the payments, one by one, to the individuals entitled to share in the prize.

Prize List Submitted by Captain of the *Gertrude*

When a prize was captured, it was sent to a port where a U.S. District Court was located for the legal proceedings to take place. The prize crew carried along an official listing of the men on the ship's books (Figure 50). All men assigned to the ship along with their rank and pay appeared on this list. The officers making up the chain of command also received prize money, and their names were specified separately from the ship's crew in the present document. The ship's captain and the officers in the chain of command received at fixed fraction of the total prize money. After deducting this amount, the payments to the rest of the ship's officers and men were calculated in the proportion of each man's salary to the total payroll of the ship.

[FIRST SHEET OF PRIZE LIST][77]

PRIZE LIST *of the U.S.* Str. Gertrude *at the time of the* capture *picking up of the* Fifty (50) bales of cotton *on the* Twentieth (20th) *day of* April, 1865, *this vessel being at the time attached to the* West Gulf *Squadron, Actg. Rear Admiral* H. K. Thatcher *Commanding.* Fleet Capt. G. Simpson, Lieut. Com'd'r. Capt. Benj. F. Sands Com'd'g. 3rd Division

I CERTIFY that I have PERSONALLY compared the annexed prize list with the muster roll of this vessel at the time of the capture picking up of the cotton above named and approved the same as correct.

Benj. C. Dean A. V. Lt.

---

[77] NARA RG 217. Records of the Accounting Officers of the Treasury, Records of the Navy Pay & Pension Division, Prize Lists 1862–1865, G-H (for *Gertrude*), Box No. 04. Entry 824.

Commanding U.S. Str. "Gertrude".

"The commanding officer of every vessel in the Navy entitled to, or claiming an award of, prize money, shall, as early as practicable after the capture, transmit to the Navy Department *a complete list* of the officers and men of his vessel entitled to share, inserting thereon the quality of every person rating, *on pain of forfeiting his whole share of the prize money resulting from such capture*, and suffering such further punishment as a court martial shall adjudge."

"No officer or other person who shall have been temporarily absent on duty from the vessel, on the books of which he continued to be borne while so absent, shall be deprived, in consequence of such absence, of any prize money to which he would otherwise be entitled."—(*Act of Congress, July 17, 1862.*)

Names.	Rating.	~~Monthly~~.
		Annual
Benj. C. Dean	A. V. Lt. Com'd'g.	1,875.
R. R. Brawley	A. A. Paymaster	1,300.
Adam Shirk	A. A. Surgeon	1,250.
Fredk. Newell	Acting Ensign	1,200.
Jas. (James) W. Munro	" "	1,200.
Isaac W. Goodrich	" "	1,200.
Wm. H. Brown	Actg. 2nd Asst. Engin'r.	1,200.
J. H. Nesen	" 2nd " "	1,200.
F. C. Morey	" 2nd " "	1,200.
Chas. O. Farciot	" 3nd " "	1,000.
		per month
Chas. A. Osborne	" Mate	40 $
		per annum
Isaac Jackson	Paymaster's ~~Clerk~~ Stew(ard)	400 $
(end p. 1)		
Names	Rating	Monthly Pay
Jas. R. Kelly	Yeoman	30.00
Jas. H. Marshall	Ord(inary) Seaman	16.00
Jacob A. Sharkey	Surgeon's Steward	25.00
John Hudson	M(aster) at Arms	25.00
T. H. Connors	Lands(man)	14.00
Morris Garvin	C(oal) Heaver	20.00
T. J. Alcorn	Lands.	14.00
Robt. Duffy	1st Css. (Class) Boy	10.00
Chas. Brown	1st Css. Boy	10.00
John Segel	Ship's Cook	26.00
Peter Holliman	Steerage Stew'd (Steward)	18.00
J. O. Sharon	Carpenter's Mate	30.00
A. R. Stone	Officer's Steward	35.00
T. B. Stack	Ship's Writer	18.00
Wm. Creighton	Ord. Sea.	16.00
Edwin Barton	Lands.	14.00
John Donahue	Gunner's Mate	~~30.00~~ 27.00
Edwd. Green	1st Css. Boy	10.00
Wm. Birchall	Lands.	14.00
Chas. Dillon	Lands.	14.00
Geo. Kennedy	1st Css. Boy	10.00
J. M. Germain	Ord. Seaman	16.00

Eugene Spelman	Lands.	14.00
Wm. Ball	1st Css. Fireman	30.00
Mich¹. Slattery	2nd Css. Fire.	25.00
Philip Marra	2nd Css. Fire.	25.00
Henry Masters	Seaman	20.00
Jas. T. Benson	2nd Css. Fireman	25.00
Hugh O. Donnell	1st Css. Fire.	30.00
Terence McGrath	Coal Heaver	20.00
James Fitzgibbons	C. Heaver	20.00
G. Kimberly	C. Heaver	20.00
(end p. 2)		
Austin Shea	C. Heaver	20.00
Ed. Ordway	Coxswain	25.00
He has been paid $160.00 Bounty "Army"[78]		
Jas. H. Ham	Quarter Master	25.00
He has been paid $325.00 Bounty "Army"		
Jas. Tomson	Capt. (of the) Hold	25.00
He has been paid $54.00 Bounty – 3 mos. advance		
Wm. Collins	Boat(swain's) Mate in Charge	30.00
He has been paid $54.00 Bounty – 3 mos. advance		
Francis Meeher	Seaman	20.00
He has been paid $60.00 Bounty – 3 mos. advance		
Joseph Kelly	Quarter Master	25.00
He has been paid $48.00 Bounty – 3 mos. advance		
Simeon Devine	Boat. Mate	27.00
He has been paid $60.00 Bounty – 3 mos. advance		
Owen Duffy	Quarter Gunner	25.00
He has been paid $65.00 Bounty – 3 mos. advance		
Wm. Hunt	2nd Class Fireman	25.00
He has been paid $33.33 Bounty – 1st instalment [sic]		
Jno. Halpin	2nd Class Fireman	25.00
He has been paid $33.33 Bounty – 1st instalment		
Jno. Kurnierny	1st Class Fireman	30.00
He has been paid $100.00 Bounty – 1st instalment		
Edwd. Buchanan	1st Class Fireman	30.00
He has been paid $200.00 Bounty		
Badoo McAllister	1st Class Fireman	30.00
He has been paid $100.00 Bounty		
Wm. Walters	Lands.	26.00
M. A. Wade	Lands.	14.00
Edwd. Pender	Lands.	14.00
(end p. 3)		
Alstyne Fisher	Lands.	14.00
Jas. Kelly	Lands.	14.00
Thos. Kelly	Sail Maker's Mate	25.00
He has been paid $100.00 Bounty – 1st instalment		
James Gallivan	Seaman	20.00
James Pender	Seaman	20.00
He has been paid $100.00 Bounty – 1st instalment		
Chas. Kelley	Seaman	20.00

---

[78] The reason for specifying which men had joined up for a bounty was unclear as the fact did not affect the prize money payout.

Winslow Evans	Ord. Sea.	16.00
He has been paid $100.00 Bounty – 1st instalment		
Chas. Kemp	Officer's Cook	~~25~~ 30.00
Geo. Bell	Quarter Gunner	25.00
He has been paid $100.00 Bounty – 1st instalment		
James Kelly	Capt[n.] (of the) Forecastle	25.00
Thos. Jackson	Seaman	20.00
He has been paid $100.00 Bounty – 1st instalment		
James Thompson	Seaman	20.00
He has been paid $100.00 Bounty – 1st instalment		
Wm. Thompson	Chief Quarter Mastr.	28.00
He has been paid $100.00 Bounty – 1st instalment		
Wm. Young	Capt. (of the) After Guard	25.00
Edwd. Tieniera (?)	Coxswain	25.00
Benj. Jones	Lands.	14.00
John Allen	Seaman	20.00
He has been paid $100.00 Bounty – 1st instalment		
Peter Limber (?)	Seaman	20.00
He has been paid $100.00 Bounty – 1st instalment		
Martin Wilson	Ord. Sea.	16.00
Warren Caville	Lands.	14.00
He has been paid $100.00 Bounty – 1st instalment		
Edwd. Jordan	Lands.	14.00
He has been paid $100.00 Bounty – 1st instalment		
Rich[d.] Brown	Lands.	14.00
He has been paid $100.00 Bounty – 1st instalment		
(end p. 4)		
Niel McIsaacs	Lands.	14.00
He has been paid $100.00 Bounty – 1st instalment		
Chas. Gipson	Lands.	14.00
He has been paid $100.00 Bounty – 1st instalment		
Mich[l.] (Michael) Kennedy	Lands.	14.00
He has been paid $100.00 Bounty – 1st instalment		
Pat[k.] (Patrick) Wallace	Lands.	14.00
He has been paid $100.00 Bounty – 1st instalment		
David Russell	Lands.	14.00
He has been paid $100.00 Bounty – 1st instalment		
Peter Walsh	Lands.	14.00
He has been paid $100.00 Bounty – 1st instalment		
Edmond Williams	Lands.	14.00
He has been paid $100.00 Bounty – 1st instalment		
Dan[l.] Murphy	Lands.	14.00
He has been paid $100.00 Bounty – 1st instalment		
Jos. (Joseph) Grumeblectone	Lands.	14.00
He has been paid $100.00 Bounty – 1st instalment		
Alex[r.] (Alexander) Lupton	Lands.	14.00
He has been paid $100.00 Bounty – 1st instalment		
Jno. B. Lawn	Lands.	14.00
He has been paid $100.00 Bounty – 1st instalment		
Wm. Phelan	Lands.	14.00
He has been paid $100.00 Bounty – 1st instalment		
Peter Moriarty	1st Class Boy	10.00
He has been paid $100.00 Bounty – 1st instalment		

James Reilly	1st Class Boy	10.00
He has been paid $100.00 Bounty – 1st instalment		
Patk. Fanning	1st Class Boy	10.00
He has been paid $100.00 Bounty – 1st instalment		
John Heller	2nd Class Boy	9.00
He has been paid $100.00 Bounty – 1st instalment		
(end of p. 5)		
Robt. Brown	2nd Css. Boy	9.00
He has been paid $100.00 Bounty – 1st instalment		
Thos. J. McCaber	2nd Css. Boy	9.00
He has been paid $100.00 Bounty – 1st instalment		
David More	2nd Css. Boy	9.00
He has been paid $100.00 Bounty – 1st instalment		
David Rae	2nd Css. Boy	9.00
He has been paid $100.00 Bounty – 1st instalment		
Patk. Daley	2nd Css Fire.	25.00
He has been paid $66.00 Bounty – 1st instalment		
Nathan Erskine	2nd Css. Fire.	25.00
He has been paid $66.00 Bounty – 1st instalment		
Jno. Hagerty	2nd Css. Fire.	25.00
He has been paid $100.00 Bounty – 1st instalment		
Isaac Blauvel (?)	2nd Css. Fire.	25.00
He has been paid $100.00 Bounty – 1st instalment		
Jno. Larkin	2nd Css. Fire.	25.00
He has been paid $100.00 Bounty – 1st instalment		
(end of p. 6)		
Approved (signed) Benj. C. Dean Act'g. Vol. Lieut. Com'd'g.		
	(signed) R. R. Brawley A. A. (Acting Assistant) Paymaster	

[From Prize List.]

PRIZE LIST of the U.S. Str. "Gertrude" at the time of the picking up of the Fifty (50) Bales of Cotton on the 20th day of April, 1865, this vessel being at the time attached to the West Gulf Squadron, Rear Admiral Henry K. Thatcher Acting Commanding. Fleet Capt. G. Simpson Lieut. Com'd'r. Capt Benj. F. Sands Com'd'g 3rd Division

I CERTIFY that I have PERSONALLY compared the annexed prize list with the muster roll of this vessel at the time of the capture above named, and approve the same as correct.

Benj. F. Dean A.V.Lt
Commanding U.S. Str. "Gertrude"

"The commanding officer of every vessel in the Navy entitled to, or claiming an award of, prize money, shall, as early as practicable after the capture, transmit to the Navy Department *a complete list* of the officers and men of his vessel entitled to share, inserting thereon the quality of every person rating, *on pain of forfeiting his whole share of the prize money resulting from such capture*, and suffering such further punishment as a court martial shall adjudge."

"No officer or other person who shall have been temporarily absent on duty from the vessel, on the books of which he continued to be borne while so absent, shall be deprived, in consequence of such absence, of any prize money to which he would otherwise be entitled."—(*Act of Congress, July 17, 1862.*)

	NAMES.	RATING.	MONTHLY PAY.
	Benj. C. Dean	A.V.Lt Com'd'g	annual 1875.
62.25	V. R. Brawley	A.A. Paymaster	1300.
59.86	Adam Shirk	A.A. Surgeon	1250.
	Fredk. Newell	Act'g Ensign	1200. 57.47
57.47	Jas. N. Munro	"	1200.
	Isaac H. Goodrich	"	1200.
57.46	Wm. H. Brown	Act'g 2nd Ast't Engin'r	1200.
	J. H. Nesen	" 2nd "	1200.
	F. C. Morey	" 2nd "	1200.
47.57	Chas. O'Farciot	" 3rd "	1000.
	Chas. A. Osborne	" Matt per mo	40 $ per annum 22.90
18.96	Isaac Jackson	Paymaster's Clerk	400 $

No.	Names	Rating	Monthly Pay	
45	Jas. R. Kelley ✓	Yeoman	✓ 30.00	8.85
46	Jos. H. Marshall ✓	Ord. Seaman	✓ 16.00	9.30
47	Jacob A. Sharkey ✓	Surgeon's Steward	✓ 25.00	✓
73	John Hudson ✓	M. at Arms	✓ 25.00	4.37
74		Lands		8.85
76		Lands	14.00	8.85
81	Robt. Duffy ✓	1st Cs Boy	✓ 10.00	✓
84	Chas Brown ✓	1st Cs Boy	✓ 10.00	L
86	John Seigel ✓	Ship's Cook	✓ 26.00	✓
85	Peter Holliman ✓	Steerage Stew'd	✓ 18.00	10.34
89	J. O. Sharon ✓	Carpenter's Mate	✓ 30.00	✓
90	A. R. Stone ✓	Officer's Steward	✓ 35.00	✓
91	T. B. Stock ✓	Ship's Writer	✓ 18.00	✓
92	Wm. Creighton ✓	Ord. Sea.	✓ 16.00	✓
				8.85
98	John Donahue ✓	Gunner's Mate	✓ 30.00	✓
102	Edw'd Green ✓	1st Cs Boy	✓ 10.00	✓
103	Wm. Birchall ✓	Lands	✓ 14.00	✓
104	Chas Dillon ✓	Lands	✓ 14.00	✓
105	Geo. Kennedy ✓	1st Cs Boy	✓ 10.00	5.75
106	J. M. Germain ✓	Ord. Seaman	✓ 16.00	9.69
107	Eugene Spelman ✓	Lands	✓ 14.00	✓
108	Wm. Ball ✓	1st Cs Fireman	✓ 30.00	✓
109	Mich'l Slattery ✓	2nd Cs Fire	✓ 25.00	14.35
110	Philip Marra ✓	2nd Cs Fire	✓ 25.00	✓
111	Henry Masters ✓	Seaman	✓ 20.00	✓
114	Jas. T Benson ✓	2nd Cs Fireman	✓ 25.00	✓
116	Hugh O'Donnell ✓	1st Cs Fire	✓ 30.00	7.34
118	Terence McGrath ✓	Coal Heaver	✓ 20.00	
119	James Fitzgibbons ✓	C. Heaver	✓ 20.00	✓
123	G. Kimberly ✓	C. Heaver	✓ 20.00	

#	Names	Rating	Monthly Pay
94	Austin Shear	O. Seaman	20.00 11.49
97	Ed. Ordway ✓	Coxswain ✓	25.00 ✓
	He has been paid $160.00 Bounty "Army" ✓		
98	Jas H. Ham ✓	Quarter Master ✓	25.00 14.37
	He has been paid $325.00 Bounty "Army" ✓		
~~	~~~	~~~	~~~
	He has been paid $54.00 Bounty		
87	Wm Collins ✓	Boat Mate in Charge ✓	30.00
	He has been paid $54.00 Bounty (3 mos. advance) ✓		
100	Francis Meeker ✓	Seaman ✓	20.00 ✓
	He has been paid $60.00 Bounty (3 mos. advance) ✓		
101	Joseph Kelly ✓	Quarter Master ✓	25.00 14.37
	He has been paid $48.00 Bounty (3 mos. advance) ✓		
112	Simeon Devine ✓	Boat Mate ✓	27.00 ✓
	He has been paid $00.00 Bounty (3 mos. advance) ✓		
~~	~~~	Quarter ~~~	25.00 ✓
	He has been paid $65.00 Bounty (3 mos. advance) ✓		
115	Wm Hunt ✓	2nd Class Fireman (1st installment) ✓	25.00 ✓
	He has been paid $33.33 Bounty		
117	Jno. Halpin ✓	2nd Class Fireman (1st installment) ✓	25.00 14.36
	He has been paid $33.33 Bounty		
120	Jno. Kiemierny ✓	1st Class Fireman (1st installment) ✓	30.00 17.24
	He has been paid $100.00 Bounty		
121	Edwd. Buchanan ✓	1st Class Fireman ✓	30.00 ✓
	He has been paid $200.00 Bounty		
122	Edwd McAllister ✓	1st Class Fireman ✓	30.00 17.24
	He has been paid $100.00 Bounty		
125	Wm Walters ✓	Lands. ✓	14.00 ✓
126	M. A. Wade ✓	Lands. ✓	14.00 ✓
127	Edwd. Pinder ✓	Lands. ✓	14.00 ✓
128	Alstyne Fisher ✓	Lands. ✓	14.00 ✓
129	Jas. Kelly ✓	Lands ✓	14.00 ✓

No	Names	Rates	Monthly Pay
130	Thos Kelley	Sail Maker's Mate	25.00
	He has been paid $100.00 Bounty (1st instalment)		
131	James Gallivan	Seaman	20.00
132	James Pender	Seaman	20.00
	He has been paid $100.00 Bounty (1st instalment)		
133	Chas. Kelley	Seaman	20.00
	He has been paid $100.00 Bounty (1st instalment)		
134	Winslow Evans	Ord. Sea.	16.00
	He has been paid $100.00 Bounty (1st instalment)		
135	Chas. Kemp	Officer's Cook	30. 25.00
136	Geo. Bell	Quarter Gunner	25.00
	He has been paid $100.00 Bounty (1st instalment)		
137	James Kelley	Capt. Forecastle	25.00
138	Thos. Jackson	Seaman	20.00
	He has been paid $100.00 Bounty (1st instalment)		
139	James Thompson	Seaman	20.00
	He has been paid $100.00 Bounty (1st instalment)		
140	Wm Thompson	Chief Quarter Master	28.00
	He has been paid $100.00 Bounty (1st instalment)		
141	Wm. Young	Capt After Guard	25.00 14.37
142	Edwd. Faxiers	Coxswain	25.00
143	Benj Jones	Lands.	14.00
144	Patk. Collins	Lands.	14.00
145	John Allen	Seaman	20.00
	He has been paid $100.00 Bounty (1st instalment)		
146	Peter Lumber	Seaman	20.00
	He has been paid $100.00 Bounty (1st instalment)		
147	Martin Wilson	Ord. Sea.	16.00
148	Warren Coville	Lands.	14.00
	He has been paid $100.00 Bounty (1st instalment)		
149	Edwd Jordan	Lands.	14.00
	He has been paid $100.00 Bounty (1st instalment)		
150		Lands.	14.00
	He has been paid $100.00 Bounty (1st instalment)		

No	Names	Rating	Monthly Pay
151	Niel McIsaacs	Lands	14.00
	He has been paid $100.00 Bounty (1st instalment)		
152	Chas Lipson	Lands	14.00
	He has been paid $100.00 Bounty (1st instalment)		
153	Mich'l Kennedy	Lands	14.00
	He has been paid $100.00 Bounty (1st instalment)		
154	Pat'k Wallace	Lands	14.00
	He has been paid $100.00 Bounty (1st instalment)		
155	David Russell	Lands	14.00
	He has been paid $100.00 Bounty (1st instalment)		
156	Peter Walsh	Lands	14.00
	He has been paid $100.00 Bounty (1st instalment)		
157	Edmond Williams	Lands	14.00
	He has been paid $100.00 Bounty (1st instalment)		
158	Dan'l Murphy	Lands	14.00
	He has been paid $100.00 Bounty (1st instalment)		
159	Jos Grumbleton	Lands	14.00
	He has been paid $100.00 Bounty (1st instalment)		
160	Alex'r Lupton	Lands	14.00
	He has been paid $100.00 Bounty (1st instalment)		
161	Jno B. Lown	Lands	14.00
	He has been paid $100.00 Bounty (1st instalment)		
162	Wm Phelan	Lands	14.00
	He has been paid $100.00 Bounty (1st instalment)		
163	Peter Moriarty	1st Class Boy	10.00
	He has been paid $100.00 Bounty (1st instalment)		
164	James Reilly	1st Class Boy	10.00
	He has been paid $100.00 Bounty (1st instalment)		
165	Pat'k Fanning	1st Cl Boy	10.00
	He has been paid $100.00 Bounty (1st instalment)		
166	John Biller	2nd Cl Boy	9.00
	He has been paid $100.00 Bounty (1st instalment)		

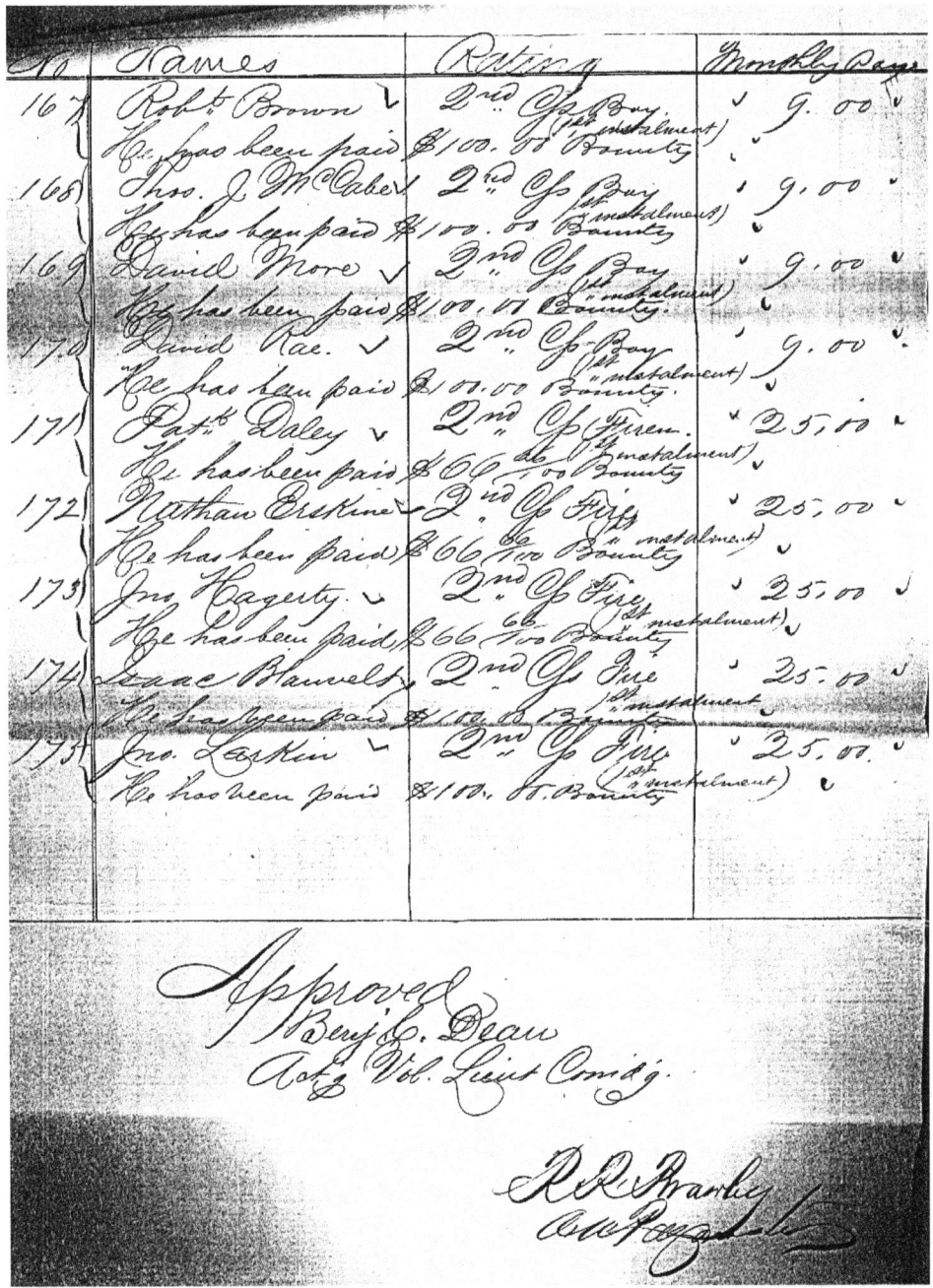

Figure 50. Prize list for the *Gertrude*, being the muster roll of the officers and crew, regarding the prize claim for floating cotton. (Page 6 of 6. This page and the prior five pages.) Dated 20 April 1865. **NARA**.

Prize Payout Ledger for the *Gertrude*

This section includes the full transcript of the prize payout ledger for the *Gertrude* vs. Fifty Bales of Cotton (Figure 51). Note the "Remarks" column recorded who collected the prize money from the navy paymaster: the sailor himself, a spouse, or a company acting as agent. The company might have been acting on behalf of the sailor, or the sailor might have assigned the proceeds to the company, essentially selling his prize for an advance payment, no doubt for a substantial discount. The meaning of "Paid N.A." was unclear, but "Paid from Navy Appropriation" was a good guess. The "Amount" column was each man's payout. The date recorded was the date of payment.

Treasury Department[79]
Fourth Auditor's Office
September 18th 1865

I certify that I have examined the claims of the officers and crew of the U.S.S. <u>Gertrude</u> for money accruing from the capture of <u>50 Bales of Cotton,</u> taken <u>April 20th</u> 1865, and find that there is due them the Sum of <u>One Thousand Nine Hundred and thirty nine dollars and thirty seven cents</u> ($ 1,939.37) to be apportioned as follows.

Names	Rank	Pay	Amount	Remarks
H. K. Thatcher	F. O.[80]	1/20	96.96	Paid order on Paymr. November 25, 1865.
E. Simpson	Fleet Captain[81]	1/100	19.39	Paid order on Paymr. January 5, 1866.
B. F. Sands	Div. Comdr.	1/50	38.78	Paid from Moiety due U.S. Boston. Nov. 15, 1865.
Benj. C. Dean	Vol. Lieut.	2/20	193.93	Paid N.A. Wash. Oct. 3, 1865. Self.
R. R. Brawley	A. A. Pay.	1,300[82]	62.25	Paid N.A. Wash. Hellon. Sept. 30, 1865.
Adam Shirk	A. A. Surg.	1,250	59.86	Paid N.A. Wash. Hellon. Sept. 30, 1865.
Frederic Newell	Ensign	1,200	57.47	Paid N.A. Boston. G. H. Newell Sept. 30, 1865.
James W. Monroe	Ensign	1,200	57.47	Paid N.A. Wash. Hellon. Sept. 30, 1865.
Isaac W. Goodrich	Ensign	1,200	57.47	Paid N.A. N.Y. Self. Nov. 2, 1865.
William H. Brown	2nd A. Engr.	1,200	57.47	Paid N.A. Boston. Andrews. Sept. 30, 1865.
H. Nesson	2nd A. Engr.	1,200	57.47	Paid N.A. Wash. Hellon. Oct. 14, 1865.
F. C. Morey	2nd A. Engr.	1,200	57.47	Paid. N.A. Wash. Hellon. Oct. 14, 1865.
Charles O. Farciot	3rd A. Engr.	1,000	47.89	Paid N.A. Hellon. Sept. 30, 1865.
Chas. A. Osborne	A. M. M.[83]	480	22.98	Paid. N.A. N.Y. Mrs. E. Osborne. Sept. 30, 1865.
S. R. Stone	Off. Stew.	420	20.11	Paid. N.A. N.Y. Kadden. Augt. 1, 1866.
Isaac Jackson	Pay. Stew.	396	18.96	Paid N.A. Wash. Hellon. Sept. 30, 1865.
James R. Kelly	Yeoman	360	17.24	Paid. N.A. Wash. Hellon. Oct. 23, 1865.
J. O. Sharon	C. M.[84]	360	17.24	Paid. N.A. Wash. Apl. 24, 1872. (Sen. 75671).
William Collins	B. M. in chge.[85]	360	17.24	See old unfsd. cl. (unfinished claims).
(End of p. 1; 373)				
Charles Kemp	Off. Cook	360	17.24	(No entry, therefore not paid?)[86]
William Ball	1st C. Fire.	360	17.24	(No entry, therefore not paid?)
Hugh O'Donnell	1st C. Fire.	360	17.24	Paid N.A. Phila. W.W.D. Sept. 30, 1865.
John Keerninny	1st C. Fire.	360	17.24	Paid N.A. Phila. Dovitt & Co. Sept. 30, 1865.
Edward Buchanan	1st C. Fire.	360	17.24	(No entry, therefore not paid?)
Zadoc McAllister	1st C. Fire.	360	17.24	Paid. N.A. Phila. Sept. 30, 1865.
William Thompson	Ch. Qr. Mr.[87]	336	16.09	Paid N.A. N.Y. Baker. Jany. 11, 1866.
John Donohue	G. M.[88]	324	15.52	(No entry, therefore not paid?)
Simeon Devine	B. M.	324	15.52	Paid N.A. N.Y. Wakefield. May 28, 1866.
John Segel	S. Cook[89]	312	14.95	(No entry, therefore not paid?)

---

[79] NARA RG 217, E 823. Abstracts of Prize Accounts, Vol. 4 of 15 (C-D) (for *Gertrude* vs. Fifty Bales of Cotton), pp. 373-376.
[80] Flag Officer.
[81] In this case meaning Flag Officer Thatcher's chief-of-staff.
[82] The rest of this column shows annual pay, which must be the figure used to make distribution share calculations.
[83] Acting Master's Mate.
[84] Carpenter's Mate.
[85] Boatswain's Mate in Charge.
[86] Note added by the present author.
[87] Chief Quartermaster.
[88] Gunner's Mate.

Jocob A. Sharkey	Surg. Stwrd.[90]	300	14.37	Paid N.A. Wash. Hellon. Jany. 26, 1866.
John Hudson	M. at A.[91]	300	14.37	Paid N.A. Boston. Andrews. Sept. 30, 1865.
Edward Ordway	Cox(swain).	300	14.37	(No entry, therefore not paid?)
James H. Ham	Qur. Mr.[92]	300	14.37	Paid. N.A. Boston. Derby. Sept. 30, 1865.
James Tomson	Capt. Hold[93]	300	14.37	(No entry, therefore not paid?)
Jos. Kelley	Qur. Mr.	300	14.37	Paid. N.A. Boston. Derby. Sept. 30, 1865.
Owen Duffy	Qr. Gr.[94]	300	14.37	(No entry, therefore not paid?)
George Bell	Qr. Gr.	300	14.37	Paid. N.A. Boston. S.S. Draw. Dec. 31, 1869.
James Kelly	C. F. C.[95]	300	14.37	(No entry, therefore not paid?)
William Young	C. A. G.[96]	300	14.37	Paid. N.A. Boston. W.W.D. Sept. 30, 1865
Ed. Tiexiera	Coxs.[97]	300	14.37	(No entry, therefore not paid?)
Michael Slattery	2nd C. Fire.	300	14.37	Paid. N.A. N.Y. W. and S. Sept. 30, 1865.
Phillip Murra	2nd C. Fire.	300	14.37	Paid. N.A. N.Y. W. and S. Oct.16, 1865.
James T. Benson	2nd C. Fire.	300	14.37	Paid. N.A. Boston. Winslow. April 2, 1866.
William Hunt	2nd C. Fire.	300	14.37	(No entry, therefore not paid?)
John Halpin	2nd C. Fire.	300	14.37	Paid. N.A. Boston. Barker. Sept. 30, 1865.
Pat. Daley	2nd C. Fire.	300	14.37	Paid. N.A. N.Y. Bascom. Oct. 24, 1865.
Nathan Erskine	2nd C. Fire.	300	14.37	(No entry, therefore not paid?)
John Haggerty	2nd C. Fire.	300	14.37	(No entry, therefore not paid?)
Isaac Blanvelt	2nd C. Fire.	300	14.37	(No entry, therefore not paid?)
John Larkin	2nd C. Fire.	300	14.37	(No entry, therefore not paid?)
Thomas Kelly	S. M. M.[98]	300	14.37	(No entry, therefore not paid?)
Henry Masters	Sea(man)	240	11.50	(No entry, therefore not paid?)
Francis Meeher	Sea.	240	11.50	(No entry, therefore not paid?)
James Gallivan	Sea.	240	11.50	(No entry, therefore not paid?)
James Pender	Sea.	240	11.50	(No entry, therefore not paid?)
(End of p. 2; 374)				
Charles Kelly	Sea.	240	11.50	(No entry, therefore not paid?)
Thomas Jackson	Sea.	240	11.49[99]	(No entry, therefore not paid?)
James Thompson	Sea.	240	11.49	(No entry, therefore not paid?)
John Allen	Sea.	240	11.49	(No entry, therefore not paid?)
Peter Limber	Sea.	240	11.49	(No entry, therefore not paid?)
Austin Shea	C. H.[100]	240	11.49	Paid. N.A. N.Y. Walden and Co. Sept. 30, 1865.
Mario (?) Garvin	C. H.	240	11.49	Paid. N.A. N.Y. Baker. April 2, 1866.
Tarence McGrath	C. H.	240	11.49	Paid. N.A. Boston Allen. Sept. 30, 1865.
James Fitzgibbons	C. H.	240	11.49	(No entry, therefore not paid?)
G. Kimberly	C. H.	240	11.49	(No entry, therefore not paid?)
Peter Holliman	Steer. Stew.[101]	216	10.34	Paid. N.A. Phila. W.W.D. Sept. 30, 1865.

---

[89] Ship's Cook.
[90] Surgeon's Steward.
[91] Master at Arms.
[92] Quartermaster.
[93] Captain of the Hold.
[94] Quarter Gunner.
[95] Captain of the Forecastle.
[96] Captain of the Afterguard.
[97] Coxswain.
[98] Sail Maker's Mate.
[99] Rounding error in the calculations required occasionally giving or taking a penny to make the grand total work out correctly. Thus Seamen Kelly and Jackson both earned $240/yr. or $20/mo., but their prize payouts were a penny different, i.e., $11.50 and $11.49 respectively.
[100] Coal Heaver
[101] Steerage Steward

T. B. Stack	S. Writer[102]	216	10.34	Paid. N.A. N.Y. Lee. April 2, 1866
Jos. H. Marshall	O. S.[103]	192	9.20	Paid. N.A. Wash. Hollen. Sept. 30, 1865.
William Creighton	O. S.	192	9.20[104]	Paid by Report March 7, 1902, Per case card #31446.
J. M. Germain	O. S.	192	9.20	Paid. N.A. Wash. Forbes. Sept. 30, 1865.
Winslow Evans	O. S	192	9.20	(No entry, therefore not paid?)
Martin Wilson	O. S	192	9.20	(No entry, therefore not paid?)
F. H. Connors	Lands(man)	168	8.05	Paid. N.A. Boston. Coffin. Sept. 30, 1865
F. J. Alcorn	Lands.	168	8.05	Paid. N.A. N.Y. Walden & Co. Sept. 30, 1865.
Edward Barton	Lands.	168	8.05	Paid. N.A. Phila. Devitt & Co. Sept. 30, 1865.
William Birchall	Lands.	168	8.05	Paid. N.A. N.Y. Swift. Sept 13, 1867.
Charles Dillon	Lands.	168	8.05	Paid. N.A. N.Y. Bascom. Oct. 24, 1865.
Eugene Spellman	Lands.	168	8.05	(No entry, therefore not paid?)
William Walters	Lands.	168	8.05	(No entry, therefore not paid?)
M. H. Wade	Lands.	168	8.05	Paid. N.A. Washn. Self. Jany 19, '86. #87170. See vou(cher) 484. 1st qr.–'86.
Edward Pender	Lands.	168	8.05	(No entry, therefore not paid?)
Alstyne Fisher	Lands.	168	8.05	(No entry, therefore not paid?)
James Kelly	Lands.	168	8.05	(No entry, therefore not paid?)
Benj. Jones	Lands.	168	8.05	(No entry, therefore not paid?)
Pat. K. Collins	Lands.	168	8.05	Paid. N.A. Boston. Derby. Nov. 23, 1867.
Warren Colville	Lands.	168	8.05	Paid. Self. Washn. July 5, 1890. #87711. Vou. 511. 1/91
Edward Jordan	Lands.	168	8.05	(No entry, therefore not paid?)
Richard Brown	Lands.	168	8.05	(No entry, therefore not paid?)
Wiel McIsaacs	Lands.	168	8.05	(No entry, therefore not paid?)
Charles Gipson	Lands.	168	8.04	(No entry, therefore not paid?)
Michael Kennedy	Lands.	168	8.04	Deserted, Aug. 11, '65, *Princeton*. Cl. Filed, Old unfed. (unfiled?) Cl.[105]
(End of p. 3, 375)				
Patk. Wallace	Lands.	168	8.04	(No entry, therefore not paid?)
David Russell	Lands.	168	8.04	Paid. Order on Paymr. Augt. 15, 1866.
Peter Walsh	Lands.	168	8.04	(No entry, therefore not paid?)
Edmond Williams	Lands.	168	8.04	(No entry, therefore not paid?)
Daniel Murphy	Lands.	168	8.04	(No entry, therefore not paid?)
Jos. Gumblestone	Lands.	168	8.04	(No entry, therefore not paid?)
Alex. Lupton	Lands.	168	8.04	Certificate #88545 issued May 14/97. Wash. Vou. 248. 4 qr. 97.
John B. Lown	Lands.	168	8.04	Paid. N.A. N.Y. Durham. Oct. 26, 1868.
William Phelan	Lands.	168	8.04	(No entry, therefore not paid?)
Robert Duffy	1st C. Boy	120	5.75	Paid. N.A. N.Y. Baker. April. 1, 1867.
Charles Brown	1st C. Boy	120	5.75	(No entry, therefore not paid?)
Edward Green	1st C. Boy	120	5.75	Paid. N.A. N.Y. W.W.D. Dec. 12, 1865.
George Kennedy	1st C. Boy	120	5.75	Paid. N.A. Wash. Brick. Sept. 30, 1865.
Peter Moriarty	1st C. Boy	120	5.75	(No entry, therefore not paid?)
James Riley	1st C. Boy	120	5.75	(No entry, therefore not paid?)
Patk. Fanning	1st C. Boy	120	5.75	Paid. List Paymr. Murray. Dec. 19, 1865.
John Heller	2nd C. Boy	108	5.18	Paid. List Paymr. Allen. March 21, 1866.
Robert Brown	2nd C. Boy	108	5.18	(No entry, therefore not paid?)
Thomas J. McCabe	2nd C. Boy	108	5.17	(No entry, therefore not paid?)

---

[102] Ship's Writer?
[103] Ordinary Seaman
[104] The prize was collected sixty-three years after the fact.
[105] Landsman Kennedy was not allowed to collect the prize money because he deserted from the navy.

David Moore	2nd C. Boy	108	5.17	Paid. N.A. N.Y. J. Hackett. Feb. 14, 1868.
David Rae	2nd C. Boy	108	5.17	Paid. List Paymr. Allen. March 21, 1866.
(End p. 4, 376)				
		Total	$ 1,959.37	
		As appears from the papers herewith transmitted to the Second		
		Comptroller of the Treasury for his revision.		
		S. J. W. Tabor		
		Auditor		
			Treasury Department,	
			Second Comptroller's office.	
			September 26th 1865.	
		I certify the foregoing apportionment to be correct.		
		J. M. Brodhead		
		Comptroller		

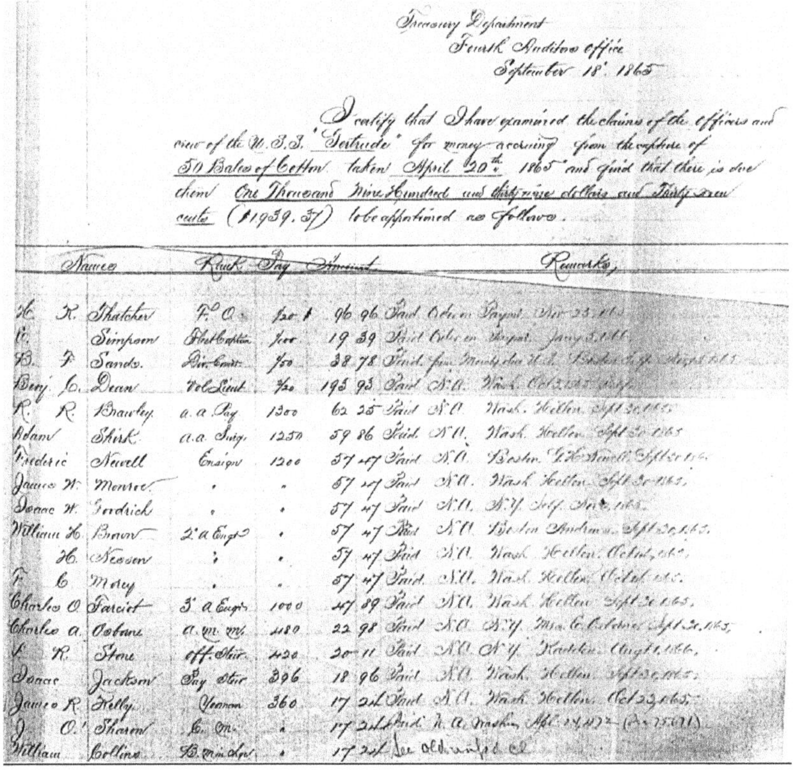

Figure 51. Treasury Department ledger of prize money paid regarding the *Gertrude* recovered floating cotton, page one. Dated 18 Sept. 1865 after which date the prize money could be paid to the officers and crew. This ledger was very large and not easily copied. Just one page was here illustrated as an example. *NARA*.

Analysis of Prize Money and Crew Ranks on the *Gertrude*

In theory, the annual salary for each rank signified the importance or level of skill. The table below was derived from the payout table above and arranged in descending order of salary and prize money awarded.

Treasury Department[106]
Fourth Auditor's Office
September 18th 1865

I certify that I have examined the claims of the officers and crew of the U.S.S. *Gertrude* for money accruing from the capture of 50 Bales of Cotton," taken April 20th 1865, and find that there is due them the Sum of One Thousand Nine Hundred and thirty nine dollars and thirty seven cents ($ 1,939.37) to be apportioned as follows.

Names	*Rank*	*Pay*	*Amount*	*Remarks*
H. K. Thatcher	Flag Officer	0.05	96.96	
E. Simpson	Fleet Captain	0.01	19.39	
B. F. Sands	Div. Comdr.	0.02	38.78	
Benj. C. Dean	Vol. Lieut.	0.1	193.93	
R. R. Brawley	A. A. Pay.	1300	62.25	
Adam Shirk	A. A. Surg.	1250	59.86	
Fred. Newell	Ensign	1200	57.47	
(End transcript and begin aggregating by rank and position.)				
2 more, total 3	Ensigns	1200	57.47	each
3	2nd A. Engr.	1200	57.47	
1	3rd A. Engr.	1000	47.89	(3rd Assistant Engineer)
1	A. M. M.	480	22.98	(Acting Master's Mate)
1	Off. Stew.	420	20.11	(Officer's Steward)
1	Pay. Stew.	396	18.96	
1	Yeoman	360	17.24	
1	C. M.	"	17.24	(Carpenter's Mate)
1	B. M. in charge	"	17.24	(Boatswain's Mate)
1	Off. Cook	360	17.24	
5	1st C. fire(man)	360	17.24	each
1	Ch. Qr. Mr.	336	16.09	(Chief Quarter Master)
1	G. M.	324	15.52	(Gunner's Mate)
1	B. M.	"	15.52	
1	S. Cook	312	14.95	(Surgeon's Cook)
1	Surg. Stwrd.	300	14.37	(Surgeon's Steward)
1	M. at A.	"	14.37	(Master at Arms)
1	Cox.	"	14.37	(Coxswain)
2	Qur. Mr.	"	14.37	each (Quartermaster)
1	Capt. of the Hold	"	14.37	(Captain of the Hold)
1	Capt. of Afterguard			
2	Qr. Gr.	"	14.37	each (Quarter Gunner)
10	2nd C. Fire(man)	"	14.37	each
1	S. M. M.	"	14.37	(Sail Maker's Mate)
9	Sea(man)	240	11.50	(5 @ 11.50 and 4 @ 11.49[107])
5	C. H.	"	11.49	(Coal Heaver)
1	Steer. Stew.	216	10.34	(Steerage Steward)
1	S(hip's) Writer	216	10.34	
5	O. S.	192	9.20	(Ordinary Seaman)
28	Lands.	168	8.05	(Landsman. 17 @ 8.05 and 11 @ 8.04)
7	1st C(lass) Boy	120	5.75	
5	2nd C(lass) Boy	108	<u>5.18</u>	(2 @ 5.18 and 3 @ 5.17)
	Total		$ 1,959.37	....

---

[106] NARA RG 217, E 823. Abstracts of Prize Accounts, Vol. 4 of 15 (C-D) (for *Gertrude* vs. Fifty Bales of Cotton), pp 373-376.
[107] For the correction of the rounding error.

The present author wished to attempt replicating the payout calculations. The process went as follows. Total crew 109 including the captain but not including the three higher officers of the chain of command, being the admiral (flag officer), fleet captain, and division commander. In figuring the crew's share, 108 officers and men were apportioned the remainder after paying the amounts specified for the admiral (1/20), fleet captain (1/100), division commander (1/50), and ship's captain (2/20). For example the captain's share was 2/20 (or 1/10 = 0.1) times 1,959.37 = $195.93. The division commander's share was paid from the U.S. government's half of the total prize money and so was not deducted from the ship's half of the prize money. Therefore to figure the amount to be shared out for the (non-command) crew, subtract the admiral's, fleet captain's, and ship's captain's shares from the ship's $1,959.37 of the total prize money leaving $1,629.43. Annual pay rates for each man were used for the calculation (total crew annual payroll not including captain $34,018).

The individual crewman's payout was successfully calculated as follows showing the example of the carpenter's mate whose annual salary was $360:

1. Annual pay of crewman (carpenter's mate) / total annual payroll of officers and crew excluding the captain = the crewman's percentage of the payroll and also his percentage of the prize money.

2. 360/34,018=0.010583

3. Crewman's (carpenter's mate's) percentage times the portion of total prize money remaining for officers and crew after deducting amounts for admiral, fleet captain, and ship's captain:
$96.96+$19.39+$193.93=$310.28 deducted from $1,939.37=$1,629.09 and then
0.010583*$1,629.09=$17.24

4. Yields $17.24 as the prize payout for the crewman (carpenter's mate) or a little more than half a month's pay. This was not life-changing money but was a nice bonus for less than half a day's work for the ship in picking up some floating bales of cotton.

The author made these calculations using a Microsoft Excel spreadsheet after determining the proper steps with the aid of a calculator. The question arose, were adding machines available to the clerks of the Civil War period? If the calculations had to be done by hand with pencil and paper, that would have required a tremendous number of man-hours. Or perhaps there was a simpler way to make the calculation?

Payout of Prize Money to the *Cornubia's* Crew and Commanders

This section included the documents dealing with the prize payout for the U.S.S. *Cornubia* vs. Eighty-nine Bales of Cotton jettisoned from the *Denbigh* in the same incident shown above for the capture by the U.S.S. *Gertrude*.

Prize List Submitted by Captain of *Cornubia*

The captain of each Union navy ship making a prize capture was required to submit a crew list showing all crewmembers on the ship's books (Figure 52). Each man's salary and the total ship's payroll (less the captain's salary) were needed to calculate prize money payouts.

[FIRST SHEET OF PRIZE LIST][108]

PRIZE LIST *of the U.S.* Steamer *Cornubia at the time of the* ~~capture~~ picking up
of the Eighty Nine (89) bales of cotton *on the* Twentieth (20th) *day of*
April, 1865, *this vessel being at the time attached to the*
West Gulf *Squadron, Actg. Rear Admiral* H. K. Thatcher
Commanding. Lieutenant Comdr. Edward Simpson, Fleet Captain
Captain B. F. Sands, Division Commander

I CERTIFY that I have PERSONALLY compared the annexed prize list with the muster roll of this vessel at the time of the ~~capture~~ picking up of the cotton above named and approved the same as correct.

John A. Johnstone
A. V. Lieut. *Commanding U.S.* Str. Cornubia

"The commanding officer of every vessel in the Navy entitled to, or claiming an award of, prize money, shall, as early as practicable after the capture, transmit to the Navy Department *a complete list* of the officers and men of his vessel entitled to share, inserting thereon the quality of every person rating, *on pain of forfeiting his whole share of the prize money resulting from such capture*, and suffering such further punishment as a court martial shall adjudge."

"No officer or other person who shall have been temporarily absent on duty from the vessel, on the books of which he continued to be borne while so absent, shall be deprived, in consequence of such absence, of any prize money to which he would otherwise be entitled."—(*Act of Congress, July 17, 1862.*)

Names.	~~Rating~~ Rank.	Monthly Pay.
John A. Johnstone	Acting Vol. Lieutenant	[109]
George A. Harriman	Acting Ensign	$ 100.00
Frank Willet	Acting Ensign	100.00
George F. Brady	Acting Ensign	100.00
Abram P. Eastlake	A. A. Paymaster	108.33
John G. Dearborn	A. A. Surgeon	104.16
William H. Thomson	Actg. ~~2d~~ 1st * Asst. Engineer	~~100.00~~ 1500[110]
Rodney F. Carter	Actg. ~~3d~~ 2d Asst. Engineer	~~83.33~~ 1200
James A. Boynton	Actg. 3d Asst. Engineer	83.33
George Altham	Actg. 3d Asst. Engineer	83.33
William Griffin	Actg. 3d Asst. Engineer	83.33
William H. Wood	Acting Master's Mate	40.00
George H. Russell	Acting Master's Mate	40.00
[111] (A note added.)		
(End p. 1)	Carried forward	$ 1,025.81
	Brought forward	$ 1,025.81
Robert Cline	Yeoman	30.00
John Calligan	Paymaster's Steward	32.00

---

[108] NARA RG 217. Records of the Accounting Officers of the Treasury, Records of the Navy Pay & Pension Division, Prize Lists 1862–1865, C-D (for *Cornubia*), Box No. 02. Entry 824.

[109] Pay not stated and not needed for calculation because ship's commander received a fixed 2/10 of ship's moiety according to statute.

[110] Rank and pay corrected by Navy Department clerk after the fact. Could it be the promotions had been approved, but the ship had not received notice when the list was prepared?

[111] Note added later: * Navy Dept. 10th June 1865. Probably the date reviewed and changed by a Navy Department official.

Alanson W. Morse	Surgeon's Steward	25.00
	Rec'd. $54 Bounty under act of Feby. 24, 1864	
Charles Smith	Boatswain's Mate in Charge	30.00
	Rec'd. $54 Bounty under act of Feby. 24, 1864	30.00
Adolph Hahn	Gunner's Mate in Charge	30.00
Albert Andrews	Carpenter's Mate	30.00
	Rec'd. $54 Bounty under act of Feby. 24, 1864	
Robert Hawston	Sail Maker's Mate	25.00
	Rec'd. $54 Bounty under act of Feby. 24, 1864	
Charles King	Ship's Corporal[112]	22.00
	Rec'd. $54 Bounty under act of Feby. 24, 1864	
William A. Leonard	Coxswain	25.00
	Rec'd. $54 Bounty under act of Feby. 24, 1864	
John A. Eayers	Quartermaster	25.00
	Rec'd. $54 Bounty under act of Feby. 24, 1864	
Edward Gorman	Quartermaster	25.00
	Rec'd. $54 Bounty under act of Feby. 24, 1864	
Daniel Hart	Quartermaster	25.00
	Rec'd. $54 Bounty under act of Feby. 24, 1864	
John Walsh	Quartermaster	25.00
	Rec'd. $54 Bounty under act of Feby. 24, 1864	
Charles Delaud	Quarter Gunner	25.00
Ezra Phinney	Captain (of the) Hold	25.00
	Rec'd. $54 Bounty under act of Feby. 24, 1864	
Daniel Sheridan	Caption of (the) Afterguard	25.00
	Rec'd. $54 Bounty under act of Feby. 24, 1864	
John Maher	2nd Class Painter	22.00
Edward Cuddy	Cabin Cook	30.00
James Jones	Ward room Steward	30.00
George Wilson	Ward room Cook	25.00
	Rec'd. $42 Bounty under act of Feby. 24, 1864	
Richard Slane	Ship's Cook	26.00
Charles Clark	Steerage Steward	20.00
	Rec'd. $42 Bounty under act of Feby. 24, 1864	
James Reynolds	Steerage Cook	18.00
Charles DeVassy	Seaman	20.00
Edward Englington	Seaman	20.00
James Hughes	Seaman	20.00
Patrick Monks	Seaman	20.00
(End p. 2)	Carried forward	$ 1,701.81
	Brought forward	$ 1,701.81
Henry Johnson	Seaman	20.00
John Ryan	Seaman	20.00
John Smith	Seaman	20.00
Harrison W. Lowell	Seaman	20.00
Thomas Pedehofft	Ordinary Seaman	16.00
Samuel Courtney	Ordinary Seaman	16.00
William Harris	Ordinary Seaman	16.00
Isaiah Mason	Ordinary Seaman	16.00
	Rec'd. $42 Bounty under act of Feby. 24, 1864	

---

[112] A petty officer who assisted the master-at-arms in his various duties.

Patrick H. O'Regan	Ordinary Seaman	16.00
James Peterson	Ordinary Seaman	16.00
Robert O. Seaver	Ordinary Seaman	16.00
	Rec'd. $42 Bounty under act of Feby. 24, 1864	
William Tylor	Ordinary Seaman	16.00
William Templeton	Ordinary Seaman	16.00
	Rec'd. $42 Bounty under act of Feby. 24, 1864	
William A. Barton	Landsman	14.00
James Brown	Landsman	14.00
Thomas Brown	Landsman	14.00
George Brown	Landsman	14.00
John Callahan	Landsman	14.00
Thomas Drummond	Landsman	14.00
Robert C. Freeman	Landsman	14.00
William Hein	Landsman	14.00
Henry Haryatt	Landsman	14.00
James Hughes	Landsman	14.00
William M. Jackson	Landsman	14.00
James Johnson	Landsman	14.00
Benjamin Jones	Landsman	14.00
(End p. 3)	Carried forward	$ 2,123.81
	Brought forward	$ 2,123.81
Michael Lear	Landsman	14.00
Edward Lenier	Landsman	14.00
Daniel McDevitt	Landsman	14.00
James McMahon	Landsman	14.00
James Redweed	Landsman	14.00
Robert Wilson	Landsman	14.00
Thomas Evans	1st Class boy	10.00
Artemas McNeal	1st Class boy	10.00
John Smith	1st Class boy	10.00
John Brown	1st Class Fireman	30.00
Greenleaf Bragg	1st Class Fireman	30.00
Peter Caghegan	1st Class Fireman	30.00
James Dixon	1st Class Fireman	30.00
Andrew Marshall	1st Class Fireman	30.00
Chris McQuade	1st Class Fireman	30.00
Patrick Phillips	1st Class Fireman	30.00
Alexander Buckly	2nd Class Fireman	25.00
Charles Brown	2nd Class Fireman	25.00
George W. Deal	2nd Class Fireman	25.00
Edward Graham	2nd Class Fireman	25.00
Michael Hoy (?)	2nd Class Fireman	25.00
Morris Hylant	2nd Class Fireman	25.00
Jonas Miller	2nd Class Fireman	25.00
James Nixon	2nd Class Fireman	25.00
Samuel B. Pratt	2nd Class Fireman	25.00
Edward Traey	2nd Class Fireman	25.00
James Stewart	2nd Class Fireman	25.00
(End p. 4)	Carried forward	$ 2,723.81
	Brought forward	$ 2,723.81

Nicholas Anderson	Coal Heaver	20.00
James Baxter	Coal Heaver	20.00
Michael Bassy	Coal Heaver	20.00
Patrick H. Connor	Coal Heaver	20.00
George Crawford	Coal Heaver	20.00
James Decker	Coal Heaver	20.00
Jeremiah Desmond	Coal Heaver	20.00
E. O. Gotham	Coal Heaver	20.00
Patrick Kelly	Coal Heaver	20.00
Phillip Lawn	Coal Heaver	20.00
Lawrence Mulcahy	Coal Heaver	20.00
Jacob Pearson	Coal Heaver	20.00
John Rae	Coal Heaver	20.00
Jon Rockford	Coal Heaver	20.00
Henry Rink	Coal Heaver	20.00
August Spitzer	Coal Heaver	20.00
Abram Smith	Coal Heaver	20.00
Patrick Sullivan	Coal Heaver	<u>20.00</u>
Total amount of Payroll for month		$ 3,082.81
	U.S. Str. *Cornubia* Off Galveston, Texas April 20th 1865 (signed) Abram P. Eastlake A. A. Paymaster	

[FIRST SHEET OF PRIZE LIST.]

PRIZE LIST of the U.S. *Steamer Cornubia* at the time of the ~~capture~~ picking up of ~~the~~ *Eighty Nine (89) bales of cotton* on the *Twentieth (20ᵗʰ)* day of *April*, 1865, this vessel being at the time attached to the *West Gulf* Squadron, Rear Admiral *H. K. Thatcher* Commanding. *Lieutenant Comdr Edward Simpson Fleet Captain Captain B. F. Sands, Division Commander*

I CERTIFY that I have PERSONALLY compared the annexed prize list with the muster roll of this vessel at the time of the ~~capture~~ picking up of the cotton above named and approve the same as correct.

John A. Johnstone
A.V. Lieut Commanding U.S. Str. *Cornubia*

"The commanding officer of every vessel in the Navy entitled to, or claiming an award of, prize money, shall, as early as practicable after the capture, transmit to the Navy Department *a complete list* of the officers and men of his vessel entitled to share, inserting thereon the quality of every person rating, *on pain of forfeiting his whole share of the prize money resulting from such capture*, and suffering such further punishment as a court-martial shall adjudge."

"No officer or other person who shall have been temporarily absent on duty from the vessel, on the books of which he continued to be borne while so absent, shall be deprived, in consequence of such absence, of any prize money to which he would otherwise be entitled."—(Act of Congress, July 17, 1862.)

NAMES.	RATING.	MONTHLY PAY.
John A. Johnstone	Acting Vol. Lieutenant	
George A. Harriman	Acting Ensign	$100.00
Frank J. Millett	Acting Ensign	100.00
George F. Praley	Acting Ensign	100.00
Abram P. Eastlake	A.A. Paymaster	108.33
John G. Dearborn	A.A. Surgeon	104.16
William A. Thomson	Actg 1st Ast Engineer	150.00
Rodney F. Carter	Actg 2d Ast Engineer	125.00
James A. Poynton	Actg 3d Ast Engineer	83.33
George Althaus	Actg 3d Ast Engineer	83.33
William Griffin	Actg 3d Ast Engineer	83.33
William A. Wood	Acting Master's Mate	40.00
George A. Russell	Acting Master's Mate	40.00
	Carried forward	$1025.81

Figure 52. Prize list, being the muster roll for the *Cornubia* when the eighty-nine bales of floating cotton were recovered. The people listed were eligible to share in the prize money. This is page one, an example of several more pages not shown. Dated 20 April 1865. *NARA.*

Prize Payout Ledger for the *Cornubia*

This section presents the complete listing of prize money as calculated for each of the crew and the command structure (Figures 53-54).

Treasury Department[113]
Fourth Auditor's Office
September 19th 1865

I certify that I have examined the claims of the officers and crew of the U.S. Steamer *Cornubia* for money accruing from the capture of "89 Bales of Cotton," taken April 20th 1865, and find that there is due them the Sum of Three thousand five hundred and ten dollars and thirty eight cents: ($3,510.38) to be apportioned as follows: viz: ___

Names	Rank	Pay	Amount	Remarks
H. K. Thatcher	Flag Offr.[114]	1/20	175.51	Paid order on Paymr. November 25, 1865.
Ed. Simpson	Fleet Capt.[115]	1/100	35.10	Paid order on Paymr. January 5, 1866.
B. F. Sands	Div. Comdr.	1/50	70.21	Paid N.A.[116] Boston. Self. Nov. 15, 1865.
To be deducted from the moiety of the U.S.				(Navy Pen. Fund).
John A. Johnston	A. Vol. Lieut.	2/20	351.03	Paid order on Paymr. Octo. 26, 1865.
Wm. H. Thompson	A. 1st A. Engr.	1500	117.97	Paid N.A. Boston. Self. Dec. 1, 1865.
Abram P. Eastlake	A. A. Paymr.	1300	102.24	Paid N.A. Boston. Self. Octo. 11, 1865.
John G. Dearborn	A. A. Surgn.	1250	98.31	Paid order on Paymr. Dec. 12, 1865.
Geo. A. Harriman	Actg. Ensign	1200	94.38	Paid N.A. Boston. Andrews.[117] Octo. 6, 1865.
Frank Millett	Actg. Ensign	1200	94.38	Paid order on Paymr. Octo. 16, 1865.
Geo. F. Brady	"	"	94.38	Paid order on Paymr. Nov. 10, 1865.
Rodney F. Carter	A. 2d A. Engr.	"	94.37[118]	Paid N.A. Phila. Self. Sept. 29, 1865.
James A. Boynton	A. 3d A. Engr.[119]	1000	78.65	Paid N.A. Boston. Self. Nov. 15, 1865.
George Altham	"	"	78.65	Paid N.A. Boston. Self. Nov. 15, 1865.
William Griffin	"	"	78.64	Paid N.A. Boston. Self. Nov. 9, 1865.
Wm. H. Wood	A. M. Mate[120]	480	37.75	Paid N.A. N.Y. Self. Jany. 4, 1865.
Geo. H. Russell	" Bounty $54.	"	37.75	Paid Order on Paymr. Nov. 15, 1865.
John Colligan	Payr. Stewd.	396	31.15	Paid List Paymr. Griffing. Feby. 3, 1866.
Robert Cline	Yeoman	360	28.32	Paid N.A. Phila. Ashman. Nov. 16, 1865.
Charles Smith	B.M.[121] in charge	"	28.32	Claim on file. See old unpaid cl. (claims?)

---

[113] NARA RG 217, E 823. Abstracts of Prize Accounts, Vol. 3 of 15 (C-D) (for *Cornubia* vs. Eighty-nine Bales of Cotton), pp. 353-356.

[114] Until shortly before the Civil War, the highest permanent grade in the U.S. Navy was that of captain, with those captains serving as squadron commanders bearing the courtesy title and flying the broad pennant of a commodore. In 1857, Congress created the title of "flag officer" and in 1862 finally authorized the first American use of the title "admiral" (2001 http://mysite.verizon.net/vzeohzt4/Seaflags/personal/fo.html).

[115] In this case, meaning Flag Officer Thatcher's chief-of-staff.

[116] Navy Appropriation? Just a guess by the present author.

[117] It seems that Andrews was paid the prize money rather than the crewmember. Andrews may have been an agent, a broker who made an advance on the payment, or perhaps the relative of a deceased or absent crewmember.

[118] 1¢ rounding errors distributed as needed to assure the correct grant total.

[119] Acting 3d Assistant Engineer.

[120] Acting Master's Mate.

[121] Boatswain's Mate.

Adolph Hahn	B. M. in charge	360	28.32	See claim old unpd.
Albert Andrews	Carp. Mate	"	28.32	Paid N.A. Boston. Self. Sept. 17, 1866.
Edward Cuddy	Cab.(in) Cook	"	28.32	Paid N.A. Boston. Copeland. Nov. 24, 1865.
James Jones	W. R. Stwrd.[122]	"	28.31	Paid N.A. Wash. Hellen. Sept. 29, 1865.
John Brown	1st Cl. fireman	"	28.31	Paid N.A. Wash. Meleges Sneed S. Nov. 6, 1865.
Greenleaf Bragg	1st Cl. fireman	"	28.31	Paid N.A. Wash. Hellen. Sept. 29, 1865.
Peter Caghegan	1st Cl. fireman	"	28.31	Paid N.A. ?. ?. Sept. 29, 1865.
James Dixon	1st Cl. fireman	"	28.31	Paid N.A. ? ?. Conklin. Sept. 29, 1865.
Andrew Marshall	1st Cl. fireman	"	28.31	Paid Dec. 14/85 #87.17 see voucher #479 Hd. Qr. 185.
Chris McQuade	1st Cl. fireman	"	28.31	Paid N.A. N.Y. ? Sept. 29, 1865.
Patrick Phillips	1st Cl. fireman	"	28.31	Paid N.A. Mound City. Preston. Sept. 29, 1865.
Michael Slane	S.(hip's) Cook	312	24.54	Paid N.A. Wash. Self present. Sept. 8, '82.
Alanson W. Morse	Surg. Stwrd.[123]	300	23.60	Paid N.A. Boston. Eastlake. Oct. 27, 1866.
Robert Harnston	Sl. M. Mate[124]	"	23.60	Paid N.A. Boston. J. Fletcher. July 29, 1870.
William A. Leonard	Coxswain	"	23.60	Paid. N.A. Boston. Self. Dec.7, 1872.
John A. Eayers	Qur. Mas.[125]	"	23.60	Claim in file old unpaid.
Edward Gorman	"	300	23.60	Paid N.A. Boston. Derby. Jany. 28, 1877.
Daniel Hart	"	"	23.60	Paid Dec 10/14. Case #45230.
John Walsh	Qur. Gunner[126]	"	23.60	[127]
Charles Delaud	Qur. Gunner	"	23.60	Paid N.A. Washn. W.A. Selden, Oct. 7, 1876.
Ezra Phinney	Capt. Hold[128]	"	23.60	Paid N.A. Boston. Damon? Jany. 6, 1866.
Daniel Sheridan	C. A. Grd.[129]	"	23.59	
George Wilson	W. R. Cook	"	23.59	Paid N.A. N.Y. Hodges McLellan. March 11, 1867.
Alexander Buckley	2d C. fireman	"	23.59	Paid N.A. Phila. Ashman. Sept. 29, 1865.
Charles Brown	"	"	23.59	
Geo. W. Deal	"	"	23.59	Paid N.A. Phila. Ashman. Sept. 29, 1865.
Edward Graham	"	"	23.59	Paid N.A. Phila. (?) W. D. Sept. 29, 1865.
Michael Cox (?)	"	"	23.59	Paid N.A. Phila. Ashman. Oct. 16, 1865.
Morris Cylant	"	"	23.59	Paid N.A. Washn. Devitt & Co. Sept. 20, 1865.
Jonas Miller	"	"	23.59	Paid N.A. N.Y. Walden & Co. Sept. 29, 1865.
James Nixon	"	"	23.59	
Saml. B. Pratt	"	"	23.59	Paid N.A. Boston. Howe. Dec. 24, 1865.
Edward Tracey	"	"	23.59	Paid N.A. N.Y. J. C. Stean. March 16, 1867.
James Stewart	"	"	23.59	Paid N.A. Phila. Ashman. Sept. 6, 1865.
Charles King	Ship's Corpl.[130]	264	20.77	Certificate #88542 issued Apr. 30/97.Wash. V.M. 2475.
John Maher	2d C. Painter	264	20.77	See old unpaid cl.
Charles Clark	Steer. Stwd.[131]	240	18.88	Paid N.A. Boston. Osiby (?). Sept. ?, 1865.
Charles De Vassy	Seaman	240	18.88	Paid N.A. N.Y. Walden & Co. Sept. 29, 1865.
Ed. Erlington	"	"	18.88	Paid N.A. N.Y. Walden & Co. Jany. 5, 1866.
James Hughes	"	"	18.88	
Patrick Monks	"	"	18.88	Paid N.A. N.Y. C.W. Bennett. Nov. 9, 1877.
Henry Johnson	"	"	18.88	
John Ryan	"	"	18.88	

---

[122] Ward Room Steward.
[123] Surgeon's Steward.
[124] Sail Maker's Mate.
[125] Quartermaster.
[126] Quarter Gunner.
[127] Perhaps the blank space indicated that the prize money never was paid.
[128] Captain of the Hold.
[129] Captain of the Afterguard.
[130] Ship's Corporal: a petty officer who assisted the master at arms in his various duties.
[131] Steerage Steward.

John Smith	"	"	18.88	
Harrison W. Sewell	"	"	18.88	
Archibald Anderson	Coal Heaver	240	18.88	Paid N.A. Boston. Andrews. Sept. 29, 1865.
James Baxter	"	"	18.88	
Michael Brany[132]	"	"	18.88	Paid N.A. N.Y. Hackett. June 26, 1866.
Patk. H. Connor	"	"	18.88	Paid N.A. N.Y. Taylor. Sept. 29, 1865.
George Crawford	"	"	18.88	Paid N.A. Phila. W.W.D. Sept. 29, 1865.
James Decker	"	"	18.88	Paid N.A. Wash. G.H.W. Octr. 21, 1865
Jeremiah Desmond	"	"	18.88	Paid N.A. Phila. Devitt & Co. Sept. 29, 1865.
E.O. Gotham	"	"	18.88	Paid N.A. N.Y. W.W.D. Sept. 29, 1865.
Patrick Kelly	"	"	18.88	Paid N.A. Phila. Devitt & Co. Nov. 6, 1865.
Phillip Lawn	"	"	18.88	
Lawrence Mulcahy	"	"	18.88	Paid N.A. Boston. Fletcher. Octr. 26, 1865.
Jacob Pearson	"	"	18.88	Paid N.A. Phila. Ashman. Sept. 29, 1865.
John Rae	"	"	18.88	Paid N.A. Washington. Self. July 31, 1867.
John Rockford	"	"	18.88	Paid N.A. N.Y. F.B. Swift. July 10, 1868.
Henry Rink	"	"	18.88	
August Spitzer	"	"	18.88	
Abram Smith	"	"	18.88	Paid N.A. N.Y. W. & S. Sept. 29, 1865.
Patrick Sullivan	"	"	18.88	Paid N.A. N.Y. W.W.D. Sept. 29, 1865.
James Reynolds	Steer. Cook	216	16.98	See old unpaid Cl.
Thomas Bedefofft	Ord. Sea.[133]	192	15.09	
Samuel Courtney	"	"	15.09	Paid N.A. Boston. Andrews. Octr. 14, 1865.
William Harris	"	"	15.09	
Isaiah Mason	"	"	15.09	
Patk. H. O'Regan	"	"	15.09	Claim in file Old unpaid.
James Peterson	"	"	15.09	Paid N.A. Wash. Loewenthal & Co. Sept. 29, 1865.
Robt. O. Leaver	"	"	15.09	
William Taylor	"	"	15.09	Paid N.A. Phila. Wakefield. Sept. 29, 1865.
William Templeton	"	"	15.09	
Theodore Trippet	"	"	15.09	Paid Apr. 11/13. Case #119416.
William A. Braston	Lands.[134]	168	13.21	Paid N.A. Boston. Benson. Nov. 2, 1865.
James Brown	"	"	13.21	Paid List. Paym. Murray. Nov. 16, 1865.
Thomas Brown	"	"	13.21	Paid List. Paym. Murray. Nov. 16, 1865.
George Berry	"	"	13.21	Paid List. Paym. Murray. Nov. 16, 1865.
John Callaghan	"	"	13.21	
Thomas Drummond	"	"	13.21	Paid N.A. N.Y. Allen & Latson. Sept. 29, 1865.
Robt. C. Freeman	"	"	13.21	Paid N.A. Phila. Allen & Latson. Sept. 29, 1865.
William Hein	"	"	13.21	Paid N.A. N.Y. Walden & Co. Octr. 16, 1865.
Henry Haryott	"	"	13.21	
James Hughes	"	"	13.21	Paid. Self. Wash. Apl. 23/92 #88,068. 4th Qr. Voucher 1897 #HNR.
Willm. M. Jackson dead	"	"	13.21	Paid N.A. Phila. Devitt & Co. Feby. 21, 1867.
James Johnson	"	"	13.21	
Benjamin Jones	"	"	13.21	Paid N.A. Phila. Devitt & Co. Sept. 29, 1865.
Michael Leary	"	"	13.21	Paid N.A. Boston. Derby. Sept. 29, 1865.
Edward Liniod	"	"	13.21	Paid List. Paymr. Murray. Nov. 16, 1865.

---

[132] Mistake in transcription of name on crew list submitted by the ship? The crew list sent in clearly shows "Brassy," not equally clear "Brany" of the prize money list. One would think the ship's clerk was more likely to have the name right.

[133] Ordinary Seaman.

[134] Landsman.

David McDevitt	"	"	13.21	
James McMahon	"	"	13.20	Paid N.A. Wash. Deforet & O'Neill. Nov. 24, 1868.
James Redwood	"	"	13.20	Paid List. Paymr. Murray. Nov. 16, 1865.
Robert Wilson	"	"	13.20	
Thomas Evans	1st C. Boy[135]	120	9.44	
Artimus McNeal	"	"	9.44	Paid A. Wash. Forbes. Sept. 29, 1865.
John Smith	"	"	9.44	
	Total		$ 3,510.38	
	As appears from the papers herewith transmitted			
	to the Second Comptroller of the Treasury for his revision.			
	S. J. W. Tabor			
		Auditor		
	Treasury Department,			
	Second Comptroller's Office. September 26th 1865.			
		J. M. Brodhead		
		Comptroller		

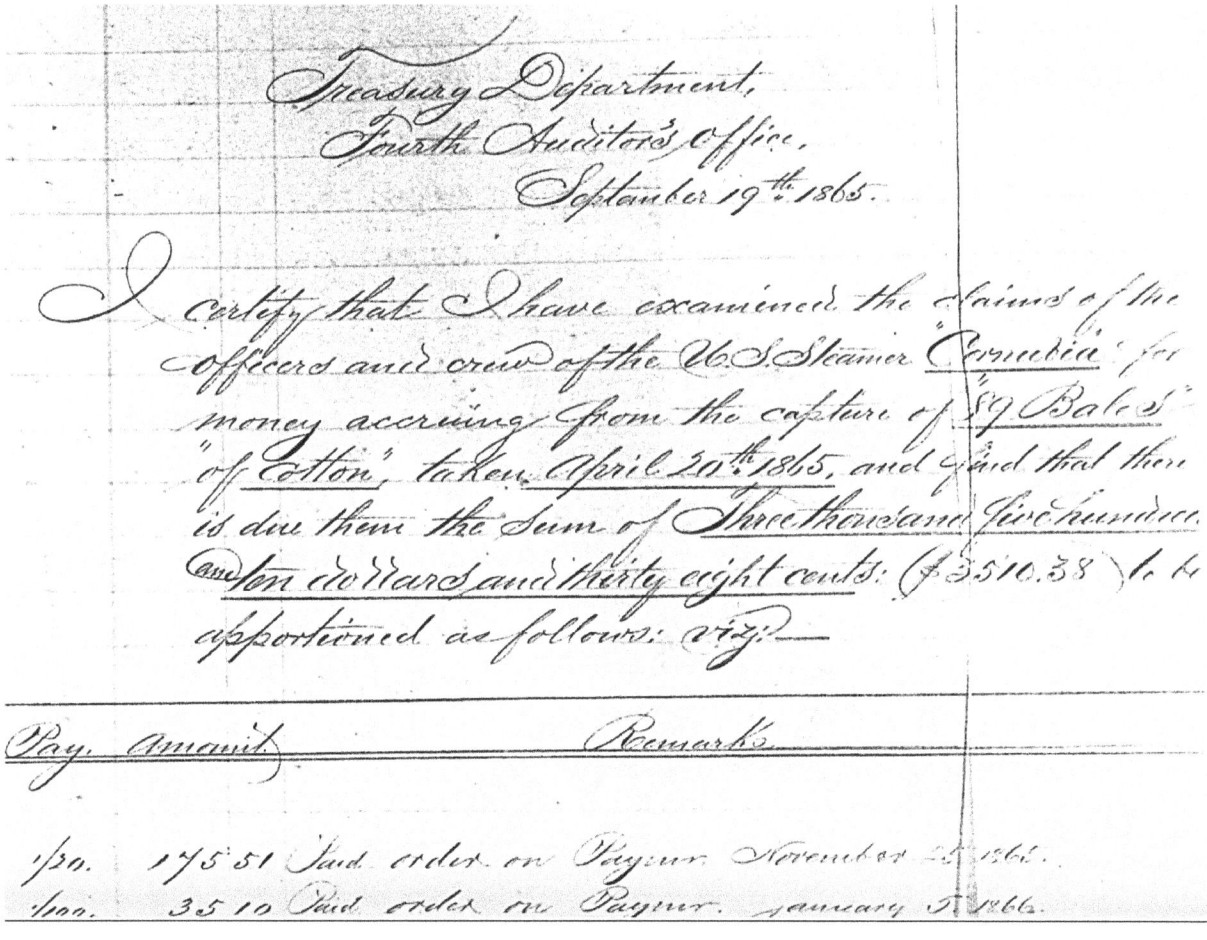

Figure 53. Heading of first page of Treasury Department ledger listing the *Cornubia's* prize money payout regarding the floating cotton. Dated 19 Sept. 1865. **NARA**.

---

[135] 1st Class Boy.

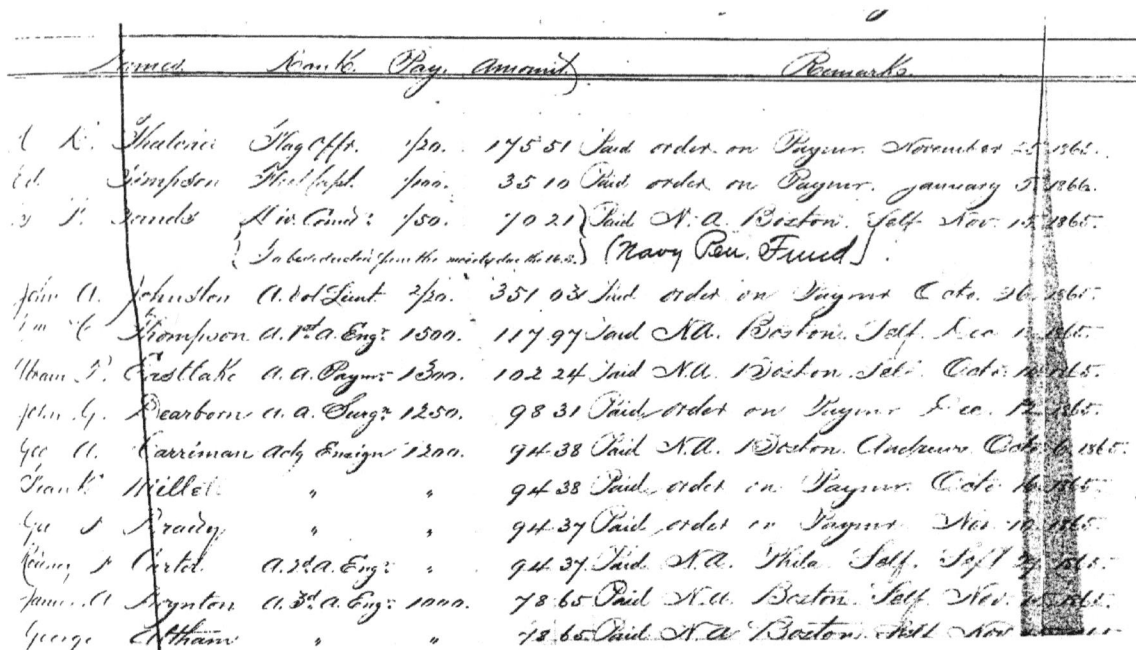

Figure 54. Top part of page 2 of the *Cornubia* prize money payout document illustrated as an example of the remainder of the document. *NARA*.

Analysis of Prize Money Payout for Each Rank on the *Cornubia*

The following table derived from the prize payout table above. Each rank was listed in descending order of pay, and the number of crewmen holding that rank was tallied.

Treasury Department
Fourth Auditor's Office
September 19th 1865

I certify that I have examined the claims of the officers and crew of the U.S. Steamer *Cornubia* for money accruing from the capture of "89 Bales of Cotton," taken April 20th 1865, and find that there is due them the Sum of Three thousand five hundred and ten dollars and thirty eight cents: ($3,510.38) to be apportioned as follows: viz: ___

Names	Rank	Pay[136]	Amount	Remarks
H. K. Thatcher	Flag Offr.[137]	1/20	175.51	Paid order on Paymr. Nov. 25, 1865.
Ed. Simpson	Fleet Capt.[138]	1/100	35.10	Paid order on Paymr. Jan. 5, 1866.
B. F. Sands	Div. Comdr.	1/50	70.21	Pd. N.A. Boston. Self. Nov. 15, 1865.
(Sands' share) to be deducted from the moiety of the U.S.				(Navy Pen. Fund).
J. A. Johnstone[139]	A. Vol. Lieut.	2/20	351.03	Pd. order on Paymr. Octo. 26, 1865.

---

[136] For the crew, this was the annual salary used to calculate each man's portion of the prize money awarded. For the chain of command, the award was specified by the law, 1/20th of the award for the squadron commander, etc. the crew's prize money came to a little less than a month's pay for each man.

[137] Commander of the West Gulf Blockading Squadron. Until shortly before the Civil War, the highest permanent grade in the U.S. Navy was that of captain, with those captains serving as squadron commanders bearing the courtesy title and flying the broad pennant of a commodore. In 1857, Congress created the title of "flag officer" and in 1862 authorized the first American use of the title "admiral" (2001 http://mysite.verizon.net/vzeohzt4/Seaflags/personal/fo.html).

[138] Flag Officer Thatcher's chief-of-staff.

[139] John A. Johnstone, Acting Volunteer Lieutenant, was the *Cornubia's* captain.

W. H. Thompson	A. 1st A. Engr.	1500	117.97	Pd. N. A. Boston. Self. Dec. 1, 1865.
A. P. Eastlake	A. A. Paymr.	1300	102.24	Pd. N. A. Boston. Self. Oct. 11, 1865
J. G. Dearborn	A. A. Surgn.	1250	98.31	Paid order on Paymr. Dec. 12, 1865.
G. A. Harriman	Actg. Ensign	1200	94.38	Pd. N.A. Boston. Andrews. Oct. 6, 1865.
(End transcript and begin aggregating by rank and position.)				
2 more, tot. 3	Actg. Ensigns	1200	94.38	each
1	A. 2d A. Engr.	1200	94.37	1¢ rounding distributed as needed.
3	A. 3d A. Engr.	1000	78.65	each (Acting 3d Asst. Engineer)
2	A. M. Mate	480	37.75	Each (Acting Masters Mate)
1	Payr. Stewd.	396	31.15	(Paymasters Steward)
1	Yeoman	360	28.32	
1	B. M. in charge	360	28.32	(Boatswains Mate)
1	G. M. in charge	360	28.32	(Gunners Mate)
1	Carp. Mate	360	28.32	
1	Cab. Cook	360	28.32	
1	W. R. Stwrd.	360	28.31	(Ward Room Steward)
7	1st Cl. fireman	360	28.31	each
1	S. Cook	312	24.54	
1	Surg. Stwrd.	300	23.60	(Surgeons Steward)
1	Sl. M. Mate	300	23.60	(Sail makers Mate)
1	Coxswain	300	23.60	
3	Qur. Mas.	300	23.60	each (Quartermaster)
2	Qur. Gunner	300	23.60	(Quarter Gunner)
1	Capt. Hold	300	23.60	(Captain of the Hold)
1	C. A. Guard	300	23.59	(Captain of the Afterguard)
1	W. R. Cook	300	23.59	
11	2d C. fireman	300	23.59	each
1	Ship's Corpl.	264	20.77	(Ship's Corporal[140])
1	2d C. Painter	264	20.77	(2d Class Painter)
1	Steer. Stwd.	240	18.88	(Steerage Steward)
8	Seaman	240	18.88	
18	Coal Heaver	240	18.88	
1	Steer. Cook	216	16.98	
10	Ord. Sea.	192	15.09	(Ordinary Seaman)
19	Lands.	168	13.21	(Landsman)
3	1st C. Boy	120	9.44	
	Total		$ 3,510.38	

Comparison of *Gertrude* and *Cornubia*

Some basic facts of ship size and navy service were extracted from the *Dictionary of American Naval Fighting Ships*[141] and Wise's *Lifeline of the Confederacy*.[142]

U.S.S. *Gertrude*: (Screw Str. t. 350; l. 156'; b. 21'; dph. 11'; a. 2 12-pdr. rifles, 6 24-pdr. howitzer). Iron steamer, a former blockade-runner (249 tons burden), built in Greenock, Scotland, in 1863. Fitted out and commissioned 22 July 1863. Decommissioned 11 August 1865 and sold a few months later.

U.S.S. *Cornubia*: (Side wheel Str: t. 589; l. 210'; b. 24' 6"; dph. 13' 3"; speed. 13 k.; complement 76; a. 1 20-pdr. rifle, 2 24-pdr. smooth bore) Formally a very successful blockade runner known as the *Lady*

---

[140] A petty officer who assisted the master at arms in his various duties.
[141] Naval Historical Center (recently renamed Naval History and Heritage Command) web site: http://www.history.navy.mil/research/histories/ship-histories/danfs.html.
[142] Wise, Stephen R. *Lifeline of the Confederacy: Blockade Running during the Civil War*. Columbia: University of South Carolina Press, 1988.

Davis (588 tons burden. Wise gave different measurements: 190' x 24.5' x 12.5'). Captured 8 Nov. 1863. Commissioned 17 March 1864. She was in the blockading squadron at Galveston when the surrender came in late May '65. Then the ship served removing harbor obstructions until 21 July when she sailed for the east coast. Decommissioned 9 August 1865 in Philadelphia.

The *Gertrude* appeared to be a smaller ship than the *Cornubia*, most apparent in the 249 and 588 tons burden[143] respectively. The crew sizes were not very different as seen in the following table:

Title	*Gertrude*	*Cornubia*
Officers	5	5
Engineers	4	5
Seamen	14	17
Firemen	15	18
Landsmen	28	19
Coal Heavers	5	18
Ship's Boys	12	3
Ship's Complement	109	112

- Officers, counting the captain and watch-keeping officers included the Master's Mate.
- Engineers, 1st, 2nd, and 3rd.
- Seamen including seaman, and ordinary seaman.
- Firemen, 1st and 2nd class.
- Landsmen, not differentiated in rank.
- Coal heavers, not differentiated in rank.
- Ship's boys, 1st and 2nd class.
- Petty officers and specialists not included in the table as they were not amenable to grouping.

Most categories of the two crews were fairly comparable except the landsmen, coal heavers, and ship's boys. The reasons for were not obvious for the marked differences in the three categories just mentioned. The slightly larger numbers of the *Cornubia's* engineers, seamen, and firemen seemed appropriate for the larger of the two ships. There were three and a half times more coal heavers on the *Cornubia* than on the *Gertrude*. One would expect more coal heavers on the larger ship, but there must have been a particular reason for the tremendous difference. Perhaps the coal storage bunkers were inconveniently located requiring a whole lot more transferring of coal. Or perhaps on the *Gertrude* some of the difference was made up by some of the extra large number of landsmen and ship's boys.

It would be interesting to know if specialists in naval history of the period might be able to explain the staffing variation in the two crews. In *Lincoln's Trident*,[144] Browning mentioned that persistent shortages in crewmen for the West Gulf Blockading Squadron ships were in part made up by enlisting Freedmen. There was perhaps a hint of this practice in the number of landsmen on the *Gertrude*.

---

[143] A measure of cargo capacity.
[144] Browning, Robert M. *Lincoln's Trident: The West Gulf Blockading Squadron During the Civil War*. Tuscaloosa: University of Alabama Press, 2015.

How the Prize Money was Paid

This section presents two types of documents dealing with the payments to each individual. The first was a voucher provided to each sailor at the conclusion of a cruise when the crew was paid off and went their separate ways. The second was an accounting document, a form recording the fact that each man had been paid.

Prize Money Pay Voucher

The purpose of the voucher was for presentation by a crewmember to the navy paymaster once the prize funds were ready for disbursement. The present author noticed several such forms for sale at a modest price online and acquired one for purposed of illustration herein (Figure 55). The form included sections for witnesses to confirm the identity of the sailor so that the paymaster could be assured he was paying out the prize money to the correct person. It seems from entries and stamps on the face of the forms that the paymaster sent the forms to the bookkeepers in Washington. When the auditors were finished recording the receipt of prize funds by the sailor, the form apparently was sent back to the sailor. Likely the return of the form was the reason this and similar forms were available for sale on the internet today.

See below the example of a payment voucher.

(Front of form.)

FORM OF OATH OF IDENTITY OF AN OFFICER, SEAMAN, ORDINARY SEAMAN, LANDSMAN, BOY, OR MARINE, ENTITLED TO PRIZE MONEY.

I, J. W. Smith, do solemnly swear that I am the identical J. W. Smith who served by that name as a Actg. Master, Vol. Lieut., & Vol. Lieut. Comdr. on board the United States ships, ~~Potomac, Connecticut,~~ & Bermuda,[145] in the years 1861 to 1866 when she captured ~~the prize~~ Cotton picked up at Sea, and who is named in the certificate dated 21st February, 1866, signed by Gideon Wells Secy. of the Navy, which is herewith presented and surrendered.

I also solemnly swear that I am now 42 years of age; am a native of New Jersey; that I ~~enlisted~~ was appointed Actg. Master; on or about the 8th day of June, 1861, in the grade of Actg. Master; ~~that my ship's number was~~ _____. I also solemnly swear that I have not made any previous assignment of or application for the prize money now claimed by me; and further, that I now reside at 75 Morton Street, and am employed as Ship Master.

(Signed by Claimant.) J. W. Smith

R. Chandler, Commander   Witness.   Late Act. Vol. Lt. Comdr.
N. (Surname illegible)   Witness.   In comd. of U.S.S. *Bermuda*

Sworn to and subscribed before me this 1st day of February, A.D. 1869. And I certify that, to my knowledge, the statement of deponent in regard to this residence and employment is true, my knowledge of deponent being derived from personal acquaintance for years. And I also certify that the above named deponent appears to be about the age stated by him; that he is about 5 feet 8 inches in height, of fair complexion, light hair, and grey eyes.

John F. Waugh
Commissioner of Deeds
(Impressed stamp with name, office,
and Kings County, State of New York.)

(Reverse side or outside cover.)

PRIZE MONEY.
APPLICATION
OF
J. H. Smith
U.S.S. *Bermuda*

(Stamped "Fourth Auditor's Office Navy. Feb. 2, 1869.")

I certify that the signature to the foregoing oath of identity is witnessed by me; that it was signed in my presence by J. W. Smith, who is well known to me as the identical person who served by that name on board the United States ships ~~Potomac~~, ~~Connecticut~~, & *Bermuda* when ~~she captured~~ the cotton claimed as being entitled to share in was picked up at Sea.

(Signed) A. E. A. Burton (?)
Comr. United States Navy

---

[145] The *Bermuda* was a famous early Civil War blockade-runner. The ship was taken into the Union navy and served as a supply ship for the West Gulf and East Gulf Blockading Squadrons. In the course of this work the *Bermuda* captured a number of runners.

FORM OF OATH OF IDENTITY OF AN OFFICER, SEAMAN, ORDINARY SEAMAN, LANDSMAN, BOY, OR MARINE, ENTITLED TO PRIZE MONEY.

I, J. H. Smith, do solemnly swear that I am the identical J. H. Smith who served by that name as a Acty Master, Vol Lieut & Ac Lt Comd on board the United States ship s. Potomac, Commodore, & Bermuda, in the years 1861 to 1866 when she captured the prize Cotton picked up at Sea, and who is named in the certificate of discharge dated 21.'" February, 1866 signed by Gideon Welles Sety of the Navy, which is herewith presented and surrendered.

I also solemnly swear that I am now 44 years of age; am a native of New Jersey; that I was appointed Acty Master, on or about the 8th day of June, 1861, in the grade of Acty Master; that my ship's number was _____. I also solemnly swear that I have not made any previous assignment of or application for the prize money now claimed by me; and, further, that I now reside at 15 Morton Street, and am employed as Ship Mayor.

(Signed by Claimant.) J. H. Smith

P. Chandler Commander Witness. Late Act Vol Lt Comdr
C H Andrews Comd'r U. S. S Bermuda

Sworn to and subscribed before me this 1st day of February, A.D. 1869. And I certify that, to my knowledge, the statement of deponent in regard to his residence and employment is true, my knowledge of deponent being derived from personal acquaintance for years. And I also certify that the above-named deponent appears to be about the age stated by him; that he is about 5 feet 8 inches in height, of fair complexion, Light hair, and Grey eyes.

John F Weugh
Commissioner of Deeds

FORM OF AFFIDAVIT OF TWO WITNESSES.

STATE OF _____
County of _____ } ss.

On this _____ day of _____, 186__, before me, a _____ in and for the State and County aforesaid, duly qualified to administer oaths, personally appeared _____, residing at _____, and employed as _____, and also _____, residing at _____, and employed as _____, who are known to me as credible witnesses, residing and employed as stated, and who, being duly sworn, depose and say: That they reside and are employed as aforesaid; that they have a personal knowledge of _____, who signed the foregoing oath of identity in their presence; and that he is the identical _____ who served on board the United States ship _____ as a _____ from _____ to _____, and who is named in the discharge dated _____, and signed by _____, which he affixed in their presence to this original receipt; that their knowledge of him was obtained

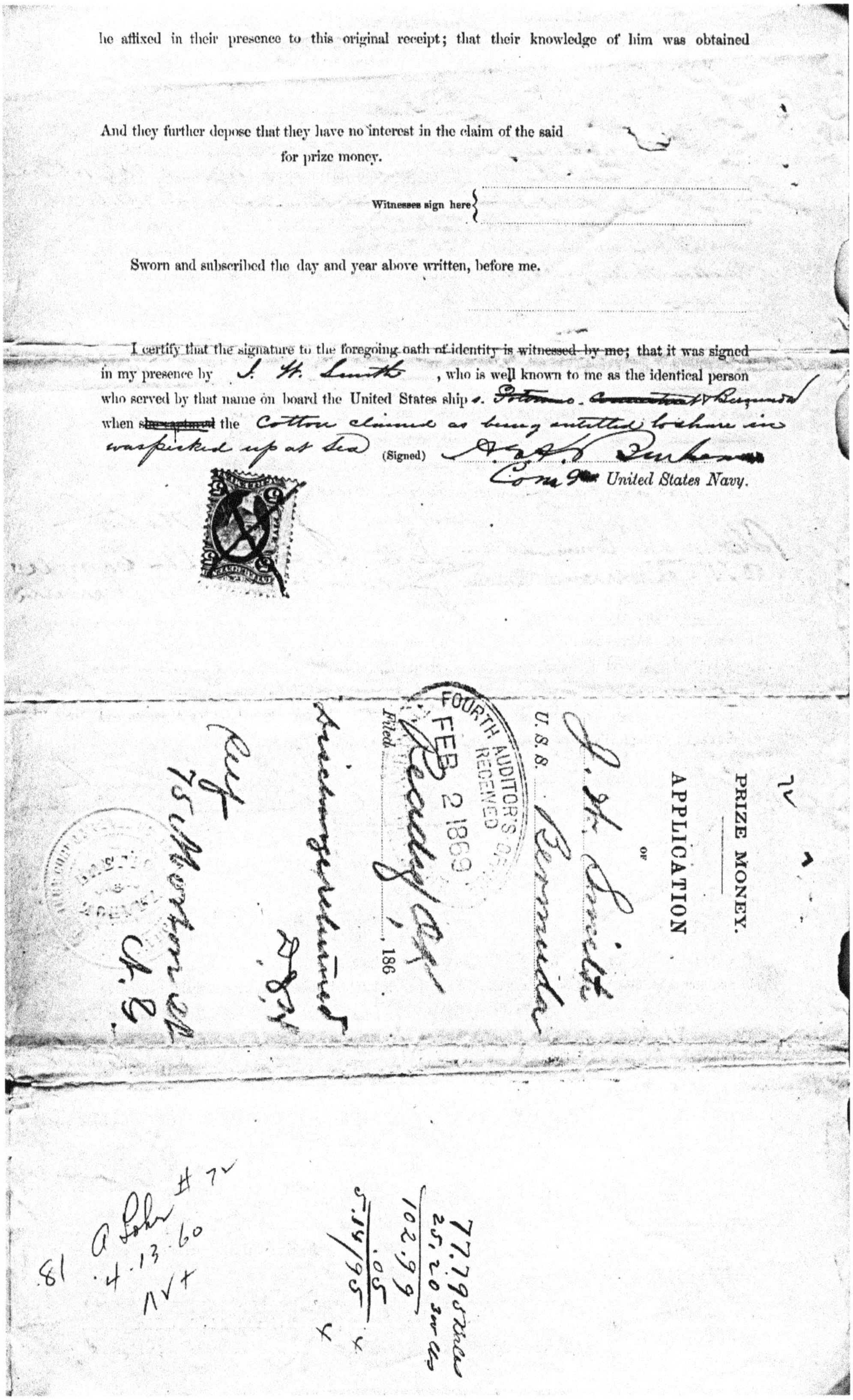

And they further depose that they have no interest in the claim of the said   for prize money.

Witnesses sign here

Sworn and subscribed the day and year above written, before me.

I certify that the signature to the foregoing oath of identity is witnessed by me; that it was signed in my presence by _J. H. Smith_, who is well known to me as the identical person who served by that name on board the United States ship _s. Potomac_ .... when she captured the _Cotton claimed as being entitled to share in was picked up at sea_ (Signed) _____
Com'g United States Navy.

Figure 55. Example of a prize money application from the captain of the U.S.S. *Bermuda*, a former blockade-running ship. (Two pages.) *From the author's personal collection.*

Appropriation Form

The below oversized documents were forms printed on heavy paper, two per page.[146] The entries were in rough order based on date approved (Figures 56-59).

No. 32160 APPROPRIATION---"PRIZE MONEY" $18.88

TAX TO BE RETAINED
BY NAVY AGENT

TREASURY DEPARTMENT,
FOURTH AUDITOR'S OFFICE.
Sept. 29th, 1865.

I CERTIFY *that I have examined and adjusted the claim of* Archibald Anderson *late* Coal Heaver *on board U.S.S.* Cornubia *and find that there is due to* him *from the United States the sum of* Eighteen 88/100 *dollars, being for his share of prize money accruing from the capture of the*

89 Bales Cotton

by the *U.S.S.* Cornubia *as appears by the accounts and vouchers herewith transmitted for the decision of the Second Comptroller of the Treasury thereon.*

To be paid to Archibald Anderson*, or order, by the* ~~*Navy Agent*~~ Paymr. *at* Boston.

S. J. W. T.
Fourth Auditor.

TREASURY DEPARTMENT,
SECOND AUDITOR'S OFFICE.
Oct. 12th, 1865.

The above claim of Eighteen *dollars and* 88 *cents is admitted.*

J. M. B.
Second Comptroller.

$ 18.88

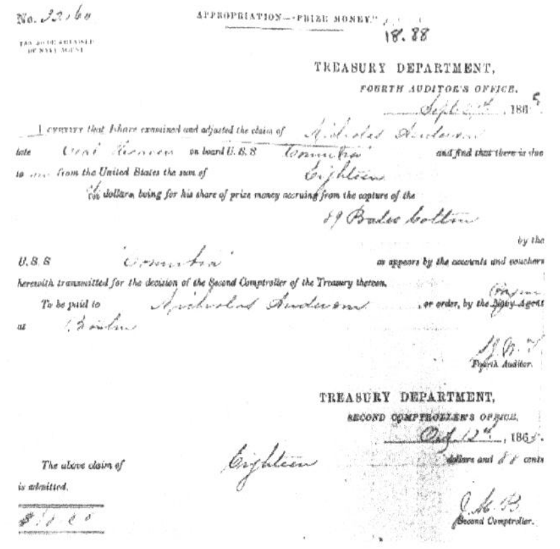
No. 32160 APPROPRIATION—"PRIZE MONEY" 18.88

TREASURY DEPARTMENT,
FOURTH AUDITOR'S OFFICE.
Sept. 29, 1865.

I certify that I have examined and adjusted the claim of Archibald Anderson late Coal Heaver on board U.S.S. Cornubia and find that there is due to him from the United States the sum of Eighteen 88/100 dollars, being for his share of prize money accruing from the capture of the 89 Bales Cotton by the U.S.S. Cornubia as appears by the accounts and vouchers herewith transmitted for the decision of the Second Comptroller of the Treasury thereon.

To be paid to Archibald Anderson, or order, by the ~~Navy Agent~~ Paymr. at Boston

S.J.W.T.
Fourth Auditor.

TREASURY DEPARTMENT,
SECOND COMPTROLLER'S OFFICE,
Oct. 12th, 1865.

The above claim of Eighteen dollars and 88 cents is admitted.

J.M.B.
Second Comptroller.

$ 18.88

Figure 56. When money was appropriated for paying prizes like the *Cornubia's* regarding the floating cotton, there was an entry for each man paid. This is an example of the multiple pages in the Treasury Department ledger for payment to Archibald Anderson, a coal heaver of the *Cornubia's* crew. Dated 29 Sept. 1865 for the approval of the prize money and 12 Oct. 1865 for the collection of the money by Anderson which was paid by the navy paymaster in Boston. *NARA*.

[146] NARA RG 217, Entry 825, Vol. 29 of 75.

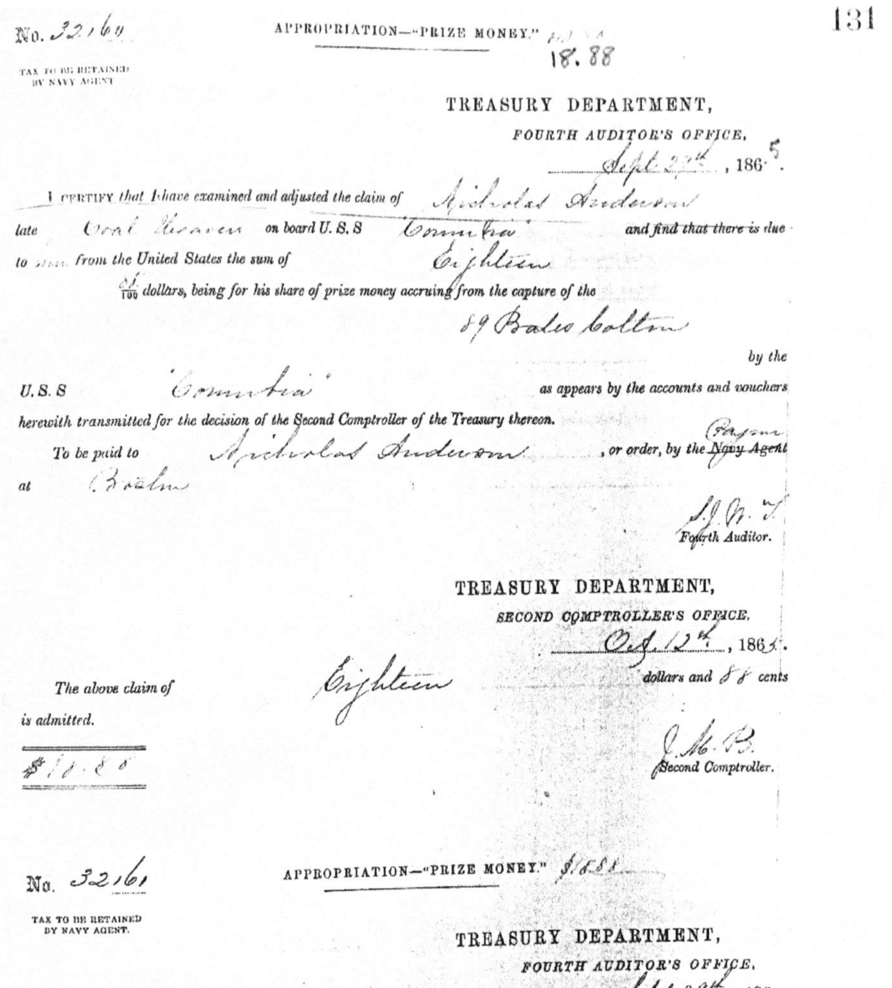

Figure 57. Each page of the prize payment ledger showed the payment details for two men, in this case for the *Cornubia*. NARA.

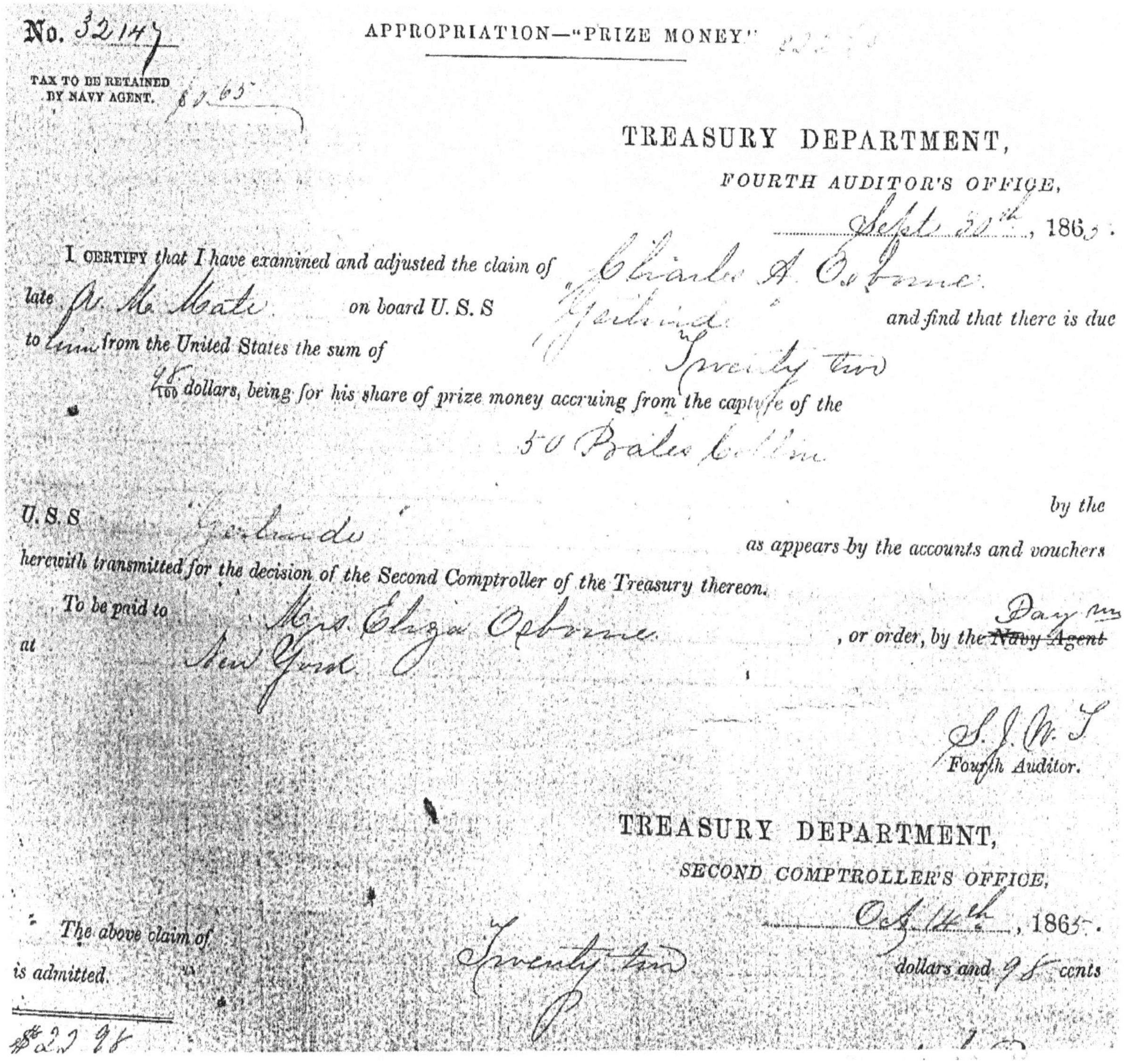

No. 32147

APPROPRIATION—"PRIZE MONEY"

TAX TO BE RETAINED BY NAVY AGENT. $0.65

TREASURY DEPARTMENT,
FOURTH AUDITOR'S OFFICE,
Sept 30th, 1865.

I CERTIFY that I have examined and adjusted the claim of _Charles A. Osborne_ late _A. M. Mate_ on board U.S.S _Gertrude_ and find that there is due to _him_ from the United States the sum of _Twenty two_ $\tfrac{98}{100}$ dollars, being for his share of prize money accruing from the capture of the _50 Bales Cotton_

U.S.S _Gertrude_ by the as appears by the accounts and vouchers herewith transmitted for the decision of the Second Comptroller of the Treasury thereon.

To be paid to _Mrs. Eliza Osborne_ at _New York_, or order, by the ~~Navy Agent~~ _Pay__

_S. J. W. T._
Fourth Auditor.

TREASURY DEPARTMENT,
SECOND COMPTROLLER'S OFFICE,
Oct 14th, 1865.

The above claim of _Twenty two_ dollars and _98_ cents is admitted.

$22.98

Figure 58. Appropriation form for a *Gertrude* crewman. **NARA**.

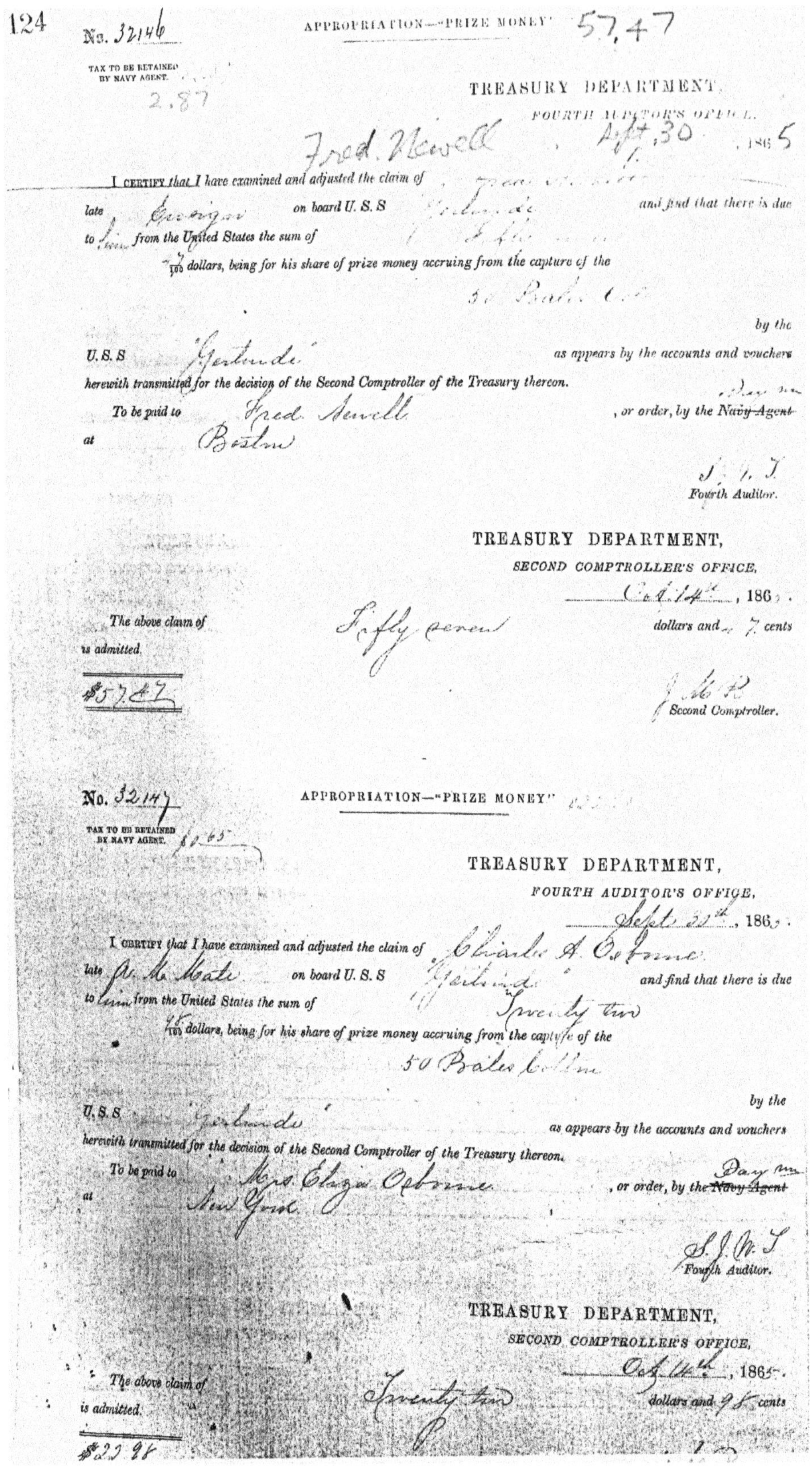

Figure 59. A page of the appropriation ledger for two *Gertrude* crewmen regarding floating cotton. *NARA*.

Postal Cover from the *Gertrude*

Internet searches for the *Gertrude* and *Cornubia* yielded an unexpected but irresistible result, the following envelope (missing its enclosure).[147] The letter probably regarding a book order showed a connection to normal life available to crewmen even when they were on station off Galveston (Figures 60-62). Undoubtedly Engineer Nesen was interested in obtaining professional reference books an example of which was advertised on the back of this envelope. Perhaps he was preparing for an examination for promotion to the next higher rank, 2nd Asst. Engineer. We know he received that promotion from the above prize list and from the below *carte-de-visite*.

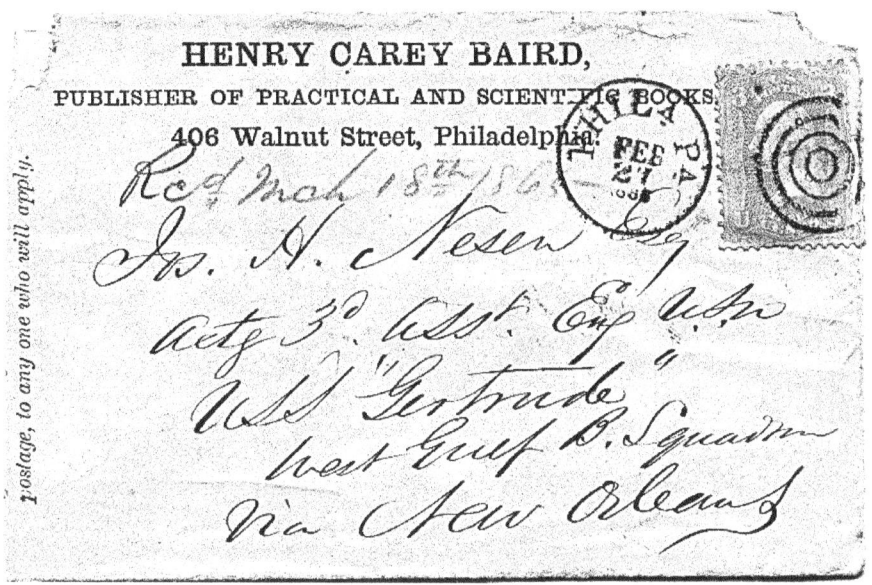

Figure 60. Bookstore letter to the blockading squadron off Galveston. *Courtesy Charles Whiting.*

Figure 61. J. H. Nesen, Acting 2nd Assistant Engineer, U.S.S. *Gertrude. Courtesy Charles Whiting.*

---

[147] Posted on the internet by Chuck Whiting, Saturday, 6 February 2010 at the following address: Bibliophemera.blogspot.com/2010/02/publisher-cover-to-USS-gertrude-civil.ht.

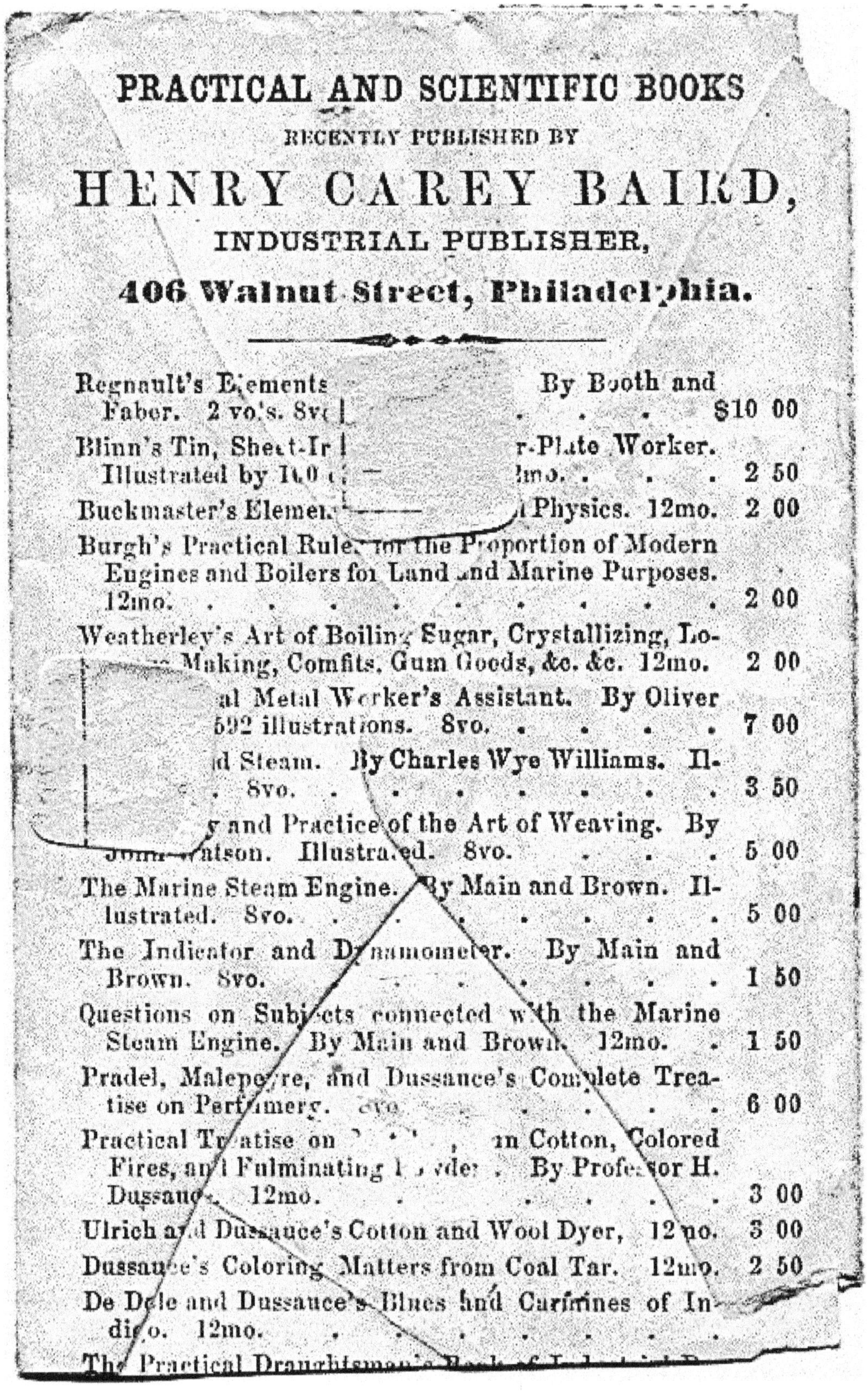

PRACTICAL AND SCIENTIFIC BOOKS
RECENTLY PUBLISHED BY
HENRY CAREY BAIRD,
INDUSTRIAL PUBLISHER,
406 Walnut Street, Philadelphia.

Regnault's Elements ... By Booth and Faber. 2 vols. 8vo.	$10 00
Blinn's Tin, Sheet-Iron ... r-Plate Worker. Illustrated by 100 ... 12mo.	2 50
Buckmaster's Elements ... Physics. 12mo.	2 00
Burgh's Practical Rules for the Proportion of Modern Engines and Boilers for Land and Marine Purposes. 12mo.	2 00
Weatherley's Art of Boiling Sugar, Crystallizing, Lozenge Making, Comfits, Gum Goods, &c. &c. 12mo.	2 00
... al Metal Worker's Assistant. By Oliver ... 592 illustrations. 8vo.	7 00
... d Steam. By Charles Wye Williams. Illustrated. 8vo.	3 50
... and Practice of the Art of Weaving. By John Watson. Illustrated. 8vo.	5 00
The Marine Steam Engine. By Main and Brown. Illustrated. 8vo.	5 00
The Indicator and Dynamometer. By Main and Brown. 8vo.	1 50
Questions on Subjects connected with the Marine Steam Engine. By Main and Brown. 12mo.	1 50
Pradel, Malepeyre, and Dussauce's Complete Treatise on Perfumery. 8vo.	6 00
Practical Treatise on ... in Cotton, Colored Fires, and Fulminating Powders. By Professor H. Dussauce. 12mo.	3 00
Ulrich and Dussauce's Cotton and Wool Dyer, 12mo.	3 00
Dussauce's Coloring Matters from Coal Tar. 12mo.	2 50
De Dele and Dussauce's Blues and Carmines of Indigo. 12mo.	
The Practical Draughtsman's ...	

Figure 62. Reverse side of the above Baird envelope. *Courtesy Charles Whiting.*

6

Papers of the Blockade-Runner *Alabama* Prize Case

When a Union navy ship captured a blockade-runner, the prize, controlled by a prize crew from the capturing ship, promptly sailed for a port having a U.S. district court. The federal district court sitting in admiralty heard a suit to condemn the prize as a lawful capture. If condemned by the court, the prize ship and cargo were sold with half the net proceeds (called a moiety) going to the U.S. government and half to the navy ship's crew and her chain of command. The court files were ultimately held by the regional branches of the NARA. These files upon investigation proved to be extraordinarily interesting and productive of many details of runner operations, Union navy operations, detailed lists of cargo, and other facts of daily life.[148]

A key question requiring investigation by the *Denbigh* Project dealt with the *Denbigh's* owner, the European Trading Company (E.T.C.). The operations of the companies that owned and operated blockade-runners turned out to be remarkably opaque. Initial research on the *Denbigh* yielded two lists of steam ships comprising the E.T.C.'s fleet of blockade-runners as shown below:

ROBERTS[149]	WISE[150]
Denbigh	*Alice*
Fanny	*Vulture*
Alabama	*Vixen*
Crescent	*Virgin*

The two lists are now thought to be not contradictory but rather sequential by time. Probably the first group was contemplated when the company was in the planning and inception phases. Three of them, the *Alabama*, the *Fanny*, and the *Crescent*, were captured in the late summer of 1863 at the same time the E.T.C. was getting off the ground. The three ships already had been sailing in the runner trade early in the Civil War under other auspices. It was not clear if arrangements were finalized for these three merchant steamers to sail for the E.T.C. The second group were added later to the E.T.C. fleet.

To investigate this question the prize case files for the *Alabama* were examined.[151] It was theorized that the ship's papers might show the ownership. While the documents disappointed in resolving the ownership question, they revealed the fascinating story of the prize ship *Alabama*. The lengthy pursuit, capture, and ensuing squabble among four Union ship captains concerning rights to the prize money did not show the Union navy in a favorable light. A perusal of the following court case from the U.S. district court in New Orleans proved highly instructive. The file was unusually long consisting of about 300 documents. A more routine prize case file might number a few dozen pages. The *Alabama* documents are shown in this chapter.

---

[148] Powell, Gerald R., Matthew C. Cordon, and J. Barto Arnold. *Civil War Blockade-Runners: Prize Claims and the Historical Record, Including the* Denbigh's *Court Documents*. College Station, Texas: Institute of Nautical Archaeology, *Denbigh* Shipwreck Project Publication 6, 2012.
[149] Roberts, Richard. *Schroders, Merchants & Bankers*. Basingstoke: Macmillan, 1992.
[150] Wise, Stephen R. *Lifeline of the Confederacy: Blockade Running during the Civil War*. Columbia: University of South Carolina Press, 1988.
[151] United States vs. The Steamer *Alabama*, Her Tackle &c. & Cargo, In Prize. U.S. District Court for the Eastern District of Louisiana. No. 7728. NARA.

Ship's Papers

The initial documents in the court file were the ship's papers taken by the captors and forwarded to the prize court (Figures 63-76). In the following transcript of the crew list column labeled "Of What Country Citizens or Subjects" was left out of the table as all crewmen were shown as "State of Alabama." The longhand script of the document proved difficult to decipher, especially the crew names. The crew for the outbound voyage to Havana numbered 32 including Captain Carrell.

Crew List

List of Persons Composing the Crew of the Steamer *Alabama*
of Mobile whereof is Master Thos. Carrell bound for Havana

NAMES	PLACE OF BIRTH	PLACE OF RESIDENCE	AGE	HEIGHT		COM-PLEXION	HAIR
				FT.	IN.		
R. Adams	Ireland	Mobile	38	5	8	Light	Red
H. Parker	Pembroke, N.H.	Do.	46	5	8	Dark	Black
Geo. Wade	Norwich, Conn.	"	54	5	10	"	"
Patrick Cole	Ireland	"	55	5	7½	Light	Grey
Rich. Goulding	Do.	"	40	5	8	"	Light
Augt. Terry	New Orleans	"	35	5	7	"	Dark
Wm. Shea	New York	"	28	7	8½	"""	Dark
John Lawless	Ireland	"	41	5	9	"	"
John Walter	Germany	"	42	5	7	"	Brown
Michael Dayle	Ireland	"	47	5	9½	Dark	Dark
Baseles John	Greece	"					
John Baillise	France						
John Emmet	Ireland						
Patrick Featherston	Ireland						
Timothy Spellman	Mobile						
Joseph Silva	Portugal	"					
W. C. Wilson	Ireland	"	46	5	8	Dark	Black
Henry Davis	Germany	Mobile	41	5	5	Light	Light
Fred. Johnson	Sweden	"	45	5	8	"	"
James Avery	Newfoundland	"	46	6	---	"	"
Criss Cillnis		"	48	5	7	Dark	Dark
Peter Spire	Iceland	"	46	5	7	Light	Light
Chas. Rohninck	Sweden	"	41	5	6	"	"
James Bragan	Ireland	"	43	5	7½	"	"
H. W. Daniels	Sweden	"	45	5	7	"	Dark
Isaac Swain	Sweden	"	45	5	11½	"	"
Saml. Coller	Sweden						
John Bayley	Virginia. Colored		30			Mulatto	Blk.
Christopher	" "		23			"	"
Wm. Caldwell	" "		35			"	"
Henry Barnett	" "					"	"
(signed) Frankn. Buchanan Admiral Aug. 21st '63	Office of Inspector Genl. Mobile Augt. 20th 1863 Capt. Carrell & the crew of the Steamer *Alabama* having complied with all the military requirements governing vessels leaving this port. Said Steamer *Alabama* is permitted to enter upon her voyage, to Havana, Cuba. (signed) C. L. Layne Major & Inspt. Genl.						

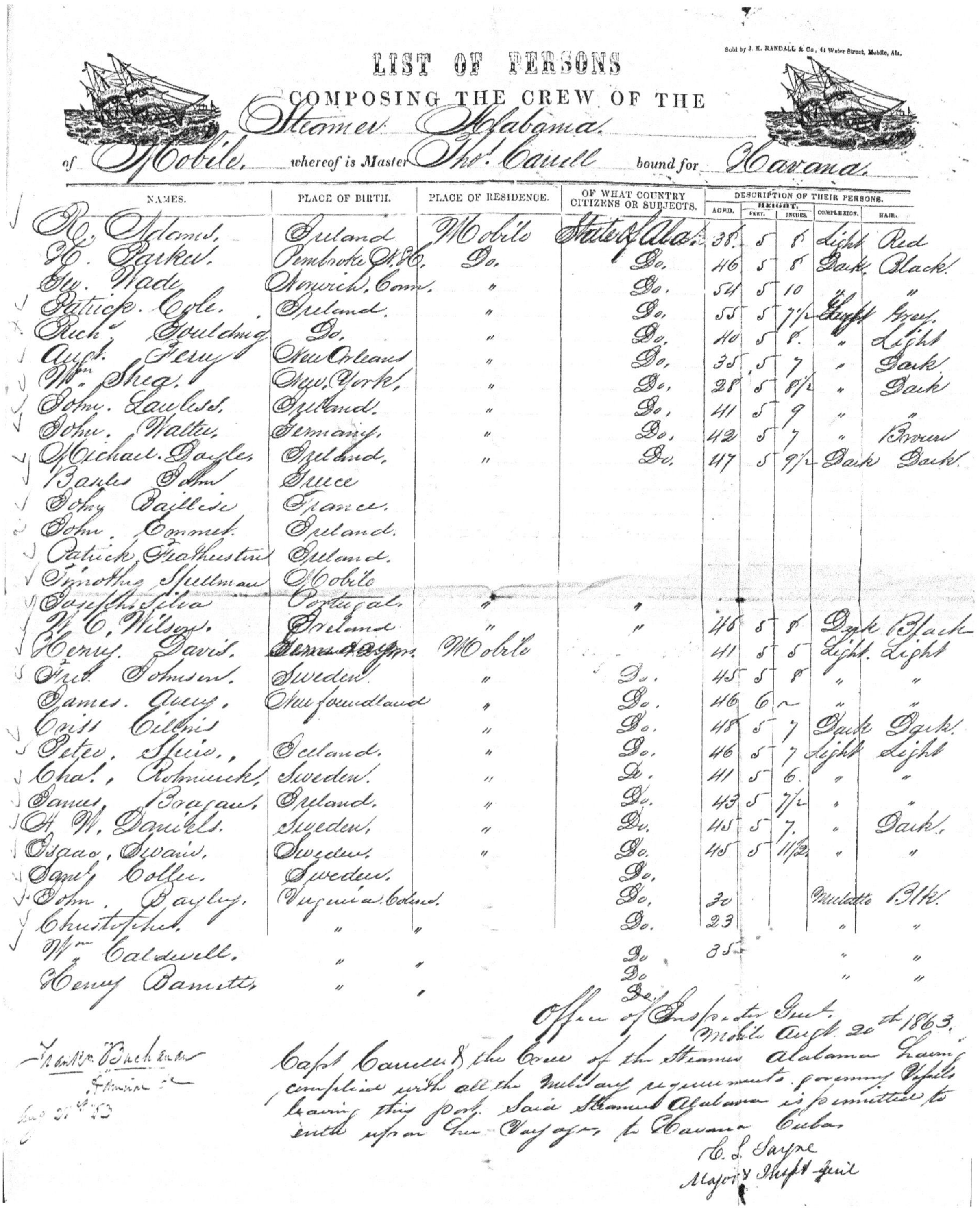

Figure 63. Crew list for Str. *Alabama*, part of the ship's papers seized when the ship was captured and filed with the prize case documents. Dated 20 August 1863 in Mobile where the proposed voyage to Havana and back commenced. Filed (in the court's case file) 19 Oct. 1863. ***NARA***.

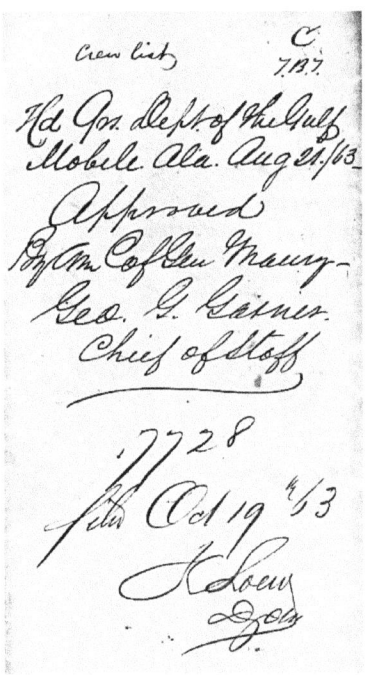

Figure 64. Notations on the reverse side of the above crew list. The notes made up an abstract added by a C.S. Army official in Mobile and by a court clerk in New Orleans. (Legible, not transcribed.) *NARA*.

Captain's Oath

    I, <u>Thomas Carroll</u> do solemnly, sincerely and truly <u>swear</u> that the within list contains the names of the crew of the <u>Str. *Alabama*</u> together with the places of their birth and residence, as far as I can ascertain the same,
    this <u>20</u>th day of <u>August</u> 18~~5~~63
    Before me <u>(signed) J. C. Colsson</u>   (signed) Thos. Carrell
        Deputy Collector

I do certify, that the within is a true copy of list of the crew of the
Steamer <u>*Alabama*</u> of <u>Mobile</u>
whereof said <u>Carroll</u> is Master, taken from the original on file in this office.

    Given under my hand and Seal of Office, at the Custom House, this <u>20</u>th day of <u>August</u> in the year of our Lord one thousand eight hundred and ~~fifty~~ <u>sixty three</u>.

    I do hereby further certify, that the within named persons who compose   (blank)   the company of the above mentioned Steamer whereof   (blank)   is at present Master, have produced to me proof, in the manner directed in the Act entitled "An Act for the Relief and Protection of American Seamen." And pursuant to the said Act, and to the Act Supplementary to the Act concerning Consuls and Vice Consuls, and for the further protection of American Seamen
    I do hereby certify, that the said <u>persons</u> are citizens of the ~~United States.~~
        Given under my hand and Seal of Office, this
<u>20</u>th day of <u>August</u> 18~~5~~63
                      (signed) J. C. Colsson
                          Deputy Collector

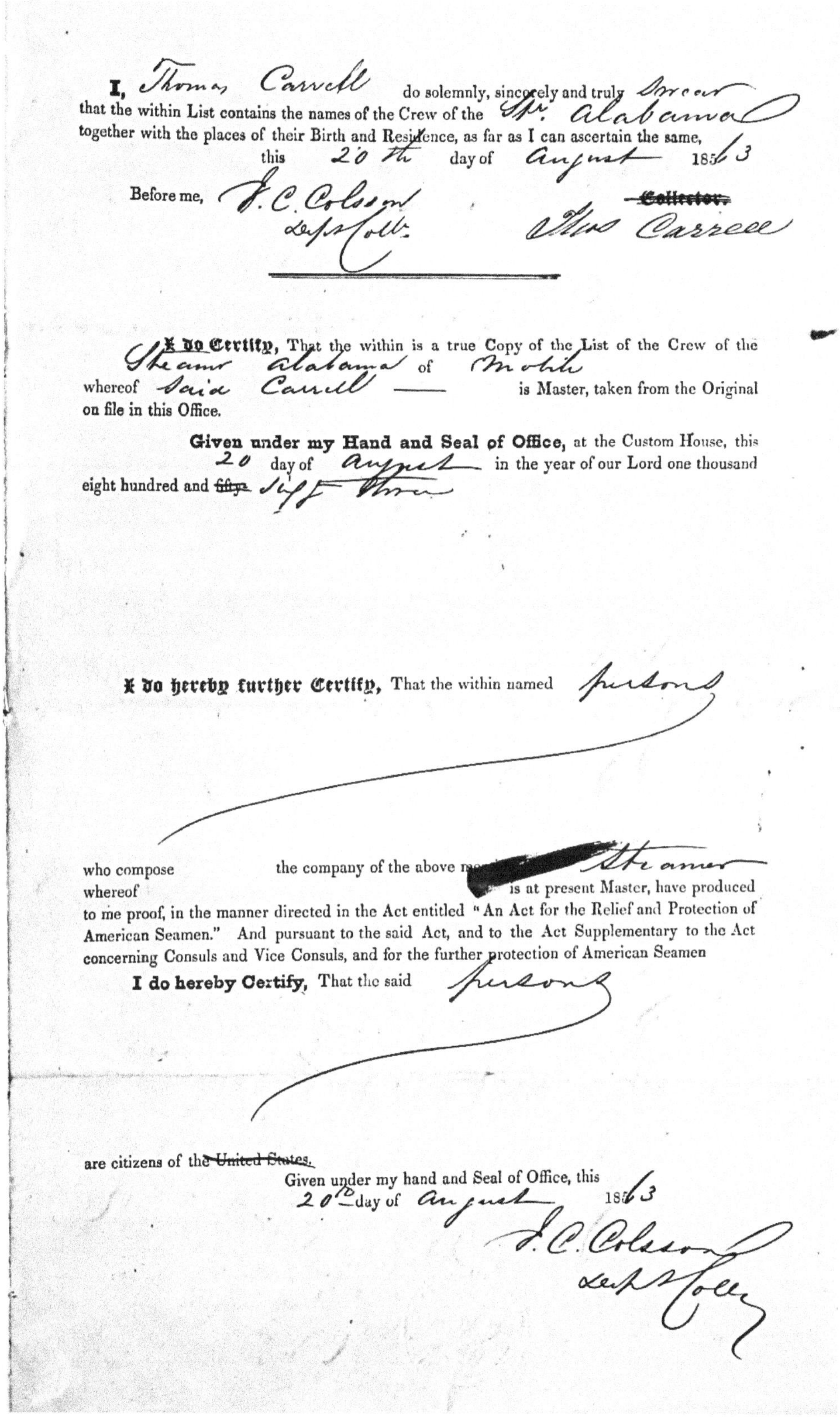

I, Thomas Carrell do solemnly, sincerely and truly Swear that the within List contains the names of the Crew of the Str. Alabama together with the places of their Birth and Residence, as far as I can ascertain the same, this 20th day of August 1863

Before me, J. C. Colson ~~Collector~~
Dept Coll.
Thos Carrell

I Do Certify, That the within is a true Copy of the List of the Crew of the Steamer Alabama of Mobile whereof said Carrell is Master, taken from the Original on file in this Office.

Given under my Hand and Seal of Office, at the Custom House, this 20 day of August in the year of our Lord one thousand eight hundred and ~~fifty~~ sixty three

I Do hereby further Certify, That the within named persons who compose the company of the above named Steamer whereof is at present Master, have produced to me proof, in the manner directed in the Act entitled "An Act for the Relief and Protection of American Seamen." And pursuant to the said Act, and to the Act Supplementary to the Act concerning Consuls and Vice Consuls, and for the further protection of American Seamen

I do hereby Certify, That the said persons are citizens of the ~~United States~~. Given under my hand and Seal of Office, this 20 day of August 1863

J. C. Colson
Dept Coll.

Figure 65. Reverse side of the *Alabama's* crew list shown above in Figures 63 and 64. Note from the dates that this form was printed in the 1850's and corrected to the 1863 date by hand when filled out. *NARA*.

Cargo Manifest for the Return Voyage

The following was a detailed listing of the cargo of the *Alabama* on her return voyage from Havana to Mobile (Figure 66). With a few exceptions, the cargo was intended for the civilian market. The left-hand column showed the shipper's or owner's marks, and those graphics were omitted from the transcript. Note the false destination intended to disguise the blockade-running nature of the voyage. Nassau was a neutral port and therefore a legal destination. The date of 8 Sept. 1863 on the document was shortly before the ship sailed from Havana.

Manifest of the Cargo of the Confederate
Steamer *Alabama*, Carrell Master, bound to Nassau

100	Kegs	Cut nails wei'g. 10,000 lbs.
20	Sacks	Puerto Rico coffee w'g. 2,447 lb.
65	Boxes	Cont'g. each 4 tin cans of 5 gals. kerosene oil, together 260 tin cans 1,300 gallons
89	Boxes	McDxB^s· Cincinnati's starch wei'g. 2,792 lbs.
10	do.	American starch 35 lbs. box, 350 lbs.
20	Barrels	Mess pork, superior
25	Boxes	of 12 bottles each French sweet olive oil
34	do.	cont'g. each 25 lbs. French comp^n· candles 850 lbs.
2	cases	wei'g. each 100 lbs. French candles 200 lbs.
100	boxes	best Castille soap wei'g. nett 9,963 lbs.
75	"	Brown soap " " 4,200 lbs.
1	case	cont'g. 18 doz. hand saw files 3x4
		6 " flat files 10, 12, 14 inches
		4 packages cont'g. 20 doz. large pocket knives
		1 do " 5 " " " "
1	case	Cont'g. 20 gross lead pencil Faber brnd.[152]
		20 do do do imitation
		20 do penholders assorted
		40 do steel pens assorted brands
		60 do " " " "
		50,000 envelopes (yellow & white)
1	case	" 25 doz. small bottles copying ink
4	cases	" viz no. (?) 3. 54, no. 4. 60, no. 5. 80, no. 6. 96 together 290 large glass bottles writing ink
1	case	" 50 lbs. powdered cantharides
		25 do do do Fly
		15 pieces adhesive plaster
		50/2 do do do
2	Barrels	Powdered linseed wei'g. Nett 263 lbs.
1	do	wei'g. 186 lbs. linseed
4	do	do 934 " Borax
3	Bales	With 50,000 fine vial corks
1	case	Cont'g. 24 doz. M. glass syringes no. 1, 2, 3, 4
		25 lbs. powdered Jalap
		12 doz. men metal P/P syringes over
		(page 2)
5	Oz.	Strychnine crystals in 1/8

---

[152] Farber is a German company that has been in business for 250 years. In 1861, Eberhard Faber set up a factory in New York that made pencils using leads from Germany. At this time most lead pencils sold in the U.S. were imported from Europe, increasingly from Germany due to an improved manufacturing process.

3	Glass	Bottles with 10 lbs. Iodide potassium
1	can	with 12 lbs. licorice est. refined
1	case	cont'g. 10 boxes assorted gum lozenges 100 lbs.
2	do	" 100 lbs. powdered gum arabic
1	do	" 100 lbs. licorice est. 1st quality
1	do	" 50 packages with 50 gross paper pill boxes
2	do	" 24 glass bottles of 4 lbs. amoniae each together 96 lbs. amoniae
2	Carboys	100 lbs. each spirits nitre fine 200 lbs.
1	Barrel	wei'g. 230 lbs. common gum arabic
1	do	wei'g. 192 " flor sulphur
1	case	cont'g. 3 gross green vials 6 oz. size
2	do	do 8 " " " 4 oz. "
2	do	do 10 " " " 2 oz. "
2	do	do 10 " " " 1 oz.
2	Bales	cont'g. 18,000 bottle corks
10	Barrels	of 42 gallons alcohol each together 420 galls.
1	case	cont'g. 61 gross Havana matches
2	do	do each 25 gross Havana colored matches together 50 gross
1	case	cont'g. 12 pieces fine colored flannel 345¾ varas[153]
1	do	do 71 doz. ladies stockings, 100 packages assorted pins
1	bale	with 50 doz. Madras hdkfs. (hand kerchief)
1	case	" 1,521½ varas colored common flannel
1	do	cont'g. 86 doz. knives & forks
10	Boxes	Colgate toilet family soap
10	do	white refining sugar wei'g. 4,241 lbs.
18/10	do	1,800 segars conchas "El Comercio"
80/10	Do	8,000 Millar "Garagozana"
20	Kegs	bicarb. of soda 112 lbs. each – 2,240 lbs.
3	Barrels	camphor of 112 lbs. each – 335 lbs.
10	Barrels	green copperas wei'g. 2,795 lbs.
10	Bags	black pepper " 1,100 "
1	Keg	wei'g. 100 lbs. balsam capaiva
1	case	with 50 lbs. calomel
1	Do	cont'g. 5 tins cans with 100 oz. quinine Over
		(page 3)
1	case	cont'g. 100 reams letter paper
1	do	do 50 reams do do
65	Boxes	cont'g. each 2 doz. pairs no. 10 Whitmore cotton cards together 130 doz.[154]
1	case	cont'g. 1654½ yards fancy silk worsted
3	cases	cont'g. 24 doz. black felt hats
1	Bundle	" 3 pieces colored flannel measuring 45½ yards fine
		Havana 8 Sept'ber. 1863

---

[153] 1 vara = 33 1/3 inches = 846.67 mm.
[154] The Confederacy had a critical shortage of cotton cards creating a bottleneck in the family production of home-spun cloth.

Manifest of the cargo of the Confederate Steamer "Alabama" Carrell Master bound to Nassau

Marks	Description
A	100 Kegs cut nails wei'g 10000 lbs
"	20 Sacks Puerto Rico Coffee w'g 2447 "
"	65 Boxes cont'g each 4 tin cans of 5 galls Kerosine oil together 260 tin cans 1300 gallons
A/B	89 Boxes Mc D & Bs Cincinati's Starch wei'g 2792 lbs
A/A	10 do American Starch 36 lb box 360 lbs
A	20 Barrels mess Pork Superior
A	25 Boxes of 12 bottles each french sweet olive oil
A/FE	34 do cont'g each 25 lbs french comp'n Candles 850 lbs
A/B	2 cases wei'g each 100 lbs french candles 200 lbs
A	100 boxes best Castille Soap wei'g nett 9963 lbs
"	75 " do Brown Soap " " 4200 "
A/c	1 case cont'g 18 doz Hand saw files 3x4
	6 do Flat files 10.12.14 inches
	4 packages cont'g 20 doz large Pocket Knives
	1 do " 5 " " "
	1 do " 9 Pocket Knives
A n°1	1 case cont'g 20 gross lead pencil Faber brand
	20 do " " imitation
	20 do Penholders assorted
	40 do Steel Pens assorted brands
	60 do " " " "
	50000 Envelopes (Yellow & white)
" n°2	1 case " 25 doz small bottles copying Ink
" n°3/6	4 cases " viz n°3. 54 - n°4. 60 n°5. 80. n°6. 96 together 290 large glass bottles writing Ink
" n°7	1 case " 50 lbs powdered cantharides
" n°8/9	25 do do do Fly
	15 pieces Adhesive Plaster
	50/2 do do do
" n°8/9	2 Barrels powdered Linseed wei'g nett 263 lbs
" n°10	1 do wei'g 186 lb Linseed
" n°11/14	4 do do 934 " Borax
" n°15/17	3 Bales with 50000 fine Vial Corks
" n°18	1 case cont'g 24 doz m. glass Syringes n° 1.2.3.4.
	25 lb powdered Jalap
	12 doz men Metal P/P Syringes

Over

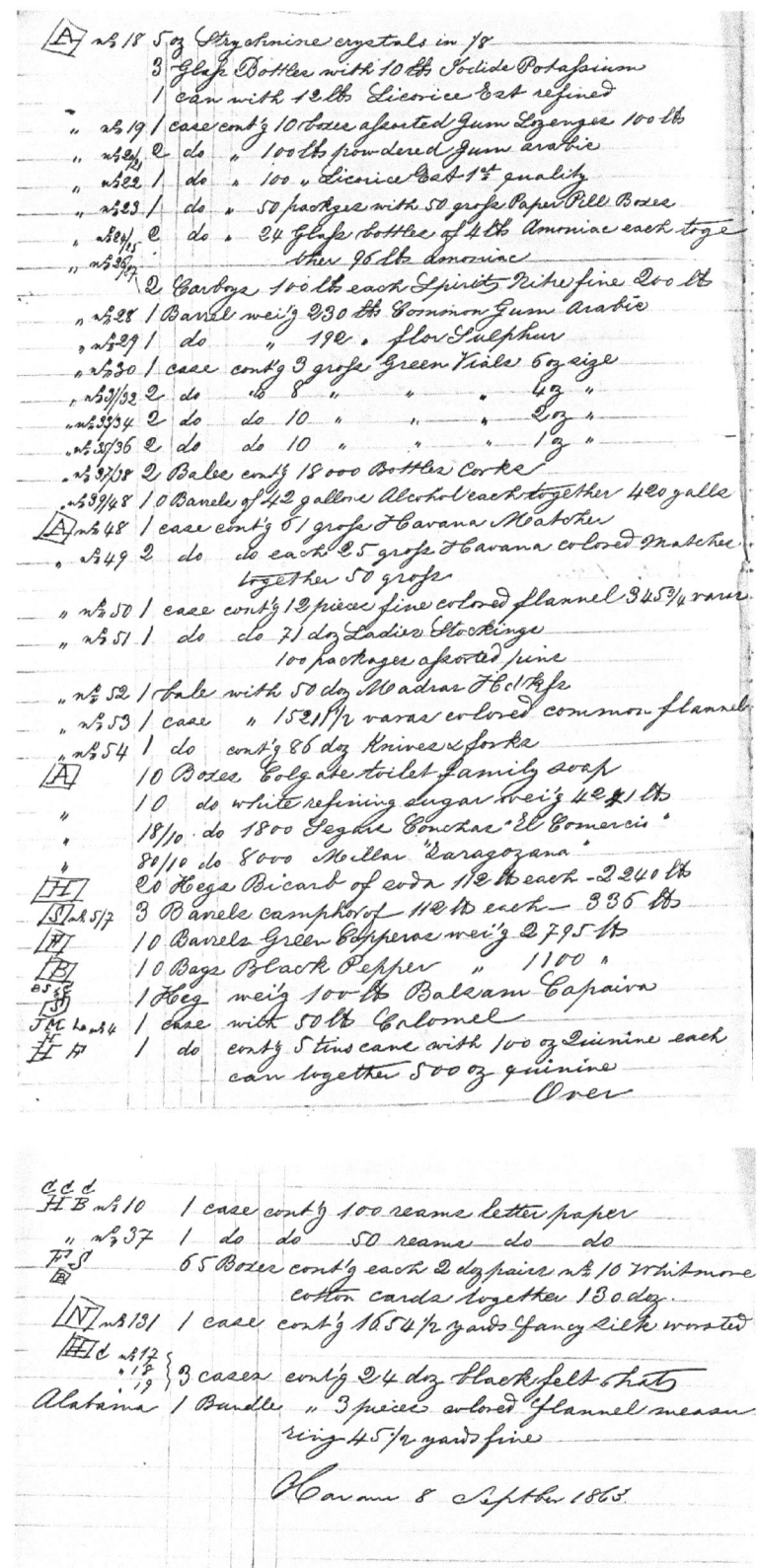

Figure 66. Cargo manifest of the Str. *Alabama* for the return voyage to Mobile from Havana, falsely claiming to be bound for Nassau. (This page and the prior page.) Dated 8 Sept. and filed 19 Oct. 1863. *NASA*.

Ship's Register Certificate

The Confederate government in Mobile issued Permanent Register No. 30 for the merchant steamer *Alabama*. It was found in the court files (Figures 67-69). Only the most pertinent text was transcribed below rather than the whole document. The dimensions and ownership of the ship as of 10 June 1863 were given. One might speculate that the ship might have been under charter to other parties or a company like the E.T.C. Such a charter might well be omitted from the register, and this would go far toward obscuring the responsible parties should the blockade-runner be captured.

In pursuance of existing laws and an act of the Congress of the Confederate States of America entitled "An act to provide for the Registration of Vessels owned in whole or in part by the Citizens of the Confederate States."

Roger A. Hiern of Mobile, State of Alabama

having taken or subscribed to the oath required by law and having sworn that –

himself owning 25/100th Sam Wolff 15/100th also of said Mobile, and R. Geddis of New Orleans 60/100th are the

only owner of the ship or vessel called the *Alabama* of New Orleans whereof R. A. Hiern is at present Master and is a citizen of the Confederate States as he hath sworn –

and that the said ship or vessel was built at Brooklyn, State of New York in the year 1859, as per Enrolment No. 103 issued at the Port of New Orleans the 14 May 1861 – now surrendered – vessel bound foreign; —

And Said Enrolment

having certified that the said ship or vessel   (blank)   has one deck and one mast and that her length is Two hundred & twenty six feet her breadth Thirty two feet (32 feet) her depth Seven 4/12 feet and that she measures Five hundred & ten 79/95 tons that she is a Steamer, has round stern, sharp bow, and no figure head. And the said R. A. Hiern having agreed to the description and admeasurement above specified and sufficient security having been given according to law the said Steamer has been duly registered at the Port of Mobile.

Given under my hand and seal at the Port of Mobile the 10th day of June One Thousand Eight Hundred and Sixty three.

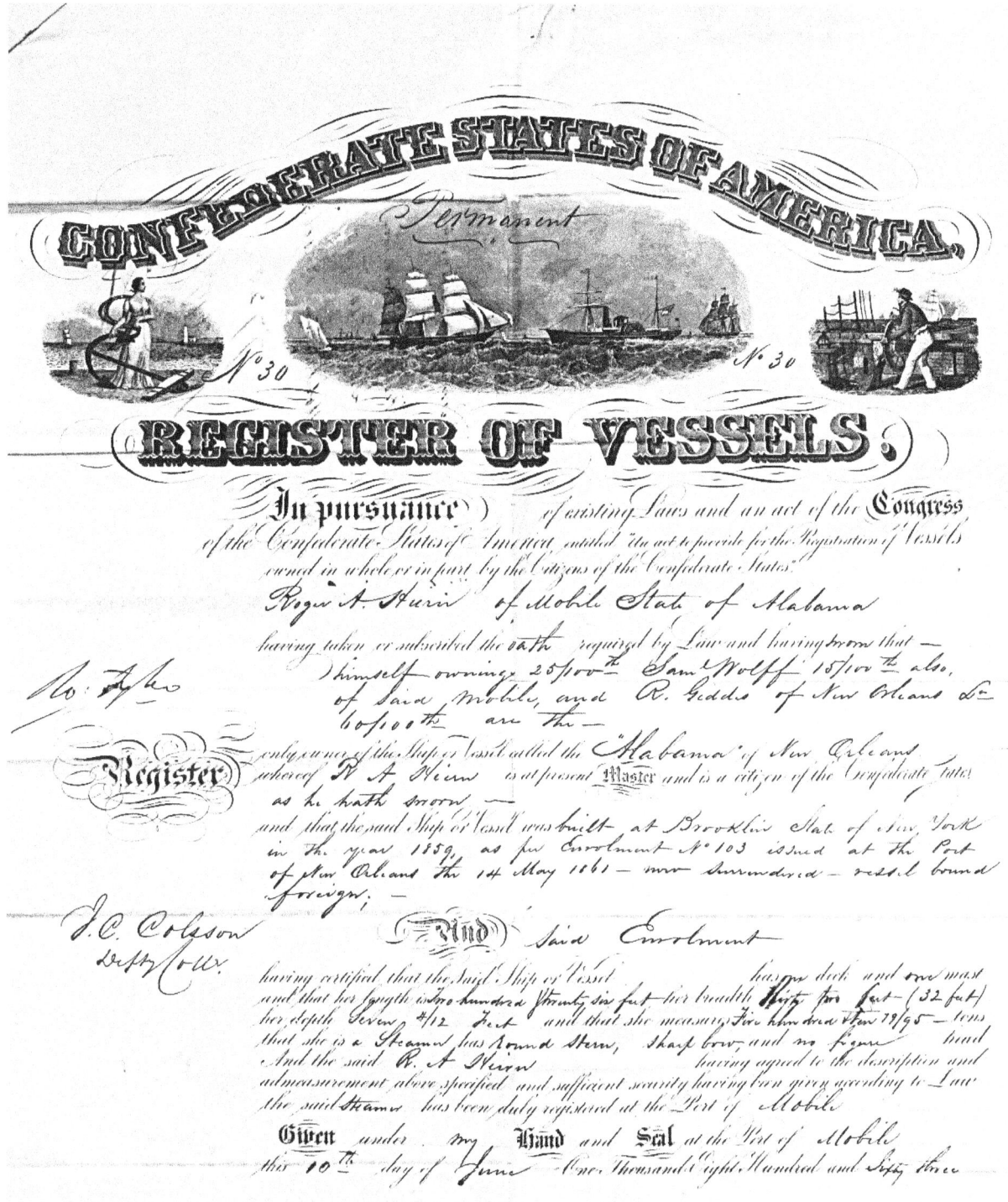

Figure 67. The official register document of the Str. *Alabama*. NARA.

In pursuance of existing Laws and an act of the **Congress** of the Confederate States of America entitled "An act to provide for the Registration of Vessels owned in whole or in part by the Citizens of the Confederate States."

Roger A. Hurn of Mobile State of Alabama having taken or subscribed the oath required by Law and having sworn that — himself owning 25/100th Sam'l Wolff 15/100th also, of said Mobile, and R. Geddes of New Orleans L— 60/100th are the — only owners of the Ship or Vessel called the "Alabama" of New Orleans whereof R A Hurn is at present **Master** and is a citizen of the Confederate States as he hath sworn —

and that the said Ship or Vessel was built at Brooklin State of New York in the year 1859, as per Enrolment N° 103 issued at the Port of New Orleans the 14 May 1861 — now surrendered — vessel bound foreign. —

And said Enrolment

having certified that the said Ship or Vessel has one deck and one mast and that her length is two hundred twenty six feet her breadth thirty two feet (32 feet) her depth Seven 4/12 Feet and that she measures Five hundred ten 79/95 — tons that she is a Steamer has Round Stern, sharp bow, and no figure head And the said R. A Hurn having agreed to the description and admeasurement above specified and sufficient security having been given according to Law the said Steamer has been duly registered at the Port of Mobile

Given under my **Hand** and **Seal** at the Port of Mobile this 10th day of June One Thousand Eight Hundred and Sixty three

Figure 68. Four enlarged sections of the *Alabama's* register. **NARA**.

The register document had small attachments pasted to the upper left corner for the purpose of up-dating changes of the ship's captain for voyages after the first voyage. The captain for the first voyage was entered on the face of the register.

DISTRICT OF MOBILE—PORT OF MOBILE.

Thomas Carrell having taken the oath required by
Law is at present Master of the Steamer *Alabama* in
Place of J. Hopkins late Master.
J. C. Colsson Depy. Collector

(Pasted on top of similar form partly obscured below.)

DISTRICT OF MOBILE—PORT OF MOBILE.

Hopkins having taken the oath required by
Law is at present Master of the Str *Alabama* in
Place of R. A. Hiern (?) late Master.
J. C. Colsson Depy. Collector

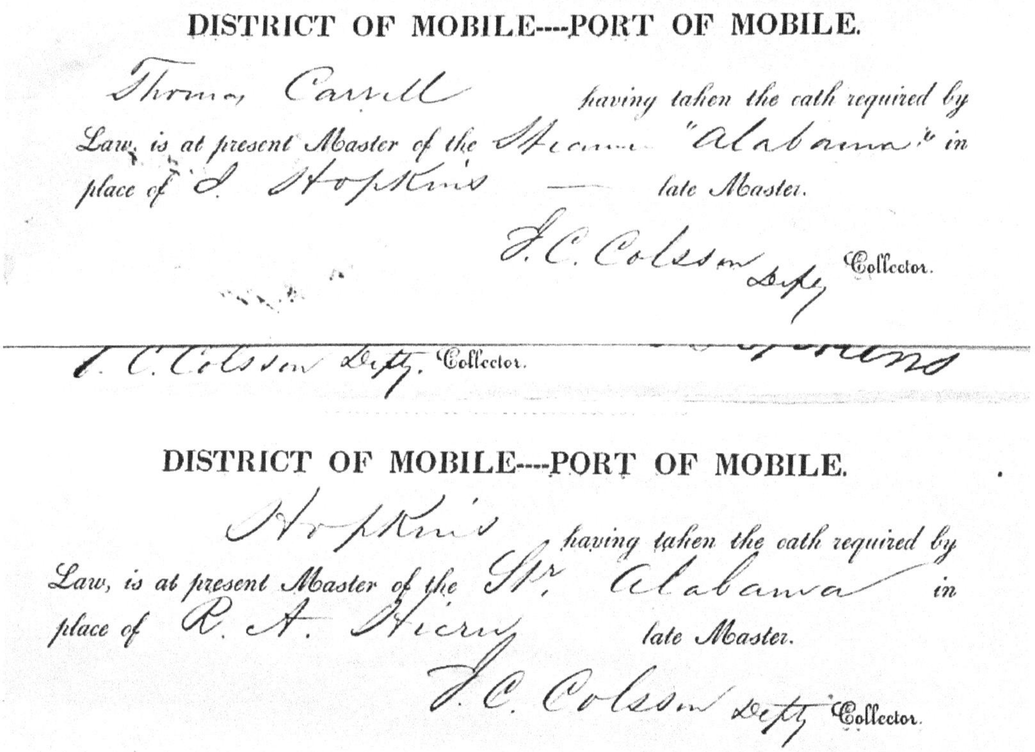
DISTRICT OF MOBILE----PORT OF MOBILE.
Thomas Carrell having taken the oath required by
Law, is at present Master of the Steamer "Alabama" in
place of J. Hopkins — late Master.
J. C. Colsson Depy Collector.

J. C. Colsson Depty. Collector.

DISTRICT OF MOBILE----PORT OF MOBILE.
Hopkins having taken the oath required by
Law, is at present Master of the Str Alabama in
place of R. A. Hiern late Master.
J. C. Colsson Depty Collector.

Figure 69. Attachments to the *Alabama's* register for updating changes of the ship's captain. *NARA*.

Articles of Agreement

The crew list specified position, rank, and pay (Figures 70-71). Note that wages were to be paid in three installments: before leaving Mobile, in Havana, and upon return to Mobile.

It is agreed between the Master and Seamen, or Mariners of the Steamer *Alabama* of Mobile, Alabama whereof   is at present Master, or whoever shall go for Master, now bound from the Port of Mobile to Havana & back to Mobile, or other Confederate Port.

Name	Place of Service	Over & Back	Advance Wages	Names	Whole Am't. of Wages	Advanced Abroad	Wages Due on return
				Aug. 27, '63			
				Receipt for wages			
				Advanced in Havana			
Thos. Carrell	Captain	3,000	1,000	Thos. Carrell	3,000	1,000	1,000
R. Adams	Purser	1,000	500	R. Adams	1,000	500	
H. Parker	Navigator	3,500	1,000	H. Parker	3,500	1,500	1,000
W. C. Wilson	Mate	1,500	750	W. C. Wilson	1,500	750	
Geo. Wade	Engineer	3,000	1,000	George Wade	3,000	1,000	1,000
Patrick Cole	1st asst. "	2,000	600	Patrick Cole	2,000	800	600
Richd. Goulding	2nd asst. "	1,500	400	R. Goulding	1,500	600	500
August Terry	Striker	800	250	August Terry	800	300	250
Wm. Shea	Carpenter	800	300	Wm. Shea	800	300	200
John Lawless	Fireman	250	100	John Lawless	250	100	50
John Walter	"	250	100	John Walter	250	100	50
M. Doyle	"	250	100	Michel Doyle	250	100	50
W. G. Thompson	"	250	100	Wm. G. Thompson	250	100	50
(6 lines left blank.)	"						
Henry Davis	Qr. Master & Seaman	225	75	H. Davis[155]	225	100	50
Fred. Johnson	"	225	75	F. Johnson	225	100	50
Criss Collins	Seaman	225	75	C. Collins	225	100	50
Peter Spire	"	225	75	P. Spire	225	100	50
Chas. Rokwick	"	225	75	C. Rokwick	225	100	50
J. Bragan	"	225	75	J. Bragan	225	100	50
A. W. Daniels	"	225	75	A. W. Daniels	225	100	50
Isaac Swain	"	225	75	Isaac Swain	225	100	50
Saml. Collier		225	75	Saml. Collier	225	100	50
James x Avely	"	225	75	James x[156] Avely	225	100	50
John Bailey	Steward Colored		150	John x Bailey		100	150
Christopher	Asst. " Colored		150			100	100
Wm. Caldwell	Cook Colored		50	Wm. x Caldwell	200	100	50
Henry Barnett	Asst. "		25	Henry x Barnett	100	50	25

Collector's Office
    Mobile   August 20th 1863
I hereby certify that the foregoing articles of agreement are a true copy of the original on file in this office. In witness whereof I have hereunto set my hand & Seal of Office
                (signed) J. C. Colsson, Depty. Collector

---

[155] At this point was squeezed in a multiple line note, probably applying from this line downward in the list: "If the vessel is detained over 25 days, $1 per day will be allowed while in Port of Havana."

[156] X in this list indicates "his mark" used when the person could not write his name.

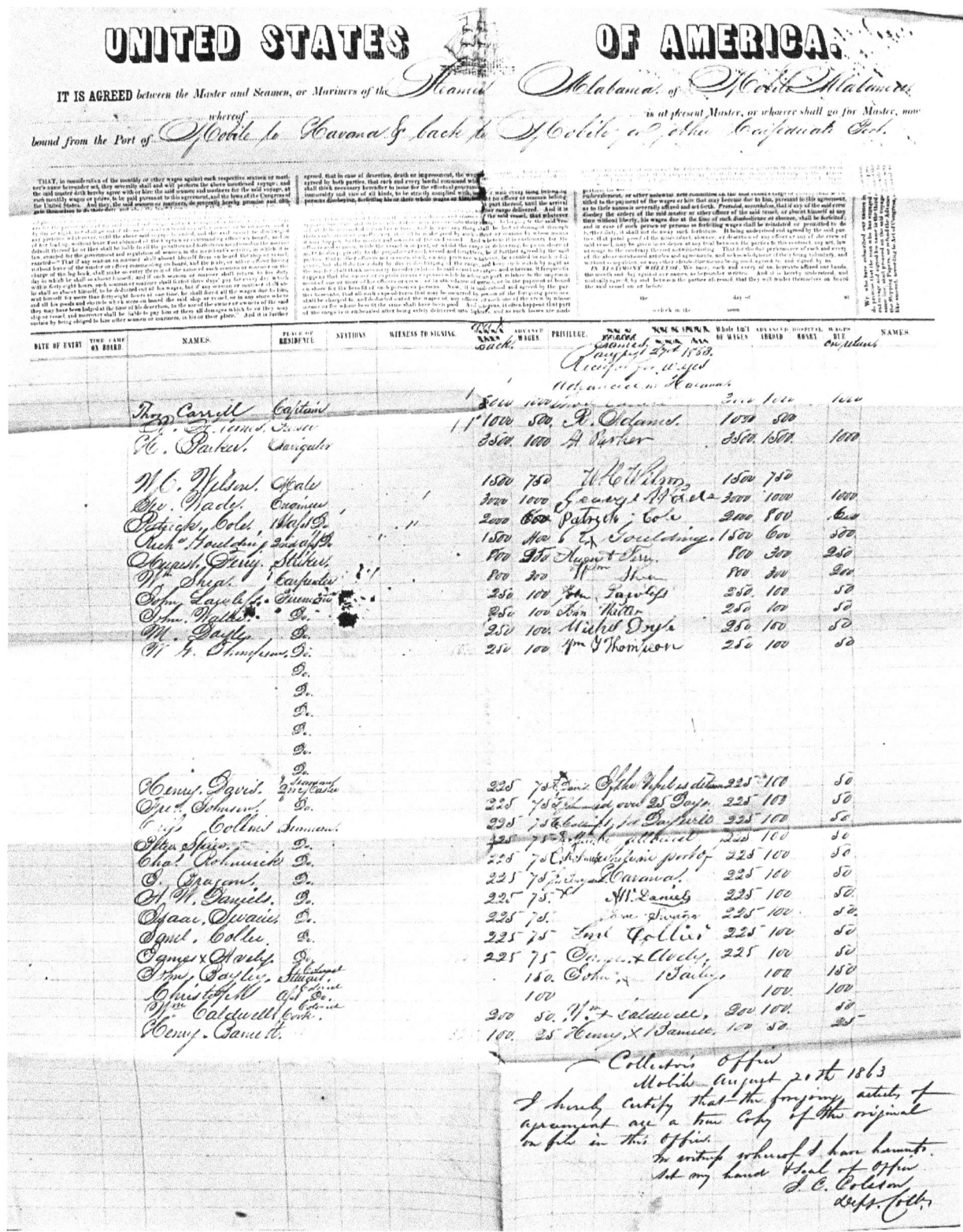

Figure 70. The *Alabama's* articles of agreement, being a list of the crew and their pay, for the voyage from Mobile to Havana and return. Dated 20 August 1863. *NARA*.

Reverse Side of Articles of Agreement

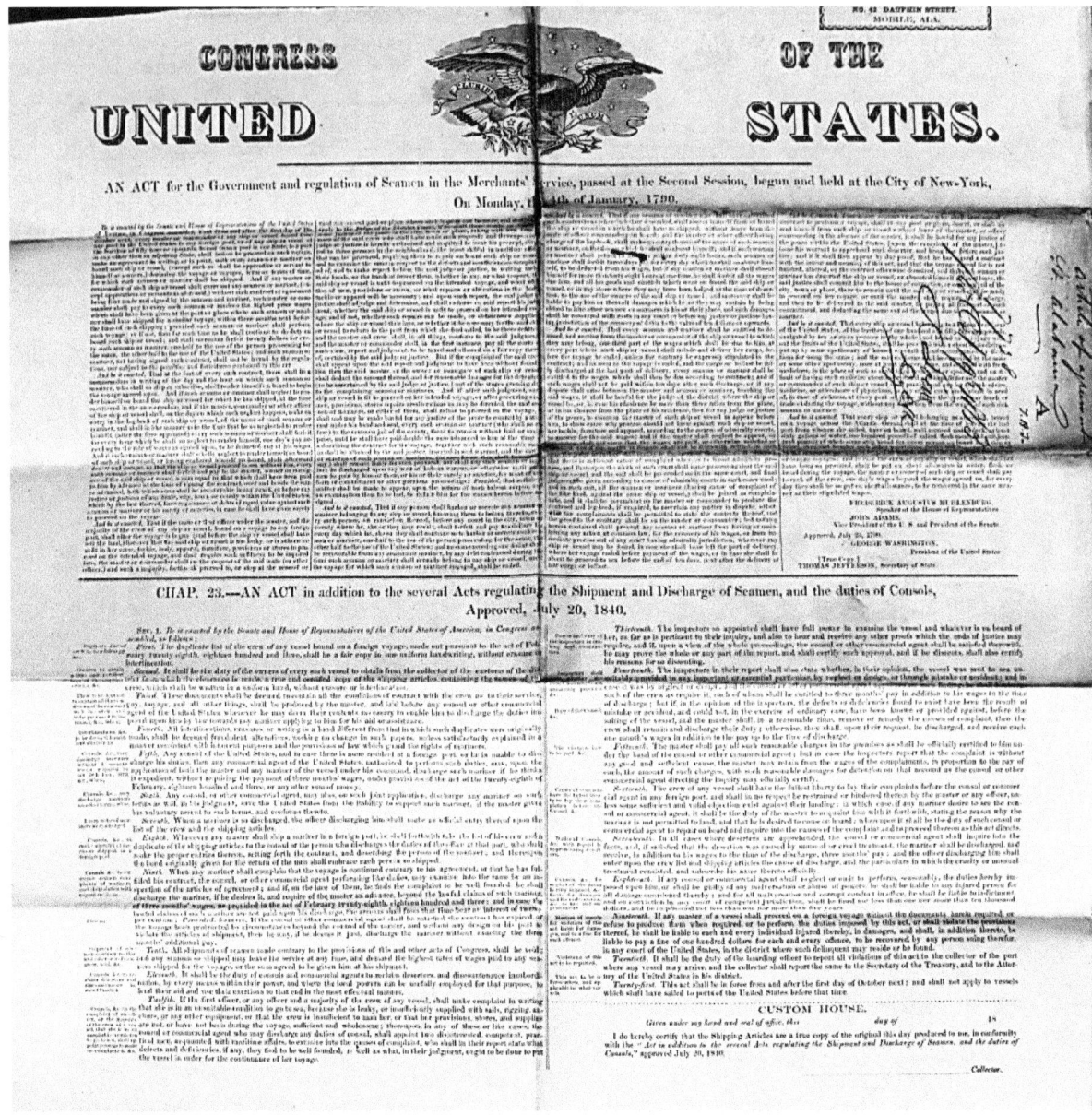

Figure 71. Reverse side of articles of agreement quoting the laws governing the terms of employment for the crew. *NARA*.

Supplementary Articles of Agreement

There was some changeover in the crew while the ship was in Havana on the previous voyage in late July and early August 1863 (Figures 72-73 and 75). The new crewmen were officially entered by recording in the following document. Note the reuse of a preexisting U.S. form by crossing out "United" and writing in "Confederate."

It is agreed between the Master, Seamen, or Mariners of the Steamer *Alabama* whereof J. Hopkins is at present Master, or whoever else shall go for Master, now bound from the Port of Havana to Nassau.

Date of Entry	Names	Stations	Birth Place	Age	Height Ft. In.		Wages/ Month	Advnc. Wages	(Signature)
1863							Receipt for wages pd. in Havana. Aug. 28, '63		
Aug. 10	Basilis (x)[157] John	Fireman	Greece	25	5	9	$ 50	$ 50	Basilis John
"	John (x) Billes	"	France	46	5	6	50	50	John Billes
"	John (x) Emmet	"	Ireland	49	5	9	50	50	John Emmet
"	Patrick (x) Featherston	"	Ireland	29	5	6	50	50	Pat Featherston
"	Timothy (x) Spelman	"	Mobile	33	5	6	50	50	T. (x) Spelman
"	Joseph (x) Silva	"	Portugal	40	5	7	50	50	Joseph (x) Silva

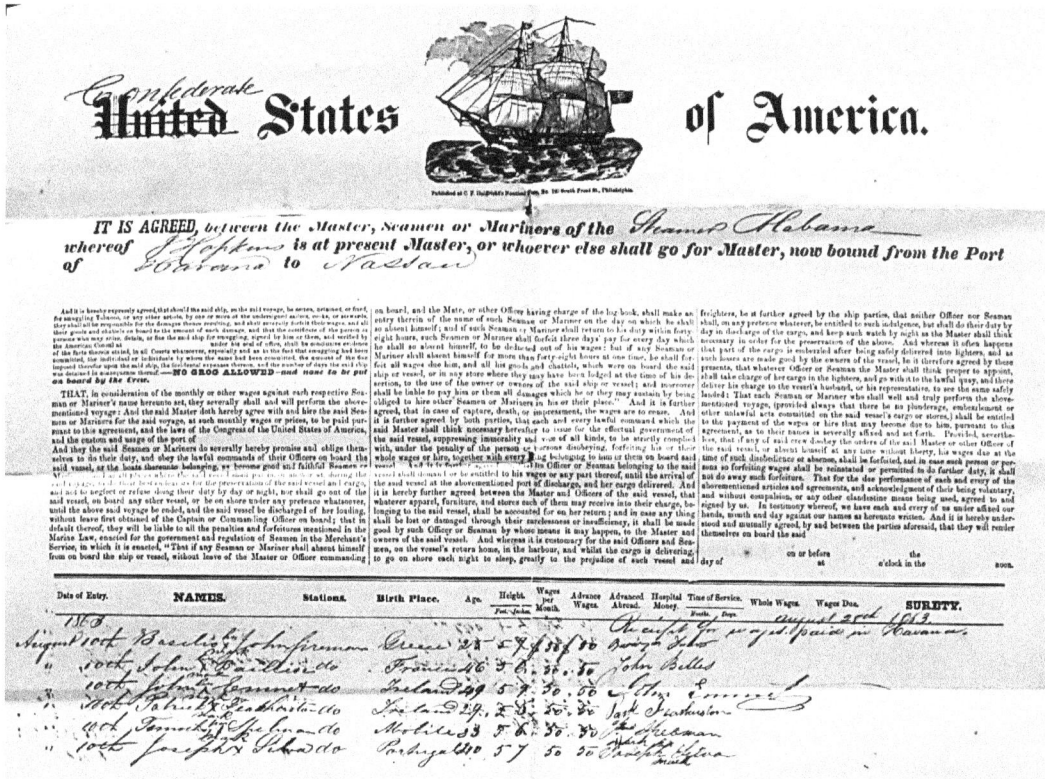

Figure 72. Crew agreement for new crewmen recruited for the *Alabama* while she was in Havana unloading cotton and loading the return cargo of general merchandise. Dated 10 August 1863. This date indicated that the document related to the previous voyage. *NARA*.

---

[157] (x) His mark. Each man in the "Name" column of this table made his mark rather than a signature, in this case not indicating illiteracy but instead that a clerk filled in this column. Under the "Advanced Aboard" column are what appear to be actual signatures with two exceptions as shown by (x).

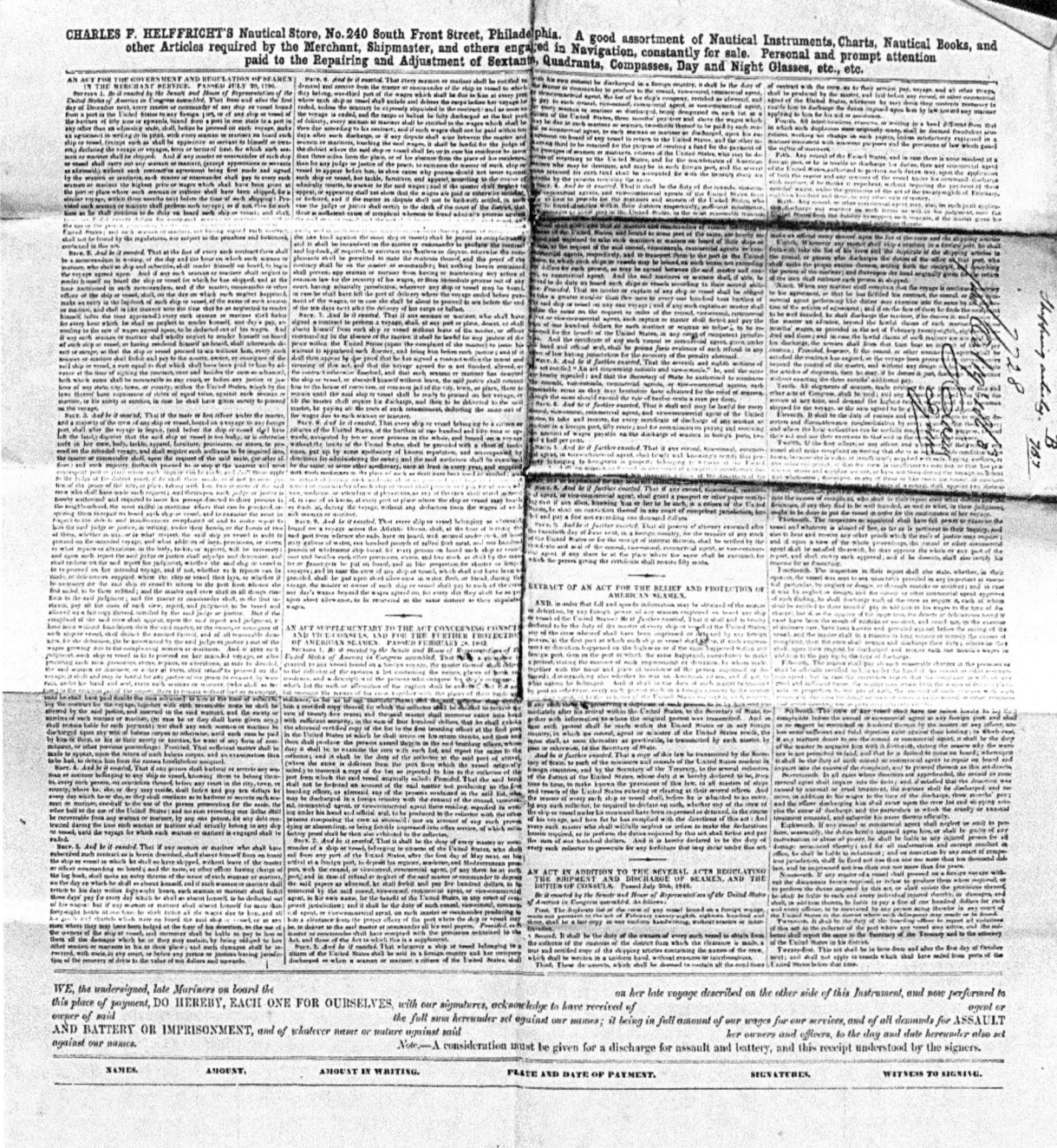

Figure 73. Reverse side of the supplementary articles of agreement for the new crewmen. The verbiage of this form constituted the terms and conditions of service. *NARA*.

Customs House Clearance, Mobile

The following document was the Mobile Customs House clearance to sail from Mobile specifying the outbound cargo of cotton and the tonnage of the *Alabama*, 510 tons (Figure 74).

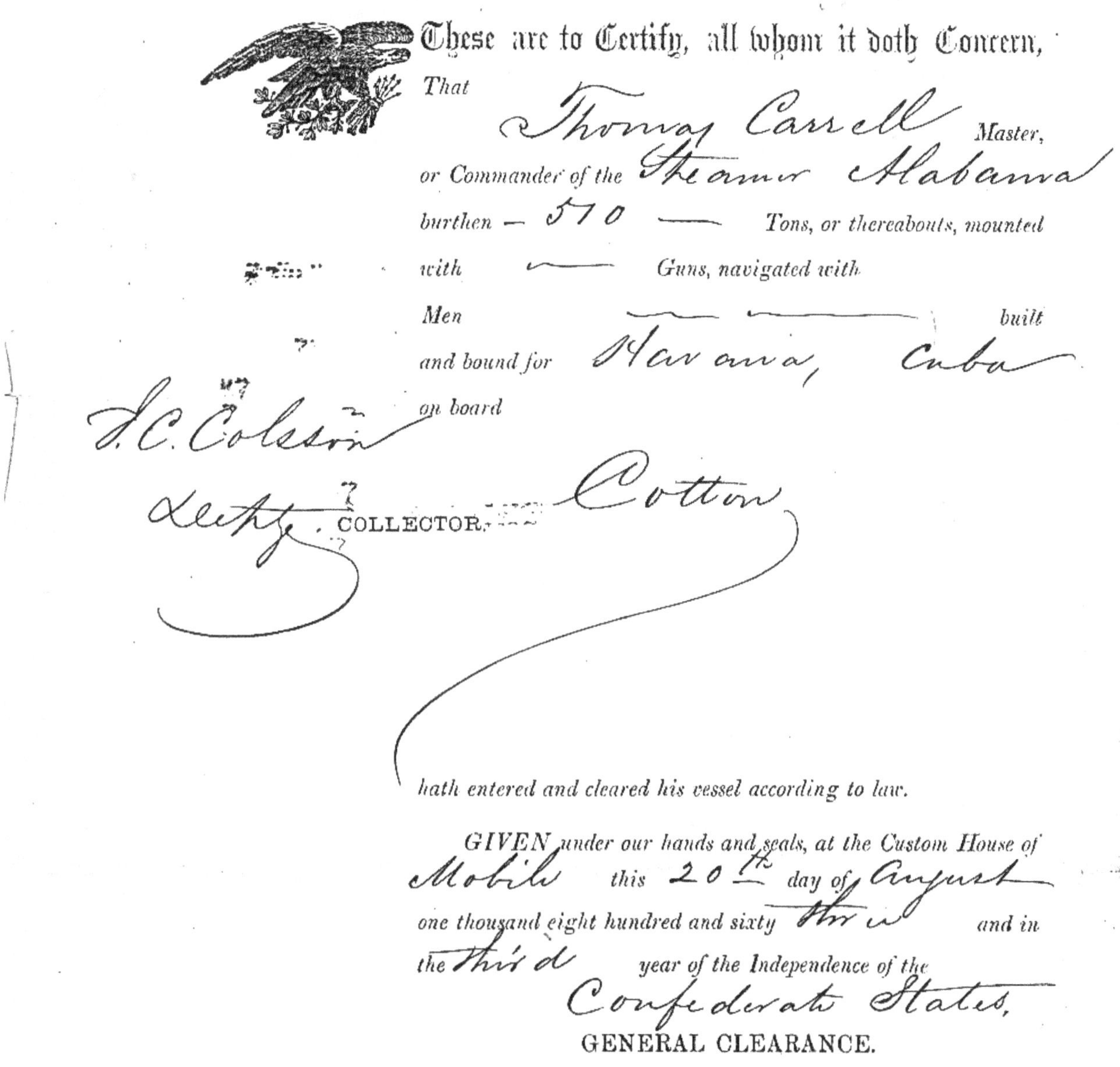

These are to Certify, all whom it doth Concern, That *Thomas Carrell* Master, or Commander of the *Steamer Alabama* burthen — *510* — Tons, or thereabouts, mounted with — Guns, navigated with — Men — built and bound for *Havana, Cuba* on board

J. C. Colson
Deputy COLLECTOR.

Cotton

hath entered and cleared his vessel according to law.

GIVEN under our hands and seals, at the Custom House of *Mobile* this *20th* day of *August* one thousand eight hundred and sixty *three* and in the *third* year of the Independence of the *Confederate States,*

GENERAL CLEARANCE.

Figure 74. Customs House clearance to sail from Mobile bound for Havana with a cargo of cotton. The *Alabama* was said to be about 510 tons burthen. Marked "E" on the reverse side. (Legible, not transcribed.) Dated 20 Aug. 1863. *NARA*.

Customs House Clearance, Havana

The following document related to the *Alabama's* prior voyage and dated 8 Aug. 1863 in Havana and likely was the clearance to sail from Havana (Figure 75).

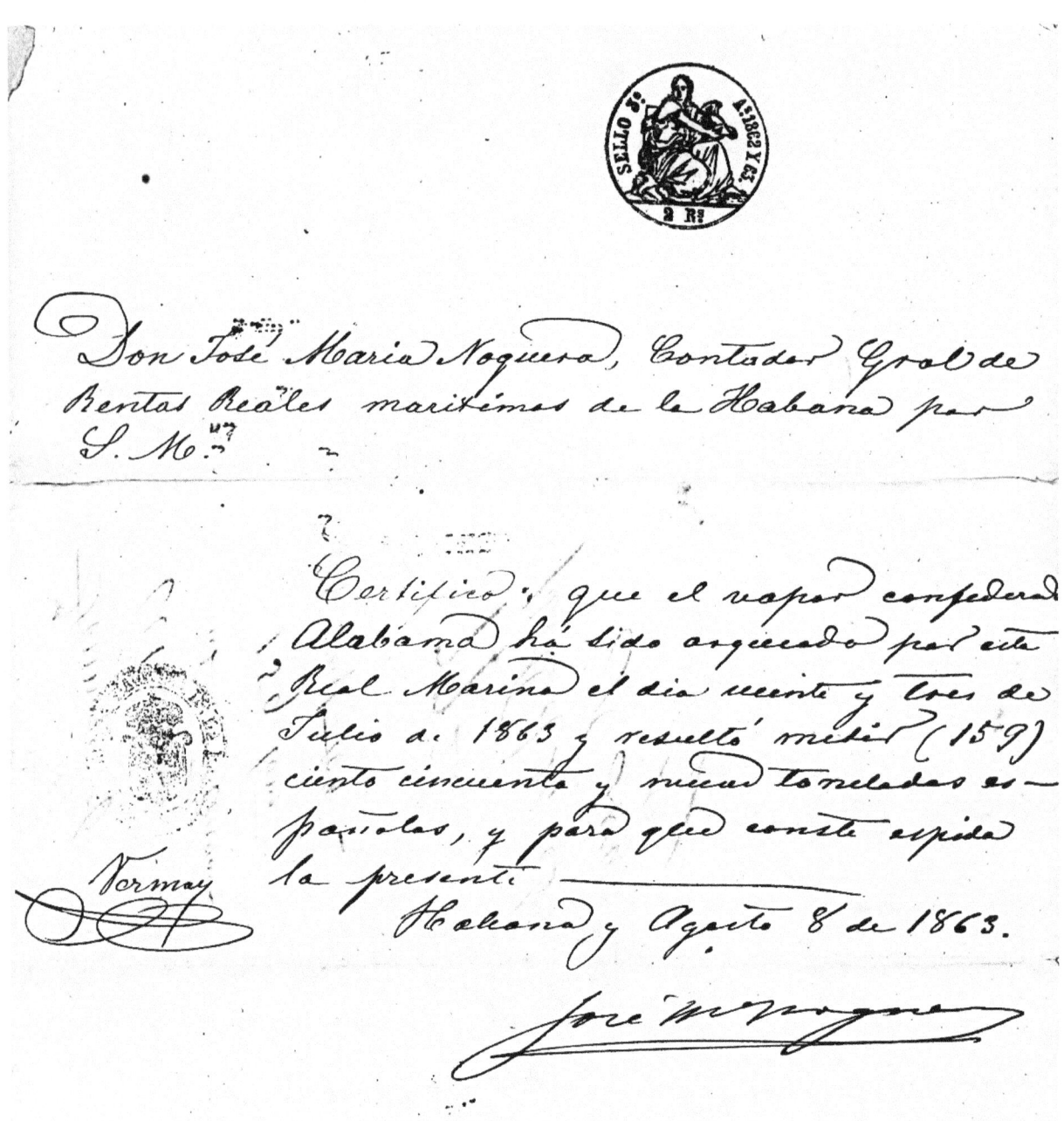

Figure 75. A customs document issued to the *Alabama* in Havana and relating to a prior voyage from Mobile. Perhaps this represented a clearance to sail from Havana. The document included the tonnage of the vessel, and about that there was something odd. The Spanish official wrote that the *Alabama* measured 159 *toneladas* (tons), a figure greatly at variance with the 510 tons on her Confederate permanent register. (Legible, not transcribed.) Dated 23 July and 8 Aug. 1863. *NARA.*

Letters

Of the packets of letters referenced in the court documents bellow, only a few scraps like the following remained in the file (Figure 76). This was a big disappointment as correspondence from other prize case files provided wonderful personal insights regarding daily life. The following notes were useful for the customs entry at Mobile, but should not be described as letters.

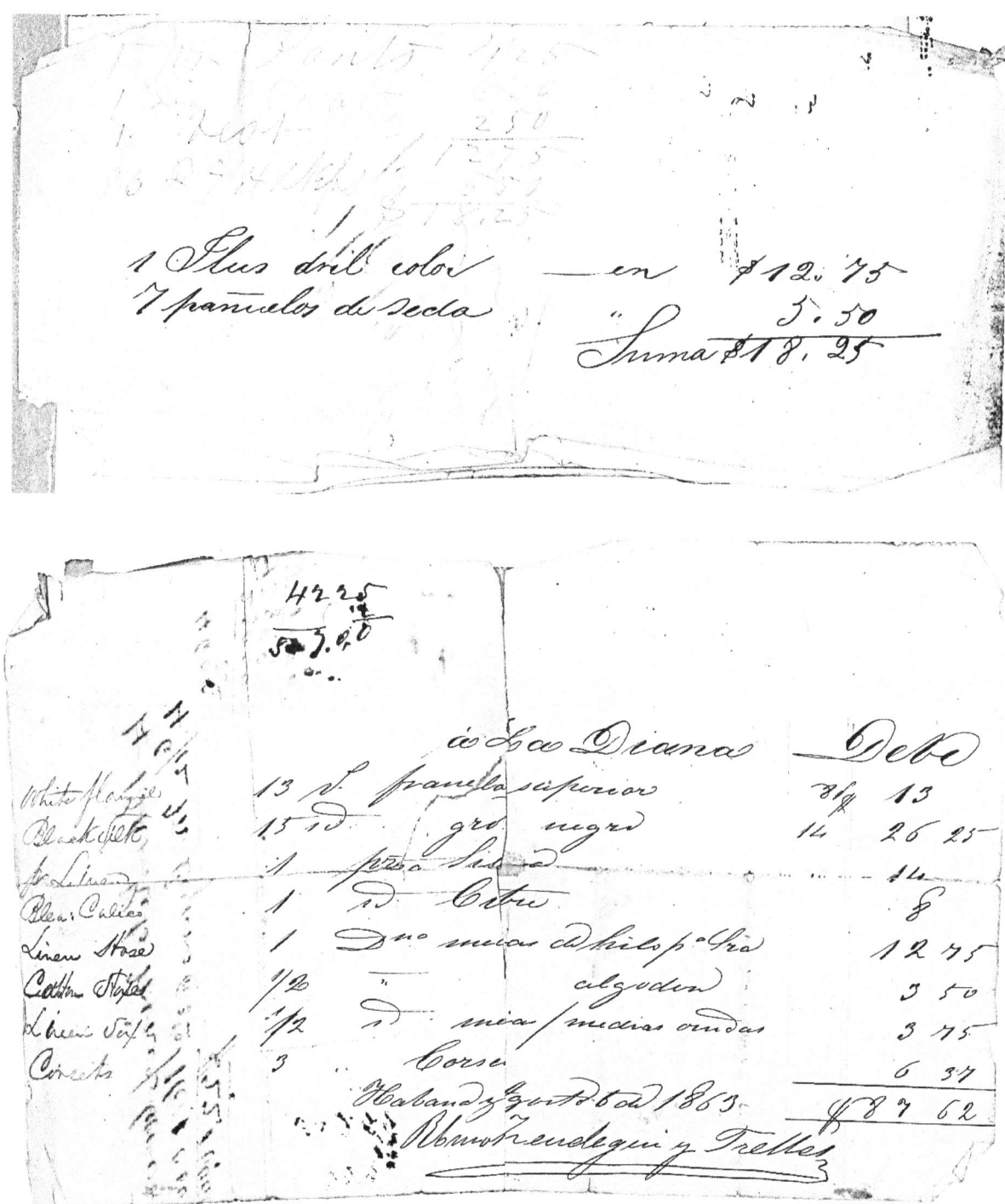

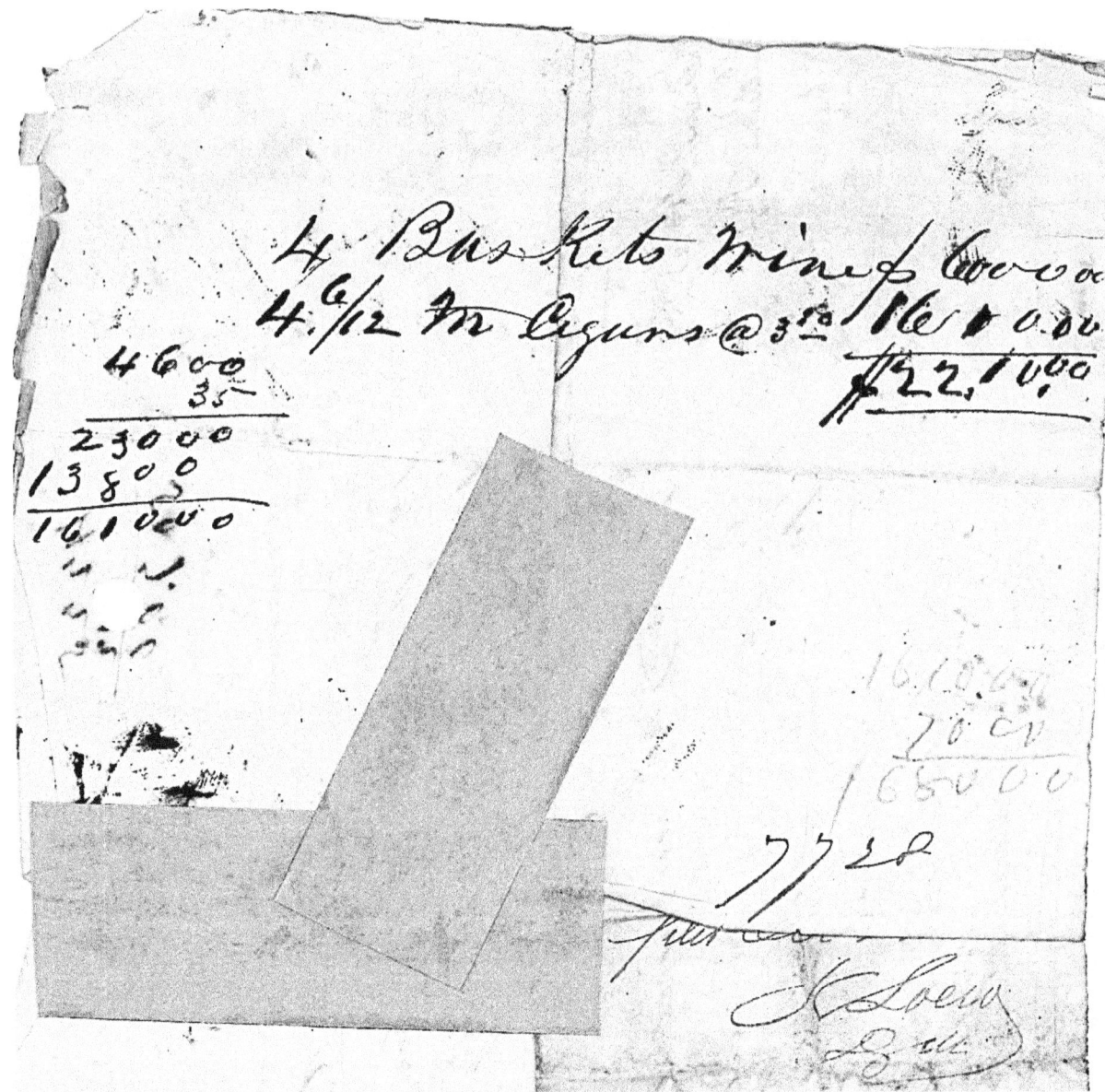

Figure 76. Informal vouchers for the purchase of small shipments. For example, one was for clothing valued at $18.25 and consisted of a pair of pants, a coat and vest, plus 36 handkerchiefs. Another listed four baskets of wine and 4½ thousand cigars. (This page and the prior page.) *NARA*.

The Legal Documents of the *Alabama's* Prize Litigation

The Monition or Libel and Admiralty Warrant

See below for the two court documents that initiated the prize case against the *Alabama* (Figures 77-80). Fortunately, the U.S. attorney's name, Rupert Waples, was typeset in some of the court forms below or the signature might easily be misread as Maples, an example of difficult, ornate capital letters often seen in the files.

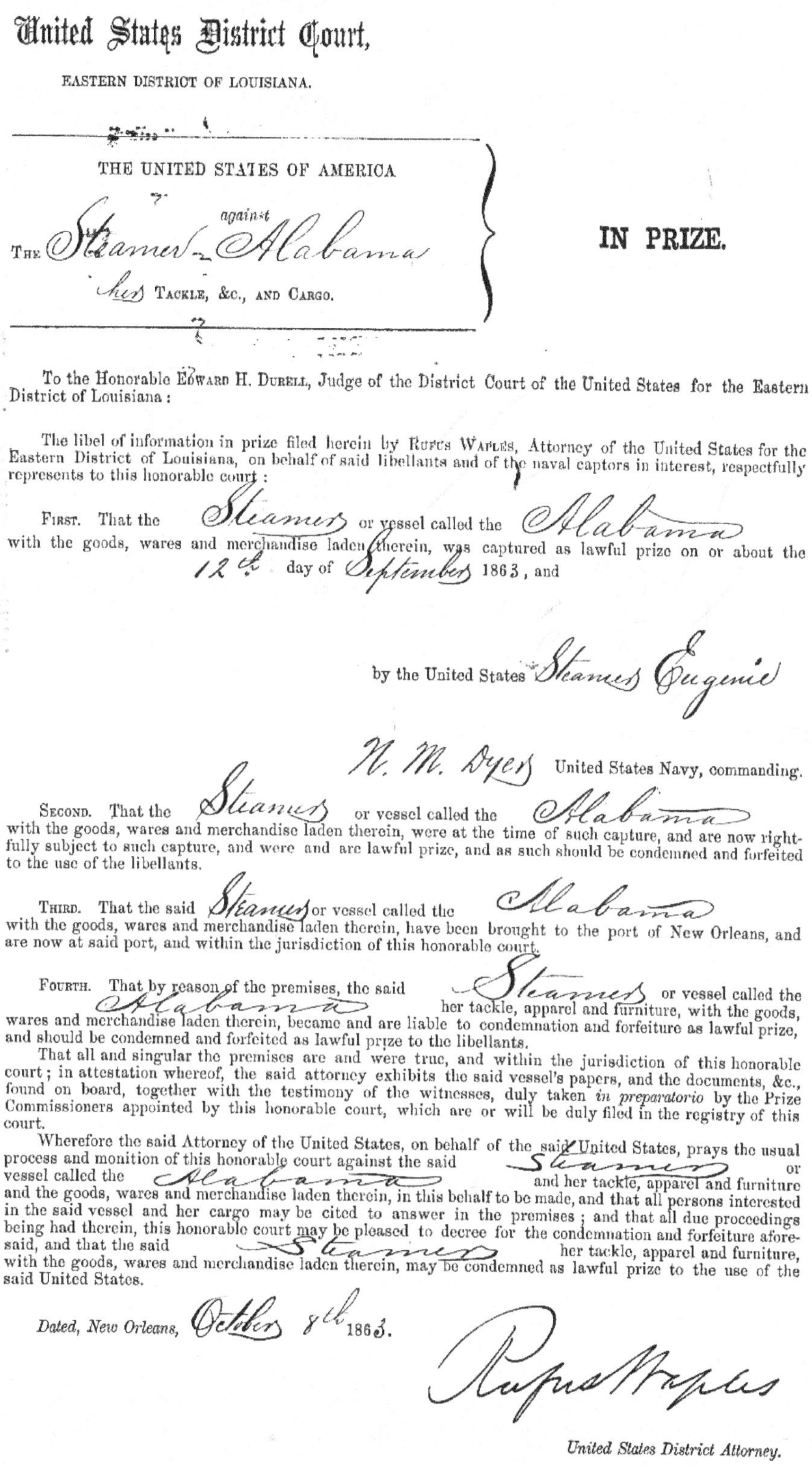

United States District Court,

EASTERN DISTRICT OF LOUISIANA.

THE UNITED STATES OF AMERICA

against

THE Steamer Alabama

her TACKLE, &c., AND CARGO.

IN PRIZE.

To the Honorable EDWARD H. DURELL, Judge of the District Court of the United States for the Eastern District of Louisiana:

The libel of information in prize filed herein by RUFUS WAPLES, Attorney of the United States for the Eastern District of Louisiana, on behalf of said libellants and of the naval captors in interest, respectfully represents to this honorable court:

FIRST. That the Steamer or vessel called the Alabama with the goods, wares and merchandise laden therein, was captured as lawful prize on or about the 12th day of September 1863, and

by the United States Steamer Eugenie

N. M. Dyer United States Navy, commanding.

SECOND. That the Steamer or vessel called the Alabama with the goods, wares and merchandise laden therein, were at the time of such capture, and are now rightfully subject to such capture, and were and are lawful prize, and as such should be condemned and forfeited to the use of the libellants.

THIRD. That the said Steamer or vessel called the Alabama with the goods, wares and merchandise laden therein, have been brought to the port of New Orleans, and are now at said port, and within the jurisdiction of this honorable court.

FOURTH. That by reason of the premises, the said Steamer or vessel called the Alabama her tackle, apparel and furniture, with the goods, wares and merchandise laden therein, became and are liable to condemnation and forfeiture as lawful prize, and should be condemned and forfeited as lawful prize to the libellants.

That all and singular the premises are and were true, and within the jurisdiction of this honorable court; in attestation whereof, the said attorney exhibits the said vessel's papers, and the documents, &c., found on board, together with the testimony of the witnesses, duly taken *in preparatorio* by the Prize Commissioners appointed by this honorable court, which are or will be duly filed in the registry of this court.

Wherefore the said Attorney of the United States, on behalf of the said United States, prays the usual process and monition of this honorable court against the said Steamer or vessel called the Alabama and her tackle, apparel and furniture and the goods, wares and merchandise laden therein, in this behalf to be made, and that all persons interested in the said vessel and her cargo may be cited to answer in the premises; and that all due proceedings being had therein, this honorable court may be pleased to decree for the condemnation and forfeiture aforesaid, and that the said Steamer her tackle, apparel and furniture, with the goods, wares and merchandise laden therein, may be condemned as lawful prize to the use of the said United States.

Dated, New Orleans, October 8th 1863.

Rufus Waples

United States District Attorney.

Figure 77. The monition or libel form asserting that the Str. *Alabama* was a lawful prize of the U.S. Str. *Eugenie.* (Sufficiently legible so not transcribed.) Dated 12 Sept. 1863 for the capture and dated and filed with the court on 8 Oct. 1863. *NARA.*

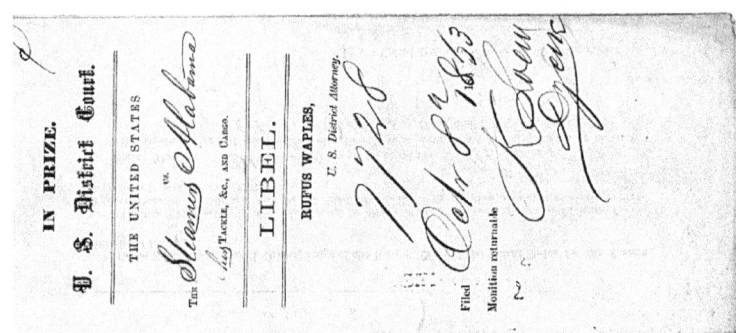

Figure 78. Reverse side of the above monition or libel providing an abstract for filing purposes. *NARA*.

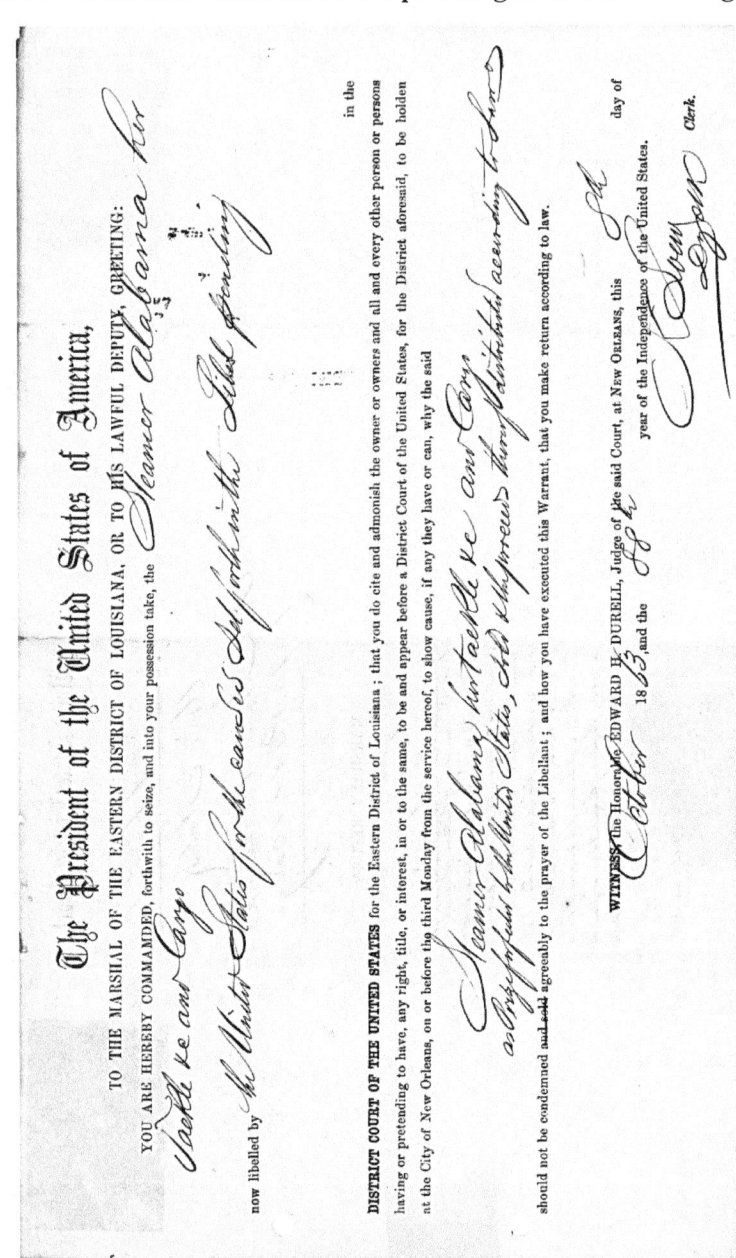

Figure 79. Admiralty warrant for the arrest or seizure of the Str. *Alabama* and contents. (Legible so not transcribed.) Dated 8 Oct. 1863. *NARA*.

Marshal's Return

No. 7728
United States
vs.
The Steamer *Alabama*
Her Tackle &c. & cargo
In Prize

Marshal's Return.

Received 9th October 1863, and on the same day seized & took into my possession the within described property. Served copies of the writ & Libel on N. M. Dyer, Prize Master, in person. Took an inventory of the same which is hereto annexed, appointed Seaborn Williams & Wm. Devlin as keepers of the same, and published Monition in the *N. O. Times* a daily newspaper printed & published in New Orleans on the 12th, 19th, 24th October 1863 returnable 26 Oct. 1863, and on the 12th inst. pursuant to an order of Court unloaded the cargo of the said steamer and placed the same in the U.S. Marshal's Warehouse No. 40 Old Levee St.

(signed) D. Christie
Dep. U.S. Marshal

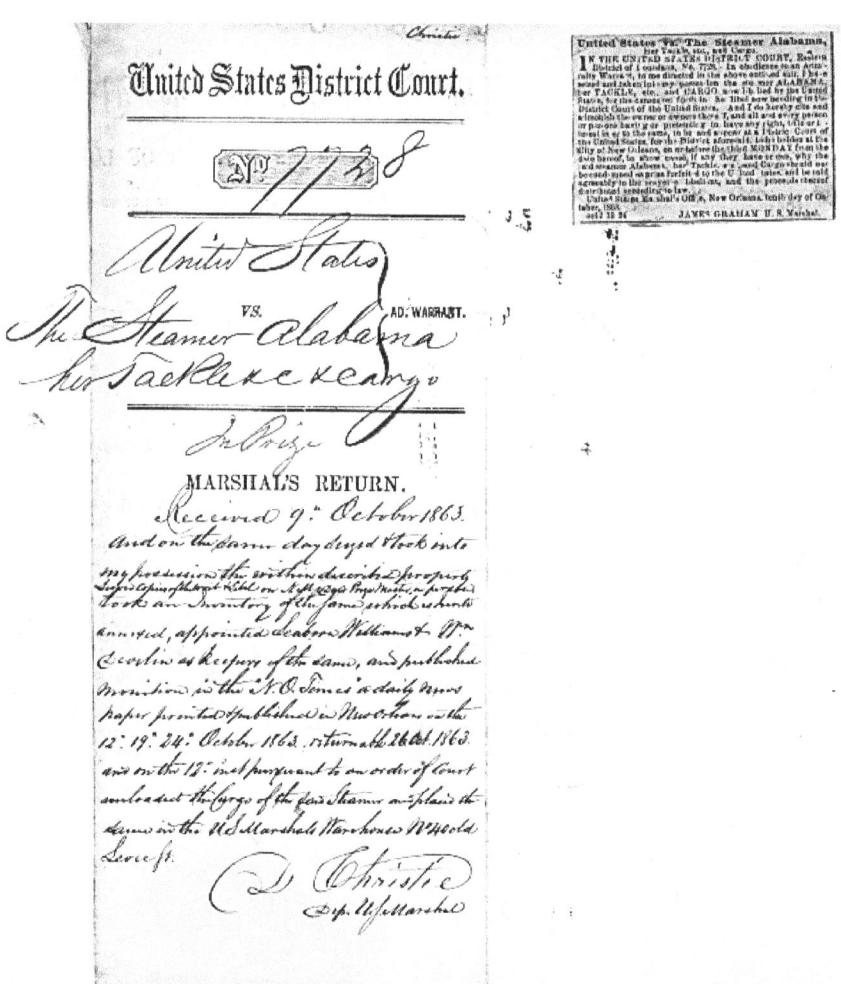

Figure 80. Reverse side of the admiralty warrant provided a form for the marshal's report to the court that he had published the initiation of the admiralty case against the *Alabama*. Dated 9 Oct. 1863. *NARA*.

Court Order to Unload the Str. *Alabama*

The document below was the court order to unload the Str. *Alabama* as moved by Rupert Waples, U.S. District Attorney, in response to a statement of T. B. Thorpe, U.S. Prize Commissioner, who said that the ship should be unloaded (Figure 81).

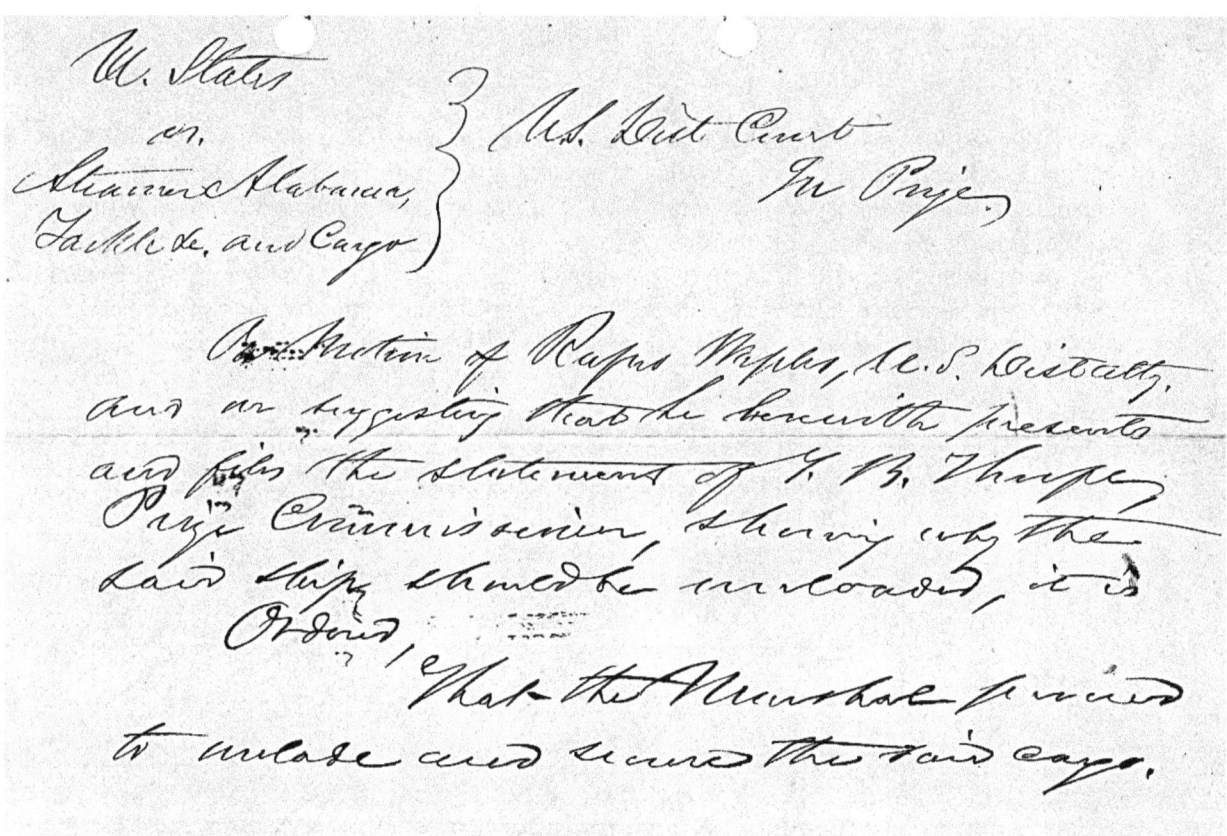

Figure 81. Order to unload the *Alabama*. (Legible, not transcribed.) *NARA*.

Appointment of Attorney for the Captors

The court appointed A. M. Buchanan as the attorney to represent the captors in the proceedings against the Str. *Alabama*, her tackle and cargo (Figure 82). The Confederates often used preexisting U.S. government printed forms, crossing out "United" and inserting a hand written "Confederate." In this case, note a Confederate form repurposed in the same way. Given the relatively quick recapture after secession, New Orleans might be one of only a few places where such reuse might have happened in 1863.

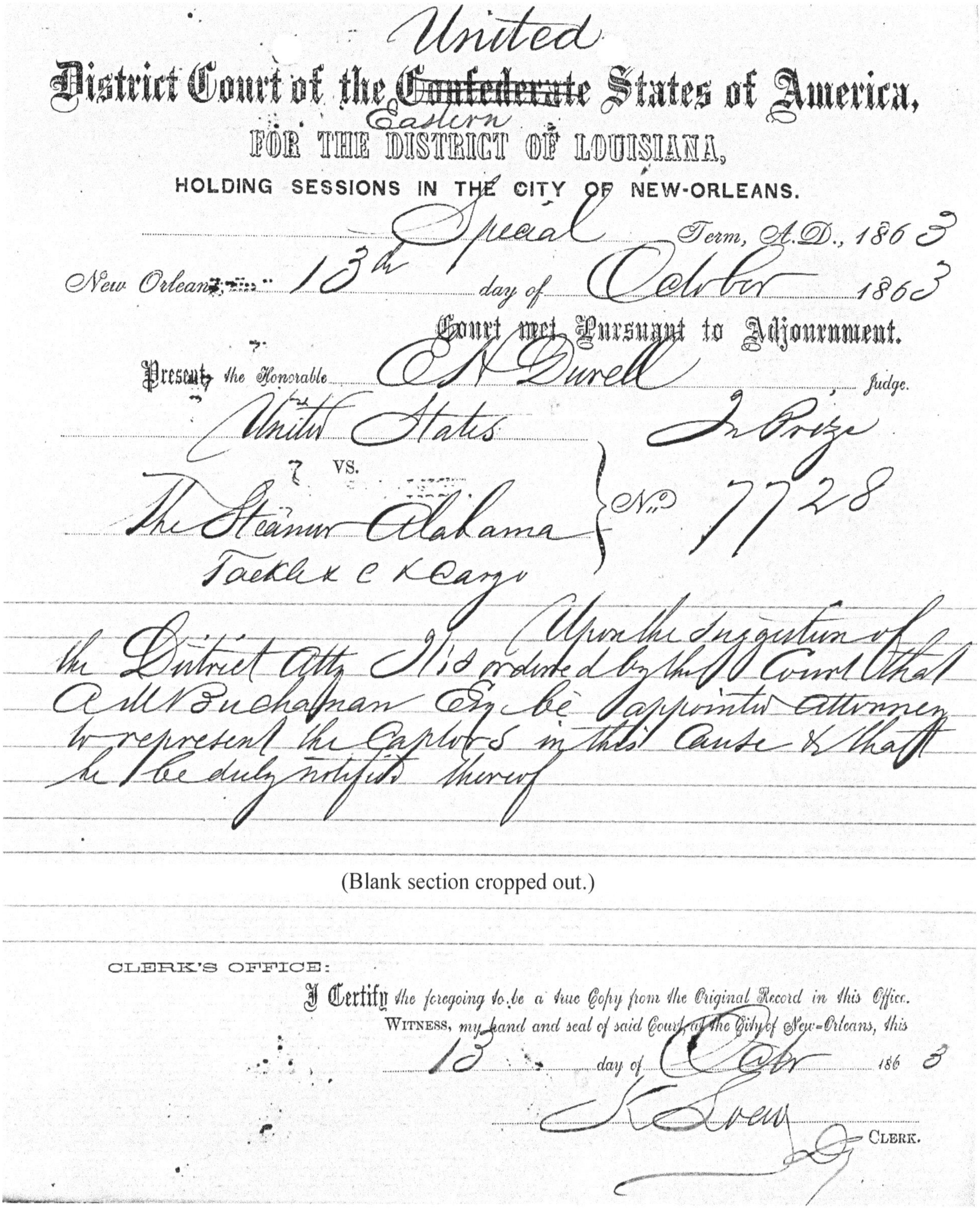

United
District Court of the ~~Confederate~~ States of America,
Eastern
FOR THE DISTRICT OF LOUISIANA,
HOLDING SESSIONS IN THE CITY OF NEW-ORLEANS.

Special Term, A.D., 1863

New Orleans, 13th day of October 1863

Court met Pursuant to Adjournment.

Present, the Honorable E H Durell Judge.

United States
vs.
The Steamer Alabama
Tackle &c & Cargo

In Prize
No. 7728

Upon the suggestion of the District Atty It is ordered by the Court that A M Buchanan Esq be appointed attorney to represent the Captors in this Cause & that he be duly notified thereof

(Blank section cropped out.)

CLERK'S OFFICE:

I Certify the foregoing to be a true Copy from the Original Record in this Office.

Witness, my hand and seal of said Court, at the City of New-Orleans, this 13 day of Octr 1863

L Loew
Clerk.

Figure 82. Appointment of A. M. Buchanan to represent the captors in the prize case. (Legible, not transcribed.) Dated 13 Oct. 1863. *NARA*.

Depositions of the Captors

Deposition of William Richardson, Acting Master, U.S.S. *San Jacinto*

The following transcript presented the deposition regarding the pursuit and capture of the *Alabama* given by William Richardson of the U.S.S. *San Jacinto*, one of several Union navy ships involved (Figure 83). The capture took place in the Chandeluer Islands about 10 miles from Ship Island (Figure 88).

> In the matter of the
> <u>Str. *Alabama*</u>
> And Cargo, a Prize
>
> District Court of the United States
> For the Eastern District of Louisiana.

Deposition of Wm. Richardson, Acting Master U.S.N. a witness produced, sworn, and examined *in preparatorio* on the ninth day of October in the year eighteen hundred and sixty three at the office of T. B. Thorpe, Special Prize Master on the standing interrogatories established by the District Court of the United States for the Eastern District of Louisiana; the said witness having been produced for the purpose of such examination, by Rufus Waples, Esq., U.S. District Attorney, in behalf of the captors of a certain vessel called the _____.

To the first interrogatory the witness answers: that about 10 A.M., Septr. 11, 1863 in Lat. 28° 30' Long. 87° 40' on board of the U.S. Steam Sloop *San Jacinto* a sail was reported from the masthead on the port quarter. We were then heading about NNW. In a few minutes discovered a smoke about 12 miles dist. Put the helm starboard & steered for her. She soon altered her course &steered more to the westward. Kept her about 2 points on the port bow. Chased her all day until sunset. She then bore about W by S & steering that course. The later 2 hours we gained on her about 1 mile per hour. At dark we were about 7½ miles dist. from her but lost sight of her. At 7 P.M. altered our course to the eastward all night.

At 6 A.M. in Lat. About 29° 00' W. Long. 88° 00' heading about N by W saw a smoke on the port bow. Headed for her. Soon after found she had altered her course to the westward. Kept her about 4 points on the starboard bow steering her with an azimuth compass all the while chasing her. At about 8 A.M. she altered her course southerly & steered to the north & eastward still chasing her & keeping her about 2 points on each bow. She apparently wanted to cross our bow. At about 11.30 A.M. she altered the course again & steered about WSW for about 1 hour, then steered to the northward again. Soon afterward steered to the westward again.

At this time a sail was discovered coming from the eastward which proved to be the Steamer *Tennessee* under sail and steam. About 1 P.M. made the south end of Chandeleur Island. The steamer then altered her course & steered to the northward. About 2 P.M. made Chandeleur Light House. At this time discovered a smoke of a steamer coming from the land. On sighting the *Tennessee* she kept in shore. The *Tennessee* passing her outside & steered for the steamer we were in chase of (which proved to be the *Alabama* of Mobile from Havana). At about 2 P.M. we were abreast of the Chandeleur Light House. The *Alabama* went inside & steered up for Breton Island passage. We fired 2 shots at her previous to her getting inside. One shot from the rifle gun went over her quarter about 11 ft. above the steamer. So the Capt. of the *Alabama* told me afterwards.

We chased her until the Light House bore ESE. Altered our course & steered out again. Signalized to the *Tennessee* to go up outside the Chandeleurs to keep the *Alabama* from escaping up Breton Island passage. The *Tennessee* not able to answer our signals, we came out again & steered up for the south passage. The *Tennessee* had gone inside the Light House. She turned back alas. *San Jacinto* turned back again & went inside and came to an anchor inside Chandeleur Light House bearing E by S in 2 fath. water.

At this time a steamer was seen coming from Ship Island. At 5 P.M. being abreast of us I knew her to be the *Eugenie*. She steered for the *Alabama*. She being in shoal water, the *Eugenie* appeared to gain on her. At 5.40 P.M. the *Alabama* appeared to be in difficulties. Altered her course a little more to the westward. At 6 P.M. she appeared to stop, not altering her bearings. At about 6.30 the *Eugenie* was up with the *Ala*. Was certain then she was aground. She bore SSE by compass, distinctly seen from the poop deck of the *San Jacinto* about 9½ or 10 miles dist. Preparations were made on board of the *San Jacinto* as soon as we came to an anchor to send our boats. The 1st cutter was the one only manned as the *Ala*. was aground & the *Eugenie* close to her.

At 6.25 the 1st cutter with howitzer, 10 men with their side arms, 3rd Asst. Engineer T. M. Emanuel, and myself in charge of Lieut. Comdg. Quackenbush left the ship & proceeded for the *Alabama*. Steered SSE when about 1½ mile from the *San Jacinto* found the compass too light, returned back and got one to ?? & proceeded for the *Ala*. again. At about 10.30 went alongside the *Eugenie*. She went ahead of the *Ala*. as a guide. Shoved off & went on board. Remained in the boat & kept the men in the boat. In a few minutes Lieut. Quackenbush told me to come on board. I went on the *Alabama's* deck toward the wheel house. Lieut. Comdg. Quackenbush introduced me to Lieut. Comdg. Green who came out in the Steamer *Eugenie*. He appeared to have charge as he was giving orders. Lieut. Quackenbush told him he was sent by Capt. Chandler to take charge of the prize Steamer *Ala*. & ask him if he would give her up to him. No he would not. He was sent here by Capt. Mayo from Ship Island. He (Green) did not care who sent ~~you~~ him (Quackenbush). He told Lieut. Quackenbush to slack the *San Jacinto's* boat astern; leave 3 men in her. He wanted to go ahead.

I left Mr. Green & Mr. Quackenbush having high words with each other, and see the boat secured. 7 men remained on board & men were put on duty in the fire room by the request of 3rd Asst. Engr. T. M. Emanuel who was on duty then in the engine room working the steamer to the anchorage. After this I came on deck towards the wheel house again found Lieut. Green & Lieut. Quackenbush in the Capt's. cabin. About 15 minutes before we came to anchor Lieut. Green ask(ed) me to tell the Engineer we were coming to anchor in a few minutes. I did so.

I came on deck again. I was ask(ed) to see the anchor all clear on the upper deck. About 11.40 P.M. saw the *San Jacinto* & Steamer *Tennessee* at anchor abreast of the Light House. Came to anchor on the starboard side of the *San Jacinto*. Lieut. Green then went on board of the *San Jacinto*. Lieut. Quackenbush went on board & told me to remain on board with 2 men. He soon returned again with orders from Capt. Chandler to take charge & not allow anything to be taken away. I had taken a globe lamp before this and went forward on the main deck & found trunks had been broken open & the contents gone. Some valises also were lying about the deck open. Boxes & demijohns were broken and lying about the deck. I immediately reported it to Mr. Quackenbush & advised him to have a guard of Marines.

Capt. Chandler sent me word they were coming. He soon came on board himself. I showed him the state of the things about (the) deck. He placed a guard at different parts of the ship & gave me particular instructions not to allow the Steamer *Ala's*. crew & passengers to take anything except their luggage and see them get in the boat with their luggage. I did so. Lieut. Smalley of the Marines on board of the *San Jacinto* was in charge of the Marines on board of the *Ala*. In the meantime the officer's & crew of the Steamers *Eugenie* & *Tennessee* were gone on board of their vessels. This was Sept. 13th about 2.30 A.M. 'till about 3 A.M.

Asst. Engr. Emanuel went on board & soon returned with 2nd Asst. Engr. H. S. David & Capt. Chandler. At day light getting ready to get under way. Ensign N. M. Dyer, in command of the *Eugenie* came on board for Lieut. Green to take him to Ship Island. The *Tennessee* had got under way for Ship Island. Ensign N. M. Dyer with Lieut. Green went on board the *Eugenie* & got under way & went to Ship Island.

About 9 A.M. the *San Jacinto* got under way. At 9.20 *Ala.* got under way, went outside the Light House steering SE. At 10.15 stopped for the *San Jacinto* to come up with us. Capt. Chandler told me he would send 2 hawsers on board & take her in tow. We were then bound for Key West. In the meantime the Steamer *Tennessee* came out to us & spoke the *San Jacinto*, sent a boat on board, & soon afterward went for Ship Island. *San Jacinto* came close to the *Ala*. Sent 2 hawsers, made them fast, went ahead, & towed her in to Ship Island.

About 5 P.M. came to anchor at Ship Island still hanging to the stern of the *San Jacinto*. At 6.15 P.M. I was ordered to let go the hawsers & come to anchor. I veered to 25 fath. chain. At about 7 P.M. Acting Master G. E. Nelson of the *Tennessee* came on board. He told me he was sent by the Capt. of the *Tennessee* (and) brought the order of Capt. Mayo. I turned the *Alabama* over to him. At about 9 Ensign N. M. Dyer, Capt. of the *Eugenie* came on board & took charge, he being the first on board from the *Eugenie* while the *Alabama* was ashore.

I then received an order from Capt. Mayo to get my clothes & have my accts. made out. I was to go to New Orleans in the *Alabama*. At about midnight we got under way & proceeded for New Orleans in tow of the *Tennessee*. About 10 A.M. Sept. 14th arrived at Pass l'Outre. At 5 P.M. came to anchor at Quarantine ???.

Octr. 9th 1863
    Wm. Richardson
        Acting Master
           U.S.N.
of the U.S. Steamer *San Jacinto*

Sworn to and signed before me T. B. Thorpe, Special Prize Commissioner this 9th day of October 1863.
    New Orleans, Louisiana
        T. B. Thorpe
    Special Prize Commissioner

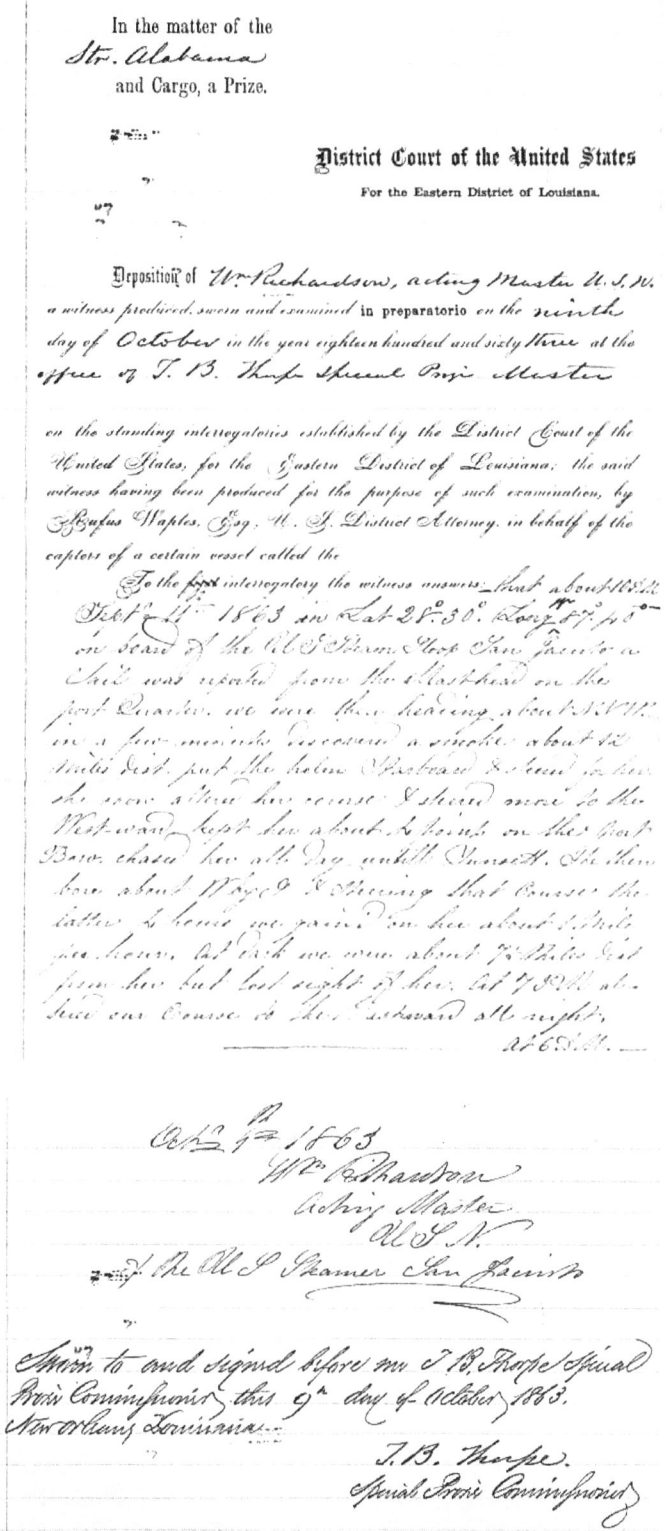

Figure 83. Deposition of Acting Master William Richardson of the U.S. Str. *San Jacinto*, the ship that first sighted and chased the *Alabama*. Shown here page 1 of 7 pages (above) and the bottom of page 7 showing the signature block. Dated 9 Oct. 1863. **NARA**.

Deposition of Prize Master N. M. Dyer, Acting Ensign and Captain, U.S.S. *Eugenie*

After extensive jockeying among the Union navy officers on site with the captured *Alabama*, a mixed prize crew with Dyer as prize master took the *Alabama* to New Orleans for adjudication (Figure 84). See below for the deposition of Captain Dyer.

The prize steamer *Alabama* was captured about 6.05 P.M. Saturday 12th September 10 miles due south from Chandeleur Light House or there abouts by the U.S.S. *Eugenie* under my command. Immediately upon coming up with her, I boarded her and transferred her crew and passengers to the *Eugenie*. At the time the *Eugenie* started for the prize, she was lying at Ship Island coaling ship. The prize was first seen from Ship Island about 3.30 P.M. at which time I received orders from Lt. Comdr. Mayo of the U.S. Gunboat *Kanawha* to get under weigh and give chase. The order being sent to me through an officer of the U.S. Gunboat *Kanawha* direct from Lt. Comdr. Mayo, Commanding the *Kanawha*, and senior officer present.

A few minutes before I started in chase, Lt. Comdr. Green of the U.S. Ship *Vincennes* came on board the *Eugenie* having received permission of Lt. Comdr. Mayo of the U.S. Gunboat *Kanawha* to go out with me, as he expressed it, to see the fun. After I had gone on board the prize and was mustering her crew and passengers, Lt. Comdr. Green of the *Vincennes* came on board the prize and by virtue of his superior rank assumed command of her. Though up to this time he had been merely a passenger on board the *Eugenie* assuming no command having received no orders to do so, but had simply been granted permission to come out in her as before mentioned "to see the fun." But by virtue of his superior rank, he did assume charge of the prize in the manner and at the time above mentioned.

The prize was then anchored, and I at once made preparations to tow her into deeper water, leaving the prize and going on board the *Eugenie* for that purpose at about 7 P.M. While on board the prize, I noticed a large amount of freight on deck consisting of cases, boxes, hams, provisions, &c. which I have since been informed by the Capt. & Super Cargo of the captured vessel was a portion of her cargo including also some private property of the crew and passengers of the prize. The value of which was supposed by them to be not less than $15,000 (Fifteen Thousand dollars) at Havana.

At about 7.30 P.M., a boat's crew from the U.S.S. *Tennessee* arrived in charge of Act. Master Nelson of that ship. This boat's crew and officers went on board the prize at my request and reported to Lt. Comdr. Green of the U.S.S. *Vincennes* and rendered valuable assistance in getting the prize into deeper water.

After nearly 5 hours of constant labor, succeeded in getting the prize finally afloat, and shaped our course for Ship Island, the prize following the *Eugenie*. At about 10.50 P.M. and after both vessels, viz.: the prize and the *Eugenie*, were on their way to Ship Island, I discovered a boat on our port beam. I immediately hailed it to come alongside and stopped our engine to allow it to do so. Upon it's coming alongside, it proved to be an armed boat's crew from the U.S.S. *San Jacinto* under charge of Lt. Comdr. Quackenbush of that ship by whom I was informed that he "came with orders from Lt. Comdr. Chandler, Commanding the *San Jacinto*, to take possession of the prize." I informed him Lt. Comdr. Green of the U.S.S. *Vincennes* was in charge of the prize at present and referred Mr. Quackenbush to him.

Lt. Comdr. Quackenbush with his boat's crew then went alongside of the prize and, after a short delay, both vessels again steamed ahead. Upon arriving near the *San Jacinto* about 12.30 A.M. Sunday 13th September, the prize was hailed from the *San Jacinto* and ordered to come to an anchor. My impression,[158] at the time of the arrival of Lt. Comdr. Quackenbush with his boat's crew alongside of the *Eugenie* and after the delivery of his order from Lt.

---

[158] And here we have what must be a world-champion run-on sentence. The officers writing these reports habitually left out punctuation and capitalization.

Comdr. Chandler, Commanding the U.S.S. *San Jacinto*, to take possession of the prize, was that the order was an arbitrary one because, as the immediate captor of the prize, I considered myself bound to retain possession of her, until with a prize crew from the *Eugenie*, she had been delivered to a competent court for adjudication, or at least until I had taken her to Ship Island and reported the facts to Lt. Comdr. Mayo of the U.S. Gunboat *Kanawha* under whose order I started in chase.

This impression I retained, and under it I went on board the *San Jacinto* immediately upon my arrival near her for the purpose of stating my case personally to Lt. Comdr. Chandler, Com'dg. the *San Jacinto*, and from whom the order originated for Lt. Comdr. Quackenbush of the *San Jacinto* to take possession of the prize. I found Lt. Comdr. Chandler in his berth and stated the circumstances to him. He appeared well disposed but requested me to anchor the *Eugenie* near the *San Jacinto* and that "in the morning he would arrange matters" manifestly feeling willing to discuss the matter then.

I accordingly left the *San Jacinto* under the impression that in the morning matters would be arranged to the satisfaction of all concerned. About 1 A.M., I went on board the prize, and found Act. Master Richardson of the U.S.S. *San Jacinto* in charge. Lt. Comdr. Green had, as I was informed, gone on board the *San Jacinto*. In a few moments, Lt. Comdr. Quackenbush again came on board the prize, and shortly after his arrival gave me a positive verbal order, purporting to come from Lt. Comdr. Chandler, Com'dg. the *San Jacinto*, to take all of the *Eugenie's* men and officers out of the prize and, about the same time, an order to the same effect was given to Act. Master Nelson of the *Tennessee* relating to the men and officers of the *Tennessee*.

I at once removed my men and officers and left myself and returned on board the *Eugenie*. At about 1.45 A.M., a boat from the U.S.S. *San Jacinto* in charge of Lt. Comdr. Quackenbush of the *San Jacinto* came alongside the *Eugenie* for all of the prisoners. They were delivered to Mr. Quackenbush, and about this time I again met Lt. Comdr. Chandler of the *San Jacinto* on board the prize, and was informed by him that he had taken possession and intended starting for Key West immediately. It was about 2 o'clock A.M. of the 13th September that I left the prize in complete possession of the *San Jacinto* at which time I noticed but little if any difference in the state of her cargo on deck.

At 5 A.M. or there abouts, I again went on board the prize for the purpose of calling Lt. Comdr. Green of the U.S.S. *Vincennes* whom I found asleep. Returning on board the *Eugenie* at once and steaming to Ship Island in company with the U.S.S. *Tennessee*. Upon a report of the facts being made to Lt. Comdr. Mayo of the *Kanawha*, senior officer present at Ship Island, he at once placed in the hands of Lt. Comdr. Green of the *Vincennes* an order to Lt. Comdr. Chandler of the *San Jacinto* to deliver the prize upon the receipt of that order to Lt. Comdr. Green of the *Vincennes* and sent the U.S.S. *Tennessee* in pursuit.

Overtook the *San Jacinto* about 25 miles at rear, and upon the delivery of the order, Lt. Comdr. Chandler decided to return to Ship Island with the prize in tow of the *San Jacinto*. Accordingly, the *Tennessee* started on her return to Ship Island followed by the *San Jacinto* towing the prize where all three vessels arrived between the hours of 2 & 6 P.M. 13th September.

Lt. Comdr. Mayo of the *Kanawha* at once gave orders to Lt. Comdr. Chandler of the *San Jacinto* to remove the men and officers belonging to the *San Jacinto* from the prize excepting Act. Master Richardson of the *San Jacinto*, and then ordered a prize crew from the *Tennessee* to proceed on board the prize, giving me about the same time written orders to take charge of the prize and proceed with her to New Orleans.

At about 9 P.M. 13th September after an absence of from 16 to 20 hours from the prize during which time the prize had been in possession of the U.S.S. *San Jacinto*, I went on board the prize and assumed command. I at once noticed that all of the cargo previously mentioned as having been on deck, with the exception of a few sacks of salt and some barrels, had been removed from the steamer. I am also informed that the prize had a quantity of silver plate which had also been taken, as well as the sextant belonging to the ship which could nowhere be found.[159]

At the time of the capture of the *Alabama* by the *Eugenie*, viz: at 6.05 P.M., to the best of my judgment the U.S.S. *San Jacinto* was beyond signal distance and, therefore, entitled to no claim in the prize money accruing from such capture.

                    N. M. Dyer

                    Act. Ensign & Prize Master

New Orleans Oct. 14th 1863

Sworn to and signed before me this the fourteenth day of October, 1863.

                    T. B. Thorpe

                    Special Prize Comr.

---

[159] Needless to say, looting a prize ship was strictly forbidden under international standards of prize law. Until adjudicated and condemned the ship and cargo remained the property of the original owners. In theory the owners might win the court case and have returned to them their property or just compensation for its value.

"At about 9 P.M. 13th September after an absence of from 16 to 20 hours from the Prize during which time, the Prize had been in possession of the U.S.S. "San Jacinto" I went on board the Prize, and assumed command. I at once noticed that all of the cargo previously mentioned as having been on deck with the exception of a few sacks of salt and some brackets had been removed from the steamer. I am also informed that the Prize had a quantity of Pilot bread, which had also been taken, as well as the sextants belonging to the ship, which could nowhere be found.—

At the time of the capture of the "Alabama" by the "Eugenie" viz: at 6.05 P.M. to the best of my judgement the U.S.S. "San Jacinto" was beyond signal distance, and therefore entitled to no share in the prize money accruing from such capture.

N. M. Dyer
Act'g Ensign Prize Master

New Orleans Oct 14th 1863
Sworn to and signed before me this the fourteenth day of October, 1863.
T. B. Thorpe
Special Prize Com'r

Figure 84. Deposition of the prize master, N. M. Dyer, Acting Ensign, commanding the *Eugenie*, page 1 of 6 (prior page) and part of page 6 (this page). Dated 12 Oct. 1863. Filed 19 Oct. 1863. *NARA*.

Prize Master's Report and Prize Commissioner's Report re. Submission of Prize Ship's Papers

Submitting and accounting for the prize ship's papers was a formal part of the adjudication and handled using the two following forms (Figures 85-86). The prize master's job was to submit the ship's papers to the U.S. prize commissioner whose duty it was to submit the papers to the court.

District Court of the United States for the Eastern District of Louisiana.

In the matter of
the Str. *Alabama*,
and her Cargo   N. M. Dyer Prize
   A Prize.   Master of the Str. *Alabama*
                           and Acting Ensign U.S.N.

detached from the U.S.S. *Eugenie* of which N. M. Dyer is commander, being duly sworn deposes and says, that the Str. *Alabama* was captured ten miles due south from Chandeleur light house, five feet and a half water, Gulf of Mexico, on Friday the 12th day of September, at about 6 o'c. 5 minutes P.M. 1863.
and said Commander Dyer deposes and says, that certain papers marked from A to I, inclusive, and further identified by the initial letters put on them by Special Prize Commissioner, T. B. Thorpe are the same that were captured with the Str. *Alabama* and that the deponent has delivered to the Prize Commissioner each and every document, paper, and writing whatsoever, received by him relating to the aforesaid captured property in the same state and condition in which he received them, without "fraud, subordination, or embezzlement," and without any addition or alterations whatsoever; and he does not know, nor does he believe that any of the documents are missing or wanting.
       And deponent further says, that if at any time hereafter, and before the final condemnation or acquital of said captured property any further or other papers relating thereto shall be found or discovered to his knowledge, they shall be delivered up, or information thereof given to the Prize Commissioners of this district, or to the Honorable Court.
                                                   (signed) N. M. Dyer,
                                                         Act. Ensign U.S.N.
       Sworn to and signed before me T. B. Thorpe Special Prize Commissioner, this the 8th day of October 1863. New Orleans, La.
                                                       (signed) T. B. Thrope
                                                       Special Prize Commissioner.

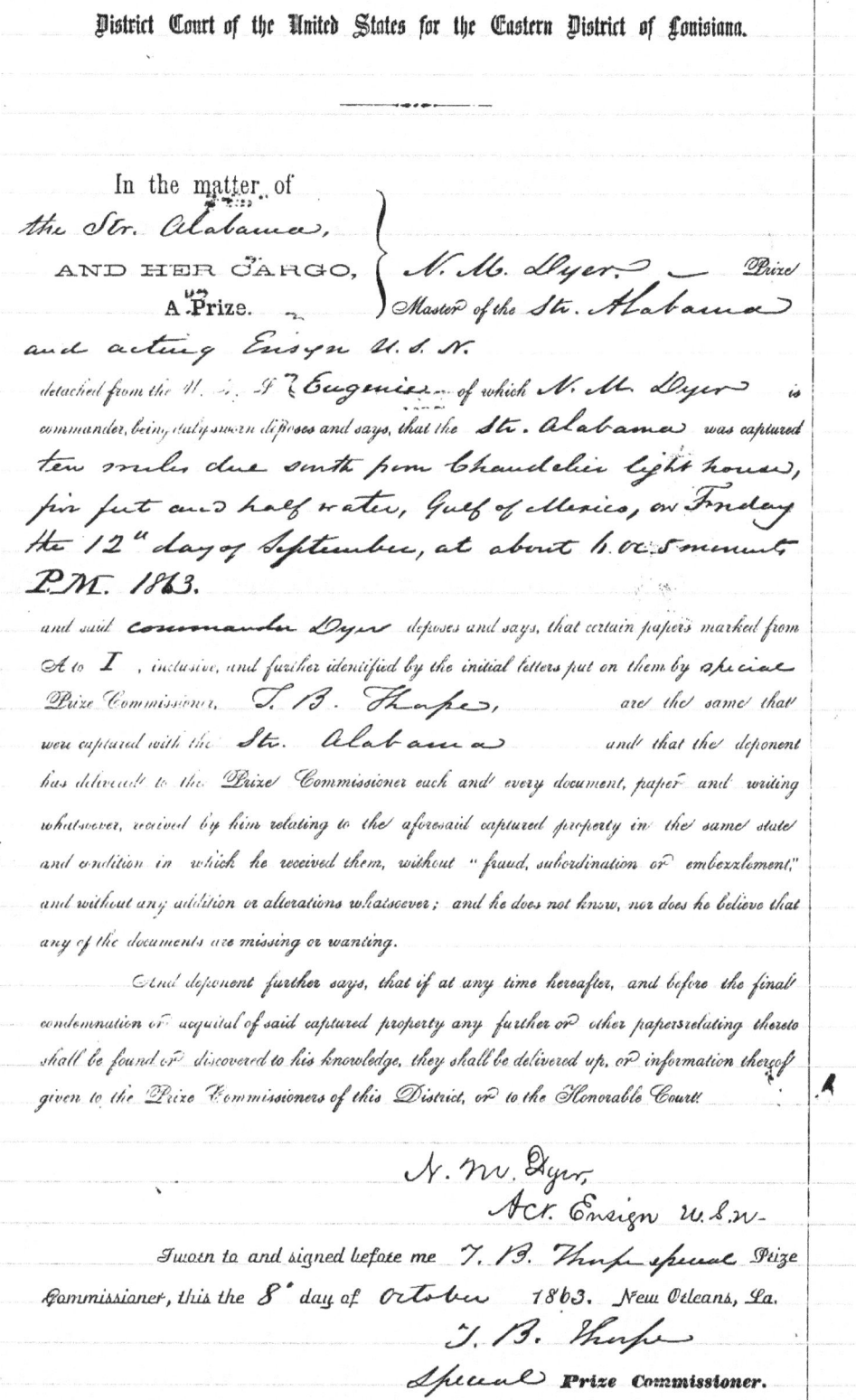

Figure 85. N. M. Dyer, commander of the *Eugenie* and captor of record of the *Alabama*, presented his deposition with bare facts of time and place of capture and testified that all the ship's papers were turned over to the prize commissioner. Dated 8 Oct. 1863. *NARA*.

(Original)

In the matter of the   Prize Commissioner's Office,
Str. Alabama,
And Cargo, a Prize.   New Orleans, October 9th 1863

To the Hon. the Judge of the District Court
of the United States for the Eastern District
of Louisiana:

    The undersigned, Prize Commissioners of the Honorable Court, respectfully report, that on the 8th day of October A.D. 1863 they received notice from N. M. Dyer Acting Ensign U.S.N. and Prize Master of the Str. *Alabama* detached from the U.S.S. *Eugenie* of which N. M. Dyer was commander that the said Acting Ensign N. M. Dyer brought with him into this City a certain vessel, called the Steamer *Alabama* and Cargo, as a Prize of the U.S.S. *Eugenie* that said Steamship had been captured on Friday the twelfth day of September 1863 on or about six o'clock and five minutes P.M. 1863, ten miles due south from Chandeleur light house, five feet and half water, Gulf of Mexico.
and the said A. [sic] M. Dyer and a prize crew had been appointed by said Dyer Comdg. the *Eugenie* to bring said Prize into this Port, and report her as required by the law and the usages in such cases observed.
    That the said Prize Master did seasonably deliver to us all such papers, passes, documents, &c. as were found on board of the captured steamship and having a reference to and a connection with the captured property, and which were in the possession, custody or power of the captors, which documents were by one of us Prize Commissioners regularly marked and numbered from

A. Articles of agreement
B. Shipping articles
C. Manifest of Cargo (Corrected to Crew List two other copies of this document not shown.)
D. (D left out of original but corrected to Manifest of Cargo in other copies not shown.)
E. Permanent register
F. Custom House clearance
G. Tonnage of Alabama
H. Package of eight letters (8)
I. Package of fourteen letters (14)

Nine
in number, and the said Prize Master made a deposition identifying said documents and complying in all respects with the law; which deposition and the said documents are herewith transmitted to your Honorable Court.
    That on the same day the said Prize Master produced to me three Witnesses, persons captured with the said Steamer to wit:

1st Thos. Carrell   Captain   of said vessel.
2d   H. Parker   Navigator   of said vessel.
3d   Robert Adams, purser   of said vessel.

    That we examined said three persons *in preparatorio* upon the standing interrogatories which have been adopted by the Honorable Court, as the same are found in Upton's work on Prize Courts, (page 479, 3d edition,) that said witnesses were examined separately on oath, and not in presence of each other and unattended by counsel, after each had taken the necessary and prescribed oath to make true answer, &c. which answers were written down as required by law, each one separately to each interrogatory: and when entirely answered the said answers were signed by each witness respectively, as the same are now transmitted to your Honorable Court.

That we proceeded aboard of said vessel, the Str. *Alabama* and examined her condition, and that it was impossible to state whether bulk had been broken or not; that the cargo consists of various merchandise, and assorted cargo and that it is impossible to ascertain exactly the nature or quality of the cargo without an unloading and undelivery [sic] of the same, which proceeding we now recommend to the Honorable Court, that we secured the hatches of said Steamer with a view to the protection of the cargo until it is undelivered.

That we took possession of the said Steamer and cargo on the 9th day of October 1863, and turned over the same to James Graham Esq. U.S. Marshal of this Court, who is to safely keep the same until further order of the Court. All of which, including the deposition of said N. M. Dyer with the documents by him identified, and the answers to the interrogatories taken as above, are now securely sealed, and are hereby submitted to this Honorable Court and deposited in its custody.

(signed) T. B. Thorpe
Special U.S. Prize Commissioners.

(Original)

In the matter of the Str. Alabama, and Cargo, a Prize.

Prize Commissioner's Office,
New Orleans, October 9.th 1863

To the Hon. the Judge of the District Court of the United States for the Eastern District of Louisiana:

The undersigned, Prize Commissioners of the Honorable Court, respectfully report, that on the 8th day of October A. D. 1863 they received notice from N. M. Dyer acting Ensign U. S. N. and Prize master of the Str. "Alabama" detached from the U. S. S. Eugenie of which N. M. Dyer was commander

that the said acting Ensign N. M. Dyer brought with him into this City a certain vessel, called the Steamer Alabama and Cargo, as a Prize of the U. S. S. Eugenie that said Steamship had been captured on Friday the twelfth day of September 1863, on or about six o'clock and five minuits P.M. 1863, ten miles due south from Chandelier light house, five feet and a half water, Gulf of Mexico.

and the said N. M. Dyer and a prize crew had been appointed by said Dyer Comdg the Eugenie to bring said Prize into this Port, and report her as required by the law and the usages in such cases observed.

That the said Prize Master did seasonably deliver to us all such papers, passes, documents, &c., as were found on board of the captured steamship and having a reference to and a connection with the captured property, and which were in the possession, custody or power of the captors, which documents were by one of us Prize Commissioners regularly marked and numbered from

A. articles of agreement
B. Shipping articles
C. Manifest of Cargo
E. Permanent register
F. Custom House clearance
G. Tonnage of "Alabama"
H. Package of eight letters (8)
I. Package of fourteen letters (14)
    Nine

in number, and the said Prize Master made a deposition identifying said documents and complying in all respects with the law; which deposition and the said documents are herewith transmitted to your Honorable Court.

That on the same day the said Prize Master produced to me three Witnesses, persons captured with the said Steamer to-wit:

1st Tho Cauell captain   of said vessel.
2d   H. Parker   Nangatu   of said vessel.
3d   Robert Adams, purser   of said vessel.
4th   of said vessel.

That we examined said three persons in preparatorio upon the standing interrogatories which have been adopted by the Honorable Court, as the same are found in Upton's work on Prize Courts, (page 479, 3d edition,) that said witnesses were examined separately on oath, and not in presence of each other and unattended by counsel, after each had taken the necessary and prescribed oath to make true answer, &c. which answers were written down as required by law, each one separately to each interrogatory; and when entirely answered the said answers were signed by each witness respectively, as the same are now transmitted to your Honorable Court.

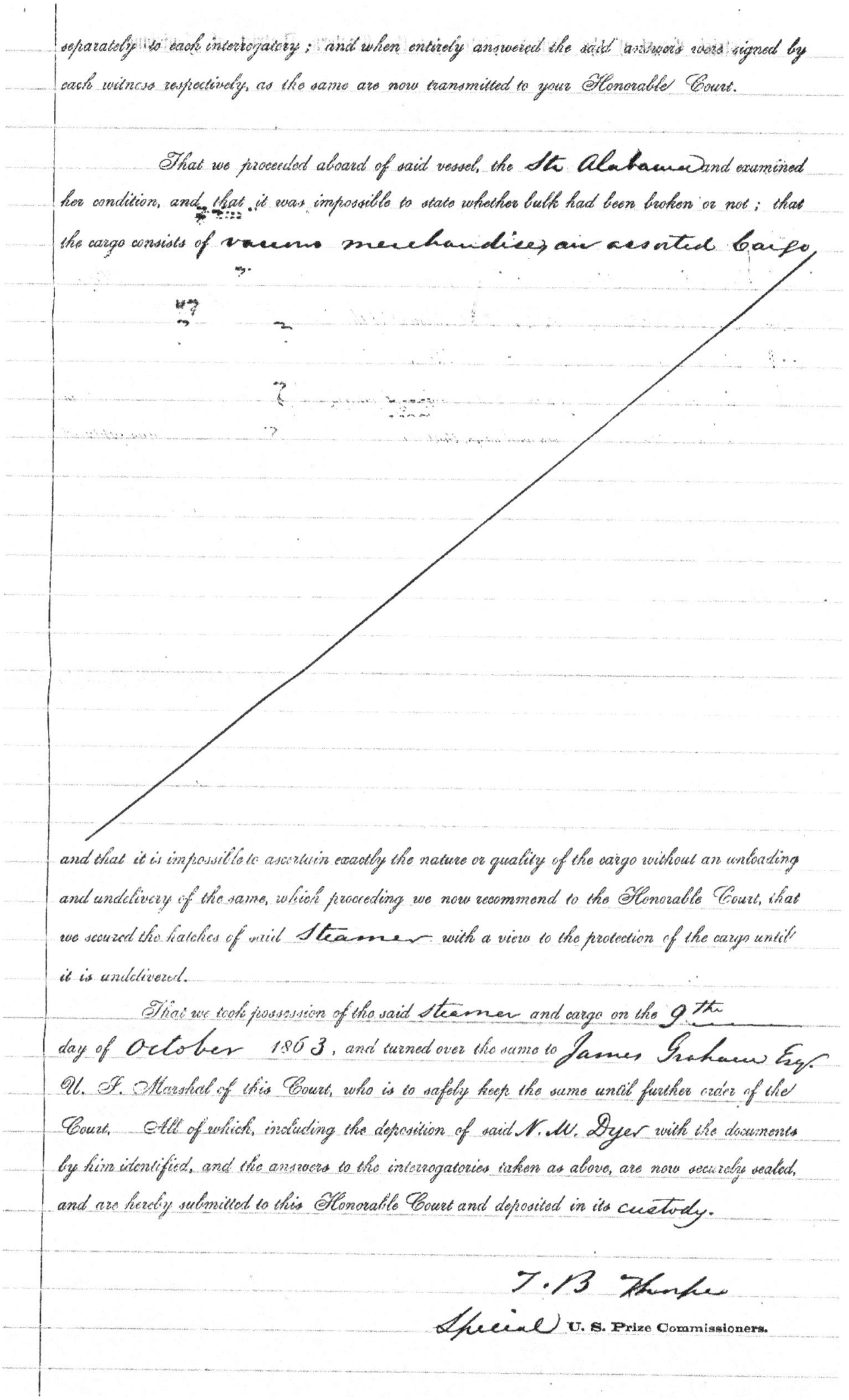

That we proceeded aboard of said vessel, the Str. Alabama and examined her condition, and that it was impossible to state whether bulk had been broken or not; that the cargo consists of various merchandises an assorted Cargo and that it is impossible to ascertain exactly the nature or quality of the cargo without an unloading and undelivery of the same, which proceeding we now recommend to the Honorable Court, that we secured the hatches of said Steamer with a view to the protection of the cargo until it is undelivered.

That we took possession of the said Steamer and cargo on the 9th day of October 1863, and turned over the same to James Graham Esq. U. S. Marshal of this Court, who is to safely keep the same until further order of the Court. All of which, including the deposition of said N. W. Dyer with the documents by him identified, and the answers to the interrogatories taken as above, are now securely sealed, and are hereby submitted to this Honorable Court and deposited in its custody.

T. B. Thorpe
Special } U. S. Prize Commissioners.

Figure 86. The prize commissioner reported to the court and gave notice of the prize's arrival, turning in the report of the prize captain who was typically an officer of the capturing ship and the depositions of participants from the captured ship. (Three pages including the two prior pages.) Dated 9 Oct. 1863. *NARA*.

Other *Alabama* Prize Case Documents in Rough Date Order

The following section included the depositions of the *Alabama's* officers presenting their depositions or answers to the standard interrogatories of a prize case (Figures 87-90). Also included were with a map showing the area of the capture (added by the present author) and an invoice for harbor dues. The testimony of Captain Carrell was particularly interesting. Important facts emerged in the interrogatory, such as the cargo was loaded in Havana 1-7 Sept. 1863, the ship sailed from Mobile on 21 Aug. and arrive in Havana on the 24th, the cargo belonged to prominent citizens of Mobile, the captain believed that the vessel's owners lived in Mobile, and the outbound cargo was cotton.

In the matter of the
Str. *Alabama*
And Cargo, a Prize

District Court of the United States
For the Eastern District of Louisiana

Deposition of <u>Thomas Carrell, Capt. of the Str. *Alabama*</u>, a witness produced, sworn and examined *in preparatorio* on the ____ day of October in the year eighteen hundred and sixty three at the office of T. B. Thorpe, Special Prize Commissioner on the standing interrogatories established by the District Court of the United States, for the Eastern District of Louisiana; the said witness having been produced for the purpose of such examination, by Rufus Waples, Esq., U.S. District Attorney in behalf of the captors of a certain vessel called the _____.

1. To the first interrogatory the witness answers: that he was born in Liverpool, England. Mobile, Ala. is now my home, consider myself a citizen of Alabama, have been raised there from a boy, married, wife, and two children living in Mobile.

2. To the second interrogatory the witness answers that he was present at the capture of the Str. *Alabama*.

3. To the third interrogatory the witness answers, that the seizure was made inside of North Point of Chandeleur Island, in sight of Ship Island, and the vessel was taken to New Orleans. Heard no reason given for the capture by the capturing officer.

4. To the fourth interrogatory the witness answers, that the vessel sailed under Confederate colors, no other colors on board. The Confederate flag was held up by one of the Quartermasters of the *Alabama*, to prevent the capturing vessel from firing on the *Montgomery* [*sic*][160] which vessel was aground at the time of her capture.

5. To the fifth interrogatory the witness answers that there was no resistance made at the time of the capture. The Steamer *Eugenie* fired two or three shots before she boarded. The capturing vessel that first boarded us was the U.S.S. *Eugenie*, there were present the U.S. Frigate *San Jacinto* and the U.S. Gunboat *Tennessee*. The *San Jacinto* sighted the *Alabama* on the 11th of September about 10 o'clock in the morning; she chased me until dark of that day. After it got so dark that I could not see the *San Jacinto* any longer, I altered my course, and hauled the vessel in due north for the land. I ran in until 3 o'clock next morning, and discovered that I had not the right soundings, found myself to the westward of Mobile point, about twenty five miles. From that point, I could not get in Mobile before day light, hauled the vessel off south again, fell in with the frigate *San Jacinto* again at six o'clock A.M., hauled to the East by S. The chase commenced again for some four or five hours, when the lookout at the mast head reported another sail on the port bow, some seven or eight miles distant. Hauled off W. by S. until I sighted the Chandeleurs, which was about 4 o'clock in the evening, ran for the

---

[160] An error by the clerk preparing the document. It should read *Alabama*.

Chandeleurs, North Point, and in going through grounded in a mud flat about 6 o'clock in the evening. The *San Jacinto*, being a heavy draft vessel, could not get in. Meantime, the *Eugenie* started from Ship Island (some ten miles from where the *Alabama* grounded) and boarded the captured vessel a little before seven o'clock, on the 13th of September 1863. The captured vessel was a merchantman; the capturing vessels were, in my opinion, U.S. vessels of war.

    6.   To the sixth interrogatory the witness answers, that the *San Jacinto* was the cause of my grounding the *Alabama* on the mud flat. The U.S.S. *Eugenie* took possession of my vessel after I was aground. The *Eugenie* never chased me at all. I must have been aground when the *Eugenie* left Ship Island. The *Alabama* has not been condemned up to the present date.

    7.   To the seventh interrogatory the witness answers that the name of the captured vessel was the *Alabama*, her captain's name was Thomas Carrell, the witness. I was appointed to command the vessel by the owners in Mobile. I live in Mobile & have my family in that city. I was born in Liverpool, England, consider Mobile my home.

    8.   To the eighth interrogatory the witness answers, that the *Alabama* is about four hundred and sixty tons. At the time of the capture there were thirty persons on the vessel, two were passengers, all came on board at Havana. The crew was shipped at Mobile.

    9.   To the ninth interrogatory the witness answers that he belonged to the ship's company as captain, had no interest in the cargo or vessel, had some few things on board for family use.

    10.   To the tenth interrogatory the witness answers, that he has known this vessel ever since she first came to Mobile in the later part of 1860, first saw her in Mobile, she carried no guns. There were thirty four men on board of her at the time of her capture. The vessel was built in New York. Never was called by any name but the *Alabama*.

    11.   To the eleventh interrogatory the witness answers, that the captured vessel was bound for Mobile when seized. The voyage began at Mobile and was to have ended at Mobile. When she was captured, she had on board an assorted cargo which was put aboard at Havana about the, or rather between the 1st and 7th of September 1863.

    12.   To the twelfth interrogatory the witness answers, that the captured vessel had no sea briefs or pass ports that I know of, she sailed on her last voyage from Mobile, was bound for Mobile when she was captured. The voyage began at Mobile and was to have ended at Mobile. She had on board a mixed cargo, which was to have been delivered at Mobile.

    13.   To the thirteenth interrogatory the witness answers, that when the *Alabama* set out on her voyage she carried cotton. She had a mixed cargo on her inward voyage. The inward bound cargo was put on board at Havana between the 1st and the 7th of September 1863.

    14.   To the fourteenth interrogatory the witness answers, that the cargo was owned by different persons living in Mobile. The owners are citizens, and born in the U.S. lived in Mobile a long time, and are citizens of the Confederate States.

    15.   To the fifteenth interrogatory the witness answers, that he knows nothing about any bill of sale of the vessel.

    16.   To the sixteenth interrogatory the witness answers, the lading of the vessel at the time she was captured was put on board in the harbor of Havana, between the 1st and seventh of September 1863. The laders have no interest in these goods. If they were registered, they would belong to citizens in Mobile.

    17.   To the seventeenth interrogatory the witness answers, that there were ten or a dozen bills of lading signed for the cargo, the last I saw of them was in the agent's office in Havana, where they were signed. The contents of these bills of lading were the cargo of the vessel in detail.

    18.   To the eighteenth interrogatory the witness answers, that he has never had any papers showing that he had any interest in the cargo, the last he saw of the bills of lading related to the cargo, they were in the hands of the agent of the vessel living in Havana.

19. To the nineteenth interrogatory the witness answers, that the vessel was captured in the latitude of the Chandeleur Islands, on the 13th of September, on Saturday between six and seven o'clock P.M. There was no charter party that I am aware of.[161]

20. To the twentieth interrogatory the witness answers, that there were no papers belonging to the vessel burnt or thrown overboard, I don't know anything about the papers, except the ship's papers. I know there were bills of lading signed as I have already stated.

21. To the twenty first interrogatory the witness answers, that he did know that the port to which he was (entering) was blockaded by the vessels of the United States, got the information from common rumor, and saw the vessels engaged in the blockade off the harbor of Mobile.

22. To the twenty second interrogatory the witness answers, that he believed that the harbor of Mobile was under blockade from the naval forces of the United States at the time he endeavored to enter it. I never received any official notice of the blockade.

23. To the twenty third interrogatory the witness answers, that he does not know that any paper or papers belonging to the *Alabama* were ever examined by any U.S. officer previous to her capture.

24. To the twenty fourth interrogatory the witness answers, that he ran out of Mobile harbor with the *Alabama* knowing the port was blockaded by vessels of the United States government. I left Mobile on the night of the 21st of August 1863 and arrived in Havana on the 24th day of the same month.

25. To the twenty fifth interrogatory the witness answers, that the *Alabama* has never before been seized as a prize.

26. To the twenty sixth interrogatory the witness answers, that he has sustained the loss of at least three or four thousand dollars, and has not, up to this date, received any indemnity for the same.

27. To the twenty seventh interrogatory the witness answers, that there was no insurance on the vessel.

28. To the twenty eighth interrogatory the witness answers, that if the *Alabama* had arrived at Mobile, the cargo would have belonged to prominent citizens of that city, the cargo was not sent to Mobile for the (chances) of a market.

29. To the twenty ninth interrogatory the witness answers, that he knows nothing of the history and production of the cargo, except that it was purchased in Havana of different stores.

30. To the thirtieth interrogatory the witness answers, that the cargo was lightered on board of the vessel from the quay or shore of the city of Havana.

31. (& 32.) To the thirty first interrogatory the witness answers: that he knows nothing of any papers that he has not already stated, to the thirty second interrogatory he makes the same reply.

33. To the thirty third interrogatory the witness answers, that the bulk was not broken during the voyage to my knowledge.

34. To the thirty fourth interrogatory the witness answers, there were two passengers on the boat, (and) four people in distress, who wished to get home. They worked as much as they could to pay their passage. No charge beyond this labor was made against them. There were no officers, soldiers, or citizens of the United States secreted on board of the vessel at the time of her capture.

35. To the thirty fifth interrogatory the witness answers: that he knows nothing of any papers belonging to the vessel not already stated.

---

[161] The item regarding a charter party is a place one might hope to find a reference to involvement by the European Trading Company which was owner or operator of the *Denbigh*. The captain's negative answer might not necessarily be decisive. It might have been fairly easy to conceal charter party information from the captain.

36. To the thirty sixth interrogatory the witness answers, that the captured vessel, at the time she was first pursued by the *San Jacinto*, was bound for Mobile. Her course was altered to escape capture by the *San Jacinto*.

37. To the thirty seventh interrogatory the witness answers, that he knows nothing of the history of the vessel, the *Alabama*, except with regard to the present owners who live in Mobile. If the vessel should be returned, it would come into the possession of owners living in Mobile.

38. To the thirty eighth interrogatory the witness answers, there were no guns on the captured vessel, no arms of any kind, no munitions of war of any kind.[162]

39. To the thirty ninth interrogatory the witness answers, the *Alabama* was loaded with cotton in the port of Mobile, that she ran the blockade on the 21th of Aug., and arrived in Havana on the 24th same month, disposed of her cotton, or landed it, if you please, in Havana. The return cargo was put on board, and the vessel was captured before she again reached the harbor of Mobile.

40. To the fortieth interrogatory the witness answers, that the captured vessel, never sailed in any convoy of ships, armed, or unarmed.

41. To the forty first interrogatory the witness answers, that he never attempted to enter any blockaded ports except that of Mobile.

42. To the forty second interrogatory the witness answers, that he had no letters of marque, or a commission to act as a privateer, or any kind of authority to cruise against the persons or property of citizens of the United States.

43. To the forty third interrogatory the witness answers, that his vessel never sailed in company or in concert with any other armed vessel or vessels, or acted in company or concert with any other armed vessel in pursuing, seizing as prizes, persons, vessels, or property of citizens belonging to the United States.

        (signed) Thomas Carrell
        Master, Str. *Alabama*

Sworn to and signed before me, Special Prize Commissioner this the ninth day of October 1863, New Orleans, La.
        (signed) T. B. Thorpe
        Special Prize Commissioner

---

[162] This statement was not quite true as can be seen below in the listing of cargo inventoried by the prize commissioner. There were 567,000 percussion caps and 24,900 cannon primers. One might speculate that other munitions in the cargo could have been jettisoned when capture was unavoidable to give the ship the impression of a peaceful merchantman.

In the matter of the
Str. "Alabama"
and Cargo, a Prize.

District Court of the United States
For the Eastern District of Louisiana.

Deposition of Thomas Carrell, Capt of the Str. "Alabama", a witness produced, sworn and examined *in preparatorio* on the ___ day of October in the year eighteen hundred and sixty three at the office of T. B. Thorpe, special prize Commissioner.

on the standing interrogatories established by the District Court of the United States, for the Eastern District of Louisiana; the said witness having been produced for the purpose of such examination, by Rufus Waples, Esq., U. S. District Attorney, in behalf of the captors of a certain vessel called the

1. To the first interrogatory the witness answers: that he was born in Liverpool, England, Mobile Ala. is now my home, consider myself a citizen of Alabama, have been raised there from a boy, married, wife and two children living in Mobile.

2. To the second interrogatory the witness answers that he was present at the capture of the Str. "Alabama".

3. To the third interrogatory the witness answers, that the siezure was made inside of North point of Chandeleur Island, in sight of Ship Island, and the vessel was taken to New Orleans. Heard no reason given for the capture by the capturing officer.

4. To the fourth interrogatory the witness answers, that the vessel sailed under Confederate

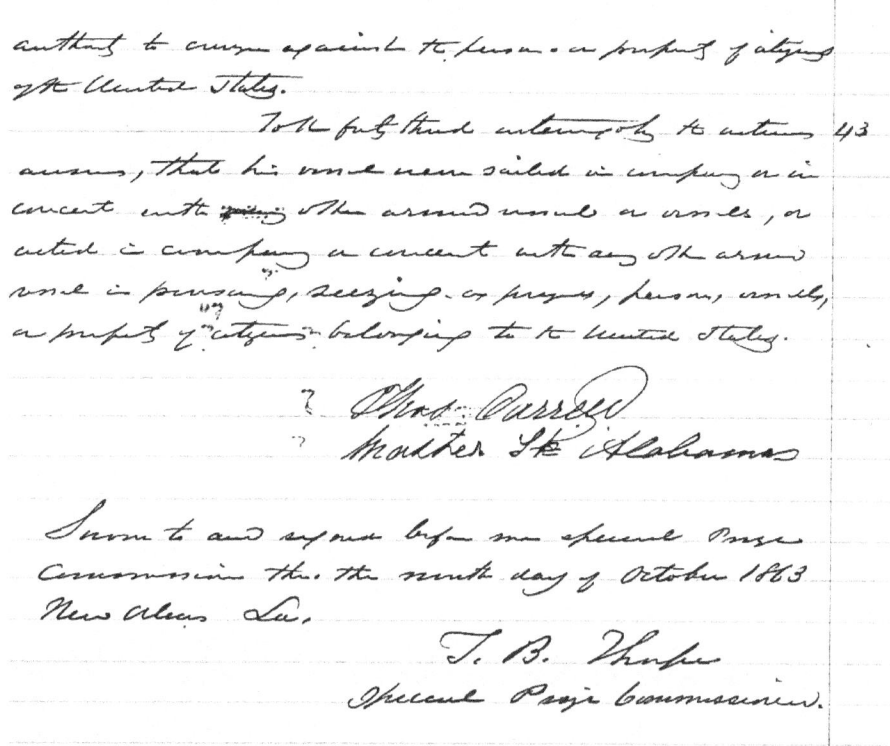

Figure 87. Interrogatory of Captain Thomas Carrell describing the *Alabama's* chase and capture by the Union navy. Shown here page 1 of 9 and the part of page 9 with the signature block. Dated 9 Oct. 1863. *NARA*.

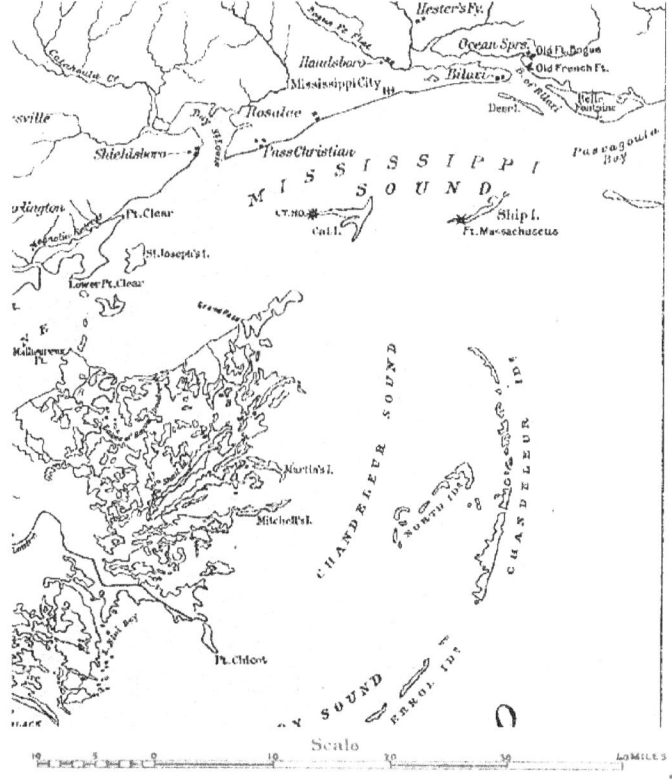

Figure 88. Chart of Chandeleur Is. and Ship I. about 10 miles apart. (North up.) From *ORN*, ser. 1, 18: 131.

Deposition of H. Parker, Navigator of the *Alabama*

The next document was Parker's response to the interrogatories in which he agreed with most of the testimony given by Captain Carrell. Parker's answers did display less detailed knowledge, and so only a small part was here transcribed (Figure 89).

There were a few new facts revealed such as in No. 14: "... he does not know the owners of the vessel and cargo. Thinks that they were owned by a stock company in Mobile, composed of citizens of Mobile." This was most interesting as there were listings of blockade-running company stock prices and notices of dividends in the *Mobile Register* newspaper during the Civil War. It was curious to note that the stock quotes for the *Alabama* continued to be shown long after the herein described capture of the ship.[163] Perhaps the company remained in business under its original name but operating a different ship or ships.

Interrogatory No. 16 also provided a new fact: that the cargo was "... put on board by a man named St. Maria, a Spaniard living in Havana." Señor Santa Maria was a name that appeared with great regularity in the *Denbigh's* invoices and customs records for cargo shipped from Havana to Mobile. The agent named Santa Maria in the context of both the *Alabama's* and the *Denbigh's* cargo perhaps hinted at the theoretical association of both ships with the European Trading Company. This was not particularly strong evidence, however. Santa Maria could easily have been a cargo agent for many ships of different ownership.

In the matter of the
Str. *Alabama*
and Cargo, a Prize

District Court of the United States
For the Eastern District of Louisiana

Deposition of H. Parker, navigator on board of the Str. *Alabama*, a witness produced, sworn, and examined *in preparatorio* on the tenth day of October in the year eighteen hundred and sixty three at the office of T. B. Thorpe, Special Prize Commissioner on the standing interrogatories established by the District Court of the United States, for the Eastern District of Louisiana; the said witness having been produced for the purpose of such examination, by Rufus Waples, Esq. U.S. District Attorney in behalf of the captors of a certain vessel called the

1. To the first interrogatory the witness answers: that he was born in New Hampshire, lives in Mobile, married, has a wife and three children in Mobile, claims to be a citizen of Alabama, has lived in Mobile over twenty years. . . .

---

[163] Arnold, J. Barto III. *The* Denbigh's *Civilian Imports: Customs Records of a Civil War Blockade Runner between Mobile and Havana.* Denbigh Shipwreck Project Publication 5. College Station, Texas: Institute of Nautical Archaeology, Texas A&M University, 2011, pp. 18-21.

In the matter of the
Str. Alabama
and Cargo, a Prize.

District Court of the United States
For the Eastern District of Louisiana.

Deposition of H. Parker, navigator on board of the Str. Alabama a witness produced, sworn and examined in preparatorio on the tenth day of October in the year eighteen hundred and sixty three at the office of T. B. Thorpe special Prize Commissioner on the standing interrogatories established by the District Court of the United States, for the Eastern District of Louisiana; the said witness having been produced for the purpose of such examination, by Rufus Waples, Esq., U. S. District Attorney, in behalf of the captors of a certain vessel called the

1. To the first interrogatory the witness answers: that he was born in New Hampshire, lives in Mobile, married, has a wife and three children in Mobile, claims to be a citizen of Alabama, has lived in Mobile over twenty years.

2. To the second interrogatory, the witness answers that he was present at the capture of the Alabama.

3. To the third interrogatory the witness answers that the capture was made back of the Chandeleur islands and the vessel was carried to New Orleans. Heard no news given for the capture.

4. To the fourth interrogatory the witness answers that the vessel sailed under confederate colors, had no other colors on board.

5. To the fifth interrogatory the witness answers that there was no resistance to the capture. The capturing vessel was a vessel of the United States, a vessel of war.

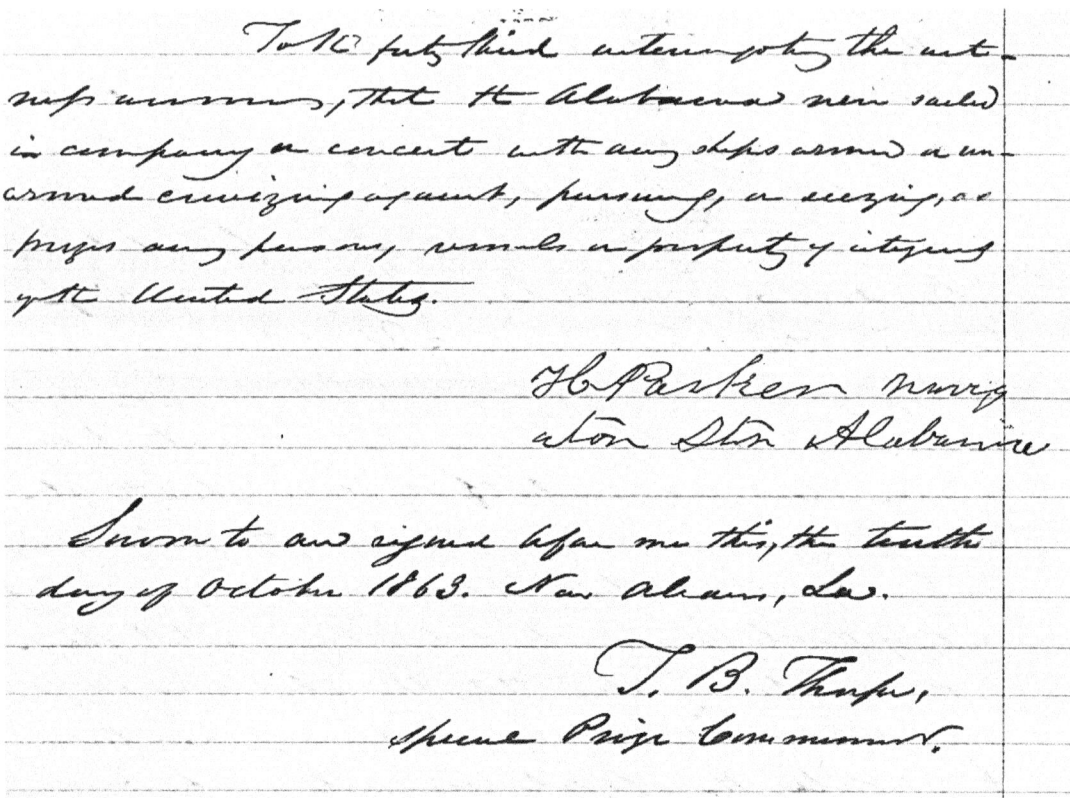

Figure 89. The deposition of H. Parker, the *Alabama's* navigator. The case file held two copies of this document and the previous one. In the second examples of each, the documents had to be redrafted in a more legible hand due to near illegibility of the originals. Dated 10 Oct. 1863. *NARA*.

Deposition of Robert Adams, Purser of the *Alabama*

As was the case with Parker, Adams added nothing of interest in addition to Captain Carrell's testimony, and so Adams' responses were not transcribed beyond the following sample (Figure 90).

> In the matter of the
> Str. *Alabama*
> and Cargo, a Prize
>
> District Court of the United States
> For the Eastern District of Louisiana

Deposition of <u>Robert Adams, Purser on Str. *Alabama*</u>, a witness produced, sworn and examined *in preparatorio* on the <u>tenth</u> day of <u>October</u> in the year eighteen hundred and sixty <u>three</u> at the <u>office of T. B. Thorpe Special Prize Commissioner</u> on the standing interrogatories established by the District Court of the United States, for the Eastern District of Louisiana; the said witness having been produced for the purpose of such examination, by Rufus Waples, Esq. U.S. District Attorney in behalf of the captors of a certain vessel called the <u>Str. *Alabama*</u>.
    1.   To the first interrogatory the witness answers: that he was born in Ireland, lives in Mobile, has lived in Mobile twenty five years. Not married. Claims to be a citizen of Alabama, not married. . . .

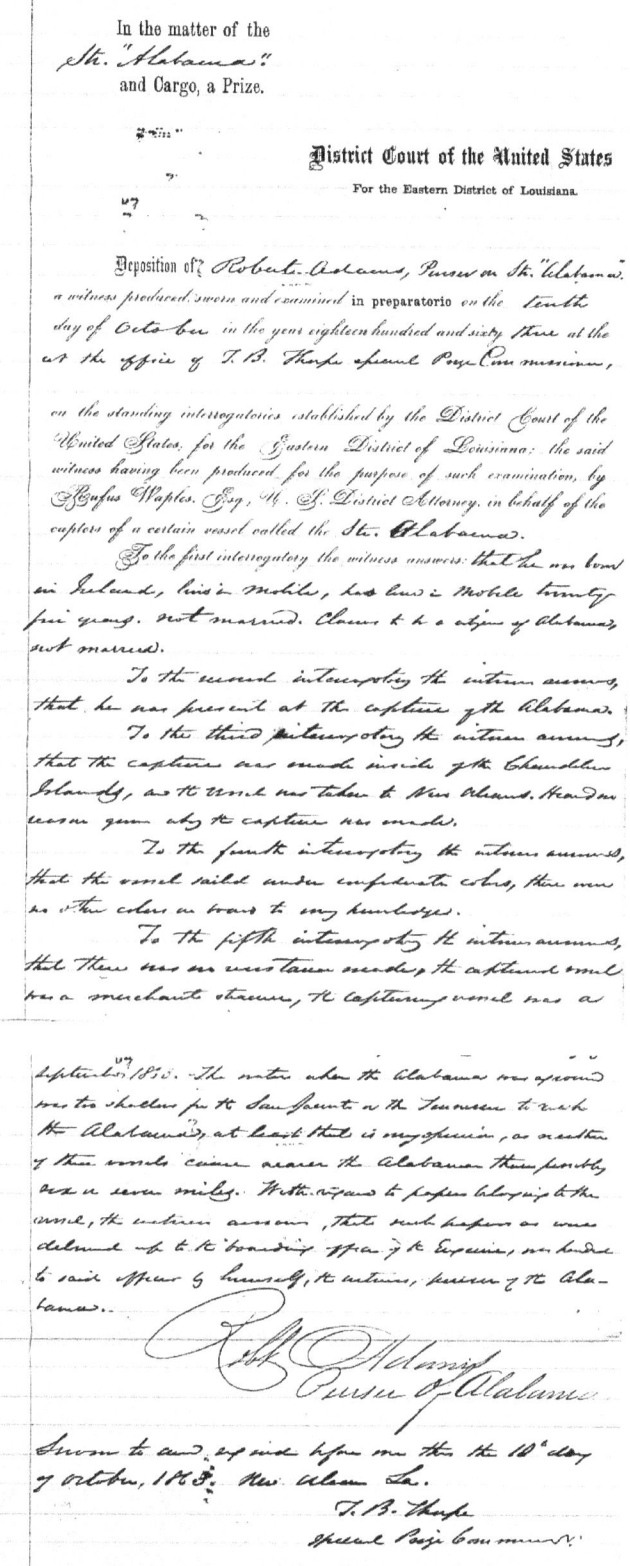

Figure 90. Deposition of Robert Adams, purser of the *Alabama*. Shown here page 1 of 7 and part of page 7 including the signature block. Dated 10 Oct. 1863. *NARA*.

Invoice for Harbor Master's Dues

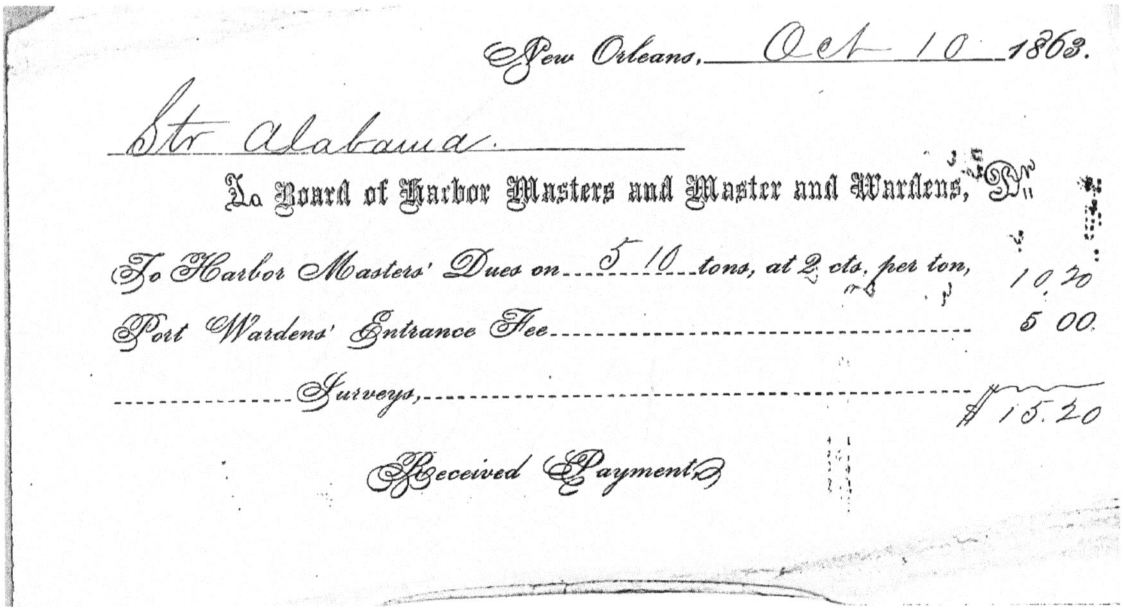
New Orleans, Oct 10 1863.

Str Alabama

To Board of Harbor Masters and Master and Wardens, Dr.

To Harbor Masters' Dues on 510 tons, at 2 cts. per ton, 10 20

Port Wardens' Entrance Fee 5 00

Surveys, $15.20

Received Payment

Figure 91. Bill for the prize ship *Alabama's* arrival fees in New Orleans. (Legible, not transcribed.) *NARA*.

Inventory of Effects on Board the Steamer *Alabama*

The following document contained a listing of every portable item of gear, equipment, and furnishings aboard the *Alabama* (not including cargo). What a fabulous archival resource for archaeologists was this inventory for a merchant steamer of this time and class (Figure 92).

Inventory of Effects on Board Steamer
Alabama.
Cabin Aft---
2 Bales Blankets
Marked G R B. 240 & 241
1 Box contents not known
2 Composition Lamps
1 astral ?? ??
1 Plated Caster & Cruets
1 Mahogany Extension Table
1 Turning Cushion Chair
19 Cane Bottom Chairs

In state Rooms
H. 1 Mattress 2 sheets 1 Pillow
G. 1 Moss Pillow
F. 1 Mattress 2 Sheets 2 Pillows
1 Blanket 1 Chronometer
E. Sealed
D. 2 Mattresses 2 Pillows 1 Blanket
21 2 Mattresses 3 Pillows
22 2 " 2 "

5 Tumblers
3 Wine Glasses
51 Plates 9 Cups 19 Saucers China Ware
2 Molasses Cups
2 Butter Dishes 3 Cake plate
1 Stop Bowl 2 Sugar Dishes
2 Butter Bowls 6 Gravy dishes
6 Pickle Dishes 1 Glass preserve Dish
1 Pair China Water Pitchers
1 Large & 2 Small Japan Waiters
1 Long Fish Dish
1 Box containing Crushed Sugar
2 Preserve Covers glass 5 Tumblers
1 Box Cayenne Peppers 1 Box Yeast
Powder
up (meaning go to top of next column to right)
1 Box Congress water
2 Kits No. 1 Mackerel
1 Meat Cleaver
1 Tin Pan 1 water Dipper
12 Sauss glasses 1 Keg faucet

9 Iron spoons 2 carving forks
1 fork 1 ?? 2 Tin Funnels
1 Oil Feeder 4 Beef Steak Dishes
1 Gravy Tureen 1 Tea Caddy
1 Large Dish pan Tin
1Tin Can 6 Demijohns
1 Containing Lemonade Vinegar
1Water Cooler 2 Water Monkey
2 Coffee Urns Plated
2 Astral Burners
1 Feeder Brass 1 Dinner Bell
2 Salt Cellar 1 Clock 1 Card Rack
   Tin
Closets of Front Table
5 Pound paper, Black Tea
3 Bibles 17 Lamp glasses
2 Baking Dickey Silver or Plated
1 Water Pitcher Plated
2 Glass Cakes Dishes
1 Large Cake Dish
2 Soap Dishes 2 plated Waiters
3 Boxes Yeast Powder 1 Cayenne
3 Cut glass salt cellars
1 Box Matches about 4 Dozen
1 Crumb Brush 1 Steel
1 Lot Pepper about 1 lb. 1 lb. ???
2 Crumb Brushes 2 Bars & ½ Bar
Castile Soap 1 Lamp to Ceiling
1 Lot Oil Cloth to Cabin floor
(End page 1)

Inventory of Steamer *Alabama* Cont'd.

Pantry
6 Boxes Vermicelli
1 Demijohn aguardiente
2 Bags Sugar about 20 lbs. ea.
1 " Coffee about 20 lbs.
1 Bag starch " 25 lbs.
1 Lot Codfish " 20 lbs.
1 Demijohn Molasses
1 Tin Can for oil
1 Hair Sifter
1 Brass Lamp 3 Copper dishes
3 Pitchers
2 Bowls
Clerk's office
1 Demijohn Caster Oil
1Broom 1 Brush 3 Old
Umbrellas 3 Brass Lamps
1 Compass Box
1 Signal Lantern
1 Starboard Boat with
Mast Mainsail & 4 Oars &
Helm Davits & falls complete
1 Larboard Boat 4 Oars

Davits & falls Complete
2 Spittoons 1 Bucket
1 Japan Tin Dust pan
Wash Room
2 ½ Barrels Jamaica Rum
1.5 Gallon Keg Molasses
On Deck Larboard Side
2 Barrels Marked <B> alum
4 "   " P <R> 1 Salt
1 "   " [H] 1222
3 "   " [R] D. M.
63–58–62
2 Barrels Marked Bread
3 Barrels Oil
2 Marked [D] <R> 1 No Mark
1 Barrel Cream Tartar
1 Sack Salt
1 Barrel <R> Salt ½ Barrel 1 Marked 4–18
1 "   <H> 12
1 ½ Barrel Marked W. G. W.
1 Barrel open Contents not Known
 1 Barrel Beans
Cook Room
16 White Stone China Dishes
1 Pitcher 1 Butter Dish
12 Tin Pans sundry Kinds
1 Coffee Mill 1 Pint Measure
2 Tin Dippers 5 Skillet
1 flour Sifter
1 Wooden Tray & Pounder
1 Soup Tureen
1 Pepper Box 6 Sheet Iron
Bake pans 6 Coffee Boilers 2 Copper
Skillets 1 Iron 1 Tin Boiler
4 large Iron Spoon 1 Grater
1 Saw 1 Tormentor 2 Egg lifters
1 Steel 1 Brass Coal Heaver
2 Wood Buckets 1 Tin
Slop Pail 3 large Knives 1 fork
1 Meat Chopper 1 Mortar
Boiler Deck
7 Coils Rope 1 Lot Old Netting
1Ax 1 Maul 1 Boat Hook
1 Oar 2 Sails 2 awnings
1 Brace & Saw 2 Barrels Rosin
34 Sacks Salt 1 Piece Leather
(End page 2)

Inventory of Steamer *Alabama* Cont'd.

Forward Deck
1 Carpenter Chest & Tools
1 Mattress 3 Pillows
2 Bow anchors & chains
1 Large spare anchor
1 " Hawser 1 Spring Line

1 Spring line mid ships
2 lengths hose
5 Lanterns forward
2 Chain Lockers
1 Containing Ropers &c 1 Empty
1 Grind Stone 2 Trunks
1 Lot Paints Oils Brushes & Pots
1 Ice Box 1 Old Hawser
2 Baskets 5 Old Brooms
1 Shovel 2 Wash tub for hose 1 Bucket
1 Barrel Beef open
1 Tin Boiler
Starboard side
1 Mast spritsail & Main sail
For small Boat 1 Tarpaulin
2 Double Blocks 1 Single
1 Bucket
Store Room
5 Barrels Bread
1   "   Tallow
1 Demijohn Coal Tar
1 ½ Barrel Tar
2 lengths Hose
2 Cans Paint in Clerk's office
1 Piece Sole Leather
up stairs
Room 17 1 Oval Table & Drawer
1 Parallel Ruler
2 Mattresses 1 Pillow
2 Sheets 1 Spread
1 Bowl & Pitcher 1 Water Pitcher
1 Zinc Lamp 1 Life Preserver
1 Cain seat Chair
Room 16
2 Mattresses 3 Blankets 2 Sheets
2 Pillows 1 Duster 1 Chair
1 Piece Leather 1 Lot Twine
1 Bolt remnants Canvas
Room 15
2 Mattresses 2 Pillows
1 Chair
Room 14
8 axes 2 Brooms
½ Bag cotton 1 Hank
Ratline stuff 1 Cotton Hook
2 Small Single Blocks
Room 13
1 Box Raisins
2 Boxes Sundries
2 Demijohns Liquor
2 Mattresses 2 Pillows
3 Blankets 2 Sheets 1 Spread
3 Lamp chimneys
1 Bag about 30 lbs. Havana Sugar
Room 12 Books & Papers
of Boat 1 Tin Box & Pen rack

1 Bill rack 1 Lot Carpet
Room R
1 Lot Carpet
Room aft Starboard Side
1 Lot Boat Linen clean & Dirty
After Deck
1 Boat
her complete Tackle
1 Slop Pail on Deck
(End page 3)

Inventory of Steamer *Alabama* Cont'd.

Room No. 9 Starboard
3 Chairs Cane seats
1 Brass Lamp
1 Stone China Pitcher
1 Tumbler
Room 7 Starboard
1 Live Preserver 1 Duster
1 Mattress 1 Sheet 1 Blanket
1 Spread 2 Pillows 1 Broom
1 Mattress Cover
Room 6 Starboard
1 Mahogany Bureau
1 Wash Stand Bowl
& Pitcher 1 Mattress
1 life Preserver
1 Broom
5 Chairs on Deck
1 Brass Oil feeder
1 Duct fan 1 Pint Measure
1 Signal Lantern
Pilot House 2 Compass Lanterns
1 Clock 1 Signal Lantern
2 Lead Line 1 Lead
1 Port 1 Starboard Signal Lanterns
Forward Upper Deck
1 Jib 1 fore Sail 1 Square Sail
2 Single Blocks & Rope
1   "   "   3 Double Blocks
& Ropes
Hurricane Deck
1 Box Containing Engineer's Blocks
& Tackle Sundry Irons
2 Tarpaulins old
2 Ladders 12 Buckets
1 lead line & Lead in
after Cabin 1 Heaving Line after

I certify the foregoing
Inventory to be correct
D. Christie
Depy. U.S. Marshal
Oct. 12th /63

Inventory of Effects on Board Steamer Alabama —

Cabin aft —

2 Bales Blankets Marked G.R.B. 240 & 241
1 Box content not known
2 Composition Saucers
1 Astral dinner Lamp
1 Plated Caster & Cruets
1 Mahogany Extension Table
1 Turning Cushion Chair
19 Cane Bottom Chairs

In State-Rooms

H — 1 Mattrass 2 Sheets 1 Pillow
G. 1 Moss Pillow
F. 1 Mattrass 2 Sheets 2 Pillows 1 Blanket — 1 Chronometer
E = Sealed
D — 2 Mattrasses 2 Pillows 1 Blanket
21 = 2 Mattrasses 3 Pillows
22 = 2   "   2   "
5 Tumblers
3 Wine Glasses
5 Plates — 9 Cups 19 Saucers China ware
2 Molasses Cups
2 Butter Dishes 3 Cake plates
1 Slop Bowl 2 Sugar Dishes
2 Butter Bowls 6 Gravy Dishes
8 Pickle Dishes 1 Glass preserve dish
1 Pair China Water Pitchers
1 Large & 2 Small Japan Waiters
1 Long Fish Dish
1 Box containing Crushed Sugar
2 Preserve Covers glass 5 Tumblers
1 Box Cayenne Pepper 1 Box Yeast Powder

1 Box Congress water
2 Kits No 1 Mackerell
1 Meat Cleaver
1 Tin Pan 1 Hot water Dipper
12 Lamp glasses 1 Keg & faucet
9 Iron Spoons 2 carving forks
1 Fork 1 Lamp 2 Tin Funnels
1 Oil Feeder 4 Buf Steak Dishes
1 Gravy Tureen 1 Tea Caddy
1 Large Dish Jean Tin
1 Tin Can 6 Demijohns
1 Containing Lemonade 1 Vinegar
1 Water Cooler 2 Water Monkeys
2 Coffee Urns Plated
2 Astral Burners
1 Feeder Brap 1 Dinner Bell
2 Salt Cellar 1 Clock Hard Rack

In Closets of Front Table

5 Pound paper Black Tea
3 Bitters 17 Lamp glasses
2 Baking Dishes Silver or Plated 1 Water Pitcher Plated
2 Glass Cake Dishes 1 Large Cake Dish
2 Soap Dishes 2 plated Waiters
3 Boxes Yeast Powder 1 Cayenne
3 Cut glass salt cellars
1 Box Matches about 4 Dozen
1 Crumb Brush 1 Steel
1 Lot Pepper about 1 lb 1 lb Nutmeg
2 Crumb Brushes 2 Bars & ½ Bar Castile Soap — 1 Lamp to Ceiling — 1 Lot Oil Cloth to Cabin floor

Inventory to Steamer Alabama Cont.

Pantry
6 Boxes Vermicella
1 Demijohn aguadiente
2 Bags Sugar about 20 lbs
1 " Coffee about 20 lbs
1 Bag starch " 25 lbs
1 Lot Codfish 20 lbs
1 Demijohn Molasses
1 Tin Can for oil
1 Hair Sifter
1 Brass Lamp 3 Copper
Dishes — 3 Pitchers
2 Bowls —
Clerks office 1 Demijohn
Castor Oil —
1 Broom 1 Brush 3 Old
Umbrellas 3 Brass Lamps
1 Compass Box —
1 Signal Lantern
1 Starboard Boat with
Mast Mainsail & 4 Oars
& Helm Davits & falls complete
1 Larboard Boat 4 Oars
Davits & falls Complete
2 Spittoons 1 Bucket
1 Japan Tin Dust pan
Wash Room
2 ½ Barrels Jamaica Rum
1.5 Gallon Keg Molasses
On Deck Larboard Side
2 Barrels Marked (B) alum
4   "   " P (R) Salts
1   "   " [H] 1222
3   "   " (R) D.M.
63-58-62-
2 Barrels Marked Bread —

3 Barrels Castor Oil
2 Marked [D] (R) 1 No Mark
1 Barrel Cream Tartar
1 Sack Salt —
1 Barrel (R) Salts ½ Barrel Marked 4-18
1   "   [H] 12 —
1 ½ Barrel Marked W.C.W.
1 Barrel open Contents not known
1 Barrel Beans —
Cook Room
16 White Stone China Dishes
1 Pitcher 1 Butter Dish
12 Tin Pans sundry Kinds
1 Coffee Mill 1 Pint Measure
2 Tin Dippers 5 Skillet
1 flour Sifter
1 Wooden Tray & Pounder
1 Soup Tureen —
1 Pepper Box 1 Sheet Iron
Bake pan 1 Coffee Boiler
3 Copper
Skillets 1 Iron 1 Tin Boiler
4 large Iron Spoons 1 Grater
1 Saw 1 Tormenter 2 Egglifters
1 Steel 1 Brass Coal Heaver
2 Wood Buckets 1 Tin
Slop Pail 3 large Knives 1 fork
1 Meat Chopper 1 Mortar
Boilers Deck
7 Coils Rope 1 Lot Old Matting
1 Ax 1 Maul 1 Boat Hook
1 Oar 2 Sails — 2 Awnings
1 Box & Saw — 2 Barrels Rosin
34 Sacks Salt 1 Piece Leather

Inventory of Steamer Alabama Cont'd

Forward Deck
1 Carpenter Chest & Tools
1 Mattress, 3 Pillows
2 Bow anchors & chains
1 Large Spar anchor
1 St. Showiser, 1 Spring
Line, 1 Spring line mid ships
2 lengths hose
5 Lanterns forward —
2 Chain Lockers
1 Containing Rope &c 1 Empty
1 Grind Stone 2 Trucks
1 Lot Paints Oils Brushes &
Pots
1 Ice Box 1 Old Hauser
2 Baskets 5 Old Bronze
1 Shovel 2 Wash Tub 1 Bucket (in Hose)
1 Barrel Beef Open
1 Tin Boiler 1 Small

Starboard Side
1 Mast spirits sail & Mainsail
for small Boat 1 Tarpaulin
3 Double Blocks 1 Single
1 Bucket

Store Room
5 Barrels Bread
1 " Tallow
1 Demijohn Coal Tar
1 ½ Barrel Tar —
2 lengths Hose
2 Cans Paint in Clerks office
1 Piece Sole Leather —
Up Stairs

Rooms
17 — 1 Oval Table & Draw
1 Parrared Ruler
2 Mattresses 1 Pillow
2 Sheets 1 Spread —
1 Bowl & Pitcher 1 Water Pitcher
1 Zinc Lamp 1 Life Preserver
1 Caw seat Chair
16 Room
2 Mattresses 3 Blankets 2 Sheets
2 Pillows 1 Duster — 1 Chair
1 Piece Leather 1 Lot Twine
1 Roll Remnants Canvass
Room
15 — 2 Mattresses 2 Pillows
1 Chair
Room 14 — 8 Axes 2 Brooms
½ Bag Cotton 1 Hank
Rollin stuff — 1 Cotton Hook
2 Small Single Blocks
Room 13 1 Box Raisins —
2 Boxes Sundries
2 Demijohns Liquor
2 Mattresses 2 Pillows
3 Blankets 2 Sheets 1 Spread
3 Lamp chimneys
1 Bag about 30 lbs Havana Sugar
Room 12 Books & Papers
of Boat 1 Tin Box & Pen
wash 1 Bill rack 1 Lot Carpet
Room R 1 Lot Carpet
Room aft Starboard Side
1 Lot Boat Linen clean
& Dirty
After Deck 1 Boat
his complete Tackle
1 Slop Pail on Deck

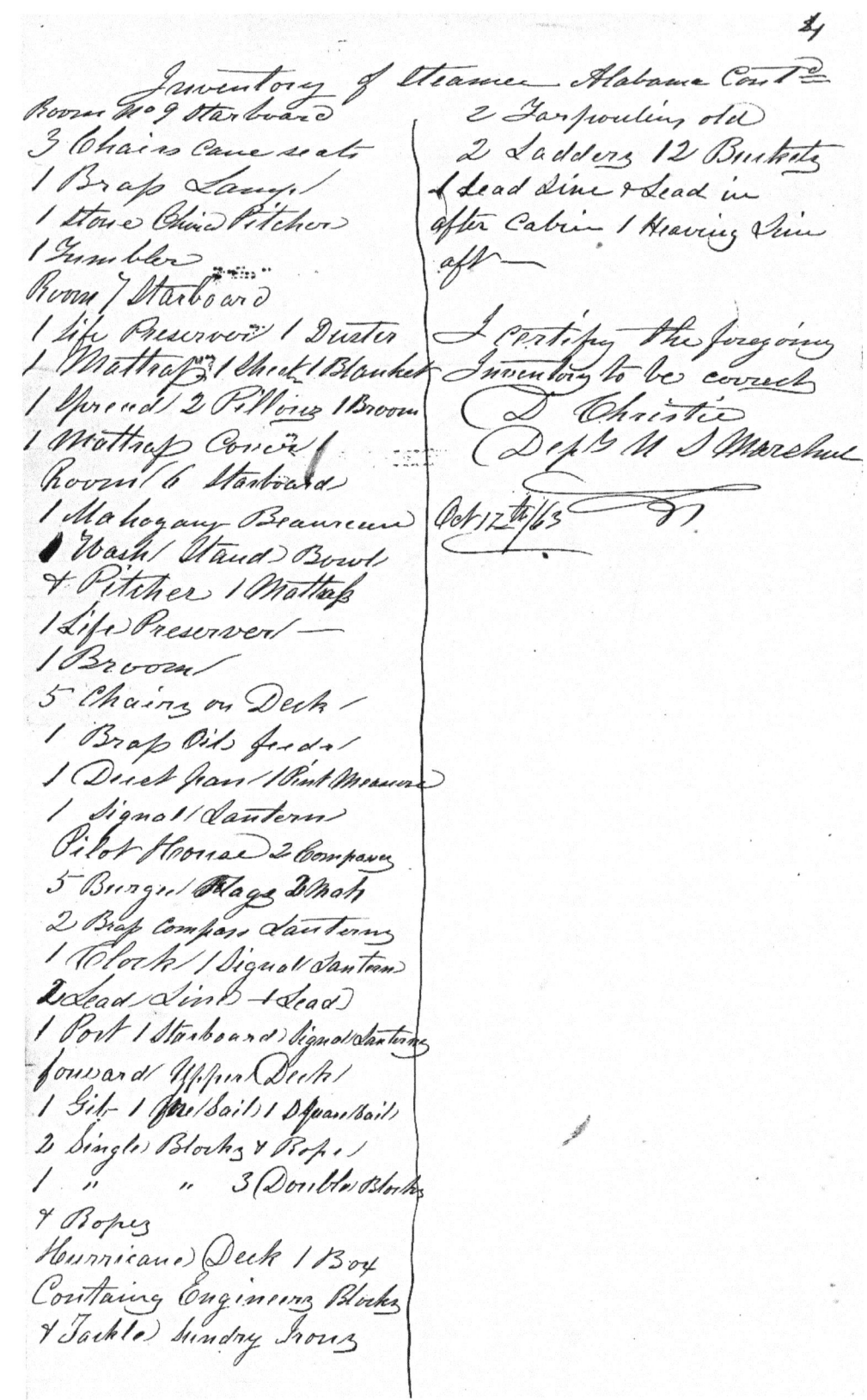

Figure 92. The inventory of effects aboard the Str. *Alabama*, all the movable gear and equipment on the ship excluding the cargo. (This page and the three preceding pages.) Dated 12 Oct. 1863. *NARA*.

Deposition of Lieutenant Commander Chandler, Captain of U.S.S. *San Jacinto*

The following document was Chandler's testimony concerning the *San Jacinto's* lengthy chase of the *Alabama* (Figure 93). One key to the story was the *San Jacinto's* assertion of being within signal distance. In fact, the *San Jacinto* was in the process of sending a boarding party to the prize when the *Alabama* was taken by the U.S.S. *Eugenie*. If within signal distance, the *San Jacinto* automatically qualified to share in the prize money. Chandler's location when filing the deposition was Key West as that was the headquarters of the squadron to which his ship was assigned. He filed his deposition with the U.S. District Court in Key West, and the court clerk forwarded it to the court in New Orleans.

Key West Florida
~~September~~ October 8th 1863

The rebel steamer *Alabama* was chased on the 11th and 12th of this month (of Sept.–letter originally dated in Sept.) by the *San Jacinto* and on the 12th inst. the said Steamer *Alabama* ran on shore in a place known in the chart as Chandeleur Sound, about six miles in a southerly direction from the light house on the north end of the Chandeleur Islands and was there taken possession of by the U.S. Steamer *Eugenia* [sic]. At the time the capture was made the U.S. Steamer *San Jacinto* was in plain sight and signal distance. To the best of my knowledge and belief, the only vessels in sight and signal distance at the time of said capture were . . . (rest of document obscured by NARA copy process).

(Scrap of paper at bottom of document showed that the submission was from Lieutenant Commander Chandler commanding the *San Jacinto*.)

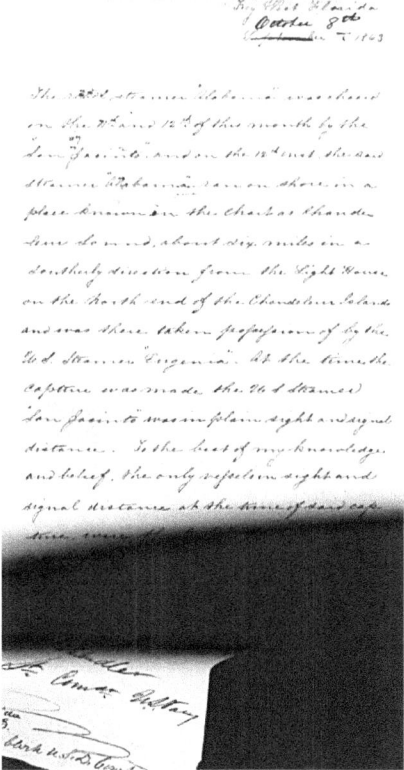

Figure 93. Deposition of Lieutenant Commander Chandler, commanding the *San Jacinto*, concerning the capture of the prize ship *Alabama*. Dated 8 Oct. 1863. *NARA*.

Deposition of Lieutenant Commander John N. Quackenbush, U.S.S. *San Jacinto*

The following document was the deposition of Lieutenant Commander Quackenbush, an officer (perhaps the executive officer) of the U.S.S. *San Jacinto* (Figure 94). He was part of the boarding party sent to the prize steamer *Alabama*.

Southern District of Florida, S. S.

In the matter of
The <u>Steamer</u>
    <u>*Alabama*</u>
        and Cargo.

Be it remembered, that on this <u>16th</u> day of <u>November</u> in the year of our Lord, one thousand eight hundred and sixty-three, before me, <u>George D. Allen, Clerk U.S. D. Court &</u> Prize Commissioner for the Southern District of Florida, personally appeared at my office in Key West, <u>John N. Quackenbush, Lt. Commander, U.S. Navy, to me personally known</u> who being duly cautioned and sworn to testify the truth, the whole truth, and nothing but the truth, did thereupon depose and say: <u>that during the eight months past he has been attached to the United States Steam Sloop of War *San Jacinto*.</u>

That in the 11th day of September 1863, the *San Jacinto* being in Lat. about 29° 2' (N) Long. about 87° 30' W. discovered a steamer about two o'clock in the afternoon and gave chase which was continued till about half past four o'clock in the afternoon of the 12th when the strange steamer ran ashore in the inside of the Chandeleur Light House, about three o'clock in the afternoon of the 12th & four guns were fired by the *San Jacinto* at the same steamer. On the *San Jacinto* getting into shoal water in the vicinity of the Chandeleur Light, the strange steamer having gone ashore, the *San Jacinto* being there about seven miles from said steamer. The first cutter of the *San Jacinto* being manned and armed was dispatched in charge of deponent toward the said steamer. That in his arrival on board found said steamer in command charge of Lt. Commander Charles H. Green, U.S. Navy, he having been taken possession of by the officers and crew of the U.S. Steamer *Eugenia* as prize of war.

That the said steamer so captured proved to be the Confederate Steamer *Alabama*, and the *San Jacinto* was in plain sight and signal distance at the time of said capture.

                (signed) J. N. Quackenbush
                Lieu. Com'dr. U.S.N.

Sworn and subscribed
before me this 16th day of
November 1863, at Key West.
    (signed) George D. Allen
        Clerk U.S. D. Court

Southern District of Florida, S. S.

In the matter of
The Steamer
Alabama
and Cargo.

BE IT REMEMBERED, that on this 16th day of November in the year of our Lord, one thousand eight hundred and sixty-three, before me, George D. Allen, clerk U.S. & acting Prize Commissioner for the Southern District of Florida, personally appeared at my office in Key West, John N. Quackenbush Lt. Commander U.S. Navy, to me personally known who being duly cautioned and sworn to testify the truth, the whole truth, and nothing but the truth, did thereupon depose and say:— that during the eight months last past he has been attached to the United States Steam Sloop of War San Jacinto.

That on the 11th day of September 1863, the San Jacinto being in Lat. about 29° 2' Long about 87° 30' W. discovered a Steamer about two o'clock in the afternoon and gave Chase, which was continued till about half past four o'clock in the afternoon of the 12th, when the Strange Steamer run ashore on the inside of the Chandeleur Light-house. About three o'clock in the afternoon of the 12th four guns were fired by the San Jacinto at the said Steamer. On the San Jacinto getting into shoal water in the vicinity of the Chandeleur Light, the Strange Steamer having gone ashore, the San Jacinto being then about seven miles from said Steamer, the First Cutter of the San Jacinto being manned and armed was dispatched in charge of deponent to board the said Steamer, that on his arrival on board found said Steamer in ~~command~~

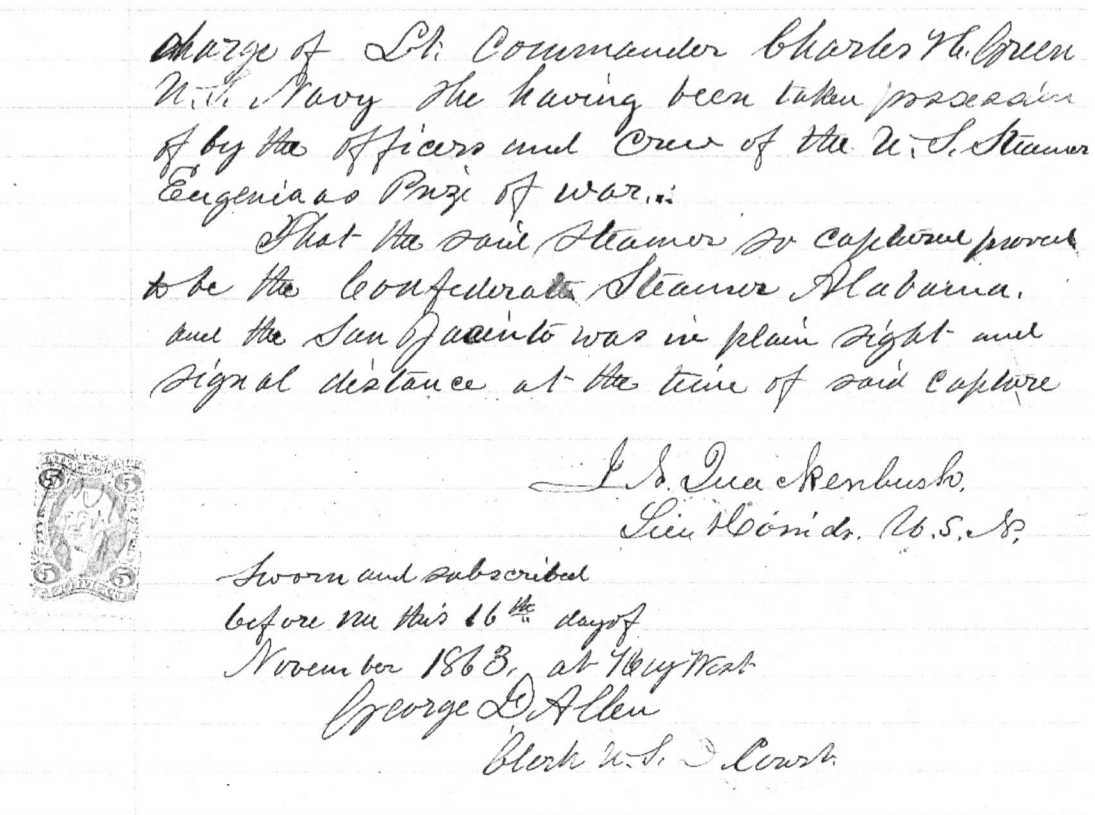

Figure 94. Deposition of Lieutenant Commander J. N. Quackenbush of the *San Jacinto*. (This page and previous.) Dated 16 Nov. 1863 at Key West. *NARA*.

Deposition of Acting Assistant Paymaster Asa C. Winter, U.S.S. *San Jacinto*

Acting Assistant Paymaster Asa C. Winter provided the following deposition concerning the capture of the *Alabama* (Figure 95).

Southern District of Florida, S. S.

In the matter of
The <u>Steamer</u>
  <u>Alabama</u>
   and Cargo.

                      Be it remembered, that on this <u>16th</u> day of <u>November</u> in the year of our Lord, one thousand eight hundred and sixty-three, before me, <u>George D. Allen, Clerk U.S. D. Court &</u> Prize Commissioner for the Southern District of Florida, personally appeared at my office in Key West, <u>Asa C. Winter, A. A. Paymaster, U.S. Navy,</u> who being duly cautioned and sworn to testify the truth, the whole truth, and nothing but the truth, did thereupon depose and say: <u>That on the 12 Sept. 1863, he was attached to the U.S. Ship *San Jacinto*.</u> That on that day the blockade runner known as the rebel Steamer *Alabama* became a prize to the U.S. Naval forces. That at the time of the capture of the said Steamer *Alabama*, the U.S. Ship *San Jacinto* was within signal distance. That the captured vessel and the captor were plainly visible from the poop deck of the *San Jacinto*. That on the 11th Sept. 1863, the said steamer was first sighted by the *San Jacinto*. That the chase was continued, by order of the Com'dg. Officer of the *San Jacinto*, until the said Steamer *Alabama* was in the possession of the U.S. Naval forces then in the vicinity of the Chandeleur Islands.

                    (signed) Asa C. Winter
                            A. A. Paymaster, U.S.N.

Sworn and subscribed
before me this 16th day of
November 1863, at Key West.
   (signed) George D. Allen
           Clerk U.S. Dist. Court & Prize Comm.

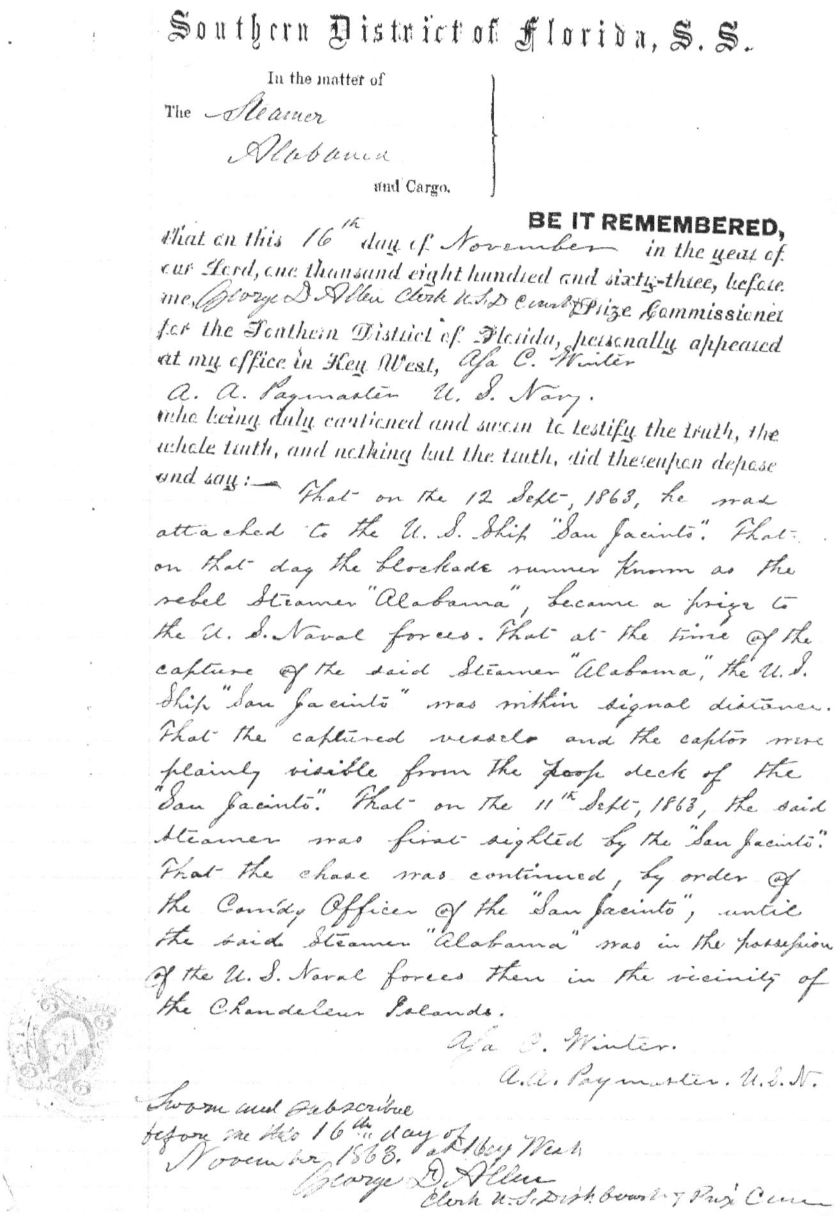

Figure 95. Deposition of Acting Assistant Paymaster Asa C. Winter of the *San Jacinto*, concerning the capture of the prize steamer *Alabama*. Dated 16 Nov. 1863 in Key West. *NARA*.

Deposition of Several Officers of the U.S.S. *San Jacinto*

A group of officers of the *San Jacinto* offered the following joint deposition regarding the capture of the *Alabama* (Figure 96).

> The undersigned, Officers on board the U.S. Sloop *San Jacinto*, do certify & make oath that, after chasing the prize Steamer *Alabama* into Chandeleur Island harbor, on the 12th of September last, and coming to anchor, we saw the *Alabama* aground, from the deck of the *San Jacinto*, and had commenced to man the first cutter with a view to the capture of that steamer, before she was boarded by the U.S. Tug *Eugenie*. At the time of the capture of the *Alabama* by the *Eugenie*, the *San Jacinto* was in plain sight.
>
> (signed) Sam$^{l.}$ B. Clark, Acting Master
> (signed) Charles Cunningham, Acting Ensign
> (signed) Thos. Turny, Acting Master's Mate
> (signed) Augustus H. Fuller, Acting Master's Mate
> (signed) Horace G. Brinker, Actg. M. Mate
>
> Southern Dist. of
> Florida S. S.
> Sworn and subscribed
> before me this 17th day of
> November 1863, at Key West.
> (signed) George D. Allen
> Clerk U.S. D. Court

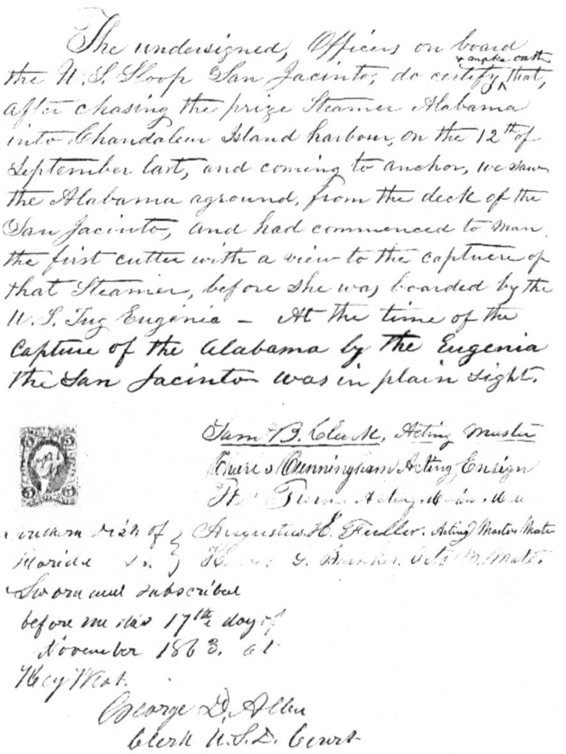
The undersigned, Officers on board the U.S. Sloop San Jacinto, do certify & make oath that, after chasing the prize Steamer Alabama into Chandaleur Island harbour, on the 12th of September last, and coming to anchor, we saw the Alabama aground, from the deck of the San Jacinto, and had commenced to man the first cutter with a view to the capture of that Steamer, before she was boarded by the U.S. Tug Eugenia – At the time of the capture of the Alabama by the Eugenia the San Jacinto was in plain sight.

Saml B. Clark, Acting Master
Charles Cunningham Acting Ensign
Augustus H. Fuller. Acting Master Mate
Sworn and subscribed before me this 17th day of November 1863, at Key West.
George D. Allen
Clerk U.S.D. Court

Figure 96. A group of the *San Jacinto's* officers made a joint deposition about the capture of *Alabama*. Dated 17 Nov. 1863. *NARA*.

The U.S.S. *Colorado* Attempted to Claim a Share of the Prize

The senior officer of the *Colorado* contacted the prize court in a far-fetched effort to claim a part of the prize money for the *Alabama* (Figure 97). The *Colorado* was on station off Mobile Bay as part of the blockading squadron. There was no valid way for the U.S.S. *Colorado* to make a valid claim as she was nowhere near within signal distance of the scene of the action.

U.S.S. Frigate *Colorado*
Blockade off Mobile
Oct. 24, 1863

Sir,

Information has been received, that the Steamer *Eugenie*, the captor of the blockade runner *Alabama*, claims to be a regularly commissioned vessel, and not a tender to the U.S. Frigate *Colorado*, and disputes the right of the *Colorado* to a share in the prize.

Capt. J. R. Goldsborough who commanded this ship at the time of the capture of the *Alabama*, and by whom prize lists were properly forwarded,[164] has since been ordered North, and I, as the officer next in rank remaining on board, believe it my duty, to act in the premises, and lay before you the following statement.

The *Eugenie* after being condemned by the prize court at Key West was sent at the request of Capt. Goldsborough to report to him; and was received by him, as a tender to the *Colorado*. A crew was put on board, taken entirely from this ship's company, and their names still held on the Paymaster's books. They are from time to time changed; no permanent crew has ever been assigned her. They comprise part of the complement of the *Colorado's* crew.

The principle part of her armament was supplied from the *Colorado's* battery and armory, water and provisions and all necessary supplies are received from her, and she moors to a hawser from the *Colorado's* stern, precisely as would one of her launches.

The names of the officers doing duty on board the *Eugenie* are carried on the books of the *Colorado*, and have always been considered as attached to her.

The *Eugenie's* duties have ever been those of a tender to the *Colorado*, and never to any other vessel, except during a short space of time she was lent to a senior vessel; but her crew have *always* been provisioned by that vessel.

At the time of the capture of the *Alabama*, the *Eugenie* was under orders to proceed to New Orleans for provisions &c. for the *Colorado*, and had on board the Paymaster, as well as a number of the crew of that vessel, more than were sufficient to man the *Eugenie*.

Until the capture of the *Alabama* I am not aware that the fact of the *Eugenie* being a *tender*, was ever doubted or questioned, and I cannot now see any just ground for releasing her from that class of vessels.

I feel it but just to the crew of the *Colorado*, many of whom have at different times served on board the *Eugenie*, and all of whom have expended much time and work upon her, and have ever considered her as part of their own vessel, that this statement should be laid before the court for their favorable consideration.

To the	Very respectfully
Hon. Rufus Waples	Your obt. Servant
U.S. District Attorney	Henry W. Miller
Eastern District of La.	Lieut.
	U.S. Navy

---

[164] The court's file on the *Alabama* prize case did not include a prize list for the U.S.S. *Colorado* or any further correspondence about the claim. Ultimately, the judge disallowed the *Colorado* claim.

U. S. Frigate Colorado
Blockade off Mobile
Oct. 24th 1863

Sir.

Information has been received, that the Steamer Eugenie, the captor of the blockade runner Alabama, claims to be a regularly commissioned vessel, and not a tender to the U. S. Frigate Colorado, and disputes the right of the Colorado to a share in the prize.

Capt. J. R. Goldsborough who commanded this ship at the time of the capture of the Alabama, and by whom prize lists were properly forwarded, has since been ordered North, and I, as the officer next in rank remaining on board, believe it my duty, to act in the premises, and lay before you the following statement.

The Eugenie after being condemned by the prize court at Key West, was sent at the request of Capt. Goldsborough to report to him; and was received by him, as a tender to the Colorado. A crew was put on board, taken entirely from this ship's company, and their names still held on the Paymaster's books. They are from time to time changed, no permanent crew has ever been assigned her. They comprise part of the complement of the Colorado's crew.

The principle part of her armament was supplied from the Colorado's battery and armory, water and provisions and all necessary supplies are received from her, and she moors to a hawser from the Colorado's stern, precisely as would one of her Launches.

The names of the Officers doing duty on board the Eugenie are carried on the books of the Colorado, and have always been considered as attached to her.

The Eugenie's duties have ever been those of a tender to the Colorado, and never to any other vessel, except during a short space of time she was lent to a senior vessel; but her crew have <u>always</u> been formed from the Colorado's ship's company and has always been provisioned by that vessel.

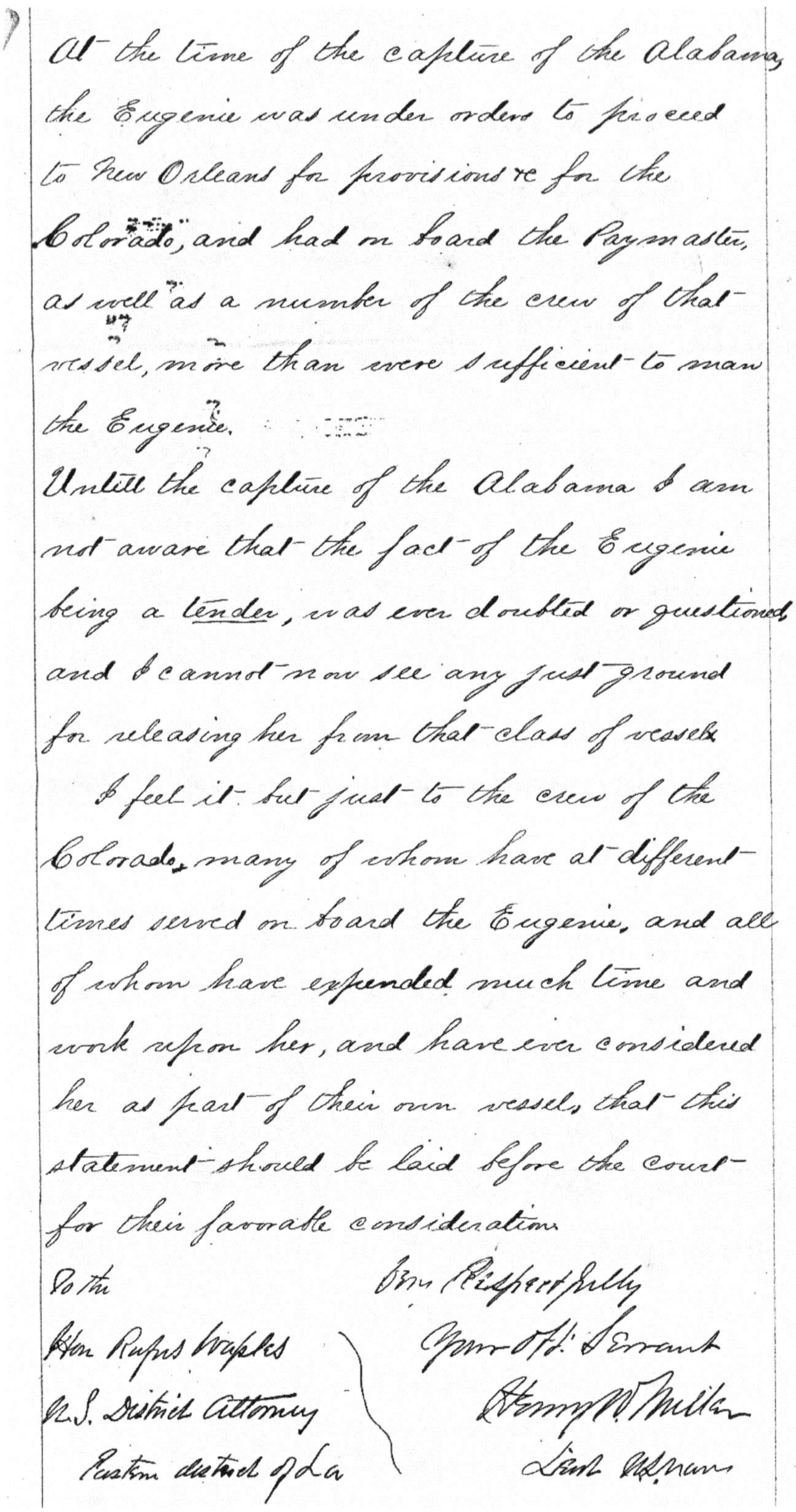

At the time of the capture of the Alabama, the Eugenie was under orders to proceed to New Orleans for provisions &c for the Colorado, and had on board the Paymaster, as well as a number of the crew of that vessel, more than ever sufficient to man the Eugenie.

Untill the capture of the Alabama I am not aware that the fact of the Eugenie being a *tender*, was ever doubted or questioned, and I cannot now see any just ground for releasing her from that class of vessels.

I feel it but just to the crew of the Colorado, many of whom have at different times served on board the Eugenie, and all of whom have expended much time and work upon her, and have ever considered her as part of their own vessel, that this statement should be laid before the court for their favorable consideration.

To the
Hon. Rufus Waples
U.S. District Attorney
Custom district of La

Very Respectfully
Wm. H. Servant
Henry W. Miller
Lieut. Adams

Figure 97. Officers of the U.S. Frigate *Colorado* made a claim for a share of the *Alabama* prize money stating that the *Eugenie* was a tender of the *Colorado* and crewed from the *Colorado's* complement. (Three pages.) Dated 24 Oct. 1863. NARA.

Commodore Bell Requested the *Alabama* for the Navy

The commander of the West Gulf Blockading Squadron wrote the prize court and requested that the prize ship be retained for service as a U.S. Navy blockading vessel with his squadron (Figure 98).

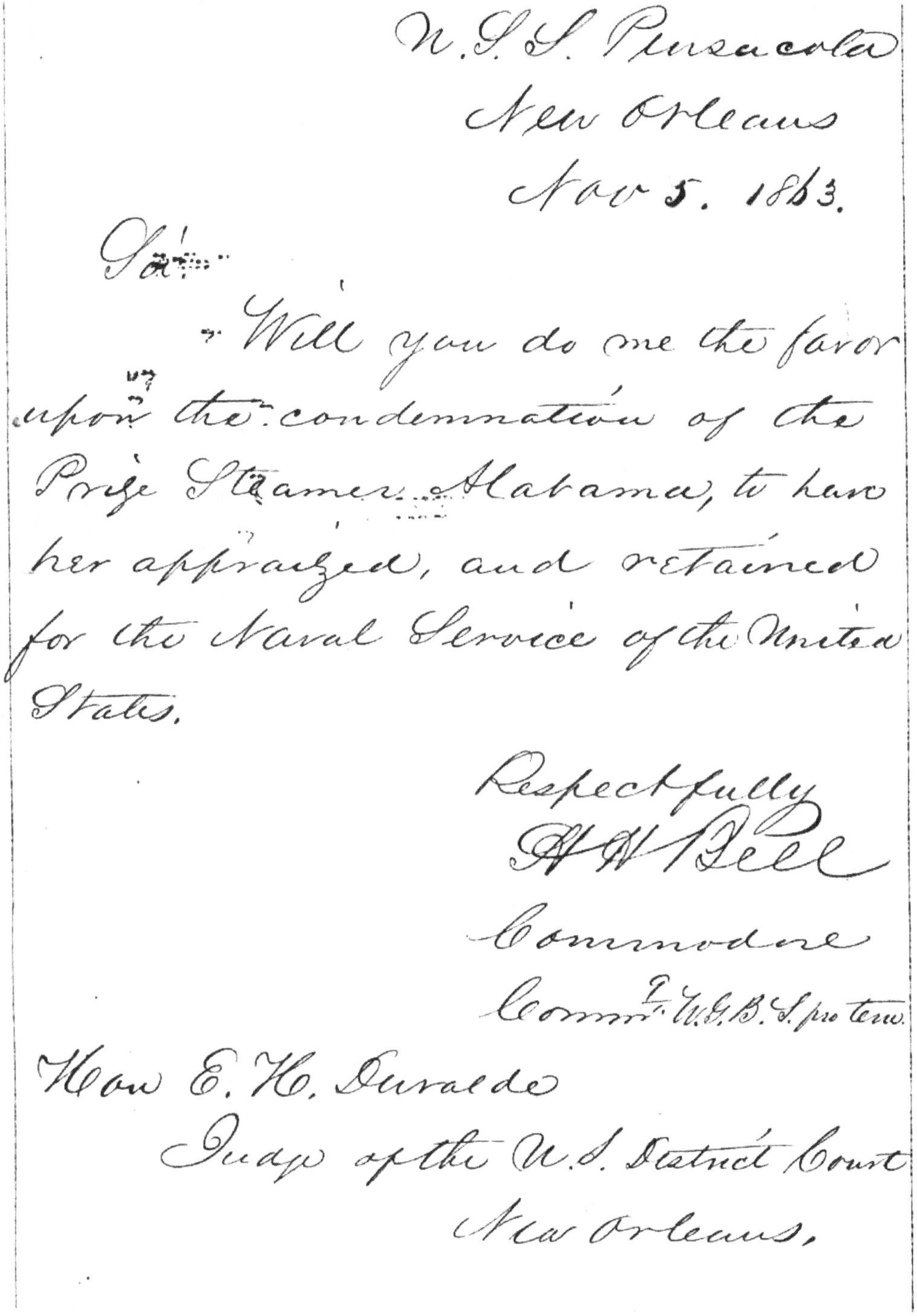

U.S.S. Pensacola
New Orleans
Nov 5. 1863.

Sir:

Will you do me the favor upon the condemnation of the Prize Steamer Alabama, to have her appraized, and retained for the Naval Service of the United States.

Respectfully
H H Bell
Commodore
Commdg. W.G.B.S. pro tem.

Hon E. H. Durell
Judge of the U.S. District Court
New Orleans.

Figure 98. Commodore Bell asked Judge Durell to turn the *Alabama* over to the navy for blockade duty. (Legible, not transcribed.) Dated 8 Nov. 1863. *NARA*.

Appointment of the Appraisers

U.S. Marshal James Graham appointed Peter Steele and L. B. Frost to be appraisers for the "*Alabama*, her tackle, apparel, furniture, and appurtenances." The appraisers signed an oath faithfully to discharge their duties (Figure 99).

    United States   U.S. District Court
        vs.
    Steamer *Alabama*   No. _____
    her tackle, &c., and Cargo   Prize

I hereby appoint Peter Steele and S. B. Frost, the appraisers, to value the Steamer *Alabama*, her tackle, apparel, furniture, and appurtenances, as she now lies at the wharf in the Port of New Orleans.

                                                                   (signed) Jas. Graham
New Orleans, Nov. 9, 1863   U.S. Marshal

We, the undersigned, having been duly appointed appraisers to value the Steamer *Alabama*, her tackle, apparel, furniture, and appurtenances, do solemnly and severally swear, that we will, to the best of our knowledge and ability, faithfully discharge the duties thus devolving on us.

                                                                   (signed) Peter Steel
New Orleans, Nov. 9, 1863   (signed) S. B. Frost

Sworn to and subscribed
before me, this 9th day
of November A.D. 1863
        (signed) Alfred Shaw
                Clerk

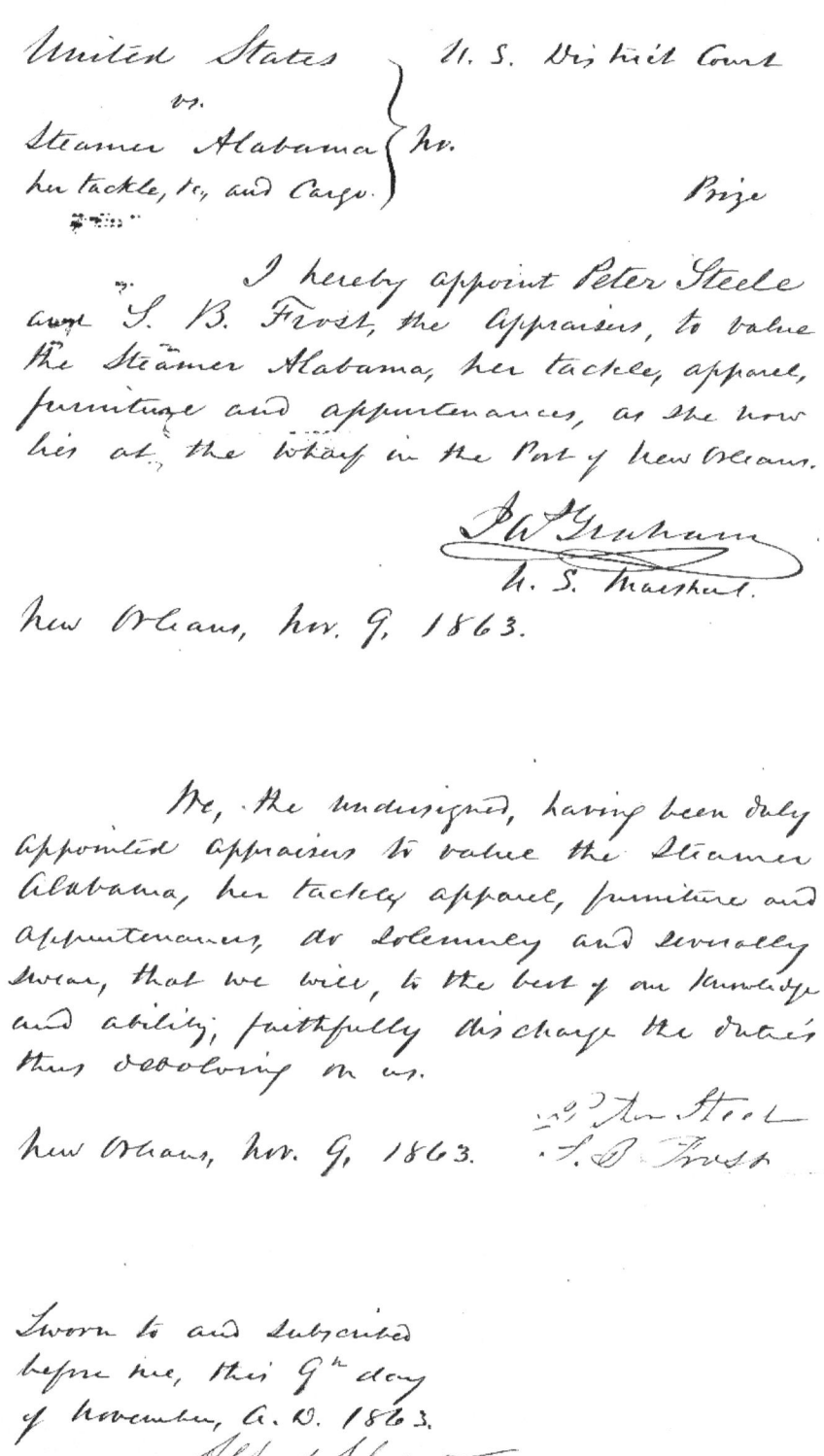

Figure 99. Appointment of the appraisers by the marshal. Dated 9 Nov. 1863. *NARA*.

Petition and Order for Appraisement of the Str. *Alabama* and Delivery to the Navy

Commodore H. H. Bell, commanding the West Gulf Blockading Squadron, petitioned Judge E. H. Durell to order an appraisal of the *Alabama* and then to turn the ship over to be taken into the naval service. The appraised price was to be paid into the register of the court by the government. The judge agreed and so ordered, dated 10 Nov. 1863 (Figure 100). Note that the ship was to be delivered to the navy only after payment to the court of the appraised value.

United States
    vs.
The Steamer *Alabama*   In Prize
her tackle, Apparel, &c.
& Cargo.

    To the Hon. E. H. Durell, Judge of the U.S. District Court, sitting in Chambers:
    The petition of Rufus Waples, U.S. District Attorney, respectfully represents:
    That he herewith presents an application of H. H. Bell, Commodore, Commanding Western Gulf Blockading Squadron, *pro tem*, addressed to your honor, asking to have the Steamer *Alabama* condemned herein as lawful prize of the United States, appraised and ~~delivered~~ retained for the Naval Service thereof;
    Wherefore the said Attorney in behalf of the United States prays for an order of appraisement and that the Marshal be directed to deliver the said Steamer to Commodore H. H. Bell, for the purposes aforesaid upon the payment of the amount of the appraisement into the Registry of the Court.
    And as in duty bound &c.

                                               (signed) Rufus Waples
                                                    U.S. Dist. Atty.

    In Chambers.
    Ordered that the Marshal shall without delay cause the Streamer *Alabama* condemned as prize to be appraised and delivered to Commodore H. H. Bell, Commanding the Western Gulf Blockading Squadron to be used in the Naval service of the United States upon the payment of the amount of the appraisement into the registry of the Court.
                                                (signed) E. H. Durell
                                                          Judge

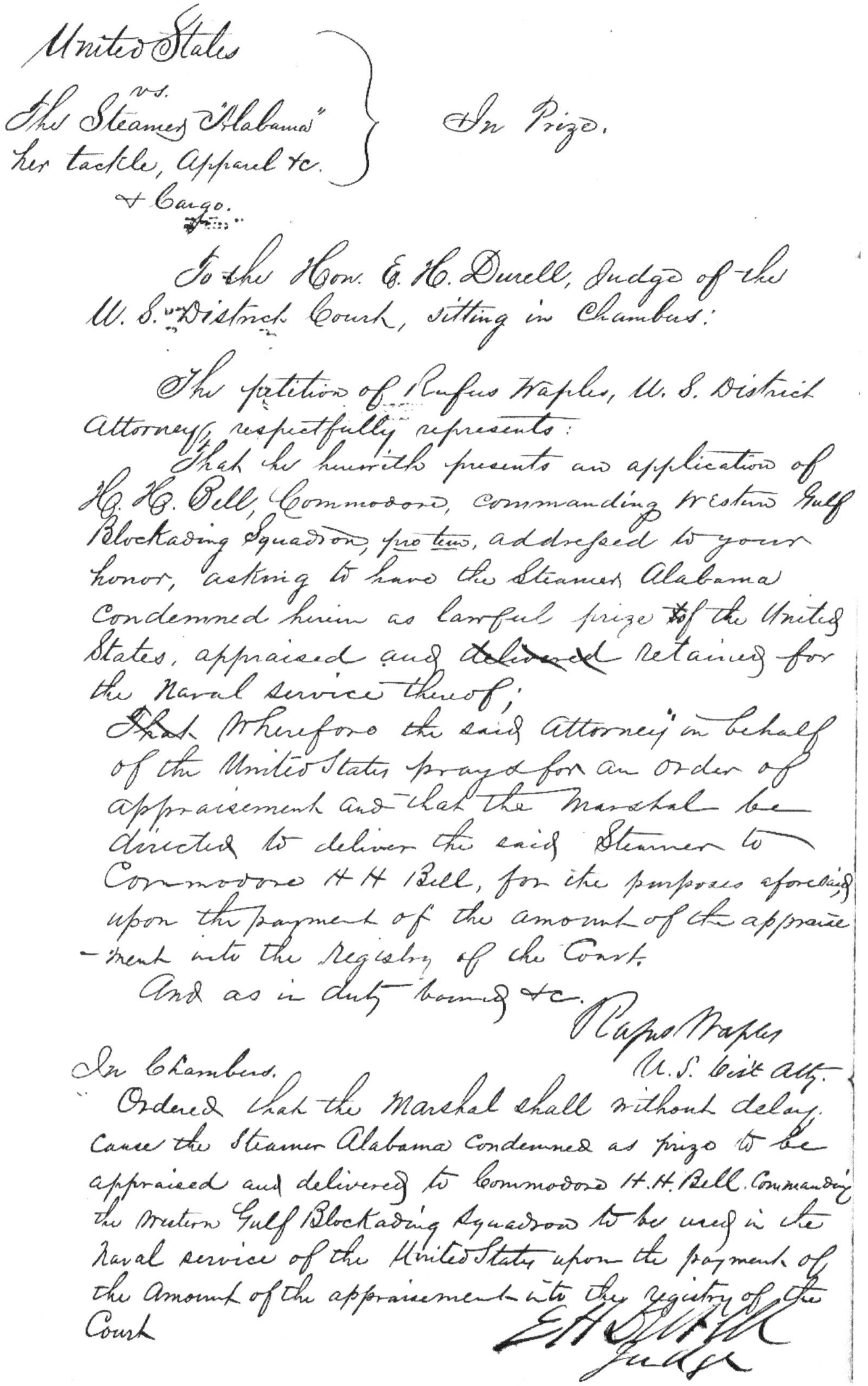

United States
    vs.
The Steamer "Alabama"
her tackle, Apparel &c.
    & Cargo.
} In Prize.

To the Hon. E. H. Durell, Judge of the U. S. District Court, sitting in Chambers:

The petition of Rufus Waples, U. S. District Attorney, respectfully represents:

That he herewith presents an application of H. H. Bell, Commodore, commanding Western Gulf Blockading Squadron, pro tem, addressed to your honor, asking to have the Steamer Alabama condemned herein as lawful prize of the United States, appraised and ~~delivered~~ retained for the Naval service thereof;

~~That~~ Wherefore the said Attorney on behalf of the United States prays for an order of appraisement and that the Marshal be directed to deliver the said Steamer to Commodore H. H. Bell, for the purposes aforesaid, upon the payment of the amount of the appraisement into the Registry of the Court.

And as in duty bound &c.

Rufus Waples
U. S. Dist. Atty.

In Chambers.

Ordered that the Marshal shall without delay cause the Steamer Alabama condemned as prize to be appraised and delivered to Commodore H. H. Bell, Commanding the Western Gulf Blockading Squadron to be used in the Naval service of the United States upon the payment of the amount of the appraisement into the registry of the Court

E. H. Durell
Judge

Figure 100. Petition and order to transfer the *Alabama* to the navy. Filed 10 Nov. 1863. *NARA*.

Testimony Regarding the Term Signal Distance

The court requested testimony on the meaning of the term "signal distance" in order to determine which U.S. Navy ships received prize money (Figures 101-102). In prize actions, the law specified that all navy ships within signal distance would share the prize money since, in theory, they were all close enough and able to be called on to assist in the capture if needed.

The U.S. prize commissioner sent Commodore Bell a requested to appear in court regarding the meaning of "signal distance," dated 7 Nov. 1863. The same letter went to Fleet Captain S. R. Franklin, U.S. Steamer *Pensacola*. A similar, but much shorter, letter went to Captain Marchand, Str. *Lackawanna*; Lt. Comdr. S. K. Mayo, *Kanawha*; Lt. Comdr. O. F. Staunton, *Panola*; and Lt. Comdr. P. C. Johnson, *Katahdin*. The commodore declined but did arrange for other officers to assist the court by testifying.

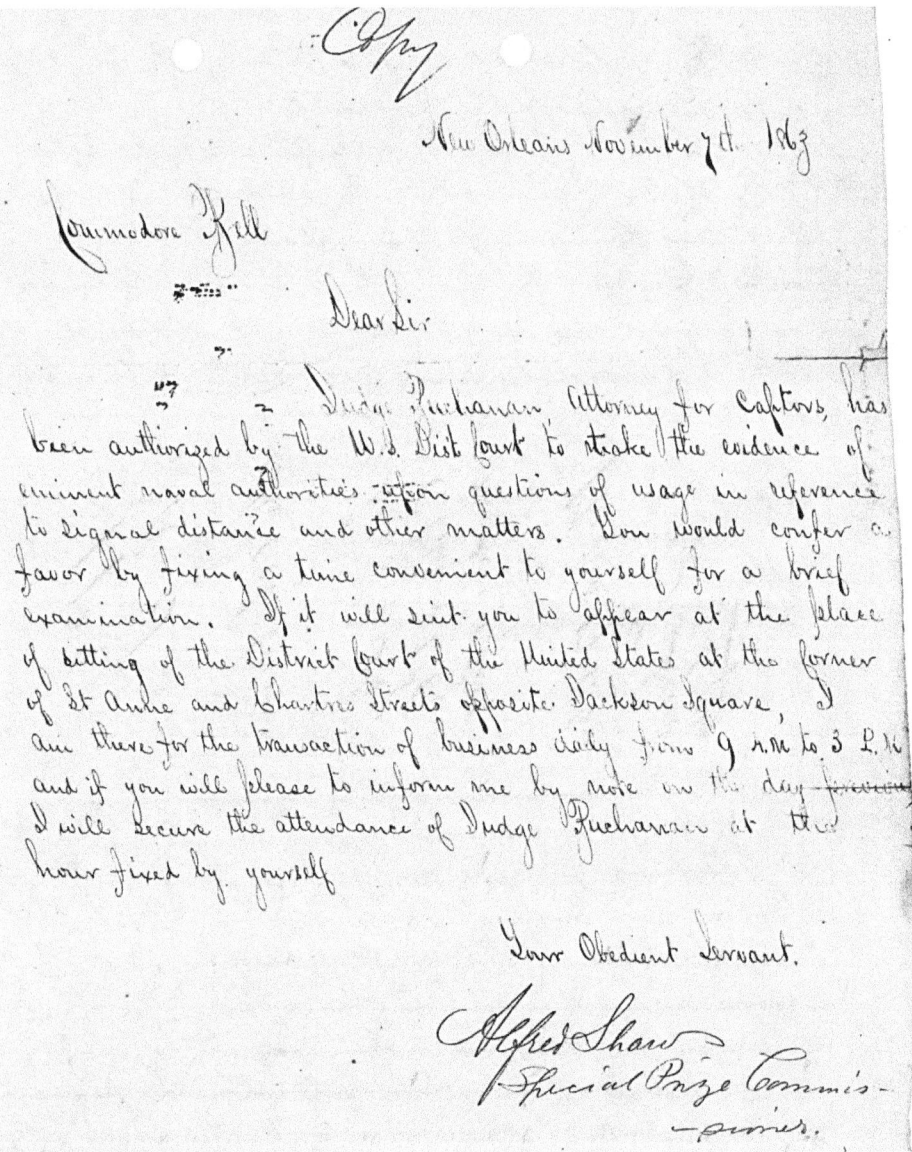

Figure 101. Letter from Special Prize Commissioner Alfred Shaw to Commodore Bell requesting his testimony about the meaning of the term "signal distance." Bell arranged for several of his ship captains to give their professional expert opinions. (Legible, not transcribed.) Dated 7 Nov. 1863. **NARA.**

Captain Jenkins began the testimony:

In the matter of	Testimony taken
The Steamer *Alabama*	before the Prize
& Cargo a Prize	Commissioner the

Witness being examined by A. M. Buchanan, Esq., Counsel for the Captors.

Captain Sherman A. Jenkins, U.S. Navy, being duly sworn, and being asked what he understands by the expression used in the 3^d Section of the Act for the Government of the Navy approved 17 July 1862, "within signal distance." Says, technically speaking, a vessel is understood to be within signal distance (and thus entitled to share in prize money generated by the capture of a blockade-runner) when she can be communicated with by any prescribed day, night, or fog signals. In this spirit of the law my opinion is that it applies to any vessel within sight, for a vessel may be considered within signal distance although the vessel making or repeating the signal may be too far from the vessel to which the signal is made to be certain of its meaning.

Being asked whether on 12 Sept. last at 5½ in the afternoon, a vessel like the *San Jacinto* being at a distance of 9½ miles, may be considered within signal distance says. Assuming from the testimony read to me in this case, that there was no impediment of foggy or hazy weather & that the several vessels including the *Alabama* were plainly seen from the deck of the *San Jacinto*, I have no hesitation in giving it as my opinion that the *San Jacinto* was within signal distance technically speaking.

Being asked who was the commander of the squadron to which the *Tennessee & Eugenie* belonged.

Answers those two vessels belonged to the Western Gulf Blockading Squadron under the command of Commodore Henry H. Bell, U.S. Navy.

Being asked who was the commander of the fleet to which the *San Jacinto* belonged at the time.

Answers Eastern Gulf Blockading Squadron under the command of Acting Rear Admiral Theodorus Baily, U.S. Navy.

                (signed) Thurston A. Jenkins
                     Capt. U.S.N.

Sworn to & subscribed before
Me Nov. 10th 1863.
  Alfred Shaw
   Special Prize Comm^r.

Captain John B. Marchand, U.S. Navy, commanding U.S.S. *Lackawanna* testified (Figure 102) to the effect that he agreed with Captain Jenkins of the U.S.S. *Richmond*. Marchand not herein transcribed.

In the matter of
the Steamer Alabama
& Cargo a
Prize
} Testimony taken
before the Prize
Commr. — the
Witnesses being examined
by A. McBuchanan Esq. Counsel for
Captors.

Captain Thornton A Jenkins U. S Navy being duly Sworn. and being asked what he understands by the expression used in the 3d Section of the act for the government of the Navy approved 17 July 1862 "within Signal distance". Says, technically Speaking, a vessel is understood to be within Signal distance when she can be communicated with, by any prescribed day, night, or fog Signals. In the Spirit of the Law my opinion is, that it applies to any vessel within sight, for a vessel may be Considered within Signal distance altho' the vessel making or repeating the signal may be too far from the vessel to which the signal is made to be certain of its meaning.

Being asked whether on 12" Septr last at 5½ in the afternoon, a vessel like the San Jacinto being at a distance of 9½ miles, may be considered within signal distance

Sept. Assuming from the testimony read to me in this case, that there was no impediment of foggy or hazy weather & that the several vessels, including the Alabama were plainly seen from the deck of the San Jacinto. I have no hesitation in giving it as my opinion that the San Jacinto was within signal distance technically speaking.

Being asked who was the commander of the Squadron to which the Tennessee & Eugenie belonged

Answer, those two vessels belonged to the Western Gulf blockading Squadron under the command pro tem of Commodore Henry H. Bell U.S. Navy

Being asked who was the Commander of the fleet to which the San Jacinto belonged at that time

Answer Eastern Gulf Blockading Squadron under the command of Acting Rear Admiral Theodorus Bailey U.S. Navy

Thornton A. Jenkins
Capt U.S.N.

Sworn to & subscribed be-
fore me Nov 10th 1863
Alfred Shaw
Special Bazeleair

Capt John B Marchand U.S. Navy being duly sworn &

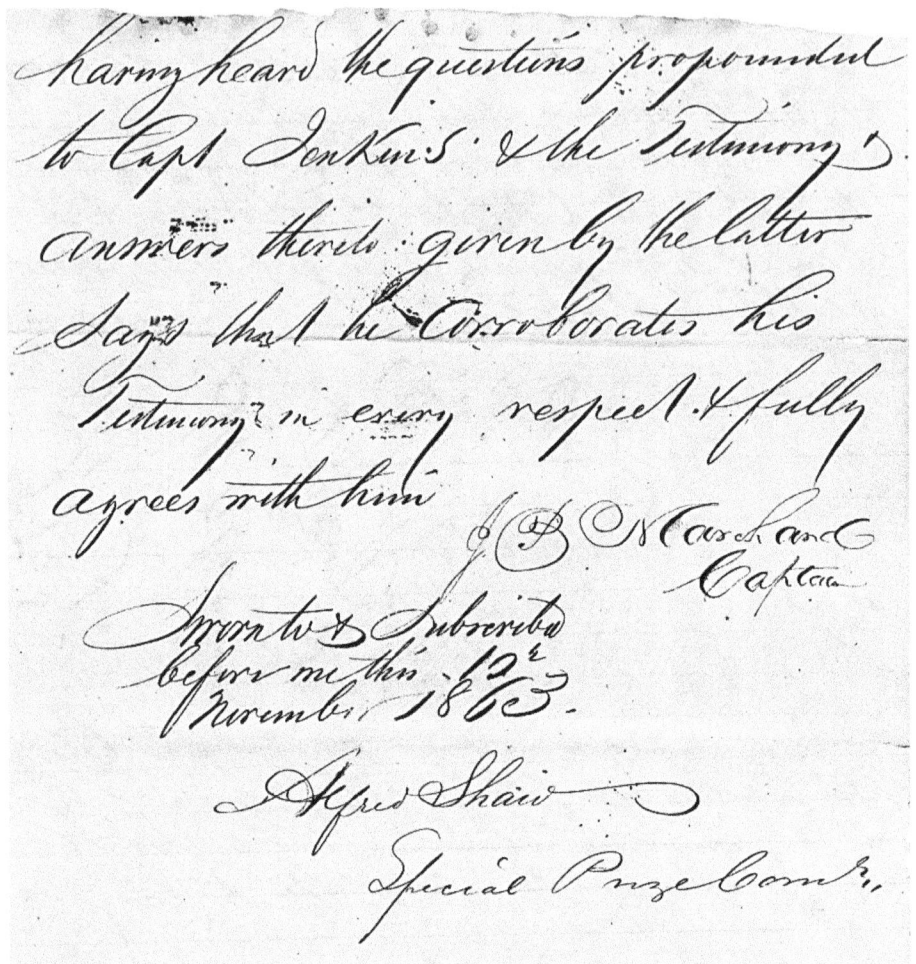

Figure 102. Testimony of Captains Jenkins and Marchand concerning the definition of the term "signal distance" as used in the prize law. (This page and two prior pages.) Dated 10 Nov. 1863. **NARA**.

Appraisal for the *Alabama*

The prize Str. *Alabama* and all contents excluding the cargo appraised for $80,000. The appraisers billed $20 for their services (Figures 103-104).

United States	U.S. District Court
vs.	
The Steamer *Alabama*	
her tackle, &c. & Cargo	

We the undersigned duly appointed to appraise the steam boat *Alabama* have performed that service and report that the said steam boat *Alabama* her tackle, apparel furniture and everything therewith belonging, is worth this day ($80,000) eighty thousand dollars and no more. There is on board of her not included in the above appraisement about twenty five tons coal worth twenty dollars per ton and three cords pine wood worth eight dollars per cord.

(signed) S. B. Frost
(signed) Peter Steel

To
Mr. James Graham
U.S. Marshal

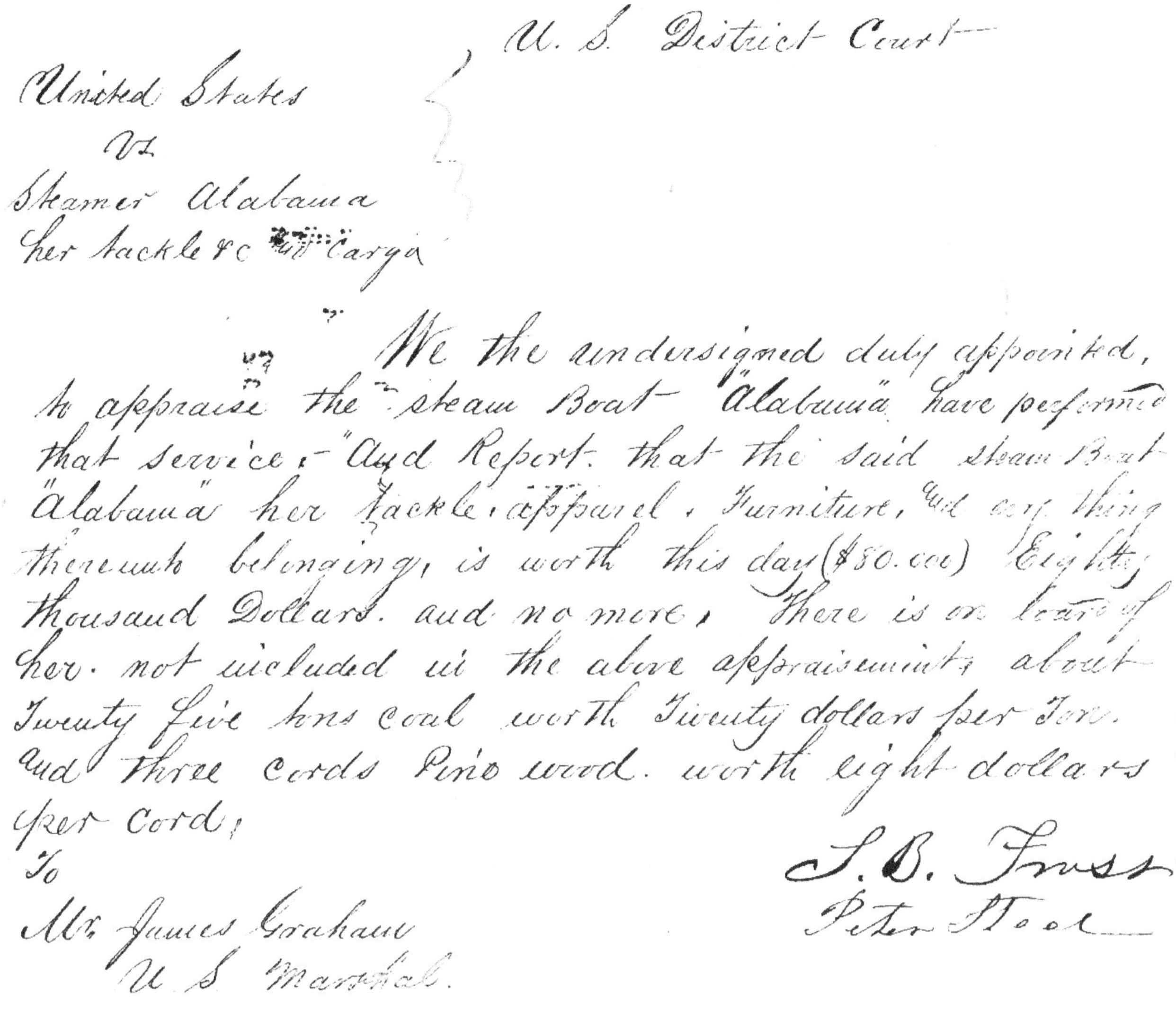

U. S. District Court

United States
vs.
Steamer Alabama
her tackle &c and Cargo

We the undersigned duly appointed, to appraise the steam Boat "Alabama" have performed that service — And Report. that the said steam Boat "Alabama" her Tackle, apparel, Furniture, and every thing thereunto belonging, is worth this day ($80.000) Eighty thousand Dollars. and no more, There is on board of her. not included in the above appraisements, about Twenty five tons coal. worth Twenty dollars per Ton. and three cords Pine wood. worth eight dollars per cord;

To
Mr. James Graham
U. S Marshal.

S. B. Frost
Peter Steel

Figure 103. The report of the appraisers. The Str. *Alabama* was valued at $80,000. Filed 10 Nov. 1863. **NARA**.

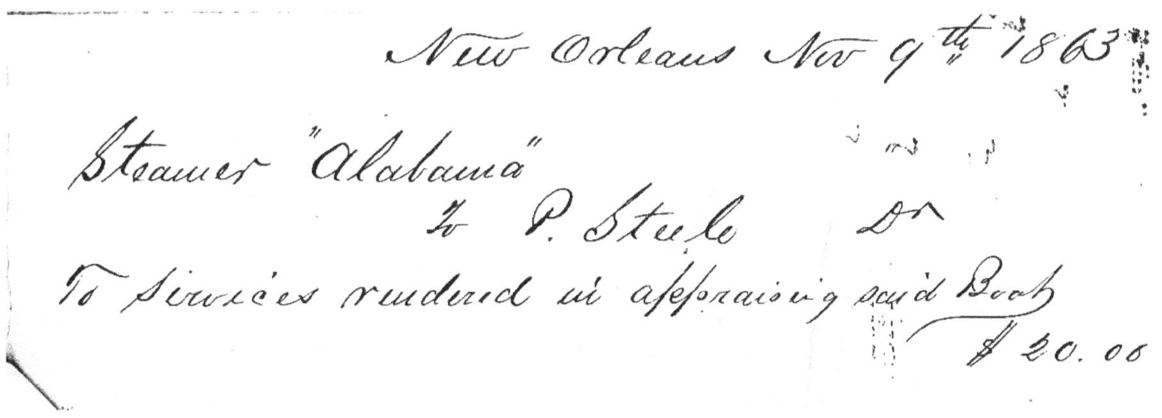

New Orleans Nov 9th 1863

Steamer "Alabama"
To P. Steele Dr
To services rendered in appraising said Boat
$20.00

Figure 104. Bill for services of the appraisers. Dated 9 Nov. 1863. **NARA**.

Attorney for the Captors Objected to the Appraisal

A. M. Buchanan, representing the captors, objected that the appraisal was too low (Figure 105).

> In the matter of the
> United States
> vs.   No. 7728
> The Steamer *Alabama*
> & Cargo

To the Honorable Edw. H. Durell, Judge of the District Court of the United States for the Eastern District of Louisiana. The petition of A. M. Buchanan, Attorney for the Captors in this case under the appointment of the Court, shows respectfully that an order made by Your Honor on the day before yesterday for an appraisement of the Steamship *Alabama*, condemned in this case, under an application made on behalf of the military authorities to take the said steam ship at the price of appraisement for account of the Government, was yesterday brought to the notice of your petitioner. Your petitioner on examining the record found therein a paper purporting to be an appraisement of said steam ship, her tackle, &c. & signed by two persons professing to be appraisers.

Now your petitioner shows that as representing the interest of the captors in the proceeds of the properly condemned, he had the right to have had a previous notice of this appraisement & to have the appointment of an appraiser, that he is dissatisfied with the appraisement, which he is informed is much less that the price for which a vessel like the *Alabama*, a first class sea going steamer, four years old & of upwards of five hundred tons burthin, could be bought as built at this time & place.

Wherefore your petitioner prays that the appraisement on file herein, above mentioned, be set aside & annulled as being irregular & that a new appraisement be ordered, to be made by two appraisers, one to be appointed by the Government & one by the Captors.

                                       And as ???, duly &c.

(signed) A. M. Buchanan
Attorney of the Captors

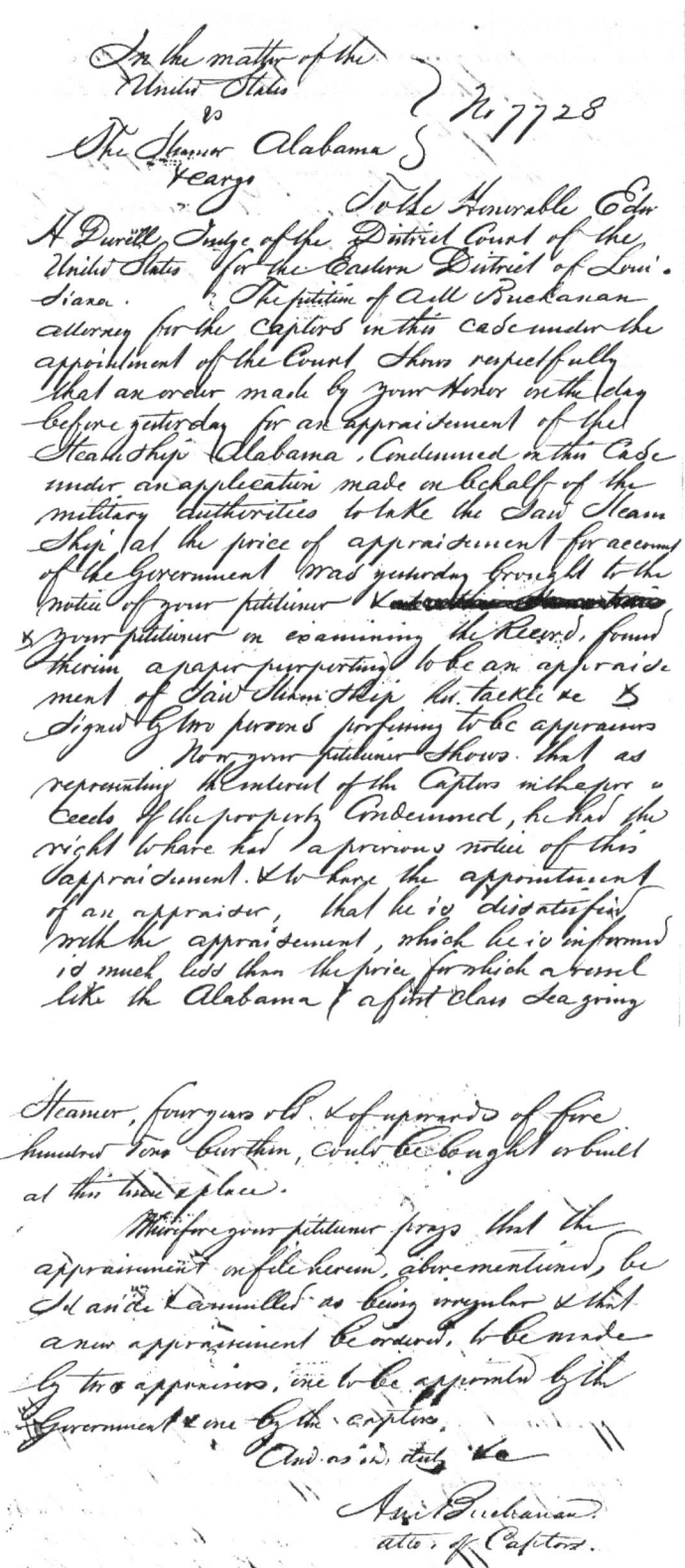

Figure 105. Attorney for the captors moved for a new appraisal based on the unreasonably low value reported for the Str. *Alabama*. Dated 12 Nov. 1863. *NARA*.

The *Alabama* Story Took an Unexpected Turn When the Union Army Unceremoniously Seized the Ship

General Charles P. Stone was chief of staff to General Nathaniel P. Banks, Union army commander of the Department of the Gulf with headquarters in New Orleans. At the time the *Alabama* case was in process, General Banks and his forces were involved in an unsuccessful incursion on the Texas coast. From the tone of Stone's communications, things were not going well in the field, and he was urgently trying to send supplies and support. General Stone simply grabbed the prize steamer *Alabama* for the army's use (Figure 106).

United States	U.S. District Court
vs.	E. H. Durell, Judge
The Steamer *Alabama*	Sitting in Chambers

    The petition of Rufus Waples, U.S. Attorney, respectfully represents that Brigadier General Charles P. Stone has forcibly taken possession of the Steamer *Alabama*, (Condemned herein as lawful prize) from the custody of the Marshal; has arrested the officers of the said Marshal while in the performance of their duty and in charge of the said steamer and in obedience of the order of the Honorable Court:

    Wherefore your petitioner prays for an order directing the said Brigadier General Charles P. Stone to show cause on Wednesday the 18th inst. (Nov. 1863) at 10 o'clock why he should not be fined and imprisoned for contempt of Court.

    (signed) Rufus Waples
    U.S. Dist. Atty.

In Chambers

Ordered that Brigadier General Charles P. Stone be cited to show cause on Wednesday the 18th inst. at 10 o'clock A.M. why he should not be fined and imprisoned for contempt of Court for the causes set forth in the foregoing petition.

    (signed) E. H. Durell

New Orleans, November 13th 1863   Judge

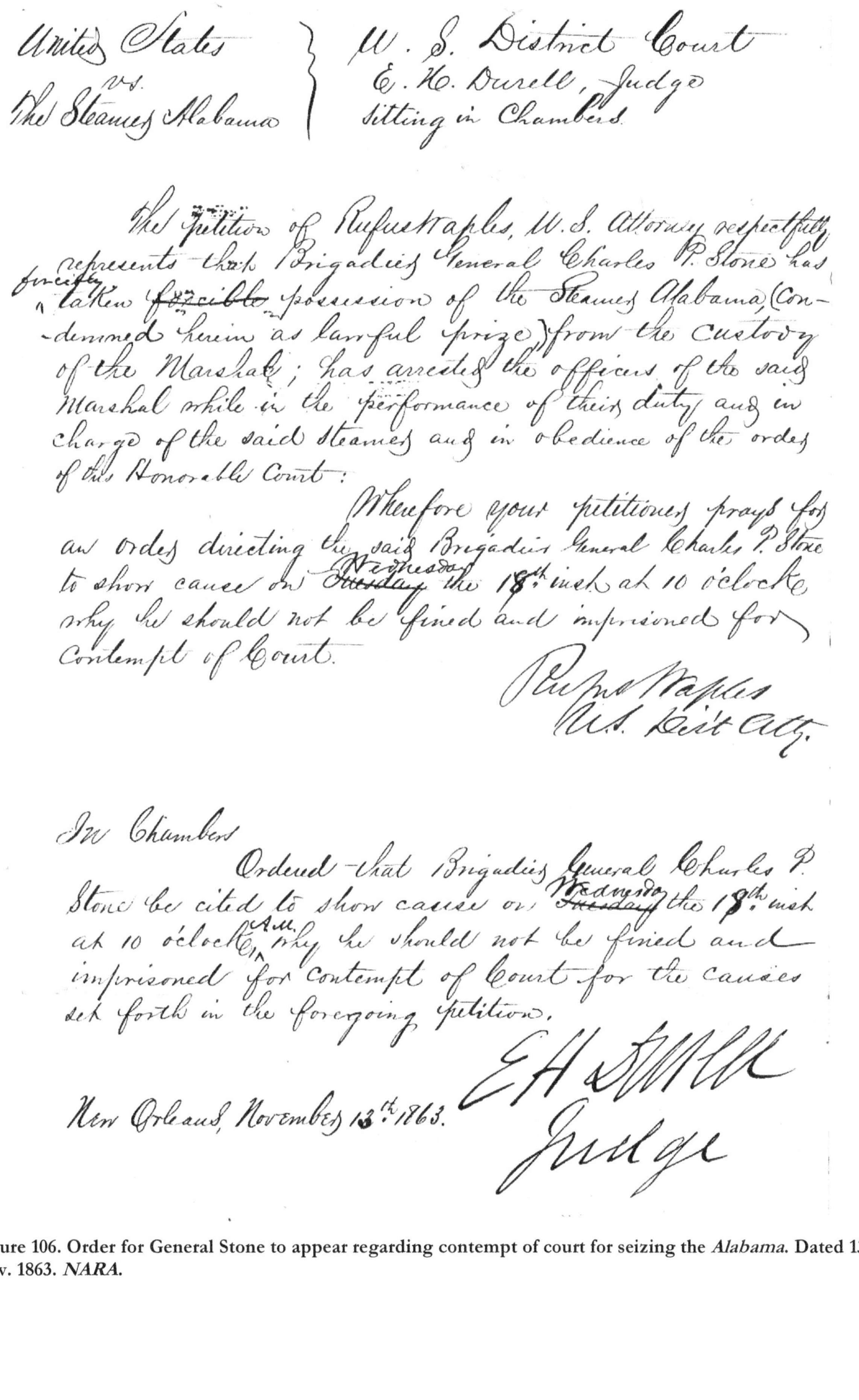

United States   U. S. District Court
     vs.   E. H. Durell, Judge
The Steamer Alabama   sitting in Chambers.

The petition of Rufus Waples, U.S. Attorney respectfully represents that Brigadier General Charles P. Stone has forcibly taken ^forcible possession of the Steamer Alabama, (Condemned herein as lawful prize) from the custody of the Marshal; has arrested the officers of the said Marshal while in the performance of their duty and in charge of the said steamer and in obedience of the orders of this Honorable Court:

Wherefore your petitioner prays for an order directing the said Brigadier General Charles P. Stone to show cause on Wednesday the 18th inst at 10 o'clock, why he should not be fined and imprisoned for contempt of Court.

                              Rufus Waples
                              U.S. Dist Atty.

In Chambers
     Ordered that Brigadier General Charles P. Stone be cited to show cause on Wednesday the 18th inst at 10 o'clock A.M. why he should not be fined and imprisoned for contempt of Court for the causes set forth in the foregoing petition.

New Orleans, November 13th, 1863.
                              E H Durell
                              Judge

Figure 106. Order for General Stone to appear regarding contempt of court for seizing the *Alabama*. Dated 13 Nov. 1863. **NARA.**

General Charles P. Stone was General Banks' chief of staff and Union army officer in charge during Banks' absence in the field. As we see below, Stone responded to the court's summons for his appearance on contempt charges (Figure 107).

> I did send an officer on board the Steamer *Alabama* with direction to arrest every person interfering with my occupation of it. Under that order Mr. Christie and two others were arrested and brought to my office. I did this with no intention of contempt of this Honorable Court but under a necessity in order that the work of preparation should be no longer delayed and that reinforcements might reach Gen. Banks in time and not too late.
>
> I admit that I believe that the U.S. Marshal was then in possession of the vessel having ejected the persons whom I had placed in charge of her. I believe all three of the persons to be officers of the Marshal and that they were in the discharge of their duty as instructed by him.
>
> I did tell the men who were arrested and brought before me on releasing them that if they attempted to retake the vessel out of my possession they must take the risk of getting shot or sent to prison as I had placed a military guard on board with instructions to defend the vessel.

> was then in possession of
> the vessel having ejected
> the persons whom I had
> placed in charge of her. I
> believe all three of the persons
> to be officers of the marshal
> and that they were in the discharge
> of their duty as instructed by him
> I did tell the men who
> were arrested and brought before
> me on releasing them that if
> they attempted to retake the vessel
> out of my possession they must
> take the risk of being shot or sent to prison
> as I had placed a military guard
> on board with instructions to defend
> the vessel

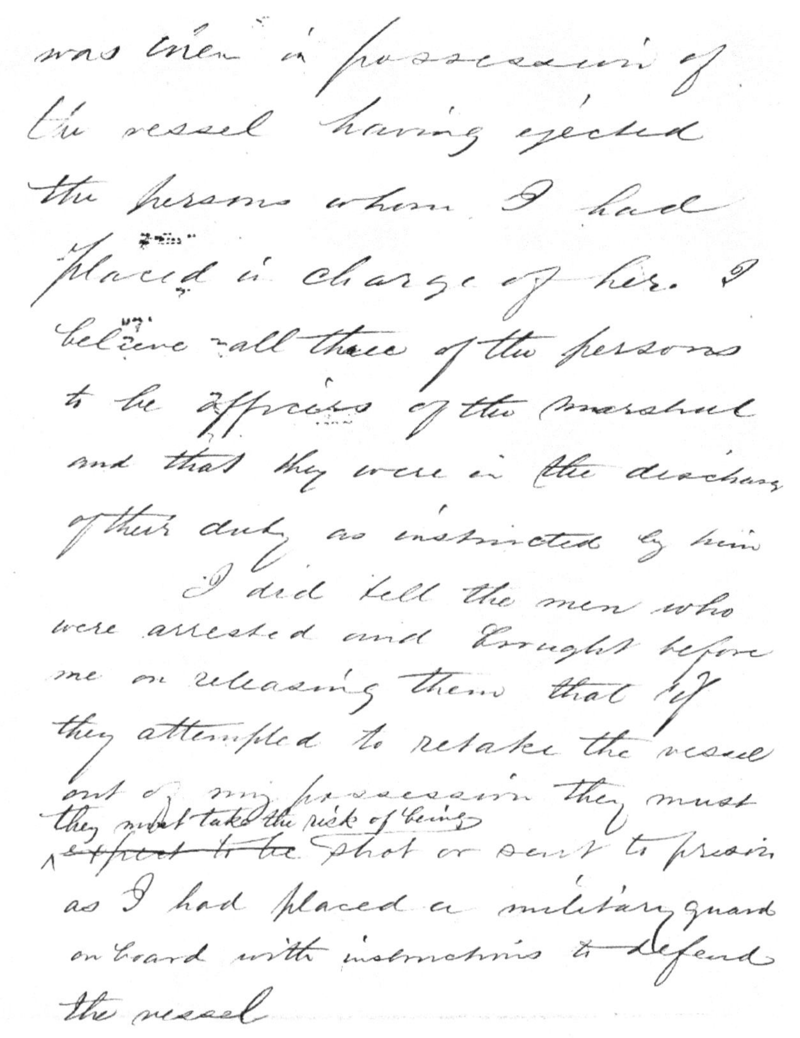

Figure 107. General Stone's statement and admission including testimony of his staff, Captain John W. Stone (below). (Two pages including this and the previous.) Filed 18 Nov. 1863. *NARA*.

Testimony of U.S. marshal's men and General Stone's men was filed 18 Nov. 1863 (Figure 108).

>Daniel Christie sworn. I was ordered on board the *Alabama*.
>Captain John W. McClure sworn. I was on board the *Alabama*, shortly after she was taken possession of by the Marshal.[165] I saw on board Mr. Christie, a Deputy Marshal. In going on board the *Alabama* I asked for the Marshal. Mr. Christie appearing I asked him to let the work proceed. He replied he would not permit a hand to be raised until he was paid the money.[166] I then informed him that Gen. Stone would give the security or I would give a script for her at the full value that she was worth. That offer was refused.
>D. Christie reexamined recalled for plaintiffs in sale (rule?).
>The testimony of Capt. McClure was now ordered to be stricken out, and Mr. Christie was not reheard.

---

[165] Taken back from Union army representatives after the army seized her from the U.S. marshal's representatives, the original caretakers.
[166] The court's representatives were bound to uphold the legal process that required that the vessel remain in the court's jurisdiction and control until funds for her sale were deposited with the court.

Daniel Christie sworn. I was ordered on board the Alabama

Captain John W. McClure sworn. I was on board the Alabama, shortly after she was taken possession of by the Marshal. I saw on board Mr Christie a Deputy Marshal. In going on board the Alabama I asked for the Marshal. Mr Christie appearing I asked him to let the work proceed. He replied he would not permit a hand to be raised until he was paid the money

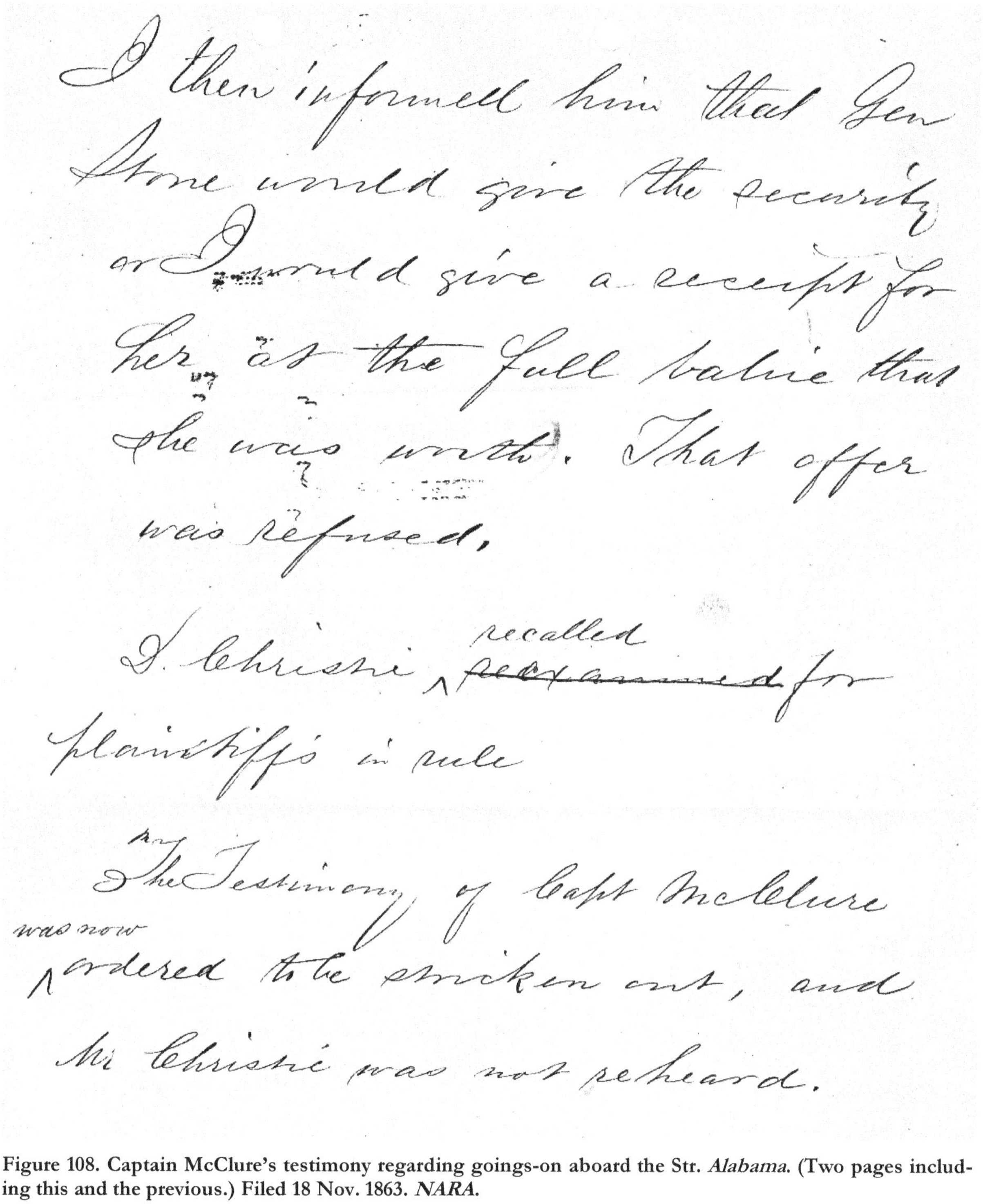

> I then informed him that Gen Stone would give the security or I would give a receipt for her, at the full value that she was worth. That offer was refused.
>
> I. Christie recalled for plaintiff's in rule
>
> The Testimony of Capt McClure was now ordered to be stricken out, and Mr Christie was not reheard.

Figure 108. Captain McClure's testimony regarding goings-on aboard the Str. *Alabama*. (Two pages including this and the previous.) Filed 18 Nov. 1863. **NARA**.

General Stone explained his view of the *Alabama* contretemps to the U.S. district attorney dated 12 Nov. 1863 (Figure 109). He adopted a (seemingly) conciliatory tone in this note.

                                        Headquarters, Department of the Gulf
                                                    New Orleans, Novr 12th 1863

Rufus Waples Esquire
    U.S. District Attorney

Sir,
    I have just had the honor to receive your letter of this date in reply to mine of even date to his Honor Judge Durell.

    As soon as the great pressure of business connected with transfer of troops & supplies to General Banks will permit I will see that the requirements of His Honor are complied with.

    The "violent measures" adopted by the Marshal, referred to in my letter, consisted in his going on board the Steamer *Alabama* with a party of deputies and suddenly arresting the work of preparation then in progress and in his ordering all the military employees off the vessel where they were busily at work in pursuance of my orders. The action taken by the Marshal will detain important reinforcements to General Banks probably one day, when his proper object, (the securing of the rights of the Honorable Court he represented) could have been obtained in a few minutes interview with me in my office.

                            I have the honor to be,
                            Sir,
                                Very respectfully
                                    Your obt. servt.
                                      Chas. P. Stone
                                      B. G. Chief of Staff

(Copy sent with dispatches to General Banks.)

Headquarters, Department of the Gulf,

New Orleans, Nov' 12th, 1863.

Rufus Waples Esquire
    U.S. District Attorney

Sir;

I have just had the honor to receive your letter of this date in reply to mine of even date to his Honor Judge Durell.

As soon as the great pressure of business connected with transfer of troops & supplies to General Banks will permit I will see that the requirements of his Honor are complied with.

The "violent measures" adopted by the Marshal, referred to in my
letter

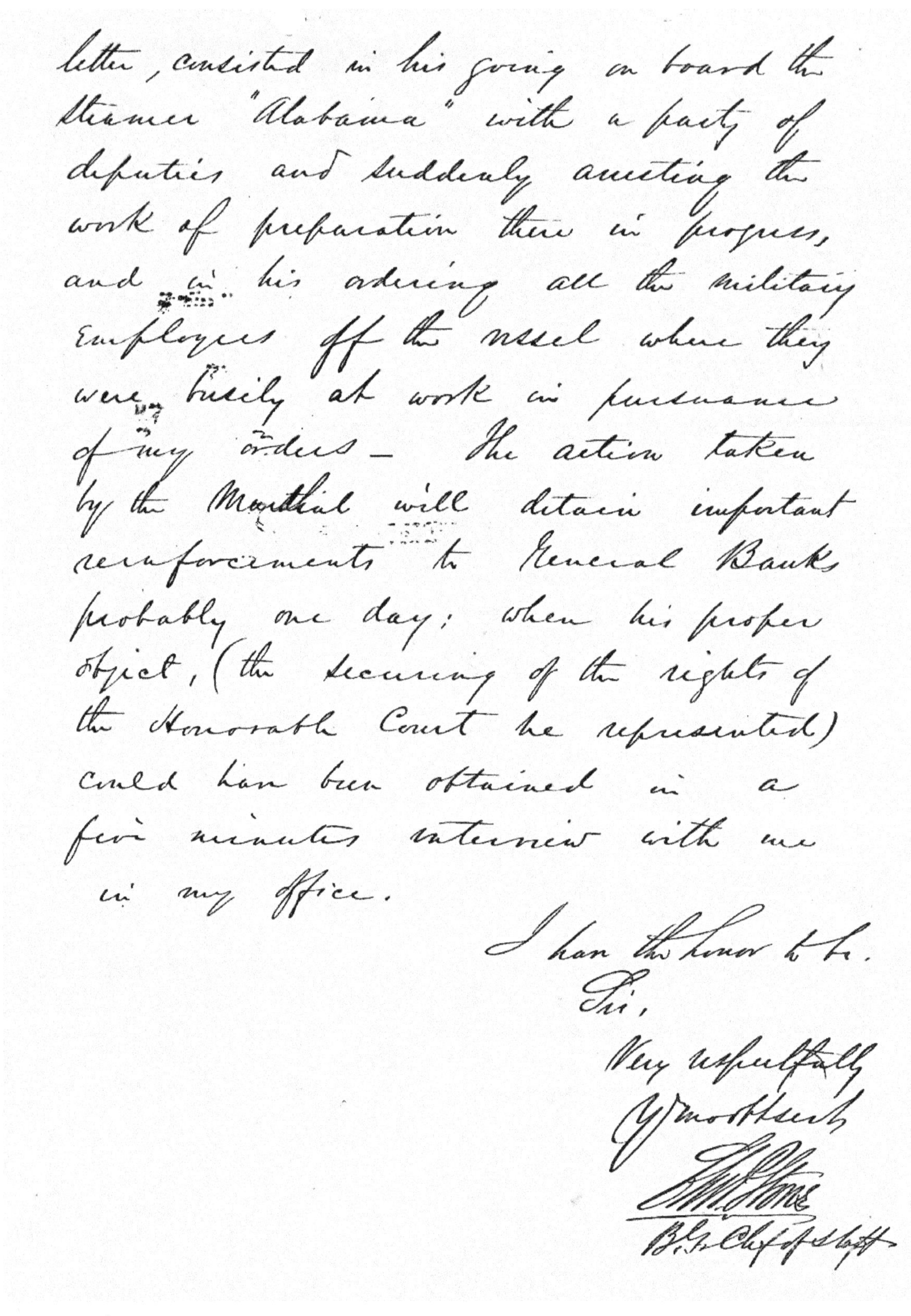

letter, consisted in his going on board the steamer "Alabama" with a party of deputies and suddenly arresting the work of preparation there in progress, and in his ordering all the military Employees off the vessel where they were busily at work in pursuance of my orders — The action taken by the Marshal will detain important reinforcements to General Banks probably one day; when his proper object, (the securing of the rights of the Honorable Court he represented) could have been obtained in a few minutes interview with me in my office.

I have the honor to be,
Sir,
Very respectfully
Y'r mo obt serv't
[signature]
B. Br. Ch'f of Staff

Figure 109. General Stone offered to pay the value of the *Alabama* to the court when allowed by the press of sending supplies to General Banks in Texas. (This page and the previous.) Dated 12 Nov. 1863. **NARA**.

General Stone explained his actions to Judge Durrell, dated 12 Nov. 1863 (Figure 110). Stone sent a copy of this letter to General Banks with dispatches. Although he sounded respectful here, General Stone revealed in the following dispatches some very high-handed attitudes and behavior toward the civil authorities.

Headquarters, Department of the Gulf
New Orleans, Novr 12th 1863
11 O'clock A.M.

To the Honorable
    The Judge, U.S. Dist. Court
        New Orleans, La.

Your Honor,

I am greatly concerned in learning that there is some difficulty about the Steamer *Alabama*, a prize under adjudication in the Court over which Your Honor so worthily presides.

The day before yesterday I was officially informed that the steamer in question had been transferred to the Navy Dept. and as her services would be of incalculable advantage to the army at the moment, I addressed myself to the Commodore Commanding the Western Gulf Squadron, asking for the use of her for one month. My request was immediately granted by the Commodore, and the necessary repairs and preparations for service were commenced that night and have been continued until the present time when she is nearly ready for sea.

It is of vital importance to the full success of military operations now prosecuting under the eye of the Major General Commanding the Dept. that the troops and supplies ordered on this steamer should reach him at once, and the vessel should sail without fail at the first moment practicable. The steamer *must sail as soon as she can be prepared*.

The public service of the county requires it.

I am ready to give any security which Your Honor may see fit to require that no pecuniary loss shall result to the Court, and I would not for any less reason than an imperative public necessity even *seem* to be disrespectful to Your Honor or your Court.

I deeply regret that the Marshal of the District should have taken violent measures to interrupt the occupation of the steamer ordered by me thirty six hours since, in a way which I supposed to be perfectly consistent with Your Honor's orders concerning her. His violent measures, adopted without first referring to me, compelled me to use prompt and efficient measures to ensure the prompt relief of the Commander of the Department.

With the highest respect
I have the honor to be,
Your Honor's
Mo. obt. servt.
Chas. P. Stone
B. G., Chief of Staff

Headquarters, Department of the Gulf,

New Orleans, Nov 12th, 1863.
11 o'clock A.M.

To the Honorable
The Judge U.S. Dist Court
New Orleans La.

Your Honor:

I am greatly concerned on learning that there is some difficulty about the Steamer "Alabama" a prize under adjudication in the Court over which Your Honor so worthily presides —

The day before yesterday I was officially informed that the Steamer in question had been transferred to the Navy Dept. and as her services would be of incalculable advantage to the Army at the moment, I addressed myself to the

the Commodore Commanding the Western Gulf Squadron, asking for the use of her for one month — My request was immediately granted by the Commodore, and the necessary repairs and preparations for Service were commenced that night and have been continued until the present time when she is nearly ready for sea.

It is of vital importance to the full success of military operations now prosecuting under the Eye of the Major General Commanding the Dept. that the troops and supplies ordered on this steamer should reach him at once, and the vessel should sail without fail at the first moment practicable — The steamer must sail as soon as she can be prepared

The Public Service of the Country requires it —

I am ready to give any security which your Honor may see fit

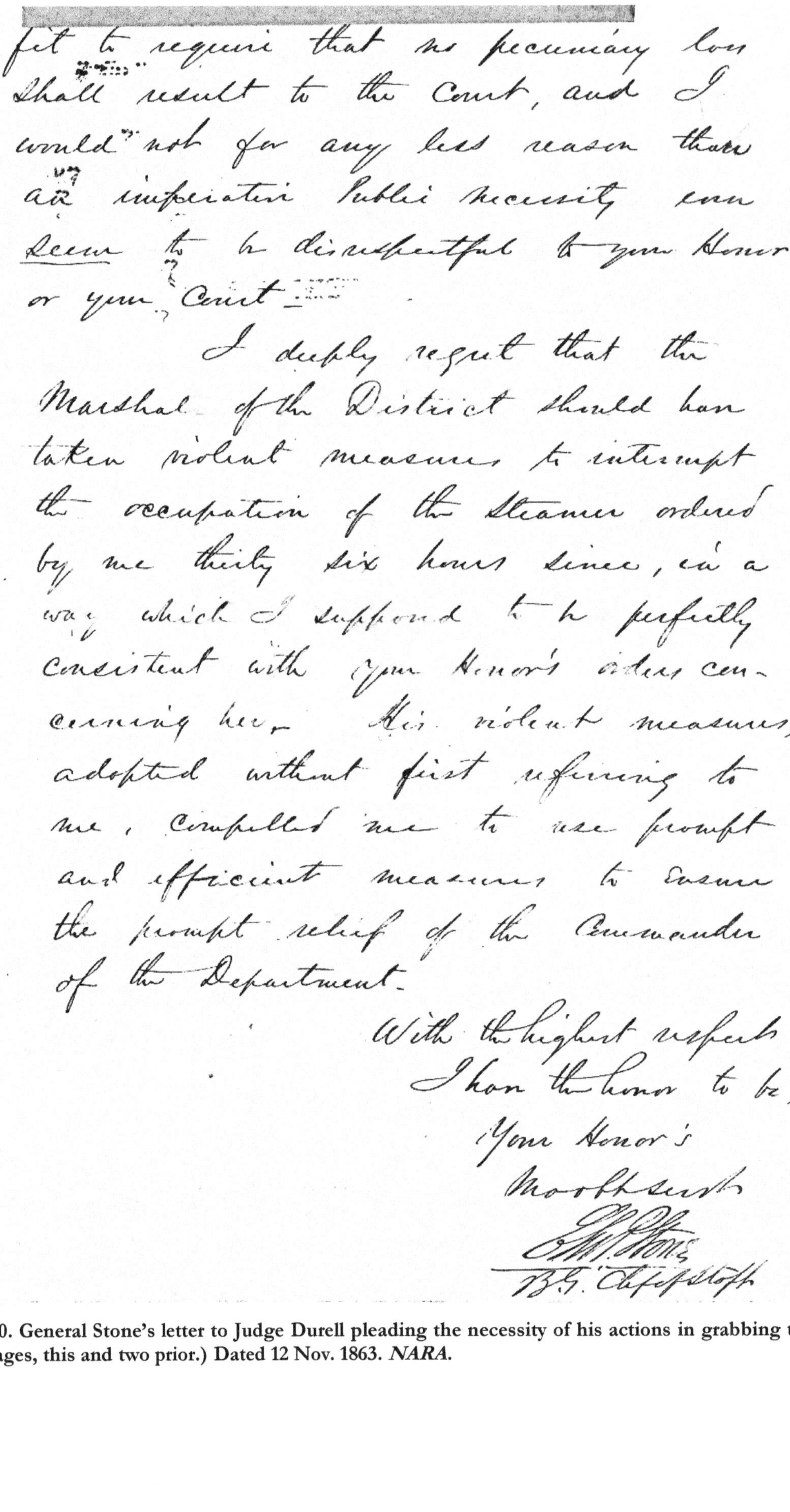

fit to require that no pecuniary loss shall result to the Court, and I would not for any less reason than an imperative public necessity even seem to be disrespectful to your Honor or your Court.

I deeply regret that the Marshal of the District should have taken violent measures to interrupt the occupation of the steamer ordered by me thirty six hours since, in a way which I supposed to be perfectly consistent with your Honor's orders concerning her. His violent measures, adopted without first referring to me, compelled me to use prompt and efficient measures to ensure the prompt relief of the Commander of the Department.

With the highest respects
I have the honor to be,
Your Honor's
most obedient
C.W. Stone
B.G. Chief of Staff

Figure 110. General Stone's letter to Judge Durell pleading the necessity of his actions in grabbing the ship. (Three pages, this and two prior.) Dated 12 Nov. 1863. *NARA*.

General Banks was campaigning on the south Texas coast capturing Brazos Island, Point Isabel, and Brownsville. His troops then proceeded up the coast as far as Matagorda Bay before withdrawing. General Stone kept him informed of things at H. Q. in New Orleans as seen in the following dispatches.[167]

NEW ORLEANS,
November 12, 1863.

Maj. Gen. N. P. BANKS,
Commanding Department of the Gulf, in the Field:
GENERAL:

The steamer *Alabama* goes to you with a full battery of artillery and some convalescent officers of the Thirteenth Army Corps.

Three schooners, which I have caused to be fitted up, will start to-morrow, to be towed to the Rio Grande. They will be loaded with extra artillery horses and forage.

I am expecting the arrival of coal from Pensacola hourly, and will cause it to be forwarded as rapidly as possible.

The *Saint Mary's* sailed from Brashear this afternoon with General Washburn on board. She takes two regiments and 30,000 rations. She was not fitted up for horses, and could not, therefore, take the battery which I had intended to send by her without great delay. I have ordered the steamer *Kate Dale* to be taken here; she is of light draught, and will be loaded to-morrow with wagons and mules, and sent to Brazos Island. She will carry about 40 wagons and 80 mules, and will carry forward some of Dana's convalescents.

In the destitution of the quartermaster's department here, I have been obliged to take all the soft coal in the private yards in the city, and, with this, hope to keep the transports going until coal shall arrive from up river.

The United States Marshal, under the orders of the district judge of the United States, this morning took possession of the *Alabama*, and drove all the workmen and coalers off, saying he would teach the military authorities to respect the courts. As I had formally borrowed the *Alabama*, and had her in possession for thirty-six hours, it is difficult to see what could have been the object of such a violent proceeding on the marshal's part. As it was no time to wait, I immediately sent Colonel Abert on board with a detachment of the First Infantry, and arrested the deputy marshals on board, caused work to be resumed, and then wrote a respectful letter to the judge on the subject. The result of the marshal's proceedings was a delay of nearly a day in getting off the battery to you, and I informed him that a repetition of such interference with my forwarding re-enforcements to you would place him in close custody. General Thomas, Adjutant-General, fully approved all that was done in the premises.

Everything is quiet at Port Hudson and along the river. Port Hudson lines, interior and exterior, are expected to be completed before the end of the month. General Andrews reports them now fully in condition to hold with his present force.

General Franklin is at Vermillion, and has recently made heavy reconnaissances to Carrion Crow Bayou and in the Mermenton (Mermentau?) country. One regiment of Texas cavalry has appeared from Niblett's Bluff, and the force in his front seems to be increasing and waiting for something.

I do not think it will be well to weaken him materially beyond the withdrawal of the 5,000 already taken, until he can withdraw to a less extended line.

Very respectfully, I am, general, your obedient servant.
CHAS. P. STONE, *Brigadier-General, and Chief of Staff*

---

[167] *OR*, ser. 1, 26.1:793-798.

The U.S. district attorney explained the legal and procedural impediments regarding the *Alabama* to General Stone, dated 13 Nov. 1863 (Figure 111).

U.S. Circ^t Court Room
Novr 13th 1863

Brig^r Gen^r Chas P. Stone
    U.S.A.
  Sir,

I have the honor to acknowledge the receipt of your note of yesterday's date, just this moment received. The order of the Court being for the delivery of the Steamer *Alabama* to Com. Bell for the Naval Service, I do not know what you have reference to, when you say that you "will see that the requirements of his honor are complied with." I suppose however, you have reference to the remark in my letter, written at the suggestion of Judge Durell, in which I stated, at my own option, that I had no doubt the Court would grant an order for the delivery of the vessel to the Army, on the same terms that it had been ordered to be delivered to the Navy, in case Com. Bell should fail to comply with the order already granted upon his application, if you should then make a similar application. That remark, however, in yesterday's letter, was written before I had knowledge that the Steamer had been taken from the possession of the Marshal. Of course, the Court can grant no order relative to the disposition of the vessel, while it is not in the custody of the Marshal, and beyond the control of the Court.

    Permit me to assure you that the Court will always be found ready and willing to facilitate as much as possible the military operations of the Army and Navy. And allow me further to assure you of my continued respect and esteem.

I have the honor to be &c.
(signed) Rufus Waples
U.S. Attorney

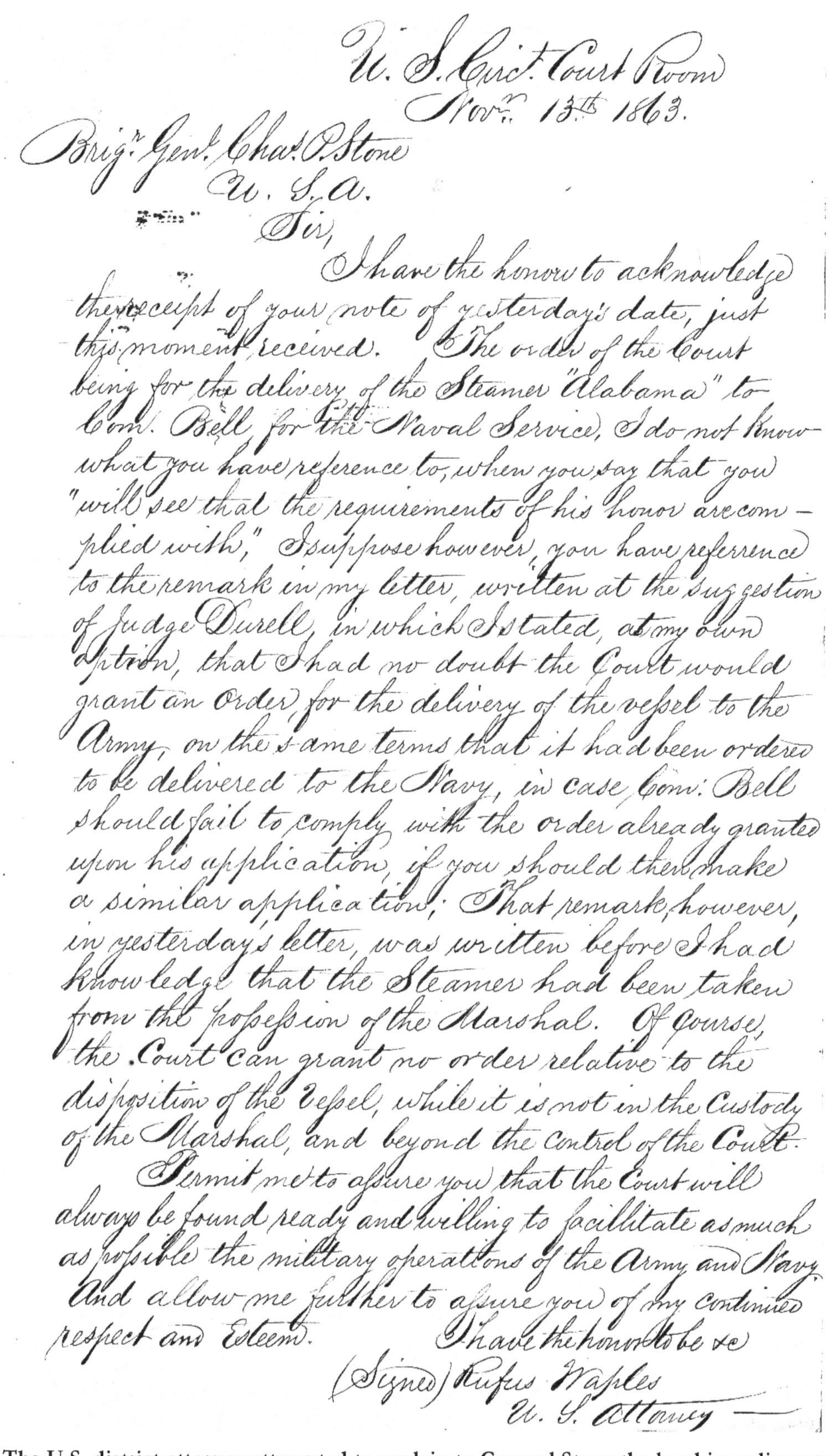

> U. S. Circt. Court Room
> Nov. 13th 1863.
>
> Brig. Genl. Chas. P. Stone
> U. S. A.
> Sir,
>
> I have the honor to acknowledge the receipt of your note of yesterday's date, just this moment received. The order of the Court being for the delivery of the Steamer "Alabama" to Com. Bell, for the Naval Service, I do not know what you have reference to, when you say that you "will see that the requirements of his honor are complied with." I suppose however, you have reference to the remark in my letter, written at the suggestion of Judge Durell, in which I stated, at my own option, that I had no doubt the Court would grant an Order for the delivery of the vessel to the Army, on the same terms that it had been ordered to be delivered to the Navy, in case Com. Bell should fail to comply with the order already granted upon his application, if you should then make a similar application. That remark, however, in yesterday's letter, was written before I had knowledge that the Steamer had been taken from the possession of the Marshal. Of course, the Court can grant no order relative to the disposition of the Vessel, while it is not in the Custody of the Marshal, and beyond the control of the Court.
>
> Permit me to assure you that the Court will always be found ready and willing to facilitate as much as possible the military operations of the Army and Navy. And allow me further to assure you of my continued respect and Esteem. I have the honor to be &c
>
> (Signed) Rufus Waples
> U. S. Attorney

Figure 111. The U.S. district attorney attempted to explain to General Stone the legal impediments to simply giving over the *Alabama* to the Union army. Dated 13 Nov. 1865. ***NARA***.

General Stone sent to General Banks a further situation report from H. Q. in New Orleans.[168] Get a load of the postscript! Stone threatened to arrest the judge and officials of the court. (Bold print added.)

NEW ORLEANS,
November 12, 1863.

Maj. Gen. N. P. BANKS,
 Comdg. Department of the Gulf, in the Field:
 (Care of Lieutenant-Colonel Chandler, Brashear.)

GENERAL: The steamer *Alabama* will sail to-night to join you, carrying a battery of artillery, men, horses, and extra ammunition.

Schooners with 80 extra artillery horses will start this evening, to be towed to you; 100 extra artillery horses have been sent to-day to Brashear, to be shipped thence to you. All quiet up the river. The enemy shows strong force in front of Franklin's troops, and there has been heavy skirmishing between Vermillion and Opelousas. General Franklin all right, but should not be weakened much at present.

One of the small steamers of the enemy was captured yesterday in Bayou Long by Lieutenant-Colonel Tarbell. All quiet in New Orleans.

Very respectfully, your obedient servant,
CHAS. P. STONE,
Brigadier-General and Chief of Staff.

NEW ORLEANS,
November 13, 1863.

Maj. Gen. N. P. BANKS,
 Commanding Department of the Gulf, in the Field:

GENERAL: The steamers *Crescent* and *Clinton* have arrived here this morning, bringing your dispatches of November 9 instant.

I shall forward the dispatches by special messenger to Washington on the steamer of tomorrow.

Permit me to tender my congratulations on the great and most important success attained. Its results cannot fail to be appreciated by the country.

The *Crescent* and *Clinton* will, if possible, be gotten off on their return to-morrow, carrying 1,000 men of First Division, Thirteenth Army Corps, and wagons and mules for General Dana's division.

Colonel Beck has been directed to send an efficient commissary of subsistence.

After conversing with Major Houston, I shall probably be able to get one or two regiments of colored troops off to you as soon as transportation can be provided. As I understand your instructions, however, I shall first send "good troops."

Very respectfully, I am, general, your obedient servant,
CHAS. P. STONE,
Brigadier-General and Chief of Staff.

P. S.–I have been this morning summoned to appear before the United States district court to show cause why I should not be fined and imprisoned for contempt of court, in having seized the steamer *Alabama*. I shall appear and treat the court with all respect, but if a disposition is shown to thwart military operations by these officers of the court, I shall arrest them.

---

[168] *OR*, ser. 1, 26.1:793-798.

S. R. Holabird, Colonel and Chief Quartermaster, wrote to Judge Durell concerning how he would secure funds to pay the court for the *Alabama* (Figure 112).

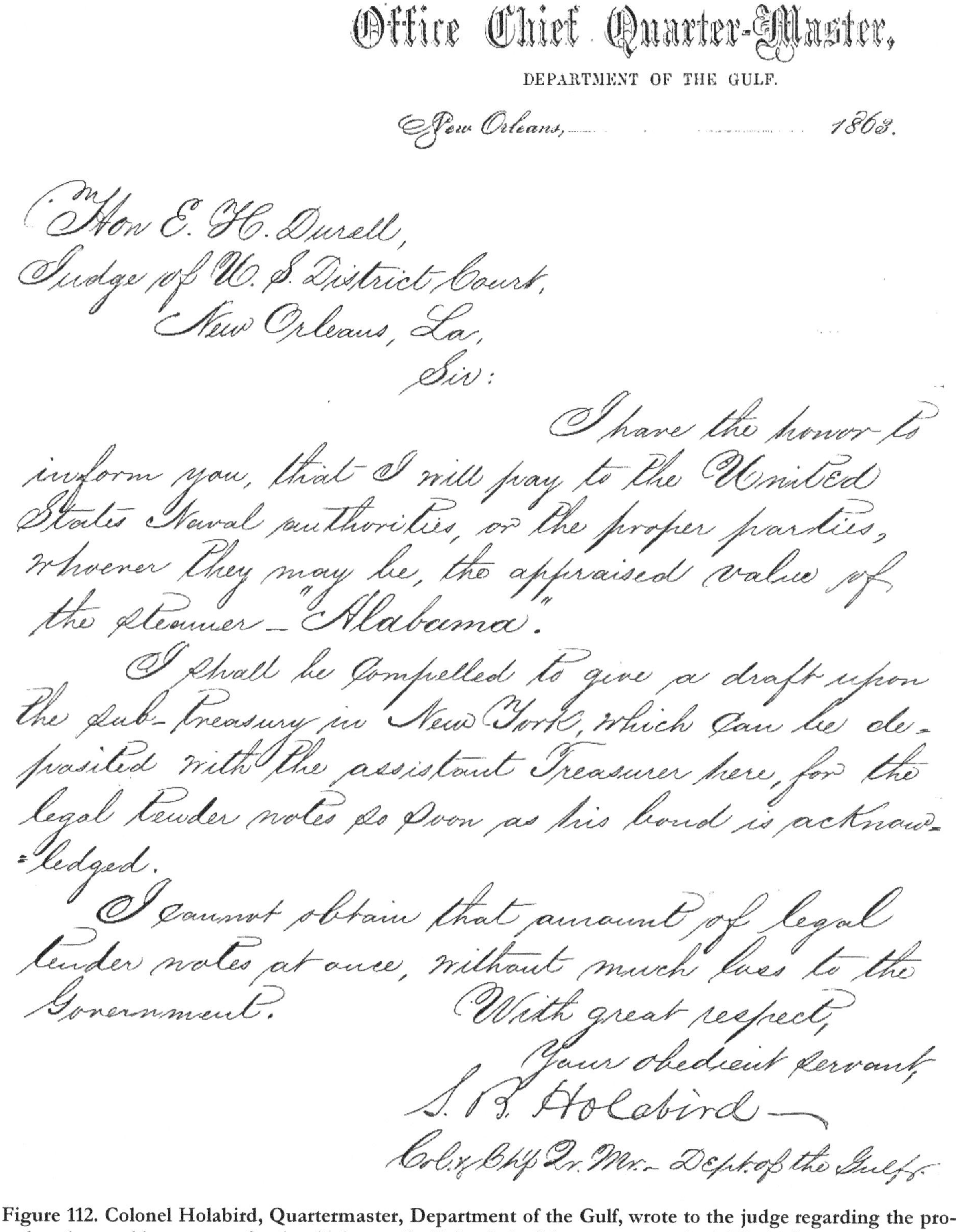

Office Chief Quarter-Master,
DEPARTMENT OF THE GULF.
New Orleans, _____ 1863.

Hon E. H. Durell,
Judge of U. S. District Court,
New Orleans, La,
 Sir:
  I have the honor to inform you, that I will pay to the United States Naval authorities, or the proper parties, whoever they may be, the appraised value of the steamer "Alabama."

I shall be compelled to give a draft upon the Sub-Treasury in New York, which can be deposited with the assistant Treasurer here, for the legal tender notes so soon as his bond is acknowledged.

I cannot obtain that amount of legal tender notes at once, without much loss to the Government. With great respect,
  Your obedient servant,
   S. B. Holabird
    Col. & Chf. Qr. Mr. Dept. of the Gulf.

Figure 112. Colonel Holabird, Quartermaster, Department of the Gulf, wrote to the judge regarding the procedure he would use to pay for the *Alabama*. (Sufficiently legible so no transcript.) Filed 1 Dec. 1863. **NARA**.

The following document was a reply from Judge Durell to Colonel Holabird. The judge instruct the colonel to pay the funds to the clerk of the court with $10,000 to be paid in cash and the remainder as a draft on the sub-treasury in New York (Figure 113).

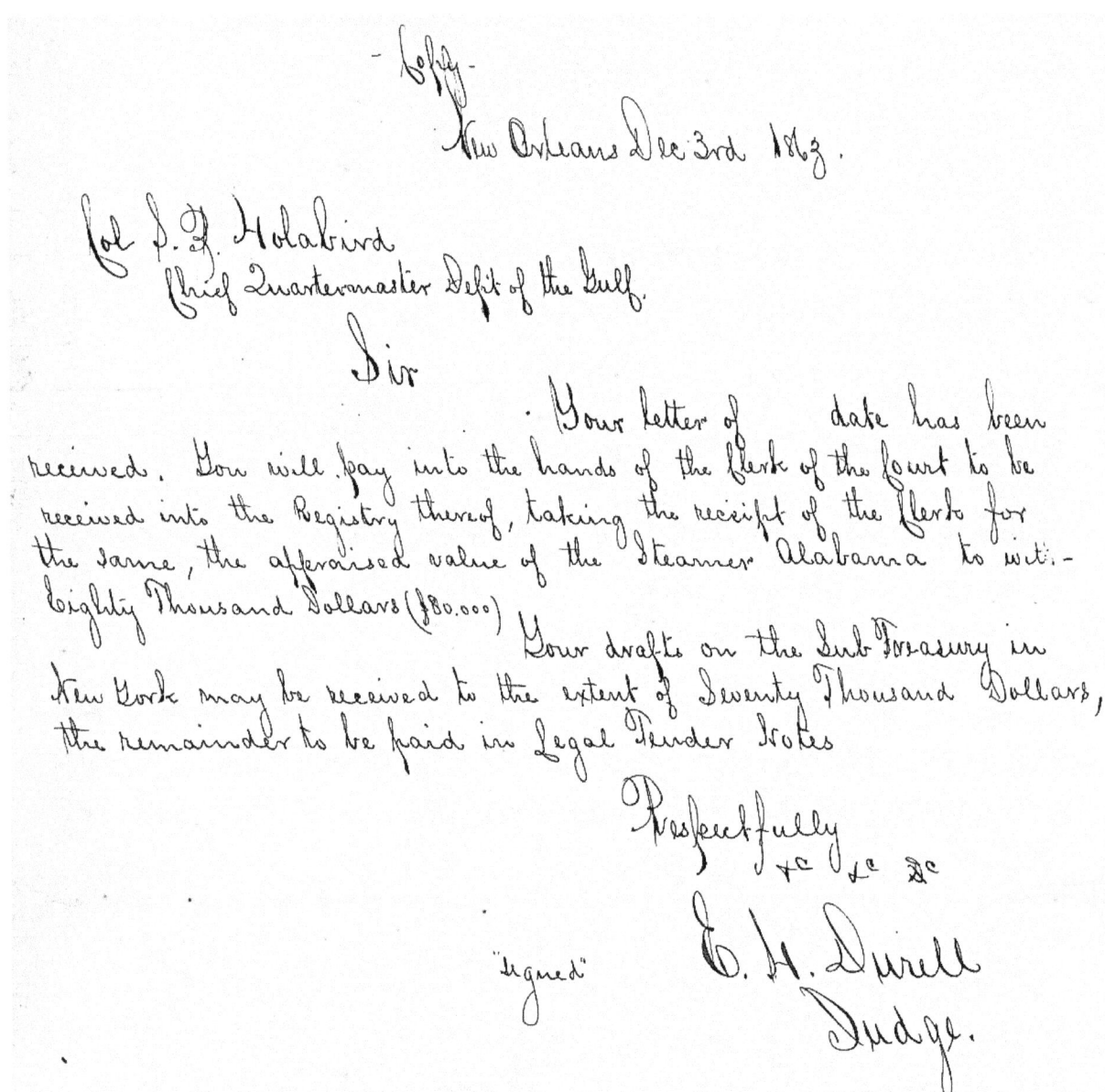

Figure 113. Judge Durell replied to Colonel Holabird with details for the payment to the court. (Sufficiently legible so no transcript.) Dated 3 Dec. 1863. *NARA*.

Two More Union Navy Captains Concerning Signal Distance

Meanwhile back at the prize court proceedings, navy ship captains continued giving testimony about the meaning of the term "signal distance" (Figure 114).

>   Wm. E. Le Roy, Commander in the U.S. Navy, commanding the U.S. Str. *Oneida*, being sworn deposes and says.
>   This matter of signal distance is not at all arbitrary; it is generally regarded as six or eight miles. The witness being questioned upon the circumstances of this case as disclosed by the depositions already on file, whether in his opinion, the Frigate *San Jacinto* being at the distance of from 10 to 11 miles from the *Alabama* when she was captured, was within signal distance of the vessel making the capture, answers. Usually among seafaring people, a vessel would be regarded with signal distance, within a distance of 10 miles, I should think. There are times from the peculiar state of the atmosphere, that signals can be read, when on other occasions of a clear day, at a much shorter distance they cannot be read. I should say under the circumstances that the *San Jacinto* can be regarded as within signal distance.

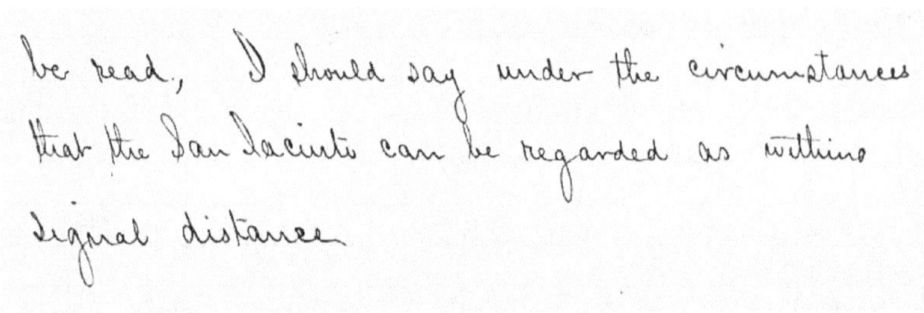

be read, I should say under the circumstances that the San Jacinto can be regarded as within signal distance

Figure 114. William E. Le Roy's testimony on the term "signal distance." (Two pages, this and the previous.) Filed 10 Dec. 1863. *NARA*.

Lieutenant Commander O. F. Staunton, commanding U.S. Str. *Panola* agreed that the *San Jacinto* was within signal distance as a gun could be fired and seen as a signal as was a common practice (Figure 115).

> Lieut. Commander O. F. Staunton, commanding the United States Steamer *Panola*, being sworn deposes and says.
>
> I consider 10 or 11 miles as "signal distance," for the reason that guns having been fired in ordinary weather, they have been seen by the other vessel and assistance could have been rendered if necessary. Firing guns have been considered as signals by the Navy, as it is often done. My opinion is, under the circumstances of this case, that the *San Jacinto* was within signal distance.

Lieut Commander O. F. Staunton, commanding the United States Steamer "Panola", being sworn deposes and says,

I consider 10 or 11 miles as "signal distance", for the reason that guns having been fired in ordinary weather, they have been seen by the other vessel and assistance could have been rendered if necessary, Firing guns have been considered as signals by the Navy, as it is often done. My opinion is, under the circumstances of this case, that the San Jacinto was within signal distance

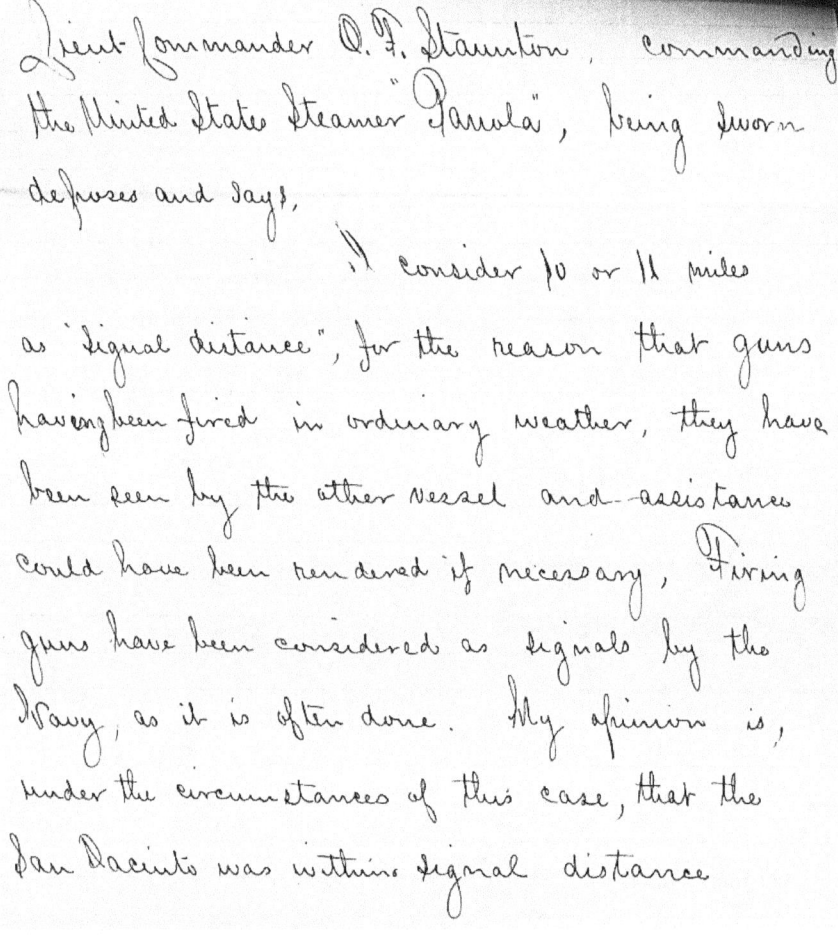

Figure 115. O. F. Staunton's testimony on "signal distance." Filed 10 Dec. 1863. *NARA*.

The *Alabama* Condemned and Ordered Sold

Judge E. H. Durell decreed the *Alabama*, her tackle, etc., and cargo, to be forfeited and ordered sale at public auction on 28 Nov. 1863 (Figures 116-117).

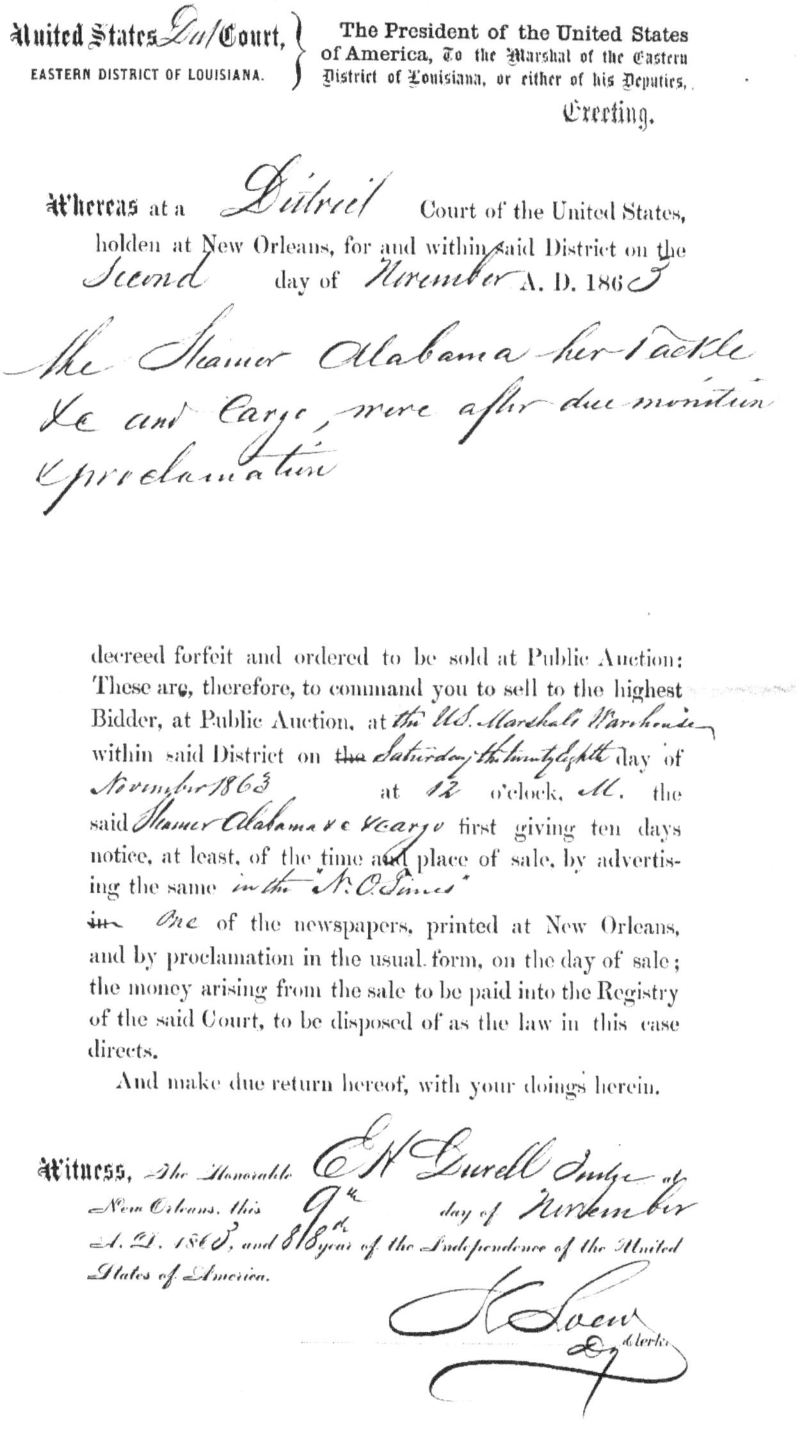

United States Dist Court, EASTERN DISTRICT OF LOUISIANA. The President of the United States of America, To the Marshal of the Eastern District of Louisiana, or either of his Deputies, Greeting.

Whereas at a District Court of the United States, holden at New Orleans, for and within said District on the Second day of November A. D. 1863 the Steamer Alabama her tackle &c and Cargo, were after due monition & proclamation decreed forfeit and ordered to be sold at Public Auction: These are, therefore, to command you to sell to the highest Bidder, at Public Auction, at the U.S. Marshal's Warehouse within said District on ~~the~~ Saturday the twenty eighth day of November 1863 at 12 o'clock, M. the said Steamer Alabama &c & cargo first giving ten days notice, at least, of the time and place of sale, by advertising the same in the "N. O. Times" ~~in~~ One of the newspapers, printed at New Orleans, and by proclamation in the usual form, on the day of sale; the money arising from the sale to be paid into the Registry of the said Court, to be disposed of as the law in this case directs.

And make due return hereof, with your doings herein.

Witness, The Honorable E H Durell Judge at New Orleans, this 9th day of November A. D. 1863, and 88th year of the Independence of the United States of America.

[signature] Dy Clerk

Figure 116. Judge Durell's order for the sale of the *Alabama* and her contents to take place beginning 28 Nov. 1863. (Sufficiently legible so no transcript.) Dated 2 Nov. 1863. *NARA*.

(Reverse of form above.)

United States of America
Eastern District of Louisiana. New Orleans, __________ 186__

Pursuant to this Precept I have caused the cargo of the Steamer *Alabama*, the Steamer having been forcibly taken out of my possession, ~~within named~~ to be sold by Public Auction, and to the highest bidder, at the U.S. Marshal's Warehouse on the day, hour and place therein directed, having given Ten days previous notice of the time, place and manner of the sale in the N. O. Times ~~and the~~ one of the newspapers printed in New Orleans and by the Crier, on the day of sale, and at such sale ~~also~~ the purchasers were as ~~under~~ specified in the Process Verbal of G. A. Hall, U.S. Prize Auctioneer here to attached & made part hereof.

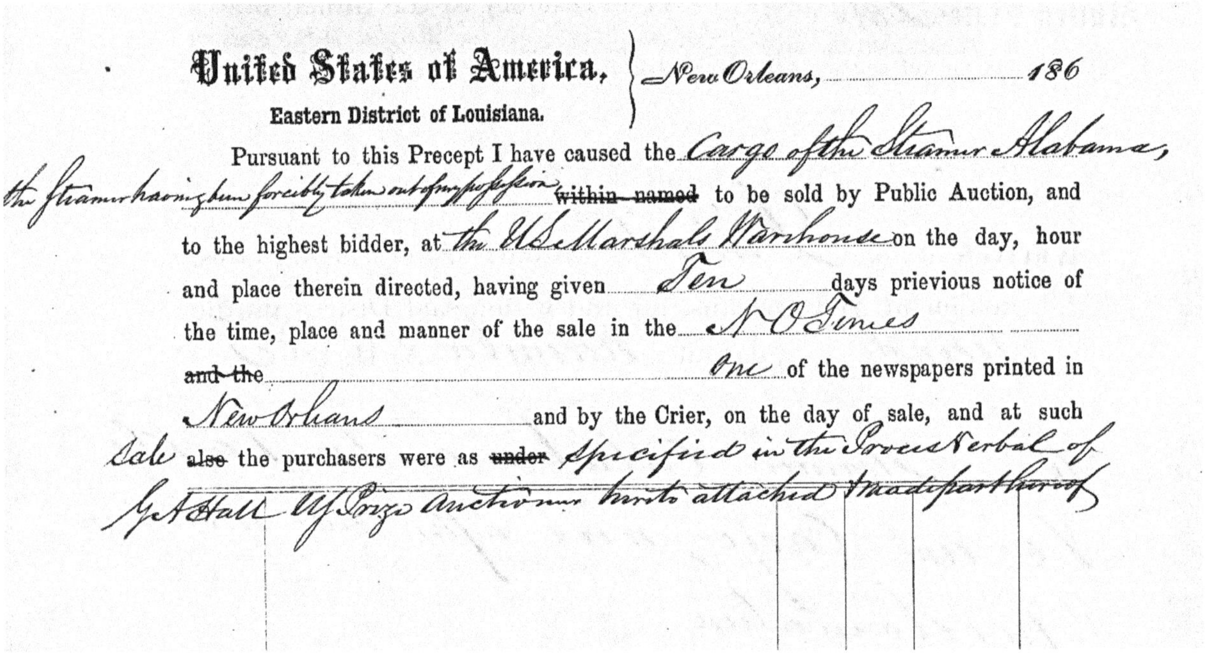

United States of America, New Orleans, __________ 186
Eastern District of Louisiana.

Pursuant to this Precept I have caused the Cargo of the Steamer Alabama, the Steamer having been forcibly taken out of my possession ~~within named~~ to be sold by Public Auction, and to the highest bidder, at the U S Marshals Warehouse on the day, hour and place therein directed, having given Ten days prievious notice of the time, place and manner of the sale in the N O Times ~~and the~~ One of the newspapers printed in New Orleans and by the Crier, on the day of sale, and at such sale ~~also~~ the purchasers were as ~~under~~ specified in the Process Verbal of G A Hall US Prize Auctioneer hereto attached & made part hereof

Figure 117. The marshal's return being the notice that the sale of the cargo took place after the required publication of notice. *NARA.*

Auctioneer's Bill for the Sale of the *Alabama's* Cargo

The following document was the invoice covering the auctioneer and his expenses relating to the sale of the *Alabama's* cargo (Figure 118).

New Orleans
Decr 4th 1863

Mr. James Graham
U.S. Marshal Eastern Dist. La.

                         To G. A. Hall
                         U.S. Prize Auctioneer   D^r

To Commission on Sale of the Cargo of Stm^r *Alabama* Novr 28th 1863			
Say one per cent on	$ 10,000.00	100.00	
" 4/10 Do. "	5,624.24	22.49	122.49
" Do. on Sale of Do. Decr 1st 1863			
Say one per cent on	$ 10,000.00	100.00	
" 4/10 Do. "	1,316.81	5.27	105.27
" Do. on Sale of Do. Decr 3rd 1863			
Say one per cent on	$ 10,000.00	100.00	
" 4/10 Do. "	14,323.05	57.29	157.29
Acc^t. as per process Verbal	$ 51,264.10		
" *N. O. Times* Advertising a/c	60.50		
" " *Era* Do. "	54.00		
" *La. St. Gazette* Do. "	51.00		
" Catalogues	70.00		
" Gauging Alcohol	1.00		236.50
			$ 621.55

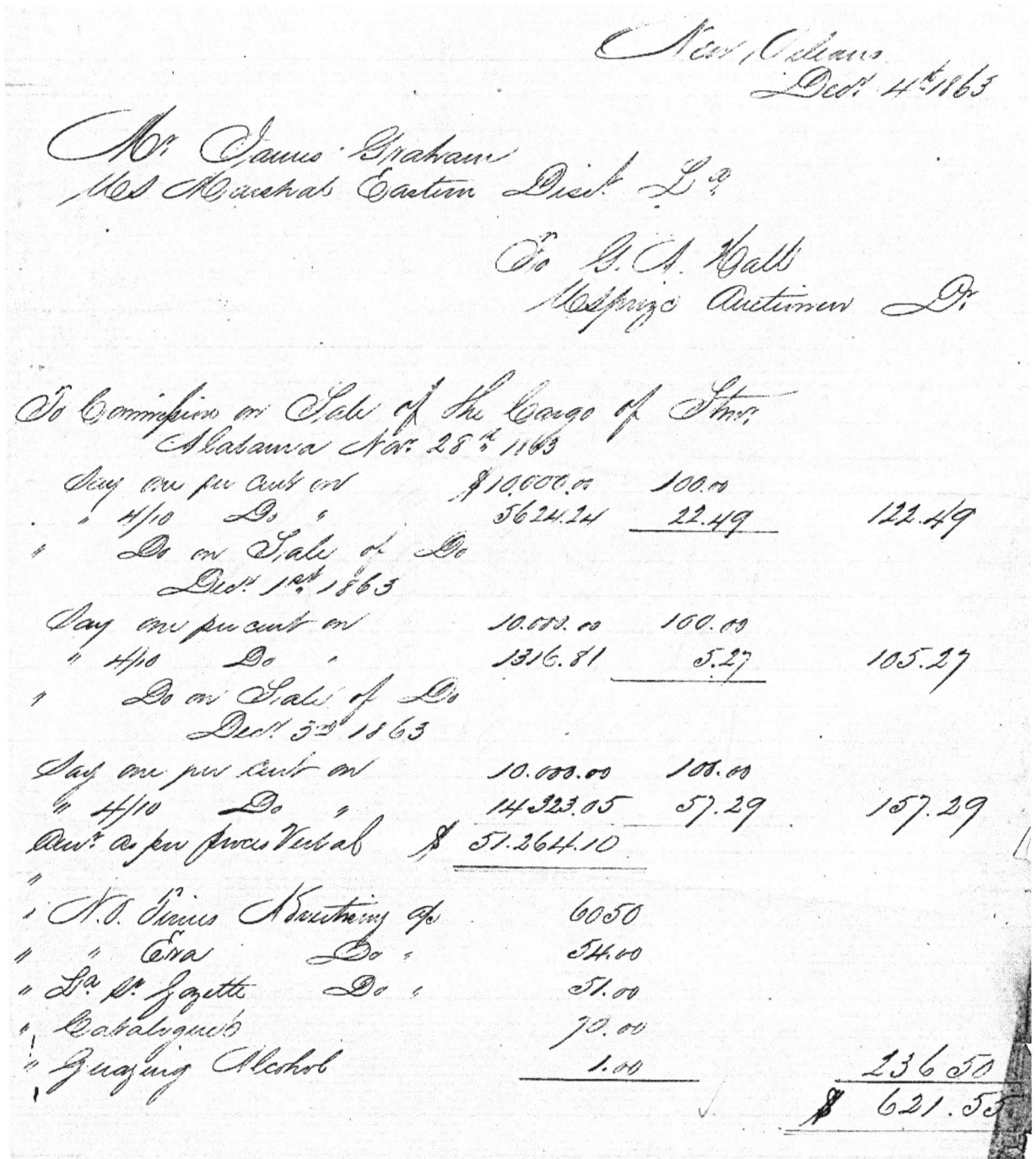

Figure 118. The auctioneer's bill regarding the sale of the *Alabama's* cargo. Dated 4 Dec. 1863. **NARA**.

U.S. District Attorney to U.S. Marshal Regarding Sale

The U.S. marshal inquired of the U.S. district attorney if he could proceed with the sale of cargo under existing documents and orders worded to cover both the ship and the cargo. The district attorney explained that the sale of the cargo could proceed in spite of the ship itself having been unceremoniously stolen by the army (Figure 119).

> This precept being for the sale of the *Alabama* and *cargo*, you certainly can sell the cargo, and make return of the sale, and then state in your return that the steamer having been forcibly taken from your passion you could not make sale of her. The *advertisement* should mention the *cargo only*.
> If you prefer a new order confined to the cargo only, you can only get it through the judge. I cannot make any change in this precept having no authority to do so.
> (to) James Graham, Esq.   (from) Rupert Waples
> Marshal   Dist. Atny.

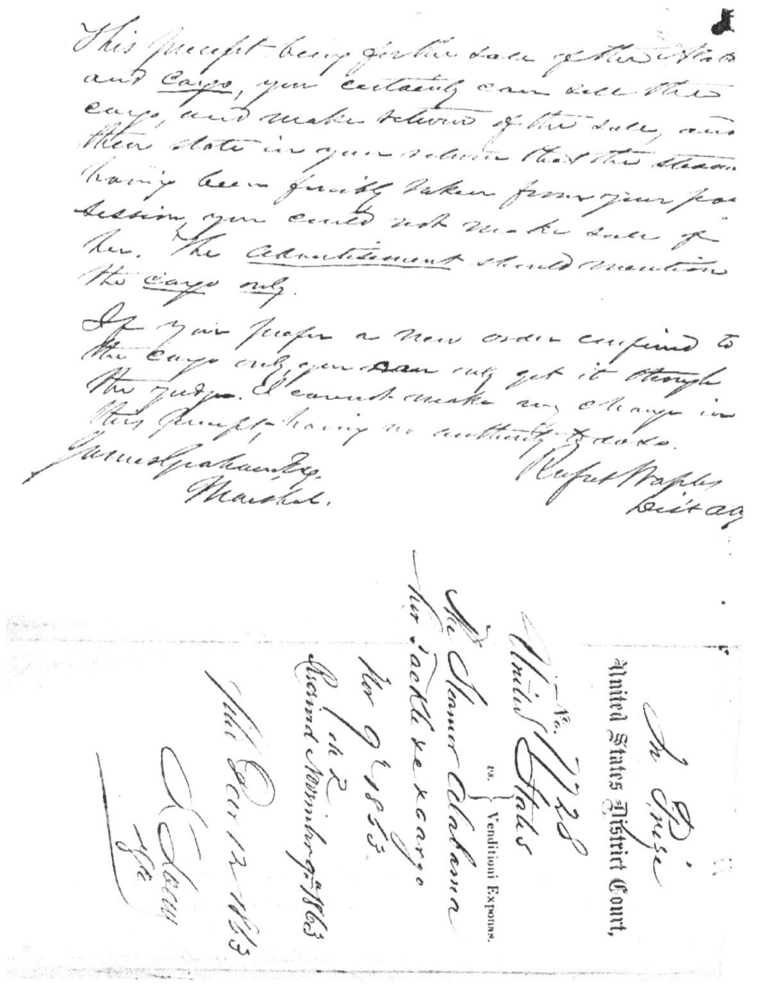

Figure 119. The cover of one of the documents reporting advertisement of the *Alabama* auction. The marshal asked the U.S. district attorney for guidance on how to proceed after General Stone confiscated the ship. Waples directed him to continue with the sale of the cargo even though the legal documents were styled as action against both ship and contents. Received 2 Nov. 1963. *NARA*.

Cargo Sale Proceeds Paid to Court Clerk

The following document was the acknowledgment that the proceeds of the *Alabama* cargo sale were deposited with the court, amounting to $53,364.10 (Figure 120).

> Whereby the ~~Net~~ Gross Sales amounted to Fifty one thousand Three hundred & Sixty four dollars ten cents, which I pay over to Alfred Shaw the Clerk of the Court aforesaid, to be disposed of as the Court directs.
> _____Attest, J. G. A. Hoit, dep. U.S. Marshal.
> Received from James Graham U.S. Marshal, Fifty one thousand Three hundred & Sixty four dollars ten cents being (as he saith) the proceeds of the aforegoing sale; the monies to be disposed of as the Court directs, and for which I have signed duplicate receipts.
> (signed) Alfred Shaw, Clerk
>
> New Orleans, 10th December 1863.

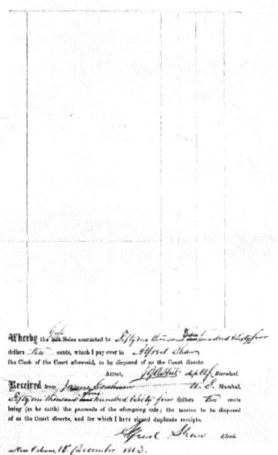

Figure 120. Funds from the auction of the *Alabama* cargo, $51,364.10, were transferred to the court by Deputy U.S. Marshal Hoit. Dated 10 Dec. 1863. *NARA*.

Report on Estimated Value Cargo

The appraisers reported to the court the estimated value of each item of the *Alabama's* cargo (Figure 121). A detailed inventory like this was of particular value to the *Denbigh* Shipwreck Project. There was little evidence of the *Denbigh's* final cargo since the wreck was burned and then spent decades in a high-energy surf zone that dispersed many of the smaller artifacts.

Office of the United States Prize Commissioners,
Appraisers' Department.

Appraisers' Report
New Orleans, November 30th 1863

Sir:

We the undersigned, duly appointed Appraisers, have examined the Cargo of Steamer *Alabama* Capt^{d.} in the Gulf of Mexico from _____ and are of opinion that the actual market value or wholesale prices of the said _____ and we do therefore appraise the same as follows:

No.	Description of Merchandise or Vessel.			Value.
10	Barrels alcohol 363½ gals.	@ $ 2.58	per gal.	908.75
24	" Jamaica rum	@ $ 120	" bar.	2,880.00
2	Half barrels do	@ $ 60	" "	120.00
4	Barrels whiskey	@ $ 160	" "	640.00
14	Cases Cognac	@ $ 6	" "	84.00
4	" "	@ $ 6	" "	24.00
16	" Claret	@ $ 5	" "	80.00
45	Bbls. Pork	@ $ 20	" "	900.00
20	" Beef	@ $ 10	" "	200.00
38	Tierces Mess beef	@ $ 10	" "	380.00
33	Sacks salt	@ $ 2	" Sack	66.00
25	Cases Olive oil	@ $ 5	" box	125.00
89	Boxes Pearl starch[169]	@ $ 3	" "	267.00
1	" " " 56 lbs.	@ $ 3	" "	3.00
10	" Starch (Kingsford & Son)[170]	@ $ 6	" "	60.00
10	" Colgate's fancy soap	@ $ 8	" "	80.00
75	" Cook's Brown soap	@ $ 3.50	" "	262.50

---

[169] Term referring to the physical form of starches dried in the old kiln type dryers that preceded the flash dryer in common use today. Referred to the shape of the finished product, not source. Used in paper making to produce a slicker surface. Source: http://www4.ncsu.edu/~hubbe/STCH.htm.

[170] Producer of cornstarch established 1848, factory located in Oswego, N.Y. http://www.amazon.com/1910-Kingsford-Oswego-Starch-Flour/dp/B005DGXS3W.

100	" Castile soap[171] 100 lbs. each	@ $16	" "	1,600.00
4	" brown "	@ $6	" "	24.00
2	Cases Candles 100 lbs. each fancy	@ $15	" "	30.00
29	Boxes "	@ $10	" "	290.00
8	" "	@ $6	" "	48.00
26	" Raisins	@ $5	" "	130.00
11	" Vermicelli & Macaroni	@ $3	" "	33.00
6	" Tea 348 lbs.	@ .80	" "	278.40
10	Bags Black Pepper 858 lbs.	@ .80	" "	257.40
	(Amount to be carried over)			$ 9,771.05
	To James Graham (signed) Indecipherable U.S. Appraisers to the U.S. Prize U.S. Marshal, (signed) Ch.s Claiborne Commissioners, New Orleans. end p. 1			
	Amount brought forward			$ 9,771.05
44	Sacks Coffee 5,409 lbs.	@ .35	" lbs.	1,893.15
18	Bbls. Bread	@ $6	" bbl.	108.00
2	Kits Mackerel	@ $2	" ea.	4.00
2	Cases dried Fish	@ $3	" "	6.00
1	Bale do do	@ $2	" bale	2.00
1	half bbl. Vinegar	@ $5	" bbl.	5.00
1	Case Tumblers	@ $4.50	" case	4.50
1	Bale Small Cordage 175 lbs.	@ .25	" lb.	43.75
3	Cases Matches	@ $40	" ea.	120.00
2	Bbls. Rosin	@ $30	" "	60.00
13	doz. Brooms	@ $3.50	" doz.	45.50
6	bbls. Lamp Oil	@ $60	" Bbl.	360.00
67	Boxes Kerosene oil—4 cans each of 5 gals.	@ $20	" Box	1340.00
76	cans White Lead	@ $7.50	" Can	570.00
1	bbl. Tallow 244 lbs.	@ .10	" lb.	24.40
1	keg Metallic Composition	@ $10		10.00
1	" Sugar	@ $15	" keg	15.00
1	Box Pickles	@ $1	" box	1.00
8	" " 24 each	@ $5	" doz.	80.00
1	" Common Picks—28	@ .75	" each	21.00
10	" Fairbank's scales,[172] different capacity	@ $30	" each	300.00
60	" *Tumba* Axes[173]—1 doz. each	@ $11.50	" doz.	690.00
2	" Hatchets—8 doz. each	@ $6	" "	96.00
2	" do No. 1, 2 doz. each	@ $4.50	" "	18.00
5	" do " 2, 2 " "	@ $6	" "	60.00
2	" do " 3, 3 " "	@ $7.50	" "	45.00
1	" do 17 " "	@ $5	" "	85.00

---

[171] A fine, hard, bland soap made from olive oil and sodium hydroxide; also: any of various similar soaps. http://www.merriam-webster.com/dictionary.

[172] Company formed in 1823 and rose to U.S. prominence located in St. Johnsbury, Vermont. Now headquartered in Kansas City, Missouri. By the Civil War period they had developed a substantial export business as well.

[173] Spanish for axes used in felling trees, "*hachas de tumba*." *Pocket Dictionary of Spanish Technical Terms*. Google Books, 1869, p. 52.

					$ 15,778.35
	(amount to be carried over)				
	end p. 2				
	Amount brought forward				$ 15,778.35
1	Box Augurs	for $ 20			20.00
99	Kegs Nails—assorted	@ $ 6	"	keg	594.00
1	Case Tacks				30.00
43	packages Knives and Forks, 3 doz. each	@ $1.25	"	doz.	161.25
1	" do do , 2 "	"	"		2.50
26	" do do , 3 "	"	"		97.50
1	" do do , 2 "	"	"		2.50
6	" do do , 1 "	"	"		7.50
9	" Scissors	@ $ 1.50	"	pk.	13.50
12	" table & 12 doz. tea spoons	@ .75	"	doz.	18.00
24	Cards Pocket Knives	@ $ 5	"	card	120.00
6	doz. large Files	@ $ 2.50	"	doz.	15.00
24	" Saw do	@ $ 1.00	"	"	24.00
4	Casks chain 1,929 lbs.	@ .06	"	lb.	115.74
2	" Solder 1,325 "	@ .70	"	"	927.50
70	Bars blister Steel 2,965 "	@ .25	"	"	741.25
44	" iron 2,095 "	@ .10	"	"	209.50
38	doz. Spades	@ $ 4.50	"	doz.	171.00
42	" Shovels	@ $ 3.50	"	"	147.00
10	Smith's Bellows	@ $ 5	"	ea.	50.00
6	red boxes Gin	@ $ 15	"	case	90.00
4	green " "	@ $ 12	"	"	48.00
37	5 gal. demijohns Gin	@ $ 15	"	ea.	555.00
3	Small " "	@ $ 6	"	ea.	18.00
222	5 gallon demijohns Rum	@ $ 10	"	"	2,220.00
7	— " " (partially full)	@ $ 6	"	"	42.00
1	5 gallon " Port Wine	@ $ 15			15.00
1	" " Annisette [sic]	@ $ 8			8.00
	(amount to be carried over)				$ 22,242.09
	end p. 3				
	Amount brought forward				$ 22,242.09
2	5 gallon demijohns Molasses	@ $ 2			4.00
1	" " Varnish	@ $ 10			10.00
6	bbls. Cream Tartar 1,109 lbs.	@ .30	"	lb.	332.70
5	" Soda 1,351 lbs.	@ .06	"	"	81.06
2	" Alum 582 lbs.	@ .05	"	"	29.10
1	" " (pulverized) 291 lbs.	@ .05	"	"	14.55
2	Casks Epsom Salts 586 lbs.	@ .02	"	"	11.72
1	bbl. Linseed 183 lbs.	@ .05	"	"	9.15
2	" " (pulverized) 260 lbs.	@ .08	"	"	20.80
4	" Borax 939 lbs.	@ .40	"	"	375.60
1	" Gum Arabic 222 lbs.	@ .50	"	"	111.00
1	" Flor Sulphur 192 lbs.	@ .13	"	"	24.96
2	" Castor Oil	@ $ 100	"	bar.	200.00
1	half barrel Balsam Cohavi (?)	@ $ 15	"	"	15.00
3	Bales 18,000 Corks	@ $ 12	"	ea.	36.00

20	kegs Carb Soda		@ $ 12	" "	240.00
10	Bbls. Copperas 2,742 lbs.		@ $ .03½	" "	95.97
1	" Turpentine		@ $ 150	" "	150.00
3	Cases Adhesive Plaster (50 half boxes) 150 pks.		@ $ 1	" pk.	150.00
3	Cans Essence Thym (Rouge) 78 lbs.		@ .32	" lb.	24.96
5	Cases Mustard (Taylor & Sons) 500 lbs.		@ .43	" "	215.00
1	" Tannin—25 bottles		@ 1.35	" bottle	33.75
1	" Hydrag Creta 50 "		@ .45	" "	22.50
2	" Aqua Ammonia, 24, 4 lb. bottles, 95 lbs.		@ .90	" "	21.60
2	" Gum Arabic 50 lb. each		@ .46	" lb.	23.00
5	boxes Extract Logwood 110 lbs.		@ .15	" "	16.50
1	Case Liquorice 100 lbs.		@ .38	" "	38.00
7	" Phials		@ $ 15	" Case	105.00
	(amount to be carried over)				$ 24,654.01
	end p. 3				
	Amount brought forward				$ 24,654.01
1	Case Pill Boxes		@ $ 50	per Case	50.00
6	bbls. Gum Camphor 666 lbs.		@ $ 2	" lb.	1,332.00
9	boxes " drops		@ $ 1	" box	9.00
1	Case assorted Drugs &c. to wit:			"	
	24 doz. Glass Syringes—12 doz. men's			"	
	Metallic Syringes—5 doz. Strychnine in		$ 175		175.00
	crystals—25 lbs. powdered Jalap—3			"	
	bottles iodine Potassium 10 lbs. each			"	
1	Case containing Adhesive Plaster &			"	
	75 lbs. Cantharides		$ 75		75.00
2	boxes Calomel—25 lbs. each		@ .86	" lb.	21.50
10	Case Pow'd. gum arabic, 575 lbs.		@ .50	" lb.	287.50
3	Boxes Quinine 100 bottles each, 294 oz.		@ $ 2.75	" oz.	808.50
1	" " 43 ", 3 broken, 40 oz.		@ $ 4	" "	160.00
2	Cases Iodine potassium 93 bottles		@ $ 3.25	" bot.	302.25
1	" 5 Cans Quinine—100 oz. each case		@ $ 4	" oz.	400.00
1	" Chloroform 98 bottles		@ $ 1	" bottle	98.00
1	Box Benzaic [sic] Acid 3 lbs.		@ $ 5	" lb.	15.00
1	Case Kreosot [sic] —24 bottles		@ $ 1.25	" bot.	30.00
2	" Pulverized Opium 48½ lbs.		@ $ 7	" lb.	339.50
2	" Cantharides 52 lbs.		@ $ 1.50	" "	78.00
2	" Ypecac [sic] 53 lbs.		@ $ 1.50	" "	79.50
7	" Flax Lint—49 packs each		@ $ 1	" pk.	343.00
1	Ship's Medicine chest		for $ 25		25.00
2	Cases Ruled paper		@ $ 100	" case	200.00
1	" Writing "		@ $ 100	" "	100.00
3	" Cartridge " 63 reams each		@ $ 63	" "	189.00
1	" containing 4 reams mapping paper &				
	14 reams cartridge paper				50.00
	(amount to be carried over)				$ 29,821.76
	end p. 4				
	Amount brought forward				$ 29,821.76
1	Case Asstd. Envelopes		@ $ 76	" case	76.00

2	" blank books	@ $ 185	" "	185.00
1	" Edward's black ink 25 doz.	@ $ 1	" doz.	25.00
4	" " " " 290 large bottles	@ .30	" ea.	87.00

1	" contg. 20 gross Lead Pencils Fabres (?)
	      20 " " Imitation
	      20 " Penholders assorted
	      40 " Steel Pens do
	      60 " do do
	      50,000 Envelopes Yellow & white
		$ 200		200.00

42	Empty Demijohns	@ .50	" ea.	21.00
1	Case assorted Stationery, containing 33 Reams folio Paper, assorted; 108 doz. Fabre's Lead Pencils; 20½ dozen Ink stands, assorted; 20 gross Pen holders, assorted; 34 boxes Steel pens.	$ 250	" " " " " "	250.00
1	Case large folio Paper	for $ 50	"	50.00
2	" French letter do	@ $ 125	" case	250.00
1	" Cottonade, 20 pieces	@ $ 18.75	" piece	375.00
1	" Denims, 30 do	@ $ 12.50	" "	375.00
2	bales grey cloth, 7 pieces each	@ $ 120	" "	1,680.00
1	" white flannel, 7 " do	@ $ 23	" "	161.00
1	" do 26 do	@ $ 27	" "	702.00
1	" Madras handkerchiefs, 48 doz.	@ $ 4	" doz.	192.00
1	Case fancy flannels 10 pieces	@ $ 12	" piece	120.00
1	" Valentia 36 pieces	@ $ 10	" "	360.00
1	" printed cotton flannel 19 pieces	@ $ 13	" "	247.00
				$ 35,177.00
	end p. 5			
	Amount brought forward			$ 351,777.00
1	Case blue flannel 4 pieces	@ $ 30	" piece	120.00
1	" red " 10 "	@ $ 25	" "	250.00
1	" contg. 24 pcks. brown linen thread and 50 pcks. Red tape	for $150		150.00
1	" containing 70 doz. Ladies Hose and 99 pkgs. Pins	for $ 250		250.00
3	" Felt hats 22¾ doz.	@ $ 6	" doz.	136.50
42	bales Army Blankets, 51 pairs	@ $ 4	" pair	8,568.00
9	pair do do	@ $ 6	" "	54.00
5	Cases cloth Cards	@ $ 225	" ea.	1,125.00
27	" Cotton " (Lory & Watson) 54 doz.	@ $ 12	" doz.	648.00
65	" " " (Whitemore)[174] 130 doz.	@ $ 12	" "	1,560.00
3	" Brogans, undressed, 330 pairs	@ $ 1.25	" pair	412.50
1	" Shoe Thread, 304 pckgs.	@ $ 1.25	" pk.	380.00
1	" asstd. Shoemaker's tools,[175] contg.			

---

[174] Properly spelled Whittemore referring to William Whittemore & Co. in Boston which manufactured and sold wool and cotton cards made with their own patented machine. See: http://quod.lib.umich.edu/c/clementsmss/umich-wcl-M-2440whi?view=text.

	1 doz. asstd. Pincers; 2 doz. Peg Awls; 48 doz. shoe tacks; 4 doz. Currier's knives	$ 75		75.00
	5 doz. Rasps; 5½ doz. hammers; 7 pks. Awls; 2 doz. Givens; 8 doz. Punches; 2 doz. Bisagles; 2 doz. Awl handles.			
4	rolls Harness Leather 661 lbs.	@ .40	" lb.	264.40
6	Kip skins[176]	@ $ 4	" kip	24.00
1	case asstd. harness Buckles & Rings, 57 pkgs.	@ $ 50		50.00
1	" Lanterns 2 doz.	@ $ 1	" piece	24.00
1	" hardware, containing 72 hand saws; 36 pks. Monkey wrenches; assd. sizes; 8 doz. Padlocks	$ 50		50.00
	(amount to be carried over)			$ 49,319.16
	end p. 6			
	Amount brought forward			$ 49,319.16
1	Military Cap	@ $ 2.80		2.80
1	box preserves	@ $ 10		10.00
2	Jars Olive oil	@ $ 1.50	" ea.	3.00
25	pkgs. Sundries	@ $ 30	" ea.	750.00
25,000	**Cannon Primers** (emphasis added)	@ $ 10	" thsand.	250.00
3	**Cases Percussion Caps**[177]	@ $ 12	" case	36.00
1	Chronometer	@ $ 100		100.00
	Lot charts & 1 Case stationery	@ $ 180		180.00
				$ 50,650.96
	H. J. Heartt (?) U.S. Appraisers to Chs. Claiborne U.S. Prize Commissioners			
	To			
	James Graham Esq.			
	U.S. Marshal			
	Eastern District of Louisiana			

---

[175] This shipment possibly replaced by a similar shipment on an early run of the *Denbigh*.

[176] Leather prepared from the skin of young or small cattle, intermediate in grade between calfskin and cowhide. From *Webster's Revised Unabridged Dictionary*.

[177] The cannon primers and percussion caps were the only obvious contraband military cargo. Such cargo alone would prove the ship a valid capture. It was easy for the author to speculate and wonder if there was more cargo of military character that was jettisoned during the daylong chase by the *San Jacinto*. Perhaps these two shipments were overlooked as the rest was dumped overboard.

Office of the United States Prize Commissioners,
APPRAISERS' DEPARTMENT.

APPRAISERS' REPORT.

NEW ORLEANS, *November 30th*, 1863.

SIR:

We, the undersigned, duly appointed Appraisers, have examined the *Cargo of Steamer Alabama Cap'td* in the *Gulf of Mexico* from _____ and are of opinion that the actual market value or wholesale prices of the said _____ and we do therefore appraise the same as follows:

MARKS.	NO.	DESCRIPTION OF MERCHANDISE OR VESSEL.		VALUE.	
	10	Barrels alcohol 363½ gals	@ $2.50 per gal	908	75
	24	" Jamaica rum	@ 120 " bar	2880	"
	2	half barrels do	@ 60 " "	120	"
	4	Barrels whiskey	@ 160 " "	640	"
	14	Cases Cognac	@ 6 " "	84	"
	4	" "	@ 6 " "	24	"
	16	" Claret	@ 5 " "	80	"
	45	Bbls Pork	@ 20 " "	900	"
	20	" Beef	@ 10 " "	200	"
	38	Tierces Mess beef	@ 10 " "	380	"
	33	Sacks salt	@ 2 " sack	66	"
	25	Cases Olive oil	@ 5 " box	125	"
	89	Boxes Pearl starch	@ 3 " "	267	"
	1	" " 56 lb	@ 3 " "	3	"
	10	" Starch (Kingsford & Son)	@ 6 " "	60	"
	10	" Colgate's fancy soap	@ 8 " "	80	"
	75	" Cook's Brown soap	@ 3.50 " "	262	50
	100	" Castile soap ~ 100 lbs each	@ 16 " "	1600	"
	4	" brown "	@ 6 " "	24	"
	2	Cases Candles ~ 100 lbs each fancy	@ 15 " "	30	"
	29	Boxes "	@ 10 " "	290	"
	8	" "	@ 6 " "	48	"
	26	" Raisins	@ 5 " "	130	"
	11	" Vermicelli & macaroni	@ 3 " "	33	"
	6	" Tea 348 lb	@ 80¢ " lb	278	40
	10	Bags Black Pepper 858 "	@ 30¢ " "	257	40
		(Amount to be carried over)		$ 9771	05

To *James Graham*
U. S. MARSHAL,
District Court, for the Eastern District of Louisiana, New Orleans.

A. F. Kearn.
Th. Claiborne
} U. S. Appraisers to the U. S. Prize Commissioners, New Orleans.

over

Appraisement of goods per Str. Alabama continued

			Amount brought forward		9721	05
44	Sacks Coffee	5409 lbs	@ 35¢ lb		1893	15
18	Bbls Bread		@ $6 bbl		108	"
2	Kits Mackerel		@ $2 ea		4	"
2	Cases dried Fish		@ $3 "		6	"
1	Bale do do		@ $2 bale		2	"
1	half bbl Vinegar		@ $5 h bbl		5	"
1	Case Tumblers		@ $4.50 Case		4	50
1	Bale Small Cordage	175 lbs	@ 25¢ lb		43	75
3	Cases matches		@ $40 ea		120	"
2	Bbls rosin		@ $30 "		60	"
13	doz. Brooms		@ $3.50 p dz		45	50
6	bbls Lamp Oil		@ $60 Bbl		360	"
67	boxes Kerosene oil – 4 cans each of 5 gals		@ $20 Box		1340	"
76	cans White Lead		@ $7.50 Can		570	"
1	bbl Tallow	244 lbs	@ 10¢ lb		24	40
1	Keg Metallic Composition		@ $10		10	"
1	" Sugar		@ $15 Keg		15	"
1	Box Pickles		@ $1 box		1	"
8	" " 24 each		@ $5 dz		80	"
1	" Common Picks – 28		@ 75¢ each		21	"
10	" Fairbank's scales, different capacity		@ $30		300	"
60	" Jumba Axes – 1 dz each		@ $11.50 p dz		690	"
2	" Hatchets – 8 dz each		@ $6 " "		96	"
2	" do No 1, 2 " "		@ $4.50 " "		18	"
5	" do " 2, 2 " "		@ $6 " "		60	"
2	" do " 3, 3 " "		@ $7.50 " "		45	"
1	" do " 17 " "		@ $5 " "		85	"
		(amount to be carried over)			15,778	35

	Amount brought forward		$15778 3
1	Box Augurs	for $20	20
99	Kegs nails – assorted	@ $6 pr Keg	594
1	Case Tacks		30
43	packages Knives and Forks, 3 dz each	⎫	161 2
1	„ do do 2 „	⎪	2 5
26	„ do do 3 „ „	⎬ $1.25 p dz	97 5
1	„ do do 2 „	⎪	2 5
6	„ do do 1 „ „	⎭	7 5
9	„ Scissors	@ $1.50 pr pkg	13 5
12	„ table & 12 dz: tea spoons	@ 75 cts „ dz	18 „
24	Cards Pocket Knives	@ $5 „ Card	120 „
6	dz large Files	@ $2.50 „ dz	15 „
24	„ Saw do	@ $1. „ „	24 „
4	Casks chain 1929 lbs	@ 6¢ „ lb	115 7
2	„ Solder 1325 „	@ 70¢ „ „	927 5
10	Bars blistered Steel 2965 „	@ 25¢ „ „	741 2
44	„ iron 2095 „	@ 10¢ „ „	209 5
38	dz Spades	@ $4.50 „ dz	171
42	„ Shovels	@ $3.50 „ „	147
10	Smith's Bellows	@ $5. ea	50
6	red boxes Gin	@ $15. pr Case	90
4	green „ „	@ $12. „ „	48
37	5 gal demijohns Gin	@ $15 ea	555
3	Small „ „	@ $6 ea	18
222	5 gallon demijohns Rum	@ $10 „	2220
7	– „ „ (partially full)	@ $6 „	42
1	5 gallon „ Port Wine	@ $15.	15
1	„ „ „ Annisette	@ $8	8
	(amount to be carried over)		$22242 0

	Amount brought forward			$22242.09
2	5 gallon demijohns Molasses	@ $2		4 "
1	" " Varnish	@ $10.		10 "
6	bbls Cream Tartar	1109 lbs	@ 30¢ pr lb	332.70
5	" Soda	1351 lbs	@ 6¢ " "	81.06
2	" Alum	582 lbs	@ 5¢ " "	29.10
1	" " (pulverised)	291 lbs	@ 5¢ " "	14.55
2	Casks Epsom Salts	586 lbs	@ 2¢ " "	11.72
1	bbl Linseed	183 lbs	@ 5¢ " "	9.15
2	" " (pulverised)	260 lbs	@ 8¢ " "	20.80
4	" Borax	939 lbs	@ 40¢ " "	375.60
1	" Gum Arabic	222 lbs	@ 50¢ " "	111 "
1	" Flor Sulphur	192 lbs	@ 13¢ " "	24.96
2	" Castor Oil		@ $100 p bar	200 "
1	half barrel Balsam Cohavi		@ $15 " h "	15 "
3	Bales 18,000 Corks		@ $12 ea	36 "
20	kegs Carb Soda		@ $12 "	240 "
10	bbls Copperas	2742 lbs	@ 3½¢ "	95.97
1	" Turpentine		@ $150 pr bbl	150 "
3	cases Adhesive Plaster (50 half boxes) 150 pks	@ $1 " pk	150 "
3	Cans Essence Thyme (Rouge) 78 lbs	@ 32¢ " lb	24.96
5	Cases Mustard (Taylor & Sons) 500 lbs	@ 43¢ " "	215 "
1	" Tannin — 25 bottles	@ $1.35 " bottle	33.75
1	" Hydrag Creta 50 "	@ 45¢ " "	22.50
2	" Aqua Ammonia, 24, 4 lb bottles, 96 lbs,	@ 90¢ " "	21.60
2	" Gum Arabic 50 lb each	@ 46¢ " lb	23 "
5	boxes Extract Logwood 110 lbs	@ 15¢ " "	16.50
1	Case Liquorice 100 lbs	@ 38¢ " "	38 "
7	" Phials	@ $15 " case	105 "
				24654.01
	(amount to be carried over)			

	Amount brought forward		$24654.01
1	Case Pill Boxes @ $50 pr Case		50
6	bbls Gum Camphor 666 lbs @ $2 " lb		1332
9	boxes " drops @ $1 " box		9
1	Case assorted Drugs &c to wit: 24 dz Glass Syringes ~ 12 dz men's metallic Syringes ~ 5 dz Strychnine in chrystals ~ 25 lb powdered Jalap ~ 3 bottles Iodine Potassium 10 lbs each	$175	175
1	Case containing Adhesive Plaster & 75 lbs Cantharides	$75	75 "
2	boxes Calomel ~ 25 lbs each @ 86¢ pr lb		21 50
10	Cases Powd'd gum arabic, 575 lbs @ 50¢ pr lb		287 50
3	Boxes Quinine 100 bottles each, 294 oz @ $2 75/100 pr oz		808 50
1	" " 43 " 3 broken, 40 oz @ $4 " "		160 "
2	Cases Iodine potassium 93 bottles @ 3 25/100 " bot		302 2.
1	" 5 Cans Quinine ~ 400 oz each @ $4 " 3		400 "
1	" Chloroform 98 bottles @ $1 " bottle		98 "
1	Box Benzoic Acid 3 lbs @ $5 " lb		15 "
1	Case Kreosot ~ 24 bottles @ $1 25/100 " bot		30 "
2	" Pulverised Opium 48½ lbs @ $7 " lb		339 50
2	" Cantharides 52 lbs @ $1 50/100 " "		78 "
2	" Ipecac 53 lbs @ $1 50/100 " "		79 50
7	" Flax Lint ~ 49 pack's each @ $1 " pk		343 "
1	Ship's Medecine chest for $25		25
2	Cases Ruled paper @ $100 " case		200
1	" Writing " @ $100 " "		100
3	" Cartridge " 63 reams each @ $1 63 " "		189
1	" containing 4 reams wrapping paper & 14 reams cartridge paper		50
	(amount to be carried over)		$29821.76

	Amount brought forward		$2982.76
1	Case Ast'd Envelopes	@ 76 pr Case	76
2	" blank books	@ 185 " "	185
1	" Edward's black ink 25 dz	@ 1 " dz	25
4	" " " " 290 large bottles	@ 30¢ ea	87
1	" con'g 20 gross Lead Pencils Fabres		
	20 " " Imitation		
	20 " Penholders assorted	$200	200
	40 " Steel Pens do		
	60 " do do		
	50,000 Envelopes Yellow & white		
42	Empty Demijohns @ 50¢ ea		21
1	Case assorted Stationery, containing		
	33 Reams folio Paper assorted; 108 doz		
	Faire's Lead Pencils; 20½ dozen	$250	250
	Ink stands, assorted; 20 gross		
	Pen holders, assorted; 34 boxes		
	Steel pens.		
1	Case large folio Paper	for 50	50
2	" French letter do	@ 125 pr case	250
1	" Cottonade, 20 pieces	@ 18¾ " piece	375
1	" Denims, 30 d°	@ 12½ " "	375
2	bales grey cloth, 7 pieces each	@ 120 " "	1680
1	" white flannel, 7 " do	@ 23 " "	161
1	" do 26 do	@ 27 " "	702
1	" Madras handkerchiefs, 48 dz	@ $4 " dz	192
1	Case fancy flannels 10 pieces	@ 12 " piece	120
1	" Valentia 36 pieces	@ 10 " "	360
1	" printed cotton flannel 19 pieces	@ 13 " "	247
			$3517.76

	Amount brought forward		$35177.74
1	Case blue flannel 4 pieces @ $30 per piece		120
1	" red " 10 " @ $25 " "		250
1	" cont'g 24 pcks brown linen thread and 50 pcks Red tape	for $150	150
1	" containing 70 dz Ladies Hose and 99 pkgs Pins	for $250	250
3	" Felt hats 22¾ dz @ $6 pr dz		136
42	bales Army Blankets, 51 pairs each @ $4 per pair		8568
9	pair do do @ $6 " "		54
5	Cases cloth Cards @ $225 ea		1125
27	" Cotton " (Lory & Watson) 54 dz @ $12 " dz		648
65	" " " (Whitemore) 130 dz @ $12 " "		1560
3	" Brogans undressed, 330 pairs @ $1.25 a pair		412
1	" Shoe Thread, 304 pckgs @ $1.25 pr pk		380
1	" assorted Shoemaker's tools, cont'g 1 dz ass'd Pincers; 2 dz Welt Cutters; 14 dz Shoe Knives; 2 dz Peg Awls; 48 dz shoe tacks; 4 dz Currier's knives; 5 dz Rasps; 5½ dz hammers; 7 pkgs Awls; 2 dz Givens; 8 doz Punches; 2 dz Bisagles; 2 dz Awl handles.	$75	75
4	rolls Harness Leather 661 lbs @ 40¢ pr lb		264
6	Kip skins @ $4 " Kip		24
1	case ass't harness Buckles & Rings, 57 pkgs $50		50
1	" Lanterns 2 dz @ $1. apiece		24
1	" hardware, containing 72 hand saws, 36 pks monkey wrenches, ass'd sizes; 8 dz Padlocks	$50	50
	(amount to be carried over)		$49319

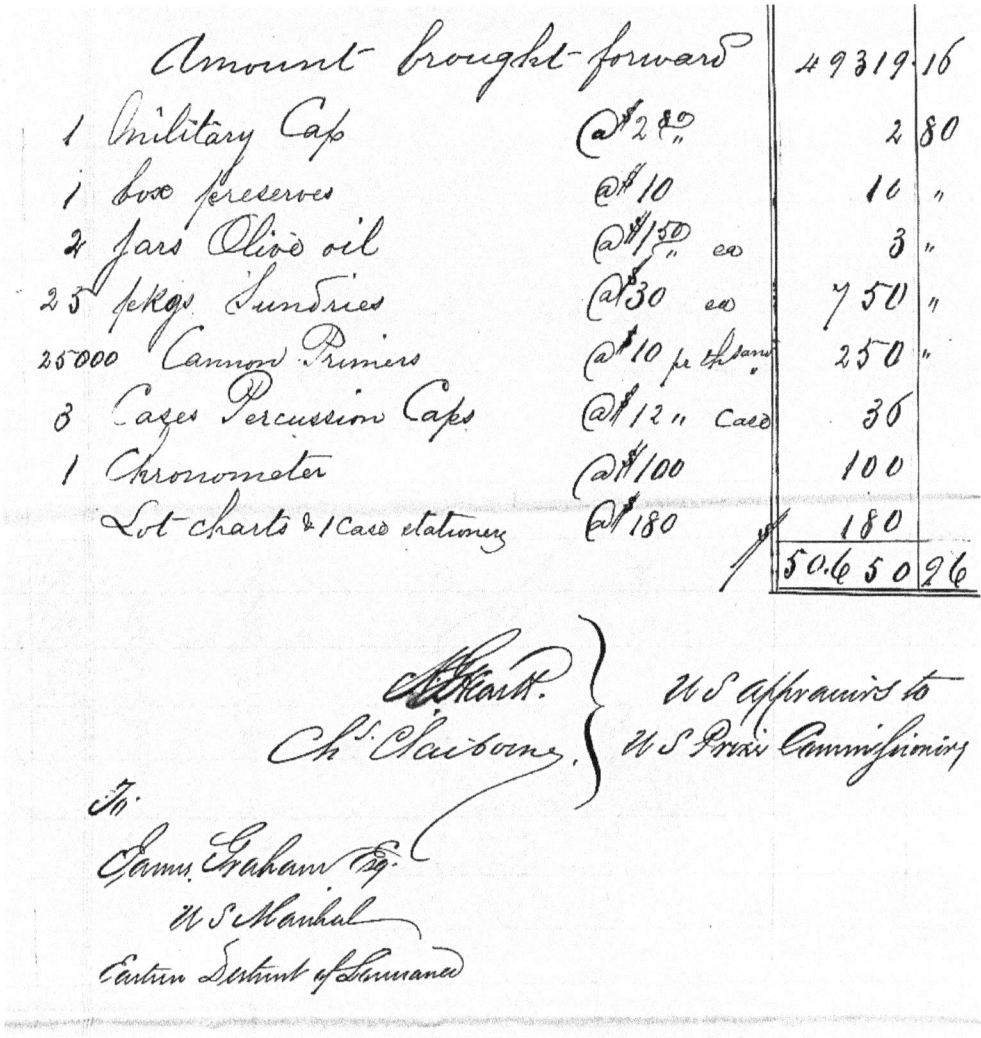

Figure 121. Appraisal of the *Alabama's* cargo. (Eight pages including this and previous seven pages.) Dated 28 Nov. 1863 and filed 21 Dec. 1863. *NARA*.

Alabama's Cargo Auction Results

The following document presented the listing of buyers, items sold, and prices realized for the *Alabama* cargo sold at auction. The auction lasted three days beginning on 28 Nov. 1863 (Figures 122-124).

Auction Day One

I, G. A. Hall, U.S. Prize Auctioneer, do hereby certify that under and by virtue of a writ of *Venditioni Exponas*[178] of the Hon. District Court of the United States, Eastern District of Louisiana directed to the U.S. Marshal of the District aforesaid, in suit by the United States vs.

---

[178] A writ of execution, directed to the sheriff or in this case the U.S. marshal, commanding him to sell goods or chattels (http://legal-dictionary.thefreedictionary.com/Venditioni+exponas).

Steamer *Alabama* her Tackle, Apparel, and Furniture and Cargo, N[o.] 7728 of the docket of the said court.

I offered the cargo of the Steamer *Alabama* for sale at public auction, on Saturday, the 28th day of November 1863, at eleven o'clock A.M. at the U.S. Marshal's Warehouse N[os.] 38 & 40 Old Levee Street, at which said sale, the same was sold for cash at the price, to the purchasers, and in the lots as hereinafter set forth, to wit:

Purchasers	Articles purchased		Amount	Total
Levi	10 Bbls. Alcohol 363½ gals. @	$ 2.25	817.88	
J. H. Carter	24 " Jamaica Rum	38.00		
Do	2 hlf. do Do Do	26.00	52.00	
Wise	3 Bbls. Whiskey	120.00	360.00	
J. H. Carter	1 Bbls. Do		135.00	
Dwyer	14 Boxes Cognac	8.00	112.00	
Stockhouse	4 " Do	11.50	46.00	2,434.88
(end p. 1)				
	Amt. forward			2,434.88
Dwyer	16 Boxes Claret	4.75	76.00	
Heath	12 Bbls. Pork	19.75	237.00	
Lepine & Ferre	20 " Do	19.50	390.00	
J. H. Carter	14 " Do	19.50	214.50	
Wise	1 " Do		14.00	
Lepine & Ferre	6 " Beef	12.50	75.00	
J. H. Carter	14 " Do	12.75	178.50	
Do	38 Tierces Mess Beef	19.50	741.00	
Wise	33 Sacks Salt	1.40	46.20	
J. H. Carter	25 cases Olive Oil	5.25	131.25	
Wise	89 boxes Starch	1.90	169.10	
McMasters	1 " Do		4.25	
Lee	10 " Do	2.65	26.50	
McMasters	10 " Soap (*Colgate*)	3.85	38.50	
Wise	75 " Do *Cook's*	3.50	262.50	
Rivet	10 " Do (Castile)	11.50	115.00	
Do	10 " Do	11.25	112.50	
Caden	10 " Do	11.00	110.00	
Carter	70 " Do	10.50	735.00	
Wise	4 " Do (Brown)	6.50	26.00	
Steers	2 Cases Candles 100 lbs. ea.	29.50	59.00	
Lepine & Ferre	29 boxes Do	7.00	203.00	
Hollander	4 " Do	7.00	28.00	
J. H. Carter	4 hlf. " Do	3.62½	14.50	
Miller	26 " Raisins dam[d.] (damaged)	2.45	63.70	
Junis	11 " Vermicelli	3.00	33.00	
Carter	6 " Tea 348 lbs.	.70	243.60	
Farrel	5 " Pepper 435 lbs.	.29	126.15	4,473.75
				6,908.63
(end p. 2)				
	Amt. forward			6,908.63

Smith Zeigler	5 Bags Pepper 423 lbs.	.29	122.67	
Spur	8 " Coffee 1,069 "	.35	374.15	
Carter	14 " Do 1,725 "	.34	586.50	
Do	12 " Do 2,625 "	.35	915.25	
Babb	4 " Do 969 "	.35½	344.00	
Carter	6 " Do 1,384 "	.35	484.40	
Wise	16 Bbls. Bread	3.35	52.00	
Do	2 " Do	5.00	10.00	
Carter	2 Kits Mackerel	1.75	3.50	
Barber	2 Cases dried Fish	4.05	8.10	
Carter	1 Bale " "		1.25	
Wise	1 hlf. bbl. Vinegar		4.50	
Currier	1 Case Tumblers 7¾ doz.	.95	7.30	
Barber	1 Bale Small Cordage 175 lbs.	.42	73.50	
Moulor	2 Cases Matches	30.00	60.00	
Wise	1 " Do		67.00	
Do	2 Bbls. Rosin	28.00	56.00	
Lepine & Ferre	13 doz. Brooms	2.30	29.90	
Barber	4 Bbls. Lamp Oil	38.00	152.00	
Wise	2 " Do Do	40.00	80.00	
Lepine & Ferre	50 Boxes Kerosene Oil	13.50	787.50	
J. Levi	10 " Do Do	16.00	160.00	
J. Tisse	3 " Do Do	11.75	35.25	
F. Barber	4 " Do Do	13.50	54.00	
Wise	76 Cans White Lead	2.50	190.00	
Carter	1 Bbl. Tallow 244 lbs.	.12	27.28	
Wise	1 Keg Metallic Comp$^{n.}$		12.00	
Moulor	1 " Sugar		15.00	4,713.05
				11,621.68
(end p. 3)				
	Amt. forward			11,621.68
Morrison	1 Box Pickles		1.75	
Do	2 " Pick Axes 48	.50	24.00	
Forney	6 " Do 144	.50	72.00	
Currier	1 " common Do 28	.40	11.20	
Keep (? Keess)	4 Fairbanks Scales	37.00	148.00	
Do	4 Do Do	29.50	118.00	
Do	1 Do Do		18.50	
Mandell	1 Do Do		25.00	
J. Graham	1 Do Do		29.50	
Fitch	30 Boxes *Tumba* Axes	8.00	240.00	
Elliot	30 " " "	7.50	225.00	
Fitch	2 Cases Hatchets 16 doz.	5.50	88.00	
Levi	2 " Do 4 "	5.50	22.00	
Currier	5 " Do 10 "	5.25	52.50	
Fitch	2 " Do 6 "	6.25	37.50	
Levi	1 " Do 16 "	5.00	80.00	
Mandell	1 " Augurs 17	.35	5.95	
Fitch	99 Kegs Nails	4.90	485.10	

Vose & West	1 Case Tacks		32.00	
Wise	43 pkgs. Knives Forks 129 doz.	1.50	232.20	
Do	1 " " " 2 doz.	2.30	4.60	
Do	33 " " " 86 doz.	1.80	154.80	
Do	9 " Scissors	1.80	16.20	
Currier	12 doz. Table Spoons	2.75	33.00	
Wise	10 " Tea "	1.35	13.50	
Samuels	23½ Cases p'ckt. (pocket) Knives	9.50	224.25	
Risset	6 doz. large Files	5.50	33.00	
Currier	18 " Small Do	1.10	19.80	2,447.35
				14,069.03
(end p. 4)				
	Amt. forward			14,069.03
Wise	4 Casks Chain 1,929 lbs.	.10½	202.55	
Walker	2 " Solder 1,325 lbs.	.47½	629.37	
McLearn	70 Bars Blist$^r.$ Steel 2,965 lbs.	.10½	311.32	
Do	44 " Iron 2,095 lbs.	.06¼	130.94	
Currier	9 Doz. Spades	4.62½	41.63	
McMasters	40 10/12 " Shovels	4.10	167.40	
Mandall	10 " Do	4.00	40.00	
Do	8 " Do	4.00	32.00	1,552.21
	Total Am't. Sold Nov. 28/63			15,624.24

At 3 o'clock P.M. Nov. 28th 1863, the above sale was adjourned by order of the U.S. Marshal, 'till Tuesday.

I, G. A. Hall, U.S. Prize Auctioneer do hereby certify, that under and by virtue of a Writ of "Venditioni Exponas" of the Hon. District Court of the United States, Eastern District of Louisiana directed to the U. S. Marshal for the District aforesaid, in Suit by the United States vs. Steamer "Alabama" her Tackle, Apparel and Furniture and Cargo, No. 7728 of the Docket of the said Court.

I offered the Cargo of the Steamer "Alabama" for Sale at Public Auction, on Saturday, the 28th day of November 1863, at Eleven o'clock A.M. at the U.S. Marshal's Warehouse Nos. 38 & 40 Old Levee Street, at which said Sale, the same was sold for Cash at the price, to the purchasers and in the lots as hereinafter set forth, to wit:

Purchasers.	Articles purchased	Amount	Total
Levi	10 Bbls. Alcohol 363½ gals. @ 2.25	817.88	
J. H. Carter	24 . Jamaica Rum $38.-	912."	
Do	2 hlf do Do Do 26.-	52."	
Wise	3 Bbls Whiskey 120.-	360."	
J. H. Carter	1 . Do	135."	
Dwyer	14 Boxes Cognac 8.-	112."	
Stackhouse	4 " Do 11.50	46."	2434.88

Purchasers	Articles purchased		Am't	Total
	Am't forward			2434 88
Dwyer	16 Boxes Claret	4.75	76 .	
Heath	12 Bbls Pork	19.75	237 .	
Lepine & Ferre	20 " Do	19.50	390 .	
I. Levi	11 " Do	19.50	214 50	
Wise	1 " Do		14 .	
Lepine & Ferre	6 " Beef	12.50	75 .	
I. H. Carter	14 " Do	12.75	178 50	
Do	38 Tierces Mess Beef	19.50	744 .	
Wise	33 Sacks Salt	1.40	46 20	
I. H. Carter	25 cases Olive Oil	5.25	131 25	
Wise	89 boxes Starch	1.90	169 10	
McMasters	1 " Do		4 25	
Lee	10 " Do	2.65	26 50	
McMasters	10 " Soap (Colgate)	3.85	38 50	
Wise	75 " Do "Cook's"	3.50	262 50	
Rivet	10 " Do (Castile)	11.50	115 .	
Do	10 " Do "	11.25	112 50	
Caden	10 " Do "	11.-	110 .	
Carter	70 " Do "	10.50	735 .	
Wise	4 " Do (Brown)	6.50	26 .	
Steers	2 Cases Candles 100 to ea	29.50	59 .	
Lepine & Ferre	29 boxes Do	7.-	203 .	
Hollander	4 " Do	7.-	28 .	
I. H. Carter	4 half " Do	3.62½	14 50	
Miller	26 " Raisins dam'd	2.45	63 70	
Tunis	11 " Vermicelli	3.-	33 .	
Carter	6 " Tea 348 lb	70¢	243 60	
Farrel	5 " Pepper 435 lb	29¢	126 15 .	4473 75
				6908 63

Purchasers	Articles purchased	Am't	Total
	Am't forward		6908 63
Smith Zeigler	5 Bags Pepper 423½ ℔ 29¢	122 67	
Spur	8 " Coffee 1069 " 35¢	374 15	
Carter	14 " Do 1725 " 34¢	586 50	
Do	12 " Do 2615 " 35¢	915 25	
Babb	4 " Do 969 " 35½	344 ..	
Carter	6 " Do 1384 " 35¢	484 40	
Wise	16 Bbl Bread 3.25	52 ..	
Do	2 " Do 5.-	10 ..	
Cash	2 Kits Mackerel 1.75	3 50	
Barber	2 Cases dried Fish 4.05	8 10	
Carter	1 Bale " "	1 25	
Wise	1 half bbl Vinegar	4 50	
Currier	1 Case Tumblers 7¾ doz 95¢	7 30	
Barber	1 Bale small Cordage 175 ℔ 42¢	73 50	
Moulor	2 Cases Matches 30.-	60 ..	
Wise	1 " Do	67 ..	
Do	2 Bbls. Rosin 28.-	56 ..	
Lepine & Ferri	13 doz. Brooms 2.30	29 90	
Barber	4 Bbl Lamp Oil 38.-	152 ..	
Wise	2 " Do Do 40.-	80 ..	
Lepine & Ferri	50 Boxes Kerosene Oil 15.75	787 50	
J. Levi	10 " Do Do 16.-	160 00	
J. Tisse	3 " Do Do 11.75	35 25	
F Barber	4 " Do Do 13.50	54 ..	
Wise	76 Cans White Lead 2.50	190 ..	
Carter	1 Brl Tallow 244 ℔ 12¢	27 28	
Wise	1 Keg Metalic Comp'n	12 ..	
Moulor	1 " Sugar	15 ..	4713 05
			11621 68

Purchasers	Articles purchased	Am't	Total
	Am't Forward		11621.68
Morrison	1 Box Pickles	1.75	
Do	2 " Pick Axes 48 – 50¢	24."	
Forney	6 " Do 144. 50¢	72."	
Currier	1 " common Do 28 – 40¢	11.20	
Keep	4 Fairbanks Scales $37	148."	
Do	4 Do Do 29.50	118."	
Do	1 Do Do	18.50	
Mandell	1 Do Do	25."	
J. Graham	1 Do Do	29.50	
Fitch	30 Boxes Jumba Axes 8.	240."	
Elliott	30 " " " 7.50	225."	
Fitch	2 Cases Hatchets 16 doz 5.50	88."	
Levi	2 " Do 4 " 5.50	22."	
Currier	5 " Do 10 " 5.25	52.50	
Fitch	2 " Do 6 " 6.25	37.50	
Levi	1 " Do 16 " 5.00	80."	
Mandell	1 " Augurs 17 35¢	5.95	
Fitch	99 Kegs Nails 4.90	485.10	
Vose & West	1 Case Tacks	32."	
Wise	43 pkgs Knives Forks 129 dz 1.80	232.20	
Do	1 " " 2 dz 2.30	4.60	
Do	33 " " 86 dz 1.80	154.80	
Do	9 " Scissors 1.80	16.20	
Currier	12 doz Table Spoons 2.75	33."	
Wise	10 " Tea " 1.35	13.50	
Samuels	23½ Cards p'ckt Knives 9.50	224.25	
Risset	6 doz large Files 5.50	33."	
Currier	18 " Small Do 1.10	19.80	2447.35
			14069.03

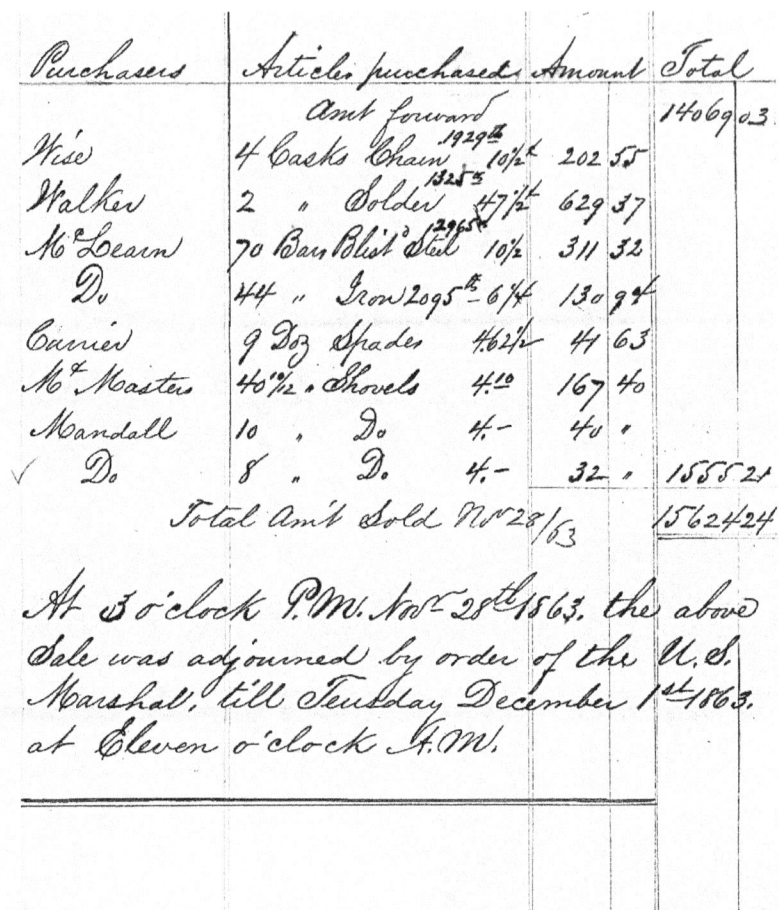

Figure 122. Auction sale of part of the *Alabama's* cargo on 28 Nov. 1863, day one. (This page and four prior.) *NARA.*

Auction Day Two

The following showed the results of the second day of the *Alabama* cargo auction, 1 Dec. 1863, beginning at 11:00 A.M. (Figure 123).

In pursuance of the aforesaid adjournment, I proceeded to sell the balance of the cargo of the Steamer *Alabama* at public auction, on Tuesday, December 1st 1863, at eleven o'clock A.M. at the U.S. Marshal's Warehouse, Nos. 38 & 40 Old Levee Street, at which said sale, the same was sold for cash at the prices, in the lots, and to the purchasers as hereinafter set forth, to wit.

Names of Purchasers	Articles purchased		Amount	Total
Cash	3 Smith's Bellows	10.00	30.00	
Wise	7　Do　　Do	6.50	45.50	
Schiller	5 red Boxes Gin	15.00	75.00	
Hearth	2　"　　"　Do	15.00	30.00	
Nobles	2　"　　"　Do	15.00	30.00	
Graham	1　"　　"　Do	15.00	15.00	
Brott & Davis	4 green " Do	14.50	58.00	
Fasnacht	37 Demijohns Do	18.00	666.00	

Durand	3 part full Demijohns Do	lot	25.50	
Wise	50 Demijohns Rum	12.00	600.00	
Cohn	10 " Do	12.50	125.00	
Priest	3 " Do	12.00	36.00	
Wise	40 " Do	12.00	480.00	
Claiborne	2 " Do	12.00	24.00	
Wise	10 " Do	12.25	122.50	
Do	40 " Do	12.00	480.00	
Do	60 " Do	12.00	720.00	3,562.50
(end p. 1)				
	Amt. forward			3,562.50
Wise	9 Demijohns Rum	9.50	85.50	
Do	7 part full " Do	lot	16.00	
Do	1 Demijohn Wine		9.00	
Schiller	1 " Anisette		10.00	
Cohn	2 " Molasses	2.50	5.00	
Do	1 " Varnish		10.50	
Marsden & Bradford	4 Bbls. Cream Tartar 1,109 lbs.	.44	487.96	
Schwab	2 " Carb. Ammonia 457 lbs.	.20	91.40	
Bontemps	5 " Sal Soda 1,351 lbs.	.03 3/8	45.60	
Pope	2 " Alum 582 lbs.	.02¾	27.84	
Schwab	1 " Pulv. Alum 291 lbs.	.13½	39.29	
Fredericks	2 " Epsom Salts 586 lbs.	.04¾	27.84	
McMasters	4 " Borax	.26	244.14	
Schwab	1 " Gum Arabic 222 lbs.	.29	64.38	
Marsden & Bradford	1 " Flor Sulphur 192 lbs.	.06	11.52	
Do	2 " Castor Oil	66.00	132.00	
E. J. Hart	1 hlf. Bale Copairs (Cohavi ?) 110 lbs.	.76	83.60	
Pope	20 Kegs Carb. Soda	6.50	130.00	
Lepine & Ferre	10 Brls. [sic] Copperas 2,742 lbs.	.02 3/8	65.12	
Chapman	1 " Turpentine, nearly empty		22.50	
Curtius (?)	3 Cases Adhi. Plaster 150 pkgs.	2.15	322.50	
Schwab	2 Cans Es. Thyme 78 lbs.	.32	24.96	
Pope	5 Cases Mustard 500 lbs.	.43	215.00	
Do	2 " Tannin 50 lbs.	1.35	67.50	
E. J. Hart	1 " Hydrag. (?) Creta 50 lbs.	.45	22.50	
Bontemps	2 " Agua Amonia [sic]	.90	21.60	
Marsden & Bradford	2 " Pulv. Gum Arabic 100 lbs.	.46	46.00	2,363.01
				5,925.51
(end p. 2)				
	Amt. forward			5,925.51
Fredericks	5 boxes Ext. Logwood 110 lbs.	.15	16.50	
Marsden & Bradford	1 Case Liquorice [sic] 100 lbs.	.38	38.00	
Deloche	1 " Phials		14.50	
Morrison	2 " Do	15.50	31.00	
Deloche	2 " Do	14.50	29.00	
Do	2 " Do	12.00	24.00	
Wise	1 Case Pill boxes		51.00	
Schwab	2 Bbls. Gum Camphor 225 lbs.	1.05	236.25	

Pope	2　”　Do　217 lbs.	1.14½	249.46	
Hart	2　”　Do　224 lbs.	1.13	253.12	
Clau	9 boxes Gum Drops	lot	25.00	
Finley	1 Case asst^d· Med. &c.		85.00	
Marsden & Bradford	65 pkg. Ad. Plaster	1.31	85.15	
Pope	Cantharide 50 lbs.	1.15	57.50	
Marsden & Bradford	Do Pulv. 25 lbs.	1.47½	36.87	
Do	2 boxes Calomel 50 lbs.	.86	43.00	
Hart	10 ” Pulv. Gum Arabic 575 lbs.	.51	293.00	
Levi	1 Case Quinine 98 lbs.[179]	2.75	269.50	
Do	1　”　Do　98 lbs.	2.75	269.50	
Do	1　”　Do　98 lbs.	2.80	274.00	
Curtius	1　”　Do　40 lbs.	2.95	118.00	
Do	2　” Iodine Potas^m· 93 lbs.	3.25	302.25	
Wise	5 Cans Quinine 100 lbs.	2.85	285.00	
Curtius	1 Case Chloroform 97 lbs.	1.00	97.00	
Frederickson	1 box Bencaic [sic] Acid 3 lbs.	11.50	34.50	
Marsden & Bradford	1　” Creosote 24 lbs.	1.25	30.00	
Do	1　” Pulv. Opium 24½ lbs.	12.00	294.00	
Curtis	1　”　”　Do　24½ lbs.	12.00	288.00	3,829.10
				9,754.61
(end p. 3)				
	Amt. forward			9,754.61
Marsden & Bradford	2 Cases Cantheride 100 lbs.	1.14	141.00	
Pope	2　” Ipecac 100 lbs.	3.95	395.00	
Marsden & Bradford	7　” Flax Lint 343 pkgs.	1.40	480.00	
Wise	1 Medicine Chest		21.00	
Do	1 Case Ruled Paper		133.00	
Do	1　”　Do　Do		80.00	
Do	1　” Writing Do		122.00	
Do	3　” Cartridge Do	50.00	150.00	
Do	1　” Do ?? Do		40.00	1,562.20
	Total Amt. sold Dec 1st/63			$ 11,316.81

At 3 o'clock P.M., Dec^r· 1st 1863, the above sale was adjourned by order of the U.S. Marshal, till Thursday, December 3rd 1863, at Eleven o'clock A.M.

---

[179] Should be 98 oz.

In pursuance of the aforesaid adjournment, I proceeded to sell the balance of the Cargo of the Steamer "Alabama" at Public Auction, on Tuesday, December 1st 1863. at Eleven o'clock. A. M. at the U.S. Marshal's Warehouse, Nos 38 & 40 Old Levee Street, at which said Sale, the same was sold for Cash, at the prices, in the lots and to the purchasers as hereinafter set forth, to wit:

Names of purchasers	Articles purchased	Amount	Total
Cash	3 Smith's Bellows 10.—	30 "	
Wise	7 Do. Do. 6.50	45 50	
Schiller	5 red Boxes Gin 15.—	75 "	
Heartt	2 " " Do. 15.—	30 "	
Noble	2 " " Do. 15.—	30 "	
Graham	1 " " Do. 15.—	15 "	
Brott & Davis	4 green " Do. 14.50	58 "	
Fasnacht	37 Demijohns Do 18.—	666 "	
Durand	3 part full " Do. Lot.	25 50	
Wise	50 " Rum 12.00	600 00	
Cohn	10 " Do 12.50	125 "	
Priest	3 " Do 12.—	36 "	
Wise	40 " Do 12.—	480 "	
Claiborne	2 " Do 12.—	24 "	
Wise	10 " Do 12.25	122 50	
Do	40 " Do 12.—	480 "	
Do	60 " Do 12.—	720 "	3562 50

Names of purchasers	Articles purchased	Amount	Total
	Amt forward		3562.50
Wise	9 Demijohns Rum 9.50	85.50	
Do	7 qrts Do lot	16.—	
Do	1 " Wine	9.—	
Schiller	1 " Annisette	10.—	
Cohn	2 " Molasses 2.50	5.—	
Do	1 " Varnish	10.50	
Marsden & Bradford	4 Bbls Cream Tartar 1109# 44¢	487.96	
Schwab	2 " Carb. Ammonia 457# 20¢	91.40	
Bontemps	5 " Sal Soda 1357# 3⅜	45.60	
Pope	2 " Alum 582. 2¾	16.—	
Marsden & Bradford	1 " Pulv. Alum 291. 13½	39.29	
Pope	2 " Epsom Salts 586. 4¾	27.84	
Schwab	1 " Linseed 183. 10¢	18.30	
Fredericks	2 " Pulv. Do 260. 10½	27.30	
McMasters	4 " Borax 26¢	244.14	
Schwab	1 " Gum Arabic 222. 29¢	64.38	
Marsden & Bradford	1 " Flr Sulphur 192. 6¢	11.52	
Do	2 " Castor Oil 66.—	132.—	
E. J. Hart	1 hlf. Bal. Copaiva 110# 76¢	83.60	
Pope	20 Kegs Carb. Soda 6.50	130.—	
Loepine & Ferre	10 Bbls Copperas 2742# 2⅜	65.12	
Chapman	1 " Turpentine nearly empty	22.50	
Curtius	3 Cases Adh. Plaster 15 pkgs 2.15	32.50	
Schwab	2 Cans Ex. Thyme 78# 32¢	24.96	
Pope	5 Cases Mustard 500# 43¢	215.—	
Do	2 " Tannin 50# 1.35	67.50	
E. J. Hart	1 " Hydrarg Creta 50# 45¢	22.50	
Bontemps	2 " Aqua Amonia 24" 90¢	21.60	
Marsden & Bradford	2 " Pulv Gum Arabic 100# 46¢	46.—	2363.01
			5925.51

Purchasers	Articles purchased	Amt	Total
	Amt forward		5925 51
Fredericks	5 boxes Ext. Liquor 110 lb 15¢	16 50	
Marsden & Bradford	1 Case Liquorice 100 lb 38¢	38 .	
Deloche	1 " Phials	14 50	
Morrison	2 " Do 15.50	31 .	
Deloche	2 Do 14.50	29 "	
Do	2 Do 12.-	24 .	
Wise	1 Case Pill boxes	51 .	
Schwab	2 Bbls Gum Camphor 225 lb 1.05	236 25	
Pope	2 " Do 217 lb 114½	248 46	
Hart	2 . Do 224 lb 1.13	253 12	
Clau	9 boxes Gum Drops lot	25 .	
Finley	1 Case asstd Medse	85 .	
Marsden & Bradford	65 pkgs Ad Plaster 1.31	85 15	
Pope	Cantharides 50 lb 1.15	57 50	
Marsden & Bradford	Do Pulv 25 . 1.47½	36 87	
Do	2 boxes Calomel 50 lb 86¢	43 .	
Hart	10 " Pulv Gum arabic 575 lb 51¢	293 .	
Levi	1 Case Quinine 98 ttt 2.75	269 50	
Do	1 . Do 98 " 2.75	269 50	
Do	1 . Do 98 . 2.80	274 "	
Curtius	1 . Do 40 . 2.95	118 .	
Do	2 . Iodin Potas 93 " 3.25	302 25	
Wise	5 Cans Quinine 100 B 2.85	285 .	
Curtius	1 Case Chloroform 97 . 1.00	97 .	
Frederickson	1 box Benzoic Acid 3 lb 11.50	34 50	
Marsden & Bradford	1 . Cresote 24 ttt 1.25	30 .	
Do	1 . Pulv Opium 24½ 12.00	294 .	
Curtius	1 " Do 24 lb 12.-	288 .	3829 10
			9754 61

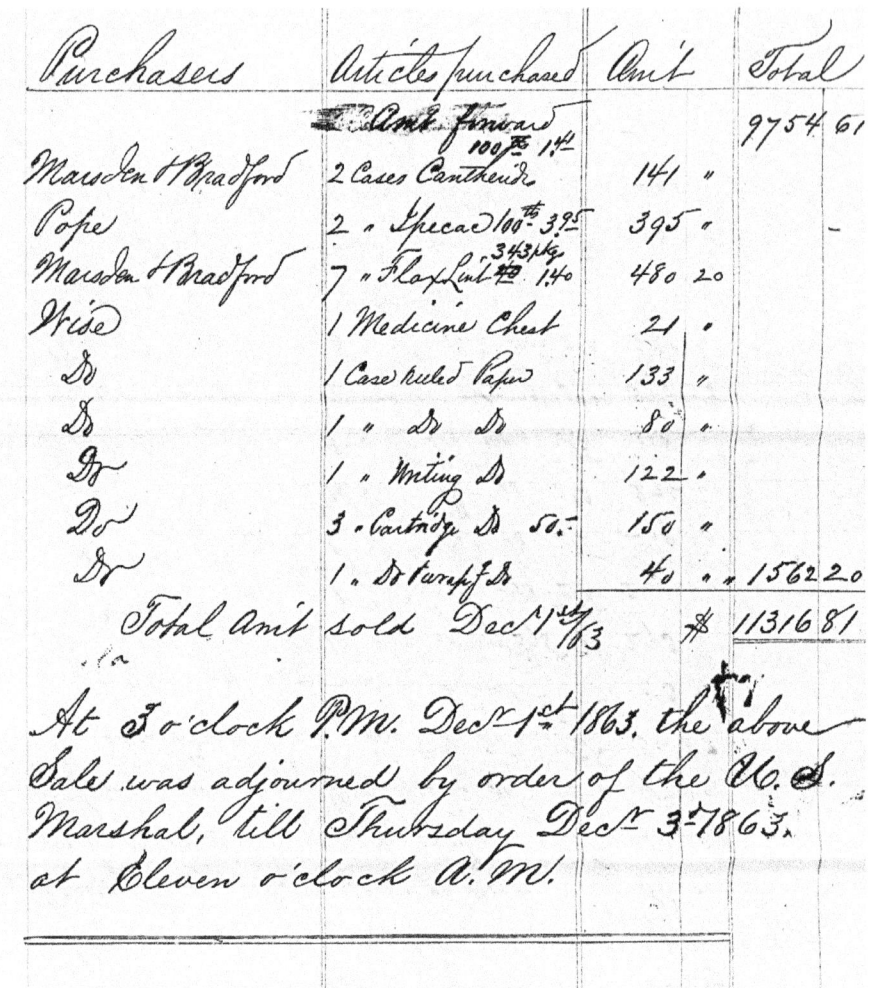

Figure 123. The second day of the *Alabama* auction sale results. (This page and the prior three pages.) Dated 1 Dec. 1863, day two. *NARA*.

Auction Day Three

The *Alabama* action resumed on 3 Dec. 1863 for the third day (Figure 124).

In pursuance of the aforesaid adjournment, I proceeded to sell the balance of the cargo of the Steamer *Alabama* at public auction, on Thursday, December 3rd 1863, at eleven o'clock A.M. at the U.S. Marshal's Warehouse, Nos. 38 & 40 Old Levee Street, at which said sale, the same was sold for cash, at the prices, in the lots and to the purchasers as hereinafter set forth, to wit:

Names of purchasers	Articles purchased		Amounts	Total
McMaster	1 Case Asstd. Envelopes		76.00	
Wise	1 " Blank Books		185.00	
Wise	1 " Ink (Edward's) 25 doz.	.60	15.00	
Hollander	4 " Do 268 bttls.	.31	83.08	
Wise	1 " Asstd. Stationary		235.00	
Ezekiel	1 " Do Do		200.00	

Wagner	1 " Folio Paper		50.00	
Wise	1 " French Do		220.00	
Lauchdnie	1 " Do Do		130.00	
Rozenberg	1 " Cottonade 20 pcs.	18.75	375.00	
Michen	30 pcs. Denims	12.50	375.00	
Mandell	2 " Do	12.50	25.00	
Girard	1 Bale Grey Cloth 7 pcs.	225.00	1,575.00	
Do	1 " Do 7 "	205.00	1,435.00	
Mandell	7 pcs. White Flannel	20.00	140.00	
Brott & Davis	26 " Do	22.00	572.00	
Forcheiman	45 doz. Mad^r· Hdkfs.	5.87½	264.37	5,955.45
(end p. 1)				
	Amt. forward			5,955.45
Morrison	2 pcs. Fancy Flannel	19.50	39.00	
Wolfe, Bro^s·	8 " Do	19.50	156.00	
Do	1 " Do		18.50	
Mandell	1 " Do		19.50	
Forchieman	34 " Valencia	15.25	518.50	
Mandell	2 " Do	15.25	30.50	
Moulor	19 " prints Cot. Flannel	12.75	242.25	
Do	5 " blue Flannel	27.50	137.50	
Brott & Davis	10 " Red Do	33.00	330.00	
Mandell	3 " " Do	35.00	99.00	
Wagner	50 pkgs. red Tape	.31	15.50	
Brott & Davis	24 " linen Thread	6.00	144.00	
Forchieman	70 doz. ladies Hose	2.87½	201.25	
Wise	98 pkgs. Pins	.55	53.35	
Forney	1 Case hats 72	.90	64.80	
French	1 " Do 67	.80	53.60	
Moulor	1 " Do 133	1.52½	202.82	
Fellman	1 Bale Blankets 51 pcs.	3.75	191.25	
Do	5 " Do 255 "	3.37½	860.63	
Goldsmith	5 " Do 255 "	3.37½	860.63	
Brott & Davis	8 " Do 408 "	3.62	1,479.00	
Levi	6 " Do 306 "	3.62½	1,109.25	
Blum & Frank	17 " Do 867 "	3.37	3,251.25	
Lepine & Ferri	9 pr. Do	Lot	37.03	
Mandell	1 " Do		3.37	
Davis & Co.	4 Cases Cloth Cards	450.00	1800.00	
Wise	1 " Do		270.00	
Levi	5 " Cotton Cards 10 doz.	18.00	180.00	
Cohn	5 " Do Do	18.00	180.00	12,548.48
				18,503.93
(end p. 2)				
	Amt. forward			18,503.93
Hollander	17 Cases Cotton Cards 34 doz.	17.50	595.00	
Withers *	10 " Do Do 20 doz.	14.00	180.00	
Levi	30 " Do Do 60 doz.	12.00	720.00	
Hoyt	25 " Do Do 50 doz.	13.00	650.00	

Frost & Co.	Case Brogans 485 prs.	1.30	630.50	
Wise	100 pkgs. Shoe thread 100 p.	.95	95.00	
Fellman	1 Case Shoe mkg. Tools		142.50	
Moulor	4 Rolls Harn^s Leather 661 lbs.	.48	317.28	
Wise	6 Kip (?) Skins	4.00	24.00	
Moulor	1 Case Rings & Buckles		58.00	
Rice	24 Lanterns	1.05	25.20	
Withers	72 hand Saws	1.30	93.60	
Ivens	16 Wrenches	1.05	16.80	
Do	8 doz. Padlocks	2.00	16.00	
Wise	1 Military Cap		3.50	
Do	2 Jars Olive Oil	3.75	7.50	
Do	1 Box Preserves		13.50	
Vose & West	1 Case Cannon Primers 24,900	7.00	174.30	
Wise	3 Cases percus^n Caps 567,000	.65	368.55	
Southern	1 Case Asst^d Stationery		180.00	
Marsden & Bradford	2 Bales Corks 18,000	3.10	55.80	
Do Do	3 " Do 50,000	.45	22.50	
McCloskey	6 part bags Coffee 376 lbs.	.34½	129.72	
Julet	1 small " & box Sugar 138 lbs.	.10¾	14.85	
Moulor	1 Trunk Sundries		70.00	4,869.30
				23,373.23
(end p. 3)				
	Amt. forward			23,373.23
Wise	1 Case Sundries		195.50	
Moulor	1 " Do		115.00	
Mandell	1 " Do		90.00	
Macon	1 " Do		77.00	
Cash	1 " Claret		2.50	
Heartt	1 Box Congress Water		1.50	
Cash	1 lot Soap &c.		3.00	
Heartt	1 box Sundries		10.00	
Cash	1 " Olive Oil		5.25	
Moulor	1 Trunk tray Sundries		3.25	
Wise	1 box Tea &c.		26.00	
Haber	1 " Crockery		15.00	
Wise	1 " Sundries		70.00	
Moulor	1 " Do		35.00	
Schiller	1 " Do		6.00	
Cash	1 " Do		14.00	
Hollander	1 " Do		13.00	
Wise	1 Keg Nails 210 lbs.	.03¾	7.87	
Cash	1 Basket Sundries		1.50	
Do	1 Box Starch		1.00	
Noble	2 Hams	1.37½	2.75	
Mandell	Lot Scrub brushes		1.50	
Lepine & Ferri	8 pr. Shoes	.85	6.80	
Levi	1 Chronometer		175.00	
Mandell	1 Set Measures		1.50	

Ja^s. Graham	1 Case Claret		4.75	
Fellman	42 Empty Demijohns		24.15	
Wise	2 Cans White Lead		5.00	913.32
				24,286.55
(end p 4)				
	Amt. forward			24,286.55
Priest	1 box Sundries		5.50	
Heard	1 " Do		5.00	
Mandell	1 pce. Denims		12.50	
Do	1 Bbl. Pork		14.00	36.50
			$	24,323.05
	Recapitulation			
	Amount of Sale—Novr. 28th		15,624.24	
	Do " Do—Decr. 1st		11,316.81	
	Do " Do—Do 3rd		24,323.05	51,264.10
	Aggregate Amt. of Sales			
	of cargo of Str. *Alabama*		$	51,264.10
* In the extension of the aforementioned amt. an error of one hundred dollars occurred, which has this 12th Day of Decr. been corrected and paid.				100.00
			$	51,364.10
N. O. La.	G. A. Hall			
Decr. 5th 1863	U.S. Prize Auctioneer			
To Jas. Graham Esq.				
U.S. Marshal				
East^n. Disct. La.				

In pursuance of the aforesaid adjournment, I proceeded to sell the balance of the Cargo of the Steamer "Alabama", at Public Auction, on Thursday, December 3d 1863, at Eleven o'clock A.M. at the U.S. Marshals' Warehouse Nos 38 & 40 Old Levee Street, at which said Sale, the same was sold for Cash, at the prices, in the lots and to the purchasers as hereinafter set forth, to wit:

Names of purchasers	Articles purchased	Amounts	Total
McMaster	1 Case Asstd Envelopes	76 .	
Wise	1 " Blank Books	185 .	
Wise	1 " Ink (Edwards) 25dz 60¢	15 .	
Hollander	4 " Do 268 btts 31¢	83 08	
Wise	1 " Asstd Stationery	235 .	
Ezekiel	1 " Do Do	200 .	
Wagner	1 " Folio Paper	50 .	
Wise	1 " French Do	220 .	
Lauchdrie	1 " Do Do	130 .	
Rozenberg	1 " Cottonade 20 pcs 18.75	375 .	
Michen	30 pcs Denims 12.50	375 .	
Mandell	2 " Do 12.50	25 .	
Girard	1 Bale Grey Cloth 7 pcs 225.-	1575 .	
Do	1 " Do 7 " 205.-	1435 .	
Mandell	7 pcs White Flannel 20.-	140 .	
Brott & Davis	26 " Do 22.-	572 .	
Forcheimer	45 dz Madr Hdkfs 5.27½	264 37	5955 45

Names of purchasers	Articles purchased	Amounts	Total
	Am't forward		5955 45
Morrison	2 pcs Fancy Flannel 19.50	39 .	
Wolfe, Bro'	8 . Do 19.50	156 .	
Do	1 . Do	18 50	
Mandell	1 . Do	19 50	
Forchieman	34 . Valencia 15.25	518 50	
Mandell	2 . Do 15.25	30 50	
Moulor	19 . print Cot. Flannel 12.75	242 25	
Do	5 . blue Flannel 27.50	137 50	
Brott & Davis	10 . Red Do. 33.-	330 .	
Mandell	3 . . Do. 33.-	99 .	
Wagner	50 pkgs red Tape .31¢	15 50	
Brott & Davis	24 . linen Thread 6.-	144 .	
Forchieman	70 doz ladies Hose 2.87½	201 25	
Wise	98 pkgs Pins 55¢	53 35	
Forney	1 Case Hats 72 - 90¢	64 80	
French	1 . Do. 67 80¢	53 60	
Moulor	1 . Do. 133 1.52½	202 82	
Fellman	1 Bale Blanket 51p. 3.75	191 25	
Do.	5 . Do 255. 3.32¼	860 63	
Goldsmith	5 . Do. 255. 3.32¼	860 63	
Brott & Davis	8 . Do. 408. 3.62½	1479 .	
Levi	6 . Do. 306. 3.62½	1109 25	
Blum & Frank	17 . Do. 867. 3.75	3251 25	
Lapine & Ferri	9 pr Do lot	37 03	
Mandell	1 . Do	3 37½	
Davis & Co.	4 Cases Cloth Cards 450.-	1800 .	
Wise	1 . Do	270 .	
Levi	5 . Cotton Cards 10 doz 18.-	180 .	
Cohn	5 . Do Do 18.-	180 .	12574 48
			18503 93

				18503	93
		Am't forward			
Hollander	17 Cases Cotton Cards 34doz	17.½	595	.	
*Withers	10 " Do Do 20doz	14.-	180	.	
Levi	30 " Do Do 60doz	12.-	720	.	
Hoyt	25 " Do Do 50doz	13.-	650	.	
Frost & Co.	Cases Brogans 485pr	1.30	630	50	
Wise	100 pkgs Shoe thread doz	95¢	95	.	
Wise	204 " Do Do	1.30	265	20	
Fellman	1 Case Shoemk's Tools		142	50	
Moulor	4 Rolls Harn.s Leath 664 lb	48¢	317	28	
Wise	6 Kip Skins	4.-	24	.	
Moulor	1 Case Rings Buckles		58	.	
Rice	24 Lanterns	1.05	25	20	
Withers	72 Hand Saws	1.30	93	60	
Ivens	16 Wrenches	1.05	16	80	
Do	8 doz Padlocks	2.00	16	.	
Wise	1 Military Cap		3	50	
Do	2 Jars Olive Oil	3.75	7	50	
Do	1. Box Preserves		13	50	
Vose & West	{1 Case Cannon Primers		—	—	
" "	{ 24.900	@ 7.00	174	30	
Wise	{3 Cases Percus.n Caps				
"	{ 567.000	65¢	368	55	
Southern	1 Case Ass't Stationery		180	.	
Marsden & Bradford	2 Bales Corks 18000	3.10	55	80	
Do Do	3 " Do 5000	45¢	22	50	
McCloskey	6 part bags Coffee 376 lb	34½	129	72	
Julet	1 smal " & box Sugar 138¾	10¾	14	85	
Moulor	1 Trunk Sundries		70	.	4869 30
				23373	23

Purchasers	Articles purchased	Amt	Total
	Amt forward		23373.22
Wise	1 Case Sundries	195.50	
Moulor	1 " Do	115.-	
Mandell	1 " Do	90.-	
Macon	1 " Do	77.-	
Cash	1 " Claret	2.50	
Haartt	1 Box Congress Water	1.50	
Cash	1 lot Soap &c	3.00	
Heartt	1 box Sundries	10.-	
Cash	1 " Olive Oil	5.25	
Moulor	1 Trunk tray Sundries	3.25	
Wise	1 box Tea &c	26.-	
Haber	1 " Crockery	15.-	
Wise	1 " Sundries	70.-	
Moulor	1 " Do	35.-	
Schiller	1 " Do	6.-	
Cash	1 " Do	14.-	
Hollander	1 " Do	13.-	
Wise	1 Keg Nails 210℔ .3¾	7.87	
Cash	1 Basket Sundries	1.50	
Do	1 Box Starch	1.-	
Noble	2 Hams 13½	2.75	
Mandell	Lot Scrub brushes	1.50	
Lepine & Ferre	8 pr Shoes 85¢	6.80	
Levi	1 Chronometer	175.-	
Mandell	1 Sett Measures	1.50	
Jas Graham	1 Case Claret	4.75	
Fellman	42 Empty Demijohns	24.15	
Wise	2 Cans White Lead	5.00	913.32
			24286.55

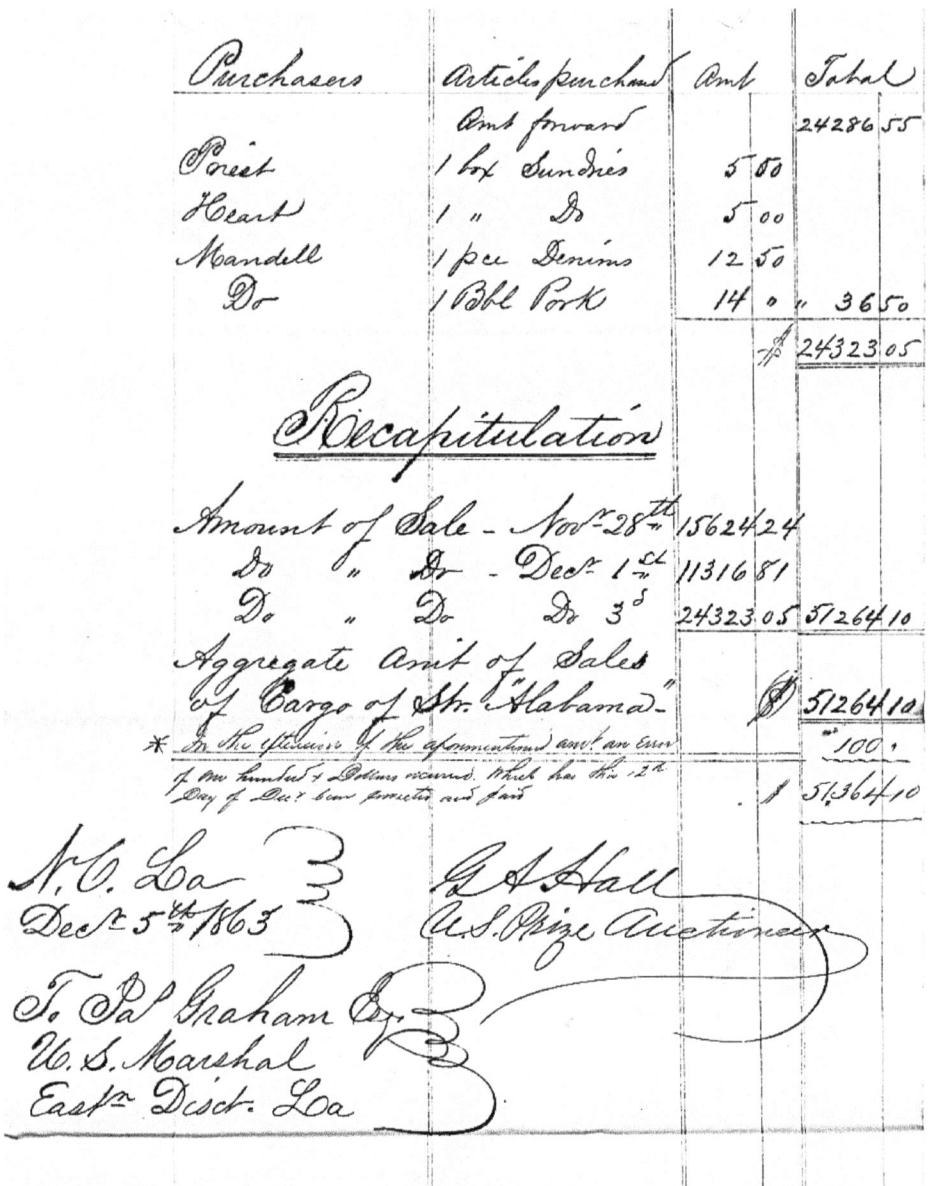

Figure 124. *Alabama* auction results, 3 Dec. 1863, the third and final day plus recapitulation of prior days totals. (This page and prior four.) *NARA*.

Estimated and Realized Prices for the Cargo Sale

The present author prepared a the following table combining the appraisers' estimated values of the *Alabama* cargo compared to the prices realized at auction. A rough comparison showed several interesting points. A number of lots were arranged differently between the two lists, probably enough to obfuscate a direct comparison. This might have been a step taken to conceal a certain amount of pilferage since a strict one to one accounting was difficult or impossible. Nevertheless, in some lots there were fewer cases or individual items present at sale time than there had been at the time the estimated values were determined. Even allowing for a very quick and not completely accurate inventory at the time of appraisal, it seemed clear that there was a certain amount of pilferage going on at the marshal's warehouse.

As for the actual values estimated and values realized at auction, there were only a few dozen lots that sold for either around half the appraisal or, by contrast, for double or more the appraisal. Further, there was a mere handful that exhibited an even greater difference between appraisal and realized price. An example of a greater difference was the lot of the three cases of percussion caps estimated at $36 but selling for $368.50. Even so, the auction sale's grand total appraised cargo value was within 1% of the grand total of prices realized. In addition, the realized total price of $51,364.10 was the larger of the two. The estimated grand total was $50,650.96. This might be taken as evidence that there was little or no price fixing going on among the buyers. Given the reputation of Union-occupied New Orleans as a den of official corruption, the seemingly upright results of the auction were a bit surprising. This was only a preliminary finding, as a more detailed economic analysis would be needed to reach a more solid conclusion.

No.	Description of Merchandise	Appraised Value.			Price Realized	
						$
10	Barrels alcohol 363½ gals.	@ $ 2.58	per gal.	908.75	$ 2.25	817.88
24	" Jamaica rum	@ $ 120	" bar.	2,880.00	38.00	912.00
2	Half barrels do	@ $ 60	" "	120.00	26.00	52.00
4	Barrels whiskey	@ $ 160	" "	640.00	3 @ 120	360.00
					1 @ 135	135.00
14	Cases Cognac	@ $ 6	" "	84.00	8.00	112.00
4	" "	@ $ 6	" "	24.00	11.50	46.00
16	" Claret	@ $ 5	" "	80.00	4.75	76.00
45	Bbls. Pork	@ $ 20	" "	900.00	14.00-19.75[180]	855.50
20	" Beef	@ $ 10	" "	200.00	6 @ 12.50 14 @ 12.75	253.50
38	Tierces Mess beef	@ $ 10	" "	380.00	19.50	741.00
33	Sacks salt	@ $ 2	" Sack	66.00	1.40	46.00
25	Cases Olive oil	@ $ 5	" box	125.00	5.25	131.25
89	Boxes Pearl starch	@ $ 3	" "	267.00	1.90	169.10
1	" " " 56 lbs.	@ $ 3	" "	3.00		3.00
10	" Starch (Kingsford & Son)	@ $ 6	" "	60.00	2.65	26.50
10	" Colgate's fancy soap	@ $ 8	" "	80.00	3.85	38.50
75	" Cook's Brown soap	@ $ 3.50	" "	262.50	3.50	262.50
100	" Castile soap 100 lbs. each	@ $ 16	" "	1,600.00	10.50-11.50[181]	1,072.00
4	" brown "	@ $ 6	" "	24.00	6.50	26.00
2	Cases Candles 100 lbs. each fancy	@ $ 15	" "	30.00	29.50	59.00
29	Boxes "	@ $ 10	" "	290.00	7.00	203.00
8	" "	@ $ 6	" "	48.00	7.00-3.62½[182]	28.00 14.50
26	" Raisins	@ $ 5	" "	130.00	2.45	63.70
11	" Vermicelli & Macaroni	@ $ 3	" "	33.00	3.00	33.00
6	" Tea 348 lbs.	@ 0.80	" "	278.40	0.70	243.60
10	Bags Black Pepper 858 lbs.	@ 0.80	" "	257.40	0.29	248.82
	(Amount to be carried over)			$ 9,771.05		

---

[180] 12 bbls. @ 19.75; 20 @ 19.50; 14 @ 19.50; 1 @ 14.00. 47 bbls. sold but 45 was the number appraised.
[181] 10 boxes @ $ 11.50; 10 @ 11.25; 10 @ 11.00; 70 @ 10.50.
[182] 8 boxes appraised but 4 whole and 4 half boxes present when sold. Pilferage in the warehouse?

				$ 9,771.05		
	Amount brought forward					
44	Sacks Coffee 5,409 lbs.	@ .35	" lbs.	1,893.15	0.34-0.35½[183]	2,704.30
18	Bbls. Bread	@ $ 6	" bbl.	108.00	16 @ 3.35 2 @ 5.00	52.00 10.00
2	Kits Mackerel	@ $ 2	" ea.	4.00	1.75	3.50
2	Cases dried Fish	@ $ 3	" "	6.00	4.05	8.10
1	Bale do do	@ $ 2	" bale	2.00		1.25
1	half bbl. Vinegar	@ $ 5	" bbl.	5.00		4.50
1	Case Tumblers	@ $ 4.50	" case	4.50		7.50
1	Bale Small Cordage 175 lbs.	@ .25	" lb.	43.75	0.42	73.50
3	Cases Matches	@ $ 40	" ea.	120.00	30.00	127.00
2	Bbls. Rosin	@ $ 30	" "	60.00	28.00	56.00
13	doz. Brooms	@ $ 3.50	" doz.	45.50	2.30	29.90
6	bbls. Lamp Oil	@ $ 60	" Bbl.	360.00	38.00 40.00	152.00 80.00
67	Boxes Kerosene oil—4 cans ea. of 5 gals.	@ $ 20	" Box	1340.00		1,036.75
76	cans White Lead	@ $ 7.50	" Can	570.00	2.50	190.00
1	bbl. Tallow 244 lbs.	@ .10	" lb.	24.40	0.12	27.25
1	keg Metallic Composition	@ $ 10		10.00		10.00
1	" Sugar	@ $ 15	" keg	15.00		15.00
1	Box Pickles	@ $ 1	" box	1.00		1.75
8	" " 24 each	@ $ 5	" doz.	80.00		[184]
1	" Common Picks—28	@ .75	" each	21.00		
10	" Fairbank's scales, different capacity	@ $ 30	" each	300.00	11 @ 18.50-37.00[185]	339.00
60	" *Tumba* Axes—1 doz. each	@ $ 11.50	" doz.	690.00	30 @ 8.00 30 @ 7.50	240.00 225.00
2	" Hatchets—8 doz. each	@ $ 6	" "	96.00	5.50	88.00
2	" do No. 1, 2 doz. each	@ $ 4.50	" "	18.00	5.50	22.00
5	" do " 2, 2 " "	@ $ 6	" "	60.00	5.25	52.50
2	" do " 3, 3 " "	@ $ 7.50	" "	45.00	6.25	37.50
1	" do 17 " "	@ $ 5	" "	85.00	5.00	80.00
	(amount to be carried over)			$ 15,778.35		
	Amount brought forward			$ 15,778.35		
1	Box Augurs	for $ 20		20.00	0.35	5.95
99	Kegs Nails—assorted	@ $ 6	" keg	594.00	4.90	485.10
1	Case Tacks			30.00		32.00
43	pckgs. Knives & Forks, 3 doz. ea.	@ $1.25	" doz.	161.25	1.50	232.20
1	" do do , 2 "	"	"	2.50	2.30	4.60
26	" do do , 3 "	"	"	97.50	33 @ 1.80[186]	154.00
1	" do do , 2 "	"	"	2.50		
6	" do do , 1 "	"	"	7.50		

---

[183] 8 bags @ 0.35; 14 @ 0.34; 12 @ 0.35; 4 @ 0.35½; 6 @ 0.35.

[184] Apparently these three lots: 8 boxes pickles and 1 box common picks did not make it to the auction. Pilferage in the warehouse?

[185] 4 @ 37.00; 4 @29.50; 1 @ 18.50; 1 @ 25.00; and 1 @ 29.50. That totals 11 scales, one scale extra not appraised. From whence came the extra set of scales?

[186] Three lots combined into one for the auction.

9	" Scissors	@ $1.50	" pk.	13.50	1.80	16.20
12	" table & 12 doz. tea spoons	@ .75	" doz.	18.00	12 @ 2.75 10 @ 1.35	33.00 13.50
24	Cards Pocket Knives	@ $5	" card	120.00	23½ @ 9.50[187]	224.00
6	doz. large Files	@ $2.50	" doz.	15.00	5.50	33.00
24	" Saw do	@ $1.00	" "	24.00	1.10	19.80
4	Casks chain 1,929 lbs.	@ .06	" lb.	115.74	0.10½	202.55
2	" Solder 1,325 "	@ .70	" "	927.50	0.47½	629.37
70	Bars blister Steel 2,965 "	@ .25	" "	741.25	0.10½	311.32
44	" iron 2,095 "	@ .10	" "	209.50	0.06¼	130.94
38	doz. Spades	@ $4.50	" doz.	171.00		281.03
42	" Shovels	@ $3.50	" "	147.00		[188]
10	Smith's Bellows	@ $5	" ea.	50.00	3 @ 10.00 7 @ 6.50	30.00 45.50
6	red boxes Gin	@ $15	" case	90.00	10 @ 15.00	150.00[189]
4	green " "	@ $12	" "	48.00	14.50	58.00
37	5 gal. demijohns Gin	@ $15	" ea.	555.00		
3	Small " "	@ $6	" ea.	18.00		2,714.50
222	5 gallon demijohns Rum	@ $10	" "	2,220.00		[190]
7	— " " (partially full)	@ $6	" "	42.00		
1	5 gallon " Port Wine	@ $15		15.00		15.00
1	" " Annisette [sic]	@ $8		8.00		8.00
	(amount to be carried over)			$22,242.09		
	Amount brought forward			$22,242.09		
2	5 gallon demijohns Molasses	@ $2		4.00	2.50	5.00
1	" " Varnish	@ $10		10.00		10.50
6	bbls. Cream Tartar 1,109 lbs.	@ .30	" lb.	332.70	0.44	487.96
5	" Soda 1,351 lbs.	@ .06	" "	81.06	0.03 3/8	45.60
2	" Alum 582 lbs.	@ .05	" "	29.10	0.02¾	27.84
1	" " (pulverized) 291 lbs.	@ .05	" "	14.55	0.13 ½	39.29
2	Casks Epsom Salts 586 lbs.	@ .02	" "	11.72	0.04¾	27.84
1	bbl. Linseed 183 lbs.	@ .05	" "	9.15		[191]
2	" " (pulverized) 260 lbs.	@ .08	" "	20.80		
4	" Borax 939 lbs.	@ .40	" "	375.60	0.26	244.14
1	" Gum Arabic 222 lbs.	@ .50	" "	111.00	0.29	64.38
1	" Flor Sulphur 192 lbs.	@ .13	" "	24.96	0.06	11.52
2	" Castor Oil	@ $100	" bar.	200.00	66.00	132.00
1	half barrel Balsam Cohavi (?)	@ $15	" "	15.00	110 lbs. @ 0.76	83.60
3	Bales 18,000 Corks	@ $12	" ea.	36.00		78.30

---

[187] One-half case missing probably from pilferage.
[188] The two lots of spades and shovels (total appraisal of $318.00) were sold as one big lot.
[189] Instead of 6 cases, the lot had 10 cases and the total cost did not add up correctly. Possibly obfuscated to cover up pilferage.
[190] The large number of demijohns of rum and gin were sold off in 11 odd lots not corresponding to the lots appraised. This obscured pilferage and made an accounting difficult or impossible if anyone happened to bother checking. For example, 7 demijohns were listed as part full when the auction came around but not when appraised.
[191] The two lots of linseed were missing by the time of the auction and perhaps had been stolen.

20	kegs Carb Soda	@ $ 12	" "	240.00	6.50	130.00
10	Bbls. Copperas 2,742 lbs.	@ $ .03½	" "	95.97	0.02 3/8	65.12
1	" Turpentine	@ $ 150	" "	150.00		22.50
3	Cases Adhesive Plaster (50 half boxes) 150 pks.	@ $ 1	" pk.	150.00	2.15	322.50
3	Cans Essence Thym (Rouge) 78 lbs.	@ .32	" lb.	24.96	0.32	24.96
5	Cases Mustard (Taylor & Sons) 500 lbs.	@ .43	" "	215.00	0.43	215.00
1	" Tannin—25 bottles	@ 1.35	" bottle	33.75	2 cases @.35[192]	67.50
1	" Hydrag Creta 50 "	@ .45	" "	22.50	0.45	22.50
2	" Aqua Ammonia, 24, 4 lb. bottles, 95 lbs.	@ .90	" "	21.60	0.90	21.60
2	" Gum Arabic 50 lb. each	@ .46	" lb.	23.00	0.46	46.00
5	boxes Extract Logwood 110 lbs.	@ .15	" "	16.50	0.15	16.50
1	Case Liquorice 100 lbs.	@ .38	" "	38.00	0.38	38.00
7	" Phials	@ $ 15	" Case	105.00	12.00-15.50	98.50
	(amount to be carried over)			$ 24,654.01		
	Amount brought forward			$ 24,654.01		
1	Case Pill Boxes	@ $ 50	per Case	50.00		51.00
6	bbls. Gum Camphor 666 lbs.	@ $ 2	" lb.	1,332.00	1.13-1.14½	738.83
9	boxes " drops	@ $ 1	" box	9.00		25.00
1	Case assorted Drugs &c. to wit: 24 doz. Glass Syringes—12 doz. men's Metallic Syringes—5 doz. Strychnine in crystals—25 lbs. powdered Jalap—3 bottles iodine Potassium 10 lbs. each	$ 175	" " " "	175.00		85.00
1	Case containing Adhesive Plaster & 75 lbs. Cantharides	$ 75	"	75.00		85.15 94.37
2	boxes Calomel—25 lbs. each	@ .86	" lb.	21.50	0.86	21.50
10	Case Pow'd. gum arabic, 575 lbs.	@ .50	" lb.	287.50	0.51	293.00
3	Boxes Quinine 100 bottles each, 294 oz.	@ $ 2.75	" oz.	808.50	2.75	813.00
1	" " 43 ", 3 broken, 40 oz.	@ $ 4	" "	160.00	2.75-2.80	118.00
2	Cases Iodine potassium 93 bottles	@ $ 3.25	" bot.	302.25	3.25	302.25
1	" 5 Cans Quinine— 100 oz. ea. case	@ $ 4	" oz.	400.00	2.85	285.00
1	" Chloroform 98 bottles	@ $ 1	" bottle	98.00	1.00	97.00
1	Box Benzaic [sic] Acid 3 lbs.	@ $ 5	" lb.	15.00	11.50	34.50
1	Case Kreosot [sic]—24 bottles	@ $ 1.25	" bot.	30.00	1.25	30.00
2	" Pulverized Opium 48½ lbs.	@ $ 7	" lb.	339.50	12.00	582.00
2	" Cantharides 52 lbs.	@ $ 1.50	" "	78.00	100 @ 1.14	141.00
2	" Ypecac [sic] 53 lbs.	@ $ 1.50	" "	79.50	3.95	395.00

---

[192] From whence came the extra case of tannin? Perhaps the appraisers' inventory work simply was not error proof.

7	" Flax Lint—49 packs each	@ $ 1	" pk.	343.00	1.40	480.00
1	Ship's Medicine chest	for $ 25		25.00		21.00
2	Cases Ruled paper	@ $ 100	" case	200.00	80.00-133.00	213.00
1	" Writing "	@ $ 100	" "	100.00		122.00
3	" Cartridge " 63 reams ea.	@ $ 63	" "	189.00	50.00	150.00
1	" containing 4 reams mapping paper & 14 reams cartridge paper			50.00		40.00
	(amount to be carried over)			$ 29,821.76		
	Amount brought forward			$ 29,821.76		
1	Case Asstd. Envelopes	@ $ 76	" case	76.00		76.00
2	" blank books	@ $ 185	" "	185.00		185.00
1	" Edward's black ink 25 doz.	@ $ 1	" doz.	25.00	0.60	15.00
4	" " " " 290 large bottles	@ .30	" ea.	87.00	268 @ 0.31[193]	83.08
1	" contg. 20 gross Lead Pencils Fabres (?) 20 " " Imitation 20 " Penholders assorted 40 " Steel Pens do 60 " do do 50K Envelopes Yellow & white	$ 200		200.00		235.00
42	Empty Demijohns	@ .50	" ea.	21.00		
1	Case assorted Stationery, containing 33 Reams folio Paper, assrted.; 108 doz. Fabre's (?) Lead Pencils; 20½ dozen Ink stands, assorted; 20 gross Pen holders, assorted; 34 boxes Steel pens.	$ 250	" " " " " "	250.00		200.00
1	Case large folio Paper	for $ 50	"	50.00		50.00
2	" French letter do	@ $ 125	" case	250.00	1 @ 220 1@ 130	350.00
1	" Cottonade, 20 pieces	@ $ 18.75	" piece	375.00		375.00
1	" Denims, 30 do	@ $ 12.50	" "	375.00		375.00
2	bales grey cloth, 7 pieces ea.	@ $ 120	" "	1,680.00	1 @ 205 1 @ 225	3,010.00
1	" white flannel, 7 " do	@ $ 23	" "	161.00	20	140.00
1	" do 26 do	@ $ 27	" "	702.00	22	572.00
1	" Madras hndkrchiefs., 48 doz.	@ $ 4	" doz.	192.00	5.71½	264.37
1	Case fancy flannels 10 pieces	@ $ 12	" piece	120.00	12 pcs. @ 19.50[194]	233.00
1	" Valentia 36 pieces	@ $ 10	" "	360.00	15.25	548.00
1	" printed cotton flannel 19 pieces	@ $ 13	" "	247.00	12.75	242.25
				$ 35,177.00		
	Amount brought forward			$ 35,177.00		
1	Case blue flannel 4 pieces	@ $ 30	" piece	120.00	5 @ 27.50	137.50

---

[193] 22 bottles of ink missing. Pilferage or breakage.
[194] 2 extra pieces indicate mixing of lots or accounting errors.

1	" red " 10 "	@ $25	" "	250.00	10 @ 33.00 3 @ 35.00[195]	566.50
1	" contg. 24 pcks. brown linen thread and 50 pcks. Red tape	for $150		150.00	6.00 0.31	144.00 15.50
1	" contng. 70 doz. Ladies Hose and 99 pkgs. Pins	for $250		250.00	2.87½ 98 @ .55[196]	201.25 53.35
3	" Felt hats 22¾ doz.	@ $6	" doz.	136.50	0.80-1.52½	321.22
42	bales Army Blankets, 51 pairs	@ $4	" pair	8,568.00	3.75	7,755.38
9	pair do do	@ $6	" "	54.00		37.03
5	Cases cloth Cards	@ $225	" ea.	1,125.00	270-450	2,070.00
27	" Cotton " (Lory & Watson) 54 doz.	@ $12	" doz.	648.00	17.50-18.00	975.00
65	" " " (Whitemore) 130 doz.	@ $12	" "	1,560.00	12.00-14.00	1550.00
3	" Brogans, undressed, 330 prs.	@ $1.25	" pair	412.50	485 pairs @ 1.30[197]	630.50
1	" Shoe Thread, 304 pckgs.	@ $1.25	" pk.	380.00	100 @ 0.95[198]	95.00
1	" asstd. Shoemaker's tools, contg. 1 doz. asstd. Pincers; 2 doz. Peg Awls; 48 doz. shoe tacks; 4 doz. Currier's knives 5 doz. Rasps; 5½ doz. hammers; 7 pks. Awls; 2 doz. Givens; 8 doz/ Punches; 2 doz. Bisagles; 2 doz. Awl handles.	$75		75.00		142.50
4	rolls Harness Leather 661 lbs.	@ .40	" lb.	264.40	0.48	317.28
6	Kip skins	@ $4	" kip	24.00		24.00
1	case asstd. harness Buckles & Rings, 57 pkgs.	@ $50		50.00		58.00
1	" Lanterns 2 doz.	@ $1	" piece	24.00	1.05	25.20
1	" hardware, contng. 72 hand saws; 36 pks. Monkey wrenches; assd. sizes; 8 doz. Padlocks	$50		50.00		
	(amount to be carried over)			$49,319.16		
	Amount brought forward			$49,319.16		
1	Military Cap	@ $2.80		2.80		3.50
1	box preserves	@ $10		10.00		13.50
2	Jars Olive oil	@ $1.50	" ea.	3.00		7.50
25	pkgs. Sundries	@ $30	" ea.	750.00		[199]
25k	Cannon Primers	@ $10	" thsnd.	250.00	24,900 @ 7.00[200]	174.30
3	Cases Percussion Caps	@ $12	" case	36.00		368.50[201]
1	Chronometer	@ $100		100.00		750.00
	Lot charts & 1 Case stationery	@ $180		180.00		180.00
				$50,650.96		

---

[195] The red and blue lots of flannel seem to have been mixed.
[196] Note one missing package of pens.
[197] More pairs of shoes auctioned than appraised, accounting problem here.
[198] About 2/3 of the packages of shoe thread went missing between appraisal and auction.
[199] This lot seems to have been sold off individually in a way that would make accounting difficult.
[200] 100 cannon primers missing.
[201] The price achieved at auction was much larger than the appraisal. Something odd here.

Court Finding on which Ships Shared in the Prize Money

The U.S. prize commissioners approved the Union navy ships *San Jacinto*, *Eugenie*, and *Tennessee* to share the prize money, but the *Colorado* claim was disallowed (Figure 125). The key factor was that these were ships within signaling distance. Furthermore, the three ships and their crews actively participated in capturing the *Alabama*. The court seemed, from the present distance, to reach a fair decision.

7728
United States United States Dist-
    vs. trict Court E. D. of La.
Steamer *Alabama*

    To the Judge of the United States District Court, Eastern District of Louisiana.

    The undersigned Prize Commissioners of your honorable Court beg leave to report:

    That with the suggestions and aid of the Attorney for Captors and the evidence taken by him before them & understanding the law to limit the right to share the proceeds to such vessels only as were within signal distance,

    They respectfully report the Steamer *Eugenie*, the U.S. Frigate *San Jacinto*, and the U.S. Gunboat *Tennessee* are entitled to share as captors and to be adjudged as such by this Honorable Court.

    They also find the fact undisputed that the capturing vessels were of superior force.

                  (signed) Michael Hahn
                  (signed) N. W. Ritchie
                        Prize Commissioners

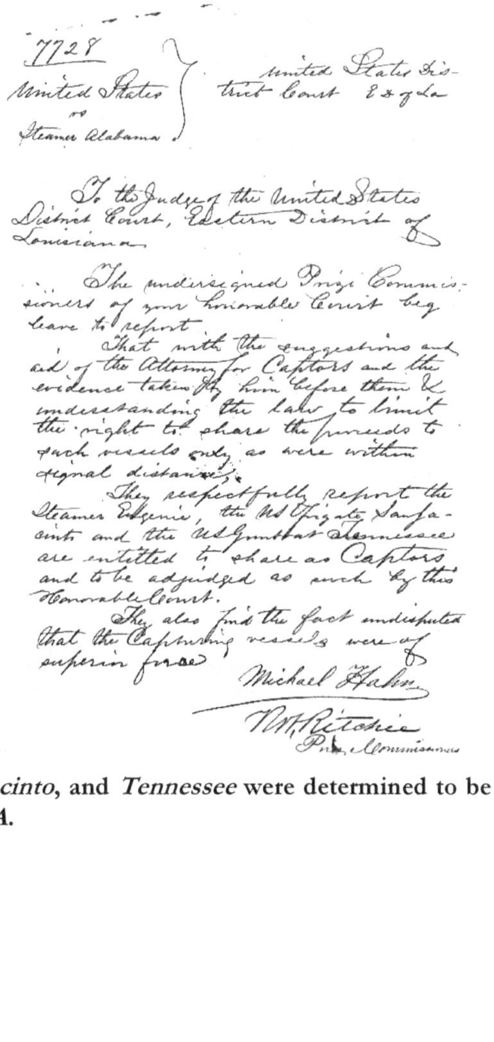

Figure 125. The *Eugenie, San Jacinto,* and *Tennessee* were determined to be eligible for the *Alabama* prize money. Filed 15 Dec. 1863. *NARA*.

Attorney for the Captors Report on Application of Prize Law to the Str. *Alabama*

A. M. Buchanan, attorney for the captors, filed with the court a report explaining in detail and with legal precedents how the new prize law applied to the blockade-running Str. *Alabama*. In so doing, he compared and contrasted the current situation with that under the prior prize act (Figure 126). It was interesting to read Judge Buchanan's legal reasoning on the *Alabama* case.

        The United States
                v.   No. 7728
     Steamer *Alabama* & Cargo

Alexander M. Buchanan,[202] appointed by the Court Attorney of Captors, has the honor to submit to the Court the following report.

The prize in this case seems to have been an unarmed vessel, and to have made no resistance. Two armed vessels of the United States were within gunshot when she was taken possession of by a boat's crew from one of them, the *Eugenie*. The other was named the *Tennessee*. A third armed vessel, of the United States, the *San Jacinto* was in sight, at a distance of 10½ to 11 miles. The first question which arises is whether the *San Jacinto* was "within signal distance." Four naval officers, disinterested in the case, have been summoned by the undersigned and have given their evidence before the Prize Commissioners, upon this point. They concurred in pronouncing that the *San Jacinto* was within signal distance. I have found no adjudged case upon the meaning of the expression in the Act of Congress except a newspaper report in the *Army and Navy Gazette* published in New York of a decision of Judge Betts in the District Court of the United States for the Southern District of New York; in which that judge rejected the claim of a ship of war, which was fifteen or sixteen miles from the vessel captured, at the time of the capture, to participate in the prize money. But it is probable that any case must rest on its own peculiar circumstances. The words on the Act of Congress, 12th Statutes at Large, page 606, are "When one or more vessels of the navy shall be within signal distance of another making a prize, all shall share in the prize, and money awarded shall be appointed among the officers and men of the several vessels according to the rates of pay of all on board who are borne upon the books, after deducting one twentieth to the flag officer, if there be any such entitled to share."

The rule, as to the distribution of prize money, before the act of July 17th 1862 was different. By the special provision of an Act of Congress, 23rd April 1800, Dodson's ??, page 281, "whenever one or more public ships or vessels are in sight at the time any one or more ships are making a prize or prizes, they shall all share equally in the prize or prizes, according to the number of men and guns on board each ship in sight." This was similar to the English rule as recognized by Lord Stovall (?) in various cases. That learned judge said in the case of the *Shackler*, 1st Dodson 360. "The rule of law upon the subject is, that a ship in sight at the time of the capture is entitled to share in the prize from that circumstance alone, *unless the* case happens to fall within one of the exceptions to that general rule, such as the circumstance of steering a distinctly different course" &c. In another case, the *Galen* 2nd Dodson 24, the same judge says that the circumstances must show that an *anisnus capiendi* existed on the part of the vessel in sight in order that she be entitled to share in the prize. In other words, she must be in chase of the vessel captured. Now, nothing can be clearer in the present case than the *anisnus capiendi* on the part of the *San Jacinto*; that vessel having been actually in chase of the *Alabama* for two whole days and one night when the latter was captured, although the capture was effected by another vessel. Another claim put in, in this case, seems to repose upon no proper foundation of law or reason. It is that of the Commander of the *Colorado* frigate,

---

[202] Associate Justice, Louisiana Supreme Court, 1853–1862.

which vessel was about one hundred miles from the scene of the capture. This claim is barred upon the assumption that the *Eugenie*, which made the capture, was a tender of the *Colorado*; and at the time of the capture was upon detached service, bound to New Orleans to procure supplies for the blockading squadron off Mobile, of which squadron the *Colorado* was the flag ship. The fact of the *Eugenie* being a tender of the *Colorado* is positively denied by Ensign Dyer, the Commander of the *Eugenie*. But however this may be, the act of Congress cited above, is in my opinion conclusive against any right of the *Colorado*, her officers, and crew, to share in the prize money in this case.

All which is respectfully submitted,
(signed) A. M. Buchanan

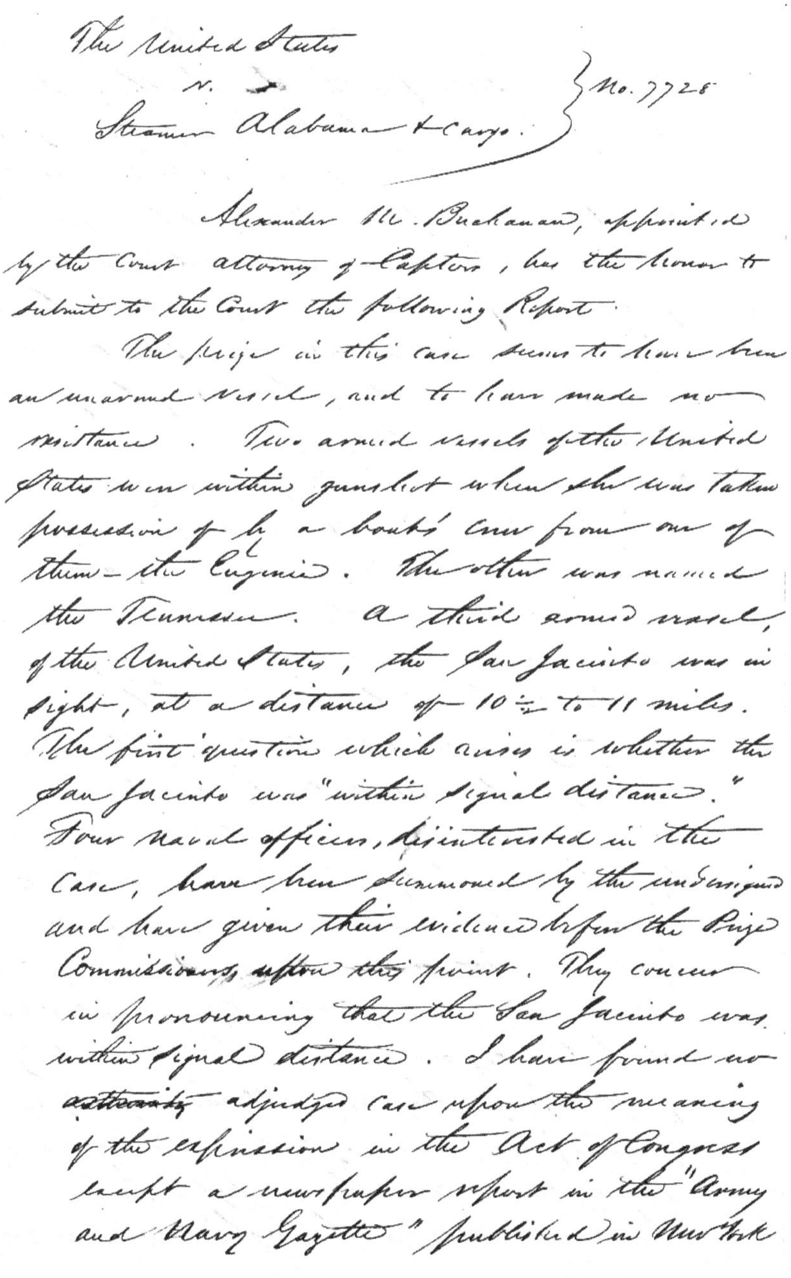

Figure 126. Report of the attorney for the captors, page one of four. To save space and as an example, only page one was here included. Filed 16 Dec. 1863. *NARA*.

Expenses of the U.S. District Attorney

The U.S. district attorney submitted his invoice to the court in a rather informal manner. The fee came to $3,940 (Figure 127).

United States	United States District Court
vs.	
The Steamer *Alabama*,	In Prize
Cargo &c.	

In all services rendered or to be rendered in this case as United States Attorney, three per centum of $131,364.00, the amount of proceeds realized upon the sale of the vessel & cargo—$3,940.

(signed) Rupert Waples
U.S. Dist. Atty.

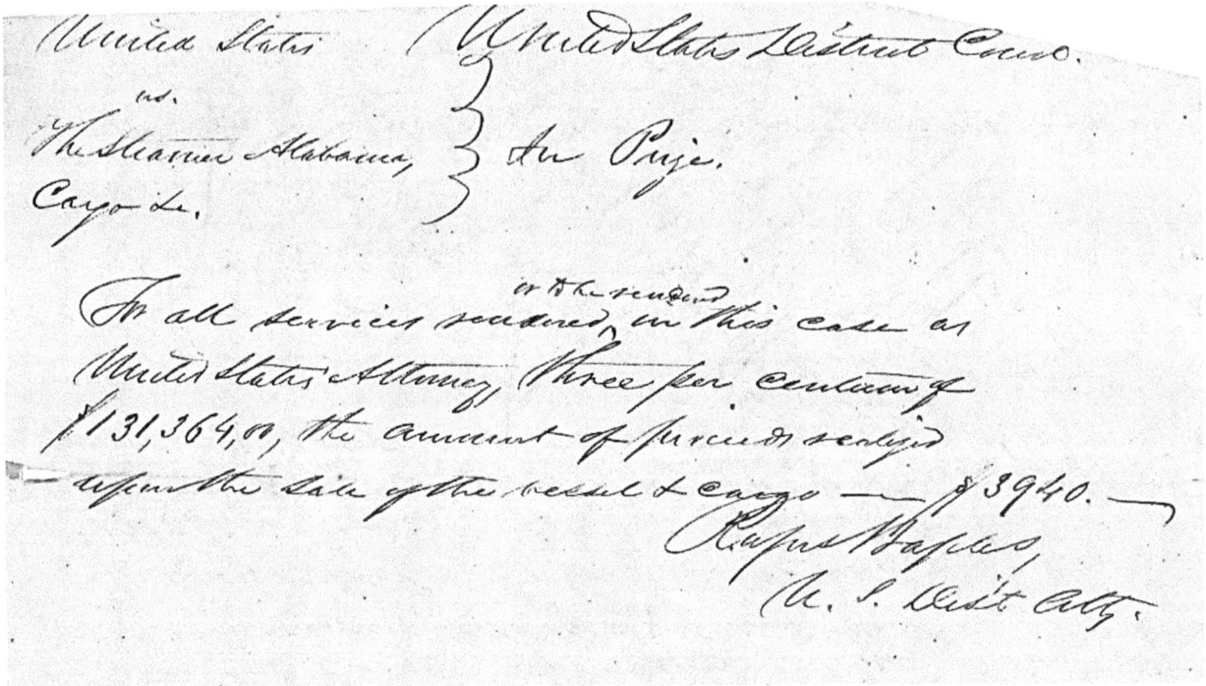

United States vs. The Steamer Alabama, Cargo &c. } United States District Court. In Prize.

For all services rendered or to be rendered in this case as United States Attorney, three per centum of $131,364.00, the amount of proceeds realized upon the sale of the vessel & cargo — $3940. —
Rupert Waples,
U. S. Dist. Atty.

Figure 127. The $3,940 invoice for the services of the U.S. district attorney in the prize case was written on the merest scrap of paper that was not even a uniform rectangular shape. This was a remarkably informal document. Filed 19 Dec. 1863. *NARA*.

U.S. Marshal's Invoice for Itemized Expenses

The $1,278.00 invoice of the U.S. marshal made an interesting contrast to the above example in that a printed form for the specific purpose was created and used. The reverse side of marshal's invoice itemized each expense regarding the prize case (Figure 128). Many of the charges were perhaps things any ship would have to pay upon entering port and discharging cargo. Generally categorized as port charges, it was instructive to see what sorts of things and scale of costs the general term "port charges" might include.

(reverse side)

Marshal's Fees	$ 23.90
Keeping (2 Keepers) 12 days from Oct. 9th to 20th Incl.	60.00
do (1 do) 40 " " " 20 to Nov. 28th	100.00
Expenses of deputy going on board	.50
Extra Services in Superintg. & dischg. Cargo	5.00
do 8 days Recg. & Storing do	40.00
do 10 do Prepg. for Sale, Selg. & delivg.	50.00
Pub. Monition & Sale N. O. Times	18.00
Expenses paid for Labor in Storing Cargo	3.00
do " do Removing do	6.25
do " Laborers Storing in Warehouse	33.25
Storage of Cargo in Warehouse	219.09
1% Commn. On Sale	250.00
Bill, Keen & Stanley dischg. Cargo	168.25
" do do Cleaning Engine	8.00
" R. Harris, Cooperage	11.50
" S. Figerio, Rating & Repairing Chronometer	37.50
" Wm. F. Levine, Pilotage	15.00
" Jas. Murphy, drayage	110.40
" Str. *Baltic*, Towage	25.00
" Wm. McDuff & Co., Weighing	19.45
" Board Health, Quarantine	23.00
" F. Pierve, hauling 21 Loads, Glassware &c.	21.00
" O. A. Pierce, dischg. Cltc. (?)	30.00
	$ 1,278.09

Figure 128. The marshal's invoice form was legible and not here transcribed. The detailed listing of expenses on the reverse side of the form was interesting. (This page and the prior page.) Dated 18 Dec. 1863. *NARA*.

Invoices for Expenses

The following invoice was for appraising the cargo of the *Alabama* prior to the auction (Figure 129).

In the Matter of the
Steam Ship *Alabama's*
and Cargo, a Prize

United States Prize Commissioner's Office,
Appraisers' Department,
New Orleans December 16th 186<u>3</u>

To. Appraisers' Office, Dr.

1863			
Decmr. 16	For appraisement of cargo of Steam Ship *Alabama* as per report attached @ $50,650.96 @ 1%.	$ 506.50	
" "	For arranging dry goods & charges of ? (drayage?) per Bills of Helert (?) & Rosseitt (?) $26 each and no charge for superintending (?) sales. 42.00.	(Fee) 42.00	<u>$ 546.50</u>
	Received (?) Payment		

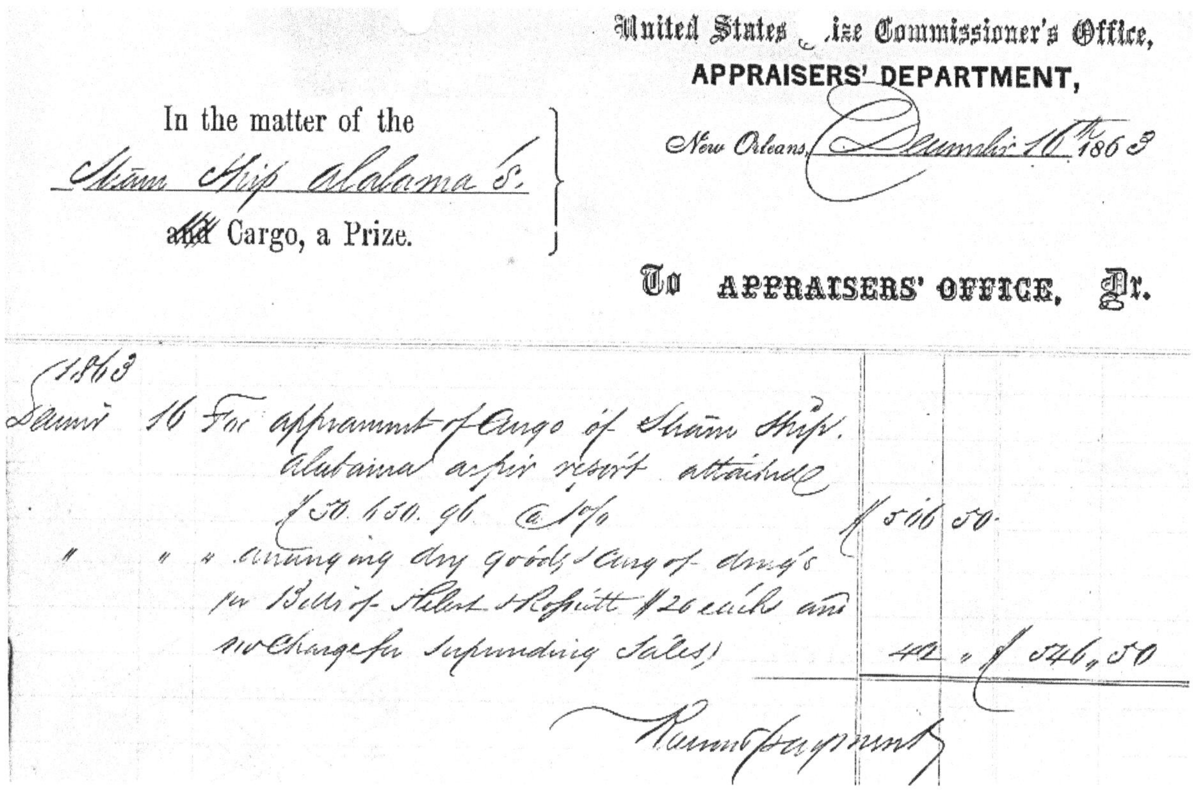

Figure 129. U.S. appraisers' invoice concerning the cargo, to be paid from proceeds of the *Alabama* sale. Dated 16 Dec. 1863 and filed 21 Dec. 1863. *NARA*.

The next bill concerned the meals of two ship keepers (Figure 130).

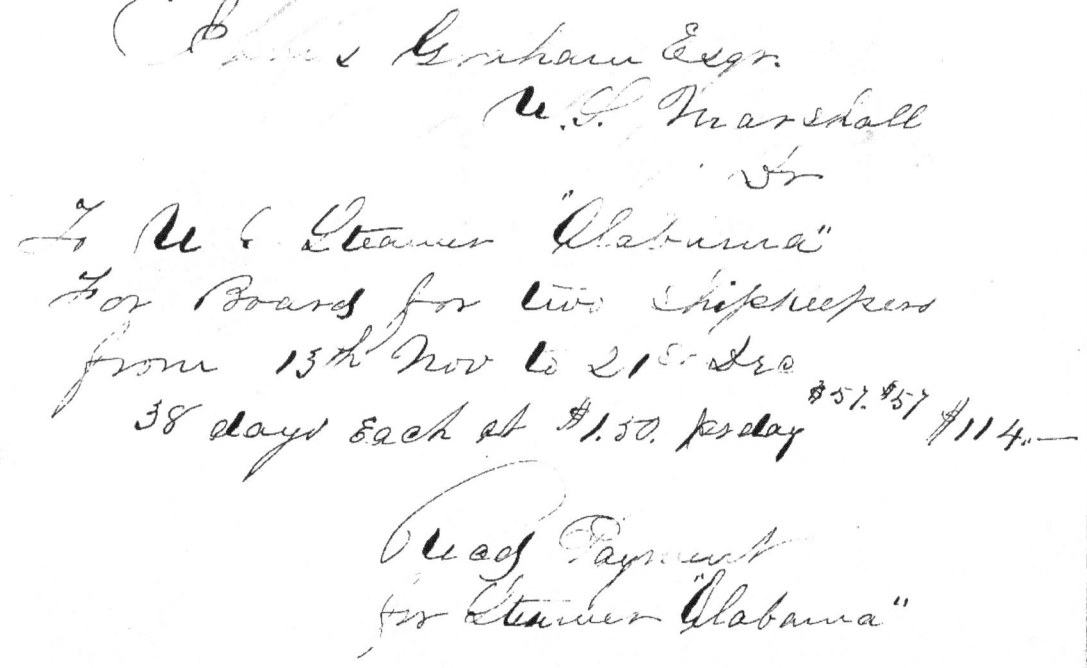

Figure 130. Invoice for the board of two ship keepers from 13 Nov. to 21 Dec., 38 days at $1.50 each for a total of $114.00. (Legible, not transcribed.) Filed 29 Dec. 1863. *NARA*.

The following were invoices for newspaper advertisements placed by the prize auctioneer announcing the *Alabama* sale (Figures 131-133).

New-Orleans, Dec 7 1863.

EVERY DESCRIPTION OF
Book and Job Printing.
EXECUTED
In the First Style of the Art.

M G. A. Hall
To THE ERA OFFICE, Dr.
No. 94 Camp Street.

1863 For advertising

Nov.	2/	Steamer Alabama & Cargo	12/	$30 00	
Dec.	1	Cargo of Str. Alabama, continued	12/	12 00	
"	3	" " " "	12/1	12 00	$54 00

Paid

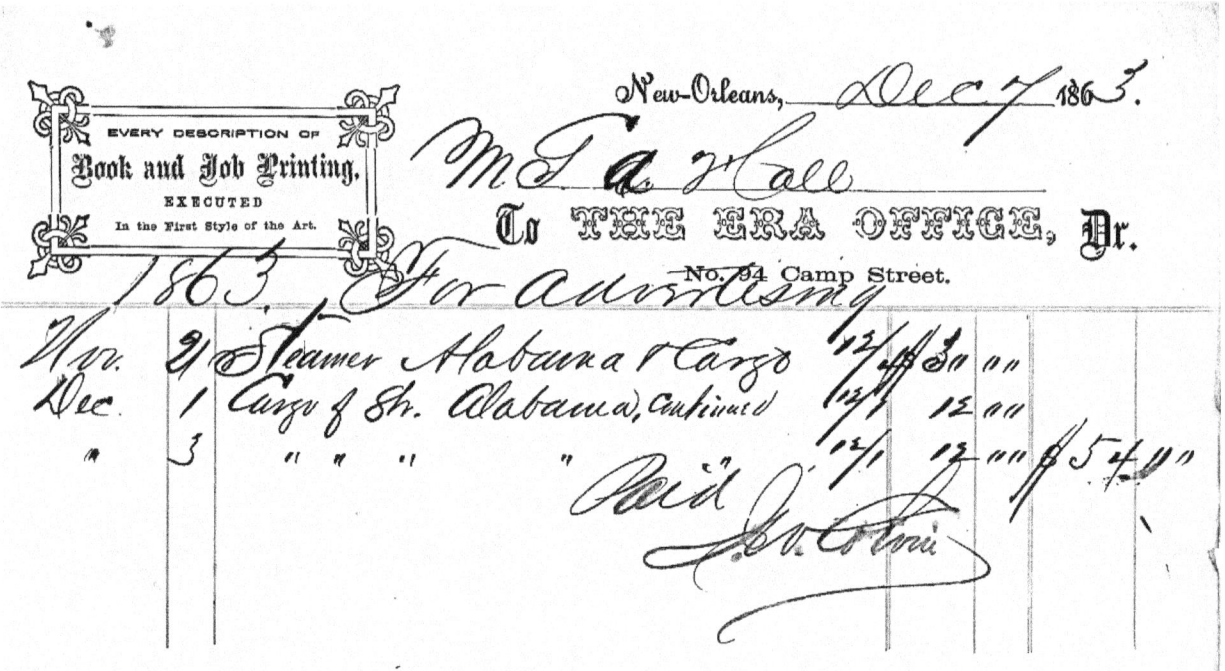

Figure 131. Invoice for advertising in *The Era* newspaper regarding *Alabama* auction. (Legible, not transcribed.) Dated 7 Dec. 1863. *NARA*.

original)

NEW ORLEANS, December 1 1863

M G. A. Hall U. S. Prize Auctioneer

LOUISIANA STATE GAZETTE. E. H. BOELITZ, Proprietor.
94 EXCHANGE ALLEY, BETWEEN CONTI AND BIENVILLE STREETS.

1863

Nov 26 For Advg U. S. Marshal's Sales on Saturday 28 Nov in German 16 squares 3 times — $35.00

Dec 3 Continuation of Sale of the Cargo of Steamer Alabama 16 sq 1 time 16.00

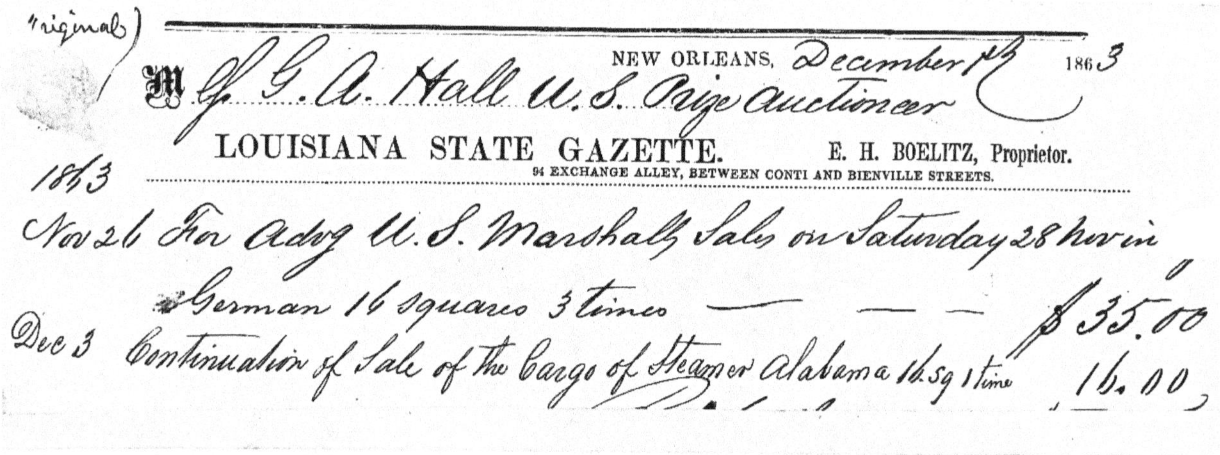

Figure 132. Invoice for advertising in the *Louisiana State Gazette* newspaper regarding the *Alabama* auction. (Legible, not transcribed.) Dated 1 Dec. 1863. *NARA*.

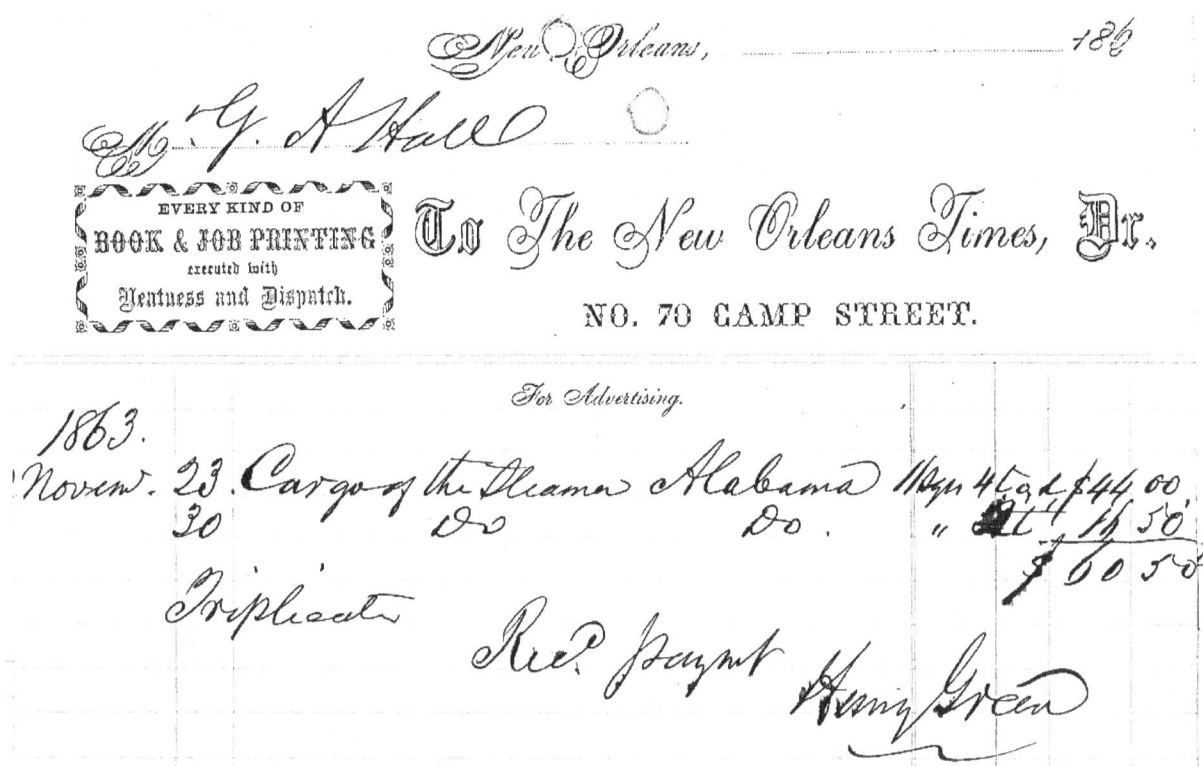

Figure 133. Invoice for advertising in the *New Orleans Times* newspaper regarding the prize case. (Legible, not transcribed.) Dated 23 Nov. 1863. **NARA**.

The barrels of alcohol in the *Alabama* cargo required an assessment of quality. The following bill was for gauging barrels of alcohol (Figure 134).

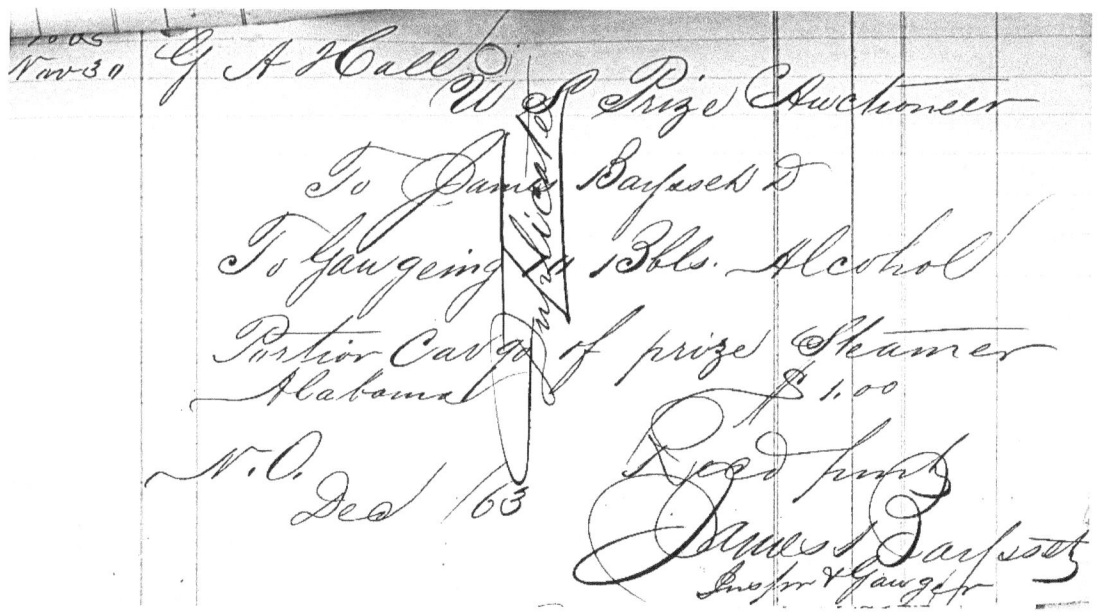

Figure 134. Invoice from James Baysset (?) who charged one dollar to gauge 10 barrels of alcohol. (Legible, not transcribed.) Dated 30 Nov. 1863. **NARA**.

The next invoice was the auctioneer's cost for producing sale catalogs for the auction of the *Alabama's* cargo (Figure 135).

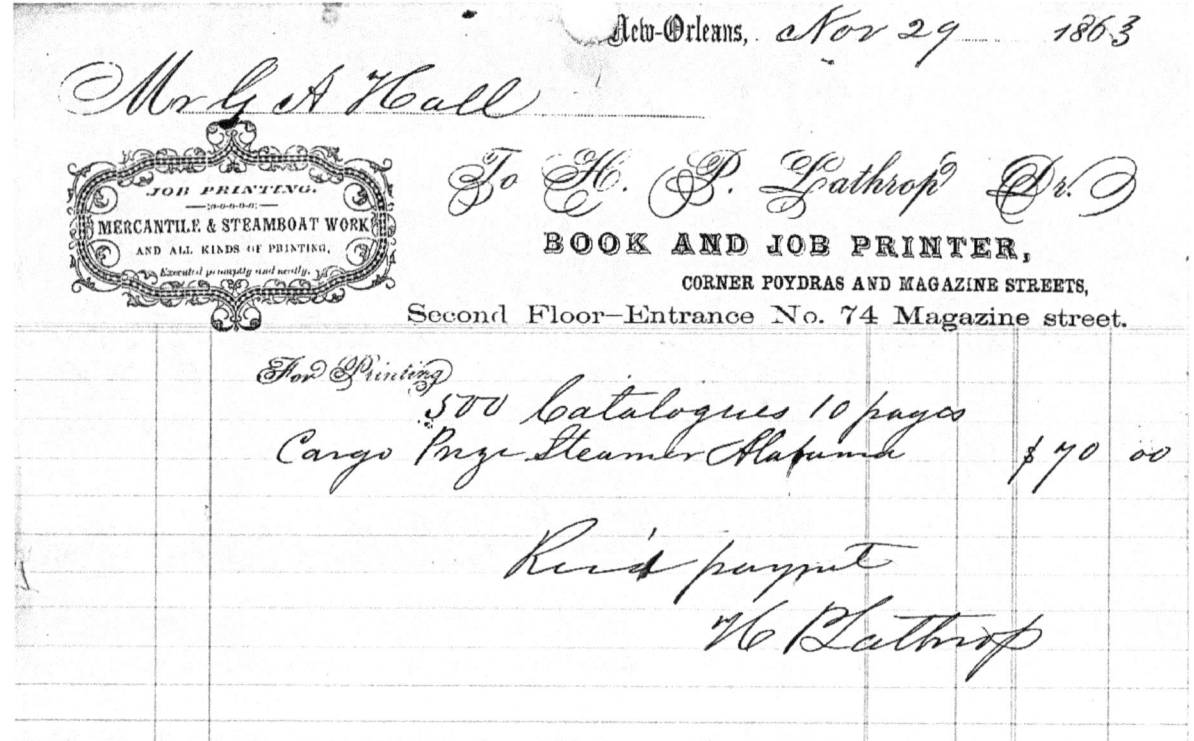

Figure 135. Invoice from printer H. P. Lathrop for producing 500 sale catalogs of ten pages each at a cost of $70.00 regarding the *Alabama's* cargo auction. (Legible, no transcript.) Dated 29 Nov. 1863. *NARA*.

U.S. Marshal's Final Invoice of Costs and Charges for the *Alabama*

The document below was marshal's bill for his final costs concerning the prize case of the *Alabama* (Figure 136).

United States Marshal's Office, Eastern District of Louisiana

United States
vs.
Steamer Alabama & Cargo.

No. 7728.

Costs & Charges.

1863		
9 Oct	Marshal's fees	$23 90
	Keeping, (Seaborn Williams) 74 days @ 2.50	185 00
	do (William Deolin) 35 do " "	87 50
	Expenses of Dep'y going on board	50
	Extra services in Sup'g and Disch'g Cargo	5 00
	Do 8 days Disch'g & Storing Cargo	40 00
	Do 10 days prep'g for sale, selling & del'g	50 00
	Pub. Monition & Sale, N.O. Times	18 00
	Expenses paid for labor in storing Cargo	3 00
	do " " " removing Do	6 25
	do " Laborers for storing in warehouse	33 25
	Bill of Keen & Stanly for disch'g Cargo	168 25
	do " do do " cleaning engine	8 00
	do " R. Harris for Cooperage	11 50
	do " L. Frigerio, for rating & repair's Chron'r	37 50
	do " Wm. T. Levine, for Pilotage	15 00
	do " Jas. Murphy " Drayage	110 40
	do " Str. Baltic " Towage	25 00
	do " W. McDuff & Co. " Weighing	19 45
	do " Board of Health (Quarantine)	23 00
	do " F. Pierce, hauling 21 loads, Glassware &c	21 00
	do " O. A. Pierce, Disch'g Clerk	30 00
	Storage of Cargo in Warehouse	219 09
	1% Commission on Sale	250 00
	Am't carried over	$1390 59

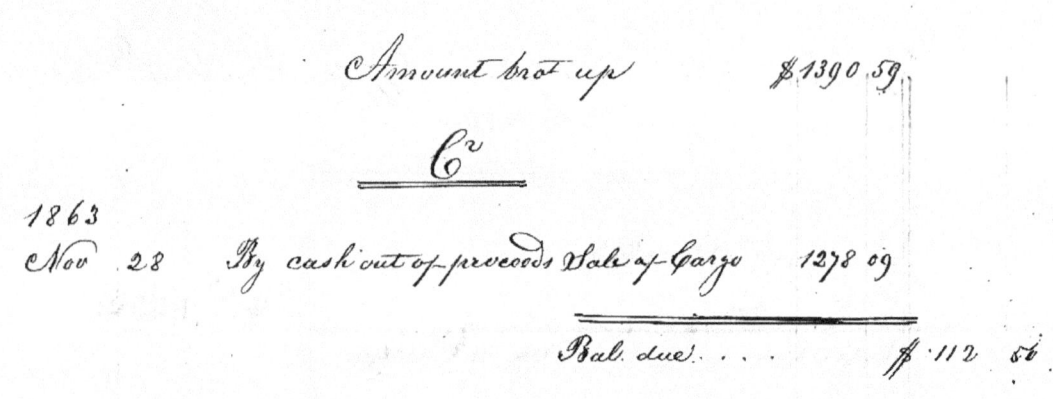

Figure 136. U.S. marshal's final *Alabama* costs and charges for 9 Oct.–28 Nov. 1863. (This page and the previous. Legible, not transcribed.) Dated 28 Nov. 1863. *NARA*.

The Court's Final Order in the Prize Case of the *Alabama*

The next document of this chapter was the final court order for the prize case against the *Alabama* finding for the captors (Figure 137).

To the present author, the court reached a reasonable and correct decision by finding in favor of the captors and determining which ships would share in the prize.

    No. 7728
    United States
        vs.
    Steamer *Alabama*
      & Cargo

The Court being satisfied that the United States Steamer *Eugenie*, the Frigate *San Jacinto*, & the Gunboat *Tennessee* are entitled to share in the Prize. That no other U.S. war vessels were within signal distance at the time of the capture & that the Prize was of inferior force to the capturing vessels. It is ordered and decreed that the aforesaid ships of war *Eugenie*, *San Jacinto*, & *Tennessee* be and they are hereby declared and adjudged to be entitled to share in the capture of the Steamer *Alabama* & cargo.
    December 15th 1863.

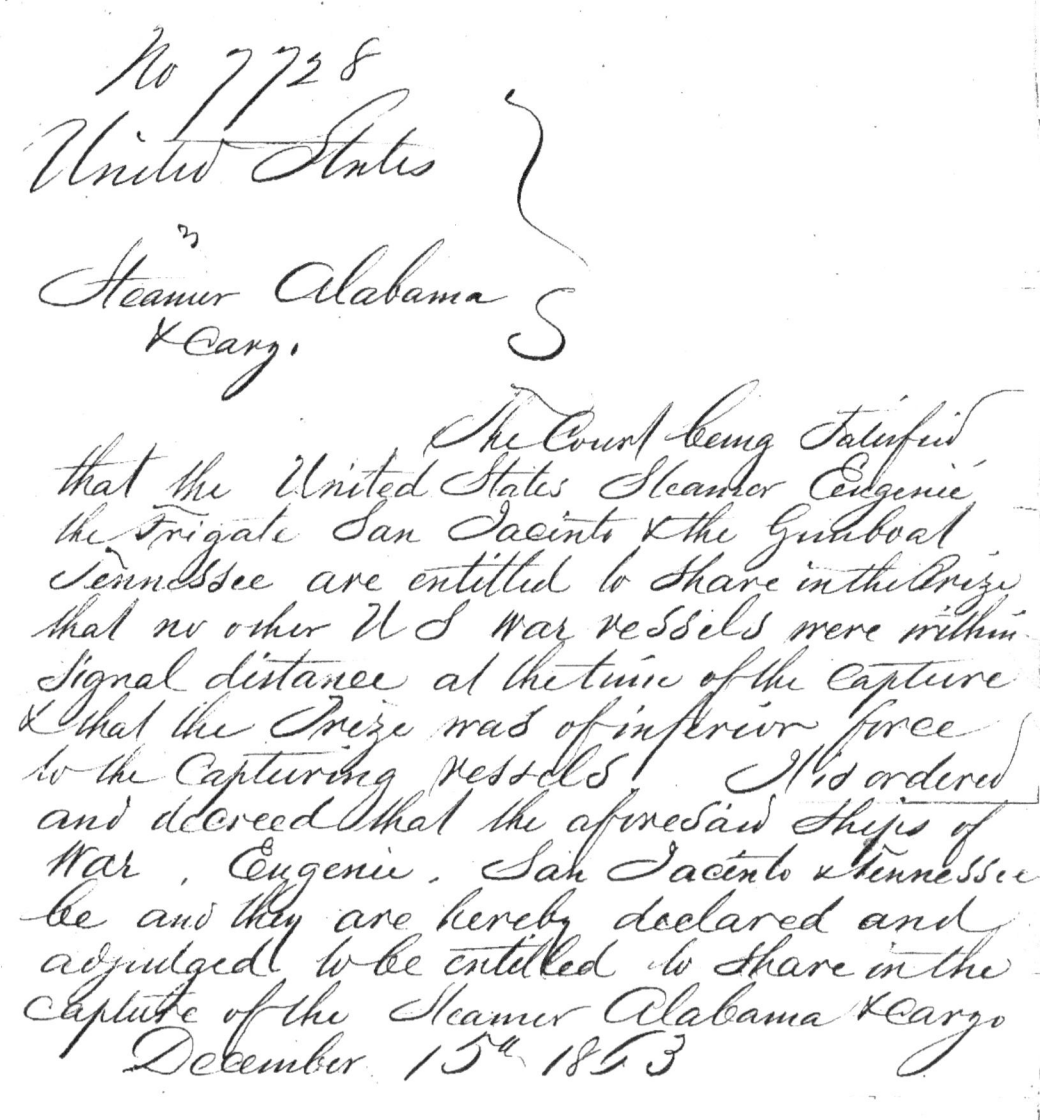

Figure 137. The court's final order for the captors regarding which ships would share in the prize. Dated 15 Dec. 1863. *NARA*.

Court Order Approving Final Costs of the *Alabama* Case

The last document of the chapter was the listing of costs approved to be deducted from the gross amount derived from sale of the ship to the U.S. Navy and the auction of the cargo (Figure 138). The total costs involved in the *Alabama* prize case were $10,298.64.

7728
U.S. vs. Str. *Alabama* & Cargo

It is ordered adjudged and decreed that the costs & charges in this case be taxed & allowed as follows:

| A. M. Buchanan | 1,970.46 | |
| Dist. Atty. | $ 3,940.00 | |

Marshal	1,278.09	
Clerk	103.80	
Registrar	1,313.64	
Prize Commrs.[203]	400.00	
Harbor Master	15.20	
Apprs.[204] Frost & Steel	40.00	vessel
Apprs. Heartt & Claiborne	546.50	
Interpreters	51.90	
G. A. Hall auctr.[205]	501.55	621.55 claimed
Reserved expenses	25.00	
add'l. fees	112.50	
board for 2 keepers	114.00	
	$ 10,412.64	
	-114.00	
	10,298.64	

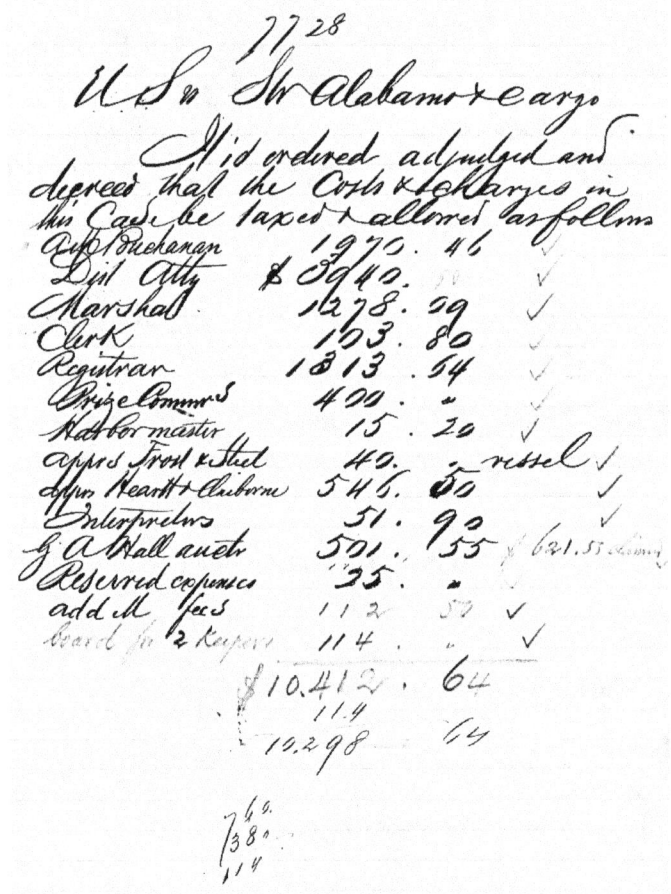

Figure 138. Final list of costs and charges for the *Alabama* prize case. *NARA*.

---

[203] Commissioners.
[204] Appraisers.
[205] Auctioneer.

7

Conclusion

This publication presented historical information on the blockade-runners as affected by the Union navy. There were both direct documentary evidence for the *Denbigh* and general context for all Gulf of Mexico blockade-runners and their navy opposition.

The *Denbigh's* documents showed how the blockading ships at Galveston, the prize court in New Orleans, and the Washington officials of the Treasury and Navy Departments processed a prize and paid the prize money to each individual sailor entitled to a share. The prize documented was a deckload of cotton unwillingly jettisoned by the *Denbigh* and picked up the following morning by two blockaders, the U.S.S. *Gertrude* and the U.S.S. *Cornubia*. The files showed the process all the way from "capture" through individual payments to each sailor, one by one. We observed the full process of capture, litigation by the prize court, public sale of the prize cotton, determination of shares, and accounting for payment of prize money.

Context for Union navy blockading efforts made up much of this book, particularly relating to areas of direct relevance to the *Denbigh*. Part I provided researchers and the public ready access to a key reference until now only available by order from a university: Robert Glover's 1974 dissertation. Glover's analysis of the West Gulf blockade was an often cited and most interesting source that now is easily obtained by its long overdue publication herein. That this landmark work was not published long ago may indicate a myopic and widespread misunderstanding of the importance of maritime matters in the broader field of Texas history.

The dive operations of the *Denbigh* Shipwreck Project provided the setting for many hours of cogitation and speculation about how and why the runners and the blockaders conducted as they did their respective missions. Examining such writings as memoirs of Union sailors and the documents generated by capture of prizes helped address some of these questions.

Careful reading of the prize case files and ship's logbooks revealed much about the behavior of the participants. The standard prize case file consisting of a few dozen documents showed a lot. The complex and controversy-filled case of the merchant steamer *Alabama* included about three hundred documents and revealed much more. The numerous depositions were particularly informative, most of all that of the *Alabama's* captain. Captain Thomas Carroll explained in detail how and why the two-day chase and capture developed as it did. This exposition on strategy and tactics of the Civil War sea chase was most valuable.

The squabbling among several Union ship captains who took part in capturing the *Alabama* suggested a chapter subtitle: "Lieutenant Commanders Behaving Badly." The case highlighted the participants' motivations and behaviors behind the blockade: get the prize money and stop at nothing. There was no letup in the grabbing when the prize was safely delivered into the custody of the prize court at New Orleans. In the middle of the legal proceedings, the Union army unceremoniously seized the *Alabama*. The Union army needed the ship for an urgent mission to deliver supplies to General Banks, who was in the field on the Texas coast conducting what turned out to be an abortive invasion. The highhanded action by General Stone, Banks' chief of staff, took the story to a new low. Stone went so far as threatening, in writing, to have his soldiers arrest the judge, U.S. marshal, U.S. district attorney, and other court officials should they impeed his plans for the *Alabama*.

For nautical archaeologists and others, there is much to learn from a patient, thorough pursuit of the documentary sources of a shipwreck like the *Denbigh*.

www.ingramcontent.com/pod-product-compliance
Lightning Source LLC
Chambersburg PA
CBHW080722300426
44114CB00019B/2463